AF622490

Making Water Smart

In Focus – Special Book Series

Making Water Smart

Editors

Matthew J Wade, Edward C. Keedwell, Jean-Philippe Steyer and M. Victoria Ruano Garcia

IWA PUBLISHING

Published by **IWA Publishing**
Unit 104-105, Export Building
1 Clove Crescent
London E14 2BA
Telephone: +44 (0)20 7654 5500
Fax: +44 (0)20 7654 5555
Email: publications@iwap.co.uk
Web: www.iwapublishing.com

First published 2021
© 2021 IWA Publishing

Apart from any fair dealing for the purposes of research or private study, or criticism or review, as permitted under the UK Copyright, Designs and Patents Act (1998), no part of this publication may be reproduced, stored or transmitted in any form or by any means, without the prior permission in writing of the publisher, or, in the case of photographic reproduction, in accordance with the terms of licenses issued by the Copyright Licensing Agency in the UK, or in accordance with the terms of licenses issued by the appropriate reproduction rights organization outside the UK. Enquiries concerning reproduction outside the terms stated here should be sent to IWA Publishing at the address printed above.

The publisher makes no representation, express or implied, with regard to the accuracy of the information contained in this book and cannot accept any legal responsibility or liability for errors or omissions that may be made.

Disclaimer
The information provided and the opinions given in this publication are not necessarily those of IWA and should not be acted upon without independent consideration and professional advice. IWA and the Editors and Authors will not accept responsibility for any loss or damage suffered by any person acting or refraining from acting upon any material contained in this publication.

British Library Cataloguing in Publication Data
A CIP catalogue record for this book is available from the British Library

ISBN: 9781789062816 (paperback)

Contents

Preface

As much as water is a vital resource for life, so data has become increasingly critical to solve the broad range of challenges faced in its use, reuse, treatment, and management. Technological and computational developments over the last half century have culminated in an exponential growth in data being generated by processes, systems, and human behaviour, providing great opportunity and challenges to acquire knowledge from the latent information stored within the diverse array of structured and unstructured datasets. However, with 'dark data' comprising between 50–90% of all data collected by industry, the value, both economically and intellectually, of improving data management and use is clear. The 4th Industrial Revolution, which strives to fuse ideologies from the physical, biological, and digital domains, provides an impetus for smarter and interconnected utilisation of data resources. For the water industry, this digital transformation is beginning to take hold as evidenced through the IWA Digital Water Programme and their White Paper series that consider emerging topics in this space, such as *Digital Twins*, *Artificial Intelligence (AI)*, *Meta-Data*, and *Uncertainty analysis*.

As the water sector embraces a 'smart water' ethos, the opportunities for reimagining and reshaping research and industry must be guided by good governance and practice, and innovation-friendly regulation. This is particularly important considering data-usage, data analysis and transformation, and advanced automation and process control. The vision of fully autonomous and integrated processes (e.g., systems of systems, Internet of Things) is appealing to many, especially to those operating in the burgeoning AI sector who look towards the water industry as the next application frontier. However, the water and wastewater fields are fundamentally driven by public health and environmental decision-making processes, established by sound engineering and science principles. It is our belief that 'smart water' will require a human-in-the-loop and that adoption of future-facing technologies and methods in this space will work best when human experts can guide their design, implementation, and application. Further, enabling a 'smart water' sector should not be limited to large, data-intensive systems typically found in advanced economies, but practical, low-cost 'smart' solutions should be established to support the myriad challenges faced by developing nations, vis-à-vis the Sustainable Development Goals. The drivers in these nations may not be the same as those in industrialised countries, but the tools and knowledge are likely to be similar such that agile and resilient design will be central to establishing the smart water paradigm for global use.

This book, part of the IWA In Focus series, comprises a series of papers highlighting the breadth of smart water technologies and applications. From the use of classical machine learning and data transformation methods for process improvement, to data management and assimilation in models for better monitoring control, a selection of these papers describes the acquisition and preparation of data as an initial step in its use as part of a digital framework for smart water applications. Importantly, several papers examine the use of smart tools and contemporary AI technologies, such as neural networks and Internet of Things, that demonstrate value in non-conventional or remote environments. This collection demonstrates the innovation possible through the entire 'pipeline' of the process of applying smart water techniques, from data measurement and collection, through initial analysis to the application of machine learning and AI techniques and finally through to system deployment. Each of these steps plays an important role in the application of data science and AI techniques to water problems. Examples of this include the provision of reliable, robust data from established sources (e.g. smart meters) and novel data collection techniques (e.g. UAVs), statistically sound and innovative data analysis through the use of neural networks and evolutionary approaches and methods that support decision makers through novel human-in-the-loop and visualisation strategies that provide complex information to users in a readily navigable format.

Collectively, these papers showcase the innovation required to leverage modern data science and AI approaches in the water sector and point the way towards a future of new measurement techniques, innovative methodologies, and intuitive human interaction to truly 'Make Water Smart'.

Editors

Matthew J. Wade
UK Health Security Agency
Environmental Monitoring for Health Protection
Windsor House
London
SW1H 0TL
United Kingdom
E-mail: matthew.wade@dhsc.gov.uk

Edward C. Keedwell
Professor of Artificial Intelligence, Dept. of Computer Science, College of Engineering,
Mathematics & Physical Sciences, Innovation Centre Phase 1,
University of Exeter, Exeter, EX4 4RN, UK
E-mail: e.c.keedwell@exeter.ac.uk

Jean-Philippe Steyer
LBE, Univ Montpellier, INRAE, 102 Avenue des etangs, 11100, Narbonne, France
E-mail: jean-philippe.steyer@inrae.fr

M. Victoria Ruano Garcia
CALAGUA -Unidad Mixta UV-UPV, Departament d'Enginyeria Química,
Universitat de València, Avinguda de la Universitat s/n, 46100,
Burjassot, València, Spain
E-mail: m.victoria.ruano@uv.es

 © 2020 The Authors

A critical review of the data pipeline: how wastewater system operation flows from data to intelligence

Jean-David Therrien IWA, Niels Nicolaï IWA and Peter A. Vanrolleghem IWA

ABSTRACT

Faced with an unprecedented amount of data coming from evermore ubiquitous sensors, the wastewater treatment community has been hard at work to develop new monitoring systems, models and controllers to bridge the gap between current practice and data-driven, smart water systems. For additional sensor data and models to have an appreciable impact, however, they must be relevant enough to be looked at by busy water professionals; be clear enough to be understood; be reliable enough to be believed and be convincing enough to be acted upon. Failure to attain any one of those aspects can be a fatal blow to the adoption of even the most promising new measurement technology. This review paper examines the state-of-the-art in the transformation of raw data into actionable insight, specifically for water resource recovery facility (WRRF) operation. Sources of difficulties found along the way are pinpointed, while also exploring possible paths towards improving the value of collected data for all stakeholders, i.e., all personnel that have a stake in the good and efficient operation of a WRRF.

Key words | data treatment, digitalization, digital twin, metadata, wastewater modelling, water resource recovery facilities

Jean-David Therrien IWA (corresponding author)
Niels Nicolaï IWA
Peter A. Vanrolleghem IWA
modelEAU, Université Laval,
1065, Avenue de la Médecine, Québec,
Canada,
QC G1 V 0A6
E-mail: *jean-david.therrien.1@ulaval.ca*

HIGHLIGHTS

- Data can be abstract: with the data pipeline concept, issues are clarified.
- Digitalization of wastewater has begun, but it is lagging behind drinking water.
- Reliable digital twins need good quality data, but reaching that goal presents unique challenges.
- Collecting data without a proper strategy leads to data graveyards.
- More collaboration between data and water experts is critical to better use wastewater data.

GRAPHICAL ABSTRACT

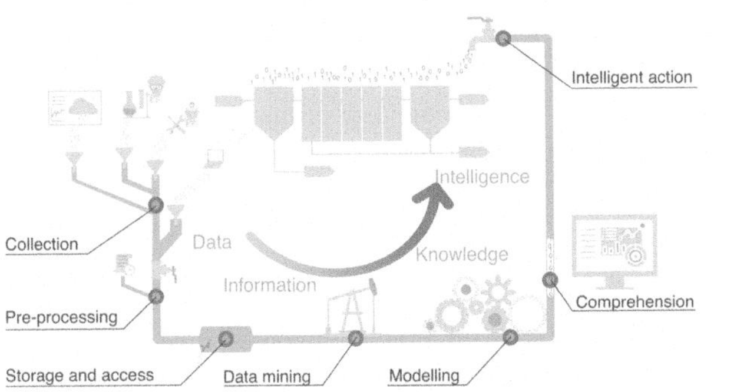

This is an Open Access article distributed under the terms of the Creative Commons Attribution Licence (CC BY 4.0), which permits copying, adaptation and redistribution, provided the original work is properly cited (http://creativecommons.org/licenses/by/4.0/).

doi: 10.2166/wst.2020.393

INTRODUCTION

Stringent limits on emissions, increasing water scarcity and rapid urbanization put a major strain on current wastewater systems. This means that the wastewater sector will have to adapt, and fast, if it is to fulfill its role as keeper of public and environmental health. In response to these stressors, WWTPs are increasingly being repurposed as water resource recovery facilities (WRRFs) (Water Environment Federation 2014). This value extraction is enabled by the deployment of a very diverse array of processes, from nutrient recovery to energy generation.

Moreover, the wastewater research community has been hard at work in recent years to develop new process models (Al-Omari *et al.* 2015; Mannina *et al.* 2016; Spérandio *et al.* 2016; Amaral *et al.* 2017; Rittmann *et al.* 2018; Varga *et al.* 2018), new process control strategies (Rieger *et al.* 2013; Jimenez *et al.* 2015; Regmi *et al.* 2015; Hernández-del-Olmo *et al.* 2016; Solon *et al.* 2017; Revollar *et al.* 2020) and new process monitoring systems (Alferes & Vanrolleghem 2016; Russo *et al.* 2019) to bridge the gap between current operation and the state-of-the-art control practices needed to efficiently and optimally run the multitude of processes involved in resource recovery given constrained capital and operational budgets.

These new developments share a need for data which has historically been difficult to collect. However, with sensor prices dropping, the increasing ubiquity of wireless communication and the proliferation of mobile devices able to ceaselessly gather information and perform sophisticated calculations, WWTPs have in recent years been exposed to an unprecedented amount of data. These trends span much wider than only the wastewater field, hence researchers from other disciplines have also dealt with this massive influx of data and have emerged with entirely new types of models leveraging machine learning and artificial intelligence. The interest in applying those new data-driven models to WRRF operation and control has recently become very strong. Some of the most discussed potential applications include the development of adaptive plant models, predictive maintenance and plant-wide control through the use of a digital twin, which have the potential to reduce costs, improve resource recovery, increase water quality and increase customer engagement (IWA and Xylem Inc. 2019). However, the pace of this digitalization has thus far been quicker in the drinking water distribution sector than in wastewater (Water Online and SWAN 2019). This could be attributed to the fact that wastewater treatment processes depend heavily on water quality sensors, which are known to be difficult to work reliably, whereas those of drinking water distribution processes rely mostly on well-established water quantity measurements.

For these newly deployed sensors and novel data-driven models to help close the digitalization gap and have a significant impact on the way WRRFs are run, they must be maintainable by the workers and professionals of the wastewater field. Simply put, the information yielded by those new techniques must be relevant enough to be looked at by these professionals; be clear enough to be understood; be convincing enough to be believed and be reliable enough to be acted upon. Regardless of the abundance of the data or the sophistication of the models, failure to attain any of those features can be a fatal blow to the adoption of even the most promising new technology. This review thus aims to provide an ensemble perspective on how data are handled in WRRFs and to point out pitfalls and possible paths for improvement to water professionals wishing to make better use of their data.

FROM DATA TO INTELLIGENCE

To have a meaningful discussion of data-driven technologies, some seemingly similar terms must be given distinct working definitions. Using the definitions put forward by Makropoulos & Savić (2019), one may categorize levels of knowledge along the following hierarchy:

- Data: Quantitative or qualitative measurement or recording of a phenomenon or process.
- Information: Fact or observation derived from the analysis of data.
- Knowledge: Insight into mechanisms that relate pieces of information together within a certain context.
- Intelligence: The ability to use knowledge of distinct aspects of a problem to develop new ideas and perspectives.

These definitions make clear the fact that different levels of understanding build on top of each other. They also show that in the pursuit of 'smart' systems, one has to attain the highest level – intelligence – by refining and analysing the available data. Makropoulos & Savić (2019) represent this process as beginning with a question about the world and continuing with all the steps required to answer it. The collection stage requires one to gather data from different stores. Once the data of interest is assembled, it should be analysed to extract information. This information is integrated into a model to capture underlying mechanisms and gain new knowledge from the studied system. This knowledge, when general enough, can be built into new tools, which then saves one from having to repeat analyses. Finally, with enough of those knowledge-based tools, intelligent decisions can be made with confidence. Of course, data interpretation is an iterative process, so higher levels of understanding gathered over time feed back into the interpretation of the same data at later stages.

From the authors' perspective, this path from data to intelligence is common to any data-driven activity. A description of each of these steps is therefore attempted here in the hope of bringing clarity to the often arduous process of distillation required by smart water systems.

Laying down the data pipeline

Using the framework outlined by Makropoulos & Savić (2019), it becomes clear that data-driven systems must be created from the ground up: without data, there is no information; without information, we know nothing; and a system that embeds no knowledge can not be intelligent. Intelligence generation can, therefore, be imagined as a pipeline with data at one end and intelligence at the other, as is pictured in Figure 1. This pipeline begins and ends at a WRRF, indicating that the intelligence extracted from the pipeline can be leveraged at later iterations of every step of the pipeline. It must also be acknowledged that different data-driven activities require different levels of complexity of analysis. Corominas *et al.* (2018) classified data-driven analysis based on their complexity in the following way:

- Basic information extraction includes simple schemes such as univariate control charts and mass, energy and stoichiometric balances.
- Advanced information extraction includes multivariate data treatment such as dimensionality reduction, feature detection and supervised machine learning.
- Human-interpretable knowledge extraction includes tools such as generalized rules, fuzzy logic, environmental decision support systems or ontologies.

This break-down emphasizes the points that lower levels of analysis may be useful on their own, and that high levels of analysis depend on the lower ones to function, which fits nicely into the pipeline framework.

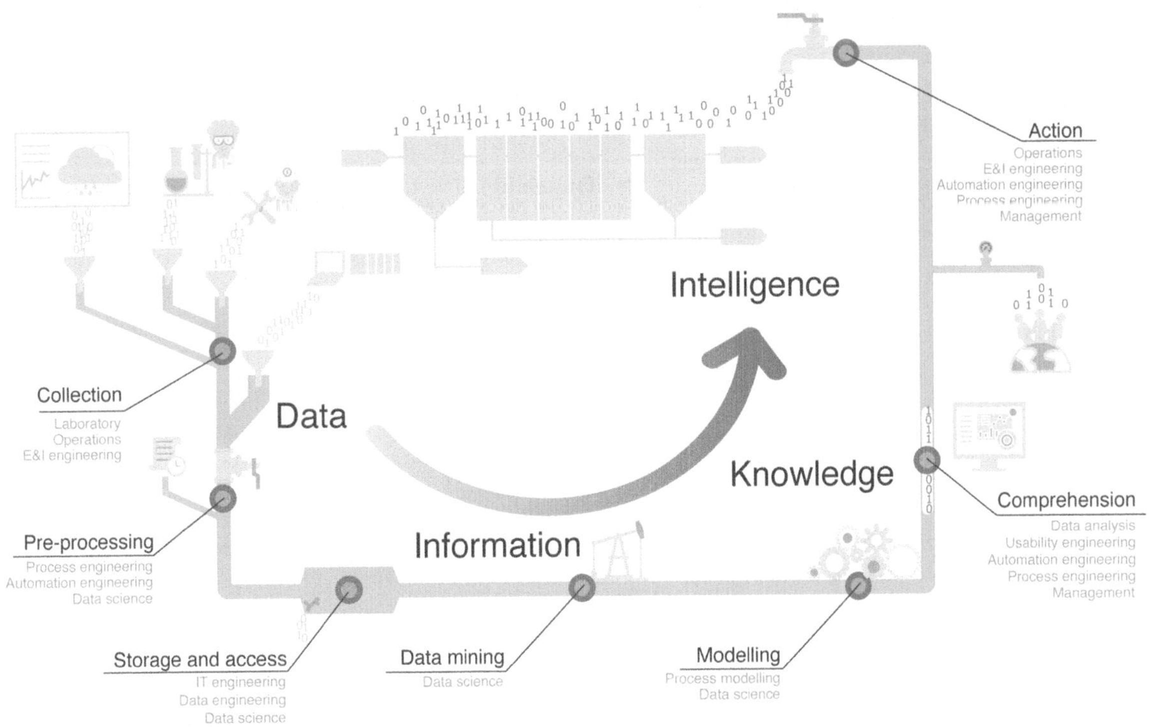

Figure 1 | From data to intelligence. For each step in the pipeline, the most essential professions are listed.

Some steps shown in the pipeline of Figure 1, data storage and data extraction, are not explicitly mentioned by Makropoulos & Savić (2019). They are, however, critical to creating usable data sets, as data collection and use may be separated by several months or years, and data collection may occur in a multitude of locations. Thus, the integration of the whole set requires a significant amount of effort and is worth discussing.

Figure 1 also lists the different actors involved in the manipulations and transformation of data along the pipeline. The variety of professions present underlines the fact that tending to the data produced by WRRFs is an inherently collaborative process and that several skill sets are required to achieve good results.

Moving away from abstractions such as 'information', 'knowledge' and 'intelligence' for a moment, one may envision the data pipeline leading to concrete, actionable insight in the following way. Let us first imagine a WRRF equipped with online sensors characterizing the quantity (flow meters) and quality (TSS, COD, COD_s, NH_4-N and NO_3-N) of the influent of a nitrifying/denitrifying biological reactor, its internal DO and TSS concentrations, as well as its secondary clarifier's recirculation line and wastage line's TSS concentrations. After having validated, cleaned and gap-filled these data, one could build a physical model of the biological treatment using one of the ASM models and a settler model. This model could be fed data series similar to what the reactor typically receives with the help of an influent generator model. Combining these two models as sources of information, one could determine whether the bioreactor would still be able to denitrify an adequate amount of nitrate if the internal recirculation rate was reduced. Equipped with this new knowledge, the operator could choose to decrease the sludge return flow to optimize pumping energy consumption.

A more involved example could include a live feed of processed sensor data into the physical model. The pre-processing needed to clean the live data is made easier by metadata, which can be either captured by automatic systems or created by the plant operators while they observe the process and inspect the sensors. This live data, when fed to an Extended Kalman filter, could help the model adjust its parameters to better reflect the current state of the reactor. When combined with meteorological observations and predictions, the model and influent generator could predict how the effluent concentration of ammonia is likely to change in the near future. Equipped with this knowledge, an operator (or an automatic controller) could adjust the dissolved oxygen setpoints with confidence that regulatory limits will not be exceeded. As one can see from these examples, the value of the data and the computational tools put in place in a digitized WRRF take all their meaning when they finally lead to a concrete, intelligent action.

Good labels make good neighbours

Before jumping into the pipeline itself, however, it is important to acknowledge that data must always be interpreted with knowledge of its history and context. That is why, ideally, metadata is also produced and collected at every step of the data pipeline. Rieger & Vanrolleghem (2008) define metadata as data about data that enables the extraction of useful information out of a signal. In other words, metadata is what describes the context in which data is created, thus making the data more easily interpreted. For industrial processes, metadata typically relates to the instrumentation producing the data (e.g. measurement unit, sampling location, equipment model etc.) and its associated quality (e.g. normal value range, raw or filtered value, measurement accuracy etc.). The metadata may be very structured (e.g. timestamps), or completely unstructured (e.g. text). Unstructured metadata is especially useful when annotating sensor signals with rich information to indicate the occurrence of an event.

When they presented the FAIR guidelines, Wilkinson *et al.* (2016), argued for general principles of data and metadata stewardship that may apply to any research area. For them, the four foundational principles that should guide data producers and publishers are:

- **Findability**: The data should be indexed, uniquely identified and contain rich metadata.
- **Accessibility**: One should be able to browse metadata using standardized protocols.
- **Interoperability**: Data should be in a widely useable format.
- **Reusability**: Metadata should reflect the needs of the domain of inquiry in which the data has been produced. The data should also be released under a clear license.

Preparing data sets such that all best practices are followed is a difficult task – especially when these best practices are still ill-defined. The FAIR guidelines thus provide a very welcome framework to incrementally improve data set quality. That is, according to Wilkinson *et al.* (2016), any of the principles can be implemented independently from the other, thus causing less friction in the beginning stages of implementation.

FLOWING DOWN THE DATA PIPELINE

Based on Figure 1, one sees that data goes through different stages throughout its lifecycle (i.e. collection, pre-processing, storage and access, mining, modelling, comprehension and action) and that the flow from one step to the other can be thought of as following the flow through a pipeline. One implication of this is that more complex data-dependent WRRF processes are vulnerable to failures anywhere along that flow. Another is that errors in early steps will influence every step downstream, potentially leading to the proverbial 'garbage in; garbage out' conundrum. This section therefore also aims at describing some of the potential failure points that are encountered in WRRF data treatment, and possible paths for improvement.

Drawing from the data well – data collection

The data collected in WRRFs is integral to their instrumentation, control and automation (ICA) systems. According to (Olsson 2012) three main objectives originally motivated the introduction of ICA to WRRFs almost 50 years ago:

1. Keep the plant running safely, i.e. get the water from influent to effluent in a controlled and reliable way.
2. Maintain good effluent quality to fulfil the plant's mission of environmental protection.
3. Optimize the plant processes to fulfil its mission while consuming as few resources as possible.

These motivations hold to this day, and data collected within the plant are indispensable in trying to accomplish these goals. Indeed, since the introduction of ICA, practically exponential growth in the amount of data being automatically digitised and stored has been observed (Olsson 2012). Common examples of such data are univariate time series coming from on-line process sensors and at-line analysers, discrete signals denoting actuator states as well as process operational settings (Vanrolleghem & Lee 2003). Because these data are sourced from dynamic systems, time series are characterised by some interdependency between sequential observations (Box & Jenkins 1970). This is in contrast to data being generated as a result of ad hoc events that do not occur deterministically in time.

Though sensors and analysers provide discrete measurements, their sampling period is usually small enough (seconds or minutes) as compared to the characteristic time constants of the treatment processes (hours or days). As such, they are invaluable tools on the path towards smart wastewater utilities (Ingildsen & Olsson 2016).

Time series data in WRRFs are either collected manually or automatically. Examples of manual data collection are offline laboratory analysis of grab and composite samples (i.e. analytically determined data), as well as observations made during visual maintenance inspections of the process itself including its field and panel-mounted measurement indicators. Note that most regulatory agencies still consider laboratory analysis to be the gold standard for water quality measurements (Yuan *et al.* 2019). However, some might say that laboratory analysis also comes with multiple drawbacks. For example, data collection is labour intensive, the measurements are delayed and infrequent (i.e., non-equidistant in time) because of extensive sample analysis and manipulation, and the measurements are prone to gross errors stemming from human distraction and fatigue.

The most essential part of any measurement instrument is, of course, the sensing element involved in capturing a physical phenomenon from the analogue world. Wastewater, however, is a harsh environment where organic and inorganic pollutants commonly cause fouling and degrading of the sensing elements (Dürrenmatt & Gujer 2012). This is eloquently demonstrated by Vanrolleghem (2014) in Figure 2.

If not maintained correctly, or compensated for during subsequent data processing, errors in data collection will directly affect the decisions taken based on the information contained in the data, and thus decrease user confidence in the measurement system (Regmi *et al.* 2019). As such, it is of utmost importance that a well-defined strategy is available and put into practice to maintain sensors, as well as the data they generate, throughout their entire life cycle. This also means that already during the conceptual phase of sensor implementation, a trade-off should be made between the value of the information being generated and the full cost of ownership (Zegers *et al.* 2019).

Another important aspect, not to be forgotten when working with automatic data collection, is the intrinsic dynamics of the measurement system itself (Rieger *et al.* 2003). These must be faster than the dynamics of the process being monitored to ensure that measurements are useful in operation and decision-making. Moreover, other low-level components involved in data acquisition, such as signal transmitters, carriers, samplers, converters and networks, will also have an impact on the data collection process and thus the final data quality. Examples include noise due to electromagnetic interference, ground loop errors, power surges, jitter, signal aliasing, quantization errors and analogue filtering (Whitt 2012). Automated data collection also requires different types of process instrumentation,

Figure 2 | Sensors in their ideal condition versus their typical operating condition. Top-left: conductivity; bottom-left: NH_4-N, K and pH; right: pH (Vanrolleghem 2014).

equipment and software applications having to communicate with each other. Whereas in the past this was typically established using custom drivers and proprietary communication protocols, the current trend is towards interoperability based on standardised and open digital communication protocols (Korodi *et al.* 2018).

By now it should be clear that the efforts of process operators, lab technicians, electrical & instrumentation engineers and automation engineers are indispensable in the data-driven operation of smart WRRFs, as these workers operate at the front end of the data pipeline.

Besides data generated by on-site measurement instrumentation, other less conventional data sources can also be integrated. A common example is when rainfall data, coming from publicly accessible weather stations, together with weather forecasts, are integrated to anticipate future operation of wastewater treatment facilities (Hernández-del-Olmo *et al.* 2019; Vezzaro *et al.* 2020). In the context of energy consumption, energy costs and load management, Aymerich *et al.* (2015) showed that energy prices, as determined by the local tariffing structure, can also affect the choice of a WRRF's operational strategy. As such, it might be of interest to also collect data from energy markets. Steel production, for example, envisions process optimisation based on real-time electricity markets where decisions can be made as frequently as every 5 minutes (Shyamal & Swartz 2019).

Cutting through the noise – data pre-processing

The process of evaluating and augmenting the quality of data, so it can be used purposefully, is often referred to as data pre-processing. This step of the pipeline is necessary since, as a result of measurement noise, errors and failures, it is impossible to collect completely accurate data. The case of on-line sensor data is especially interesting because of their high sampling frequency. Figure 3 shows WRRF online sensor data series exhibiting common problems that can be alleviated with data pre-processing. As can be seen in (a), the presence of outliers muddies the trend of the shown turbidity signal, while in (b), a sensor fault made a part of the data series unusable. In the former case, the application of an exponentially-weighted moving average outlier detection algorithm and trend smoother removed the undue noise from the data series (Alferes *et al.* 2015), while in the latter case, data filling with an average daily dry-weather TSS profile provided a substitute for the lost data (Patry 2020).

Wishing to inventory the different possible sensor states, Rosén *et al.* (2008) enumerate 7 distinct states, while Newhart *et al.* (2019) add an eighth:

1. Operational: Sensor is working properly, with normal measurement noise.
2. Excessive drift: When a sensor outputs a value progressively further from the true value.
3. Shift: When the output of the sensor is a constant amount away from its true value.
4. Fixed value: When the sensor is stuck and keeps repeating the same value.
5. Complete failure: Similar to a fixed value fault, but the sensors either give off the maximum or minimum, value, zero or no value at all.
6. Wrong gain: When signals away from the calibration point are under- or over-amplified by the sensor.

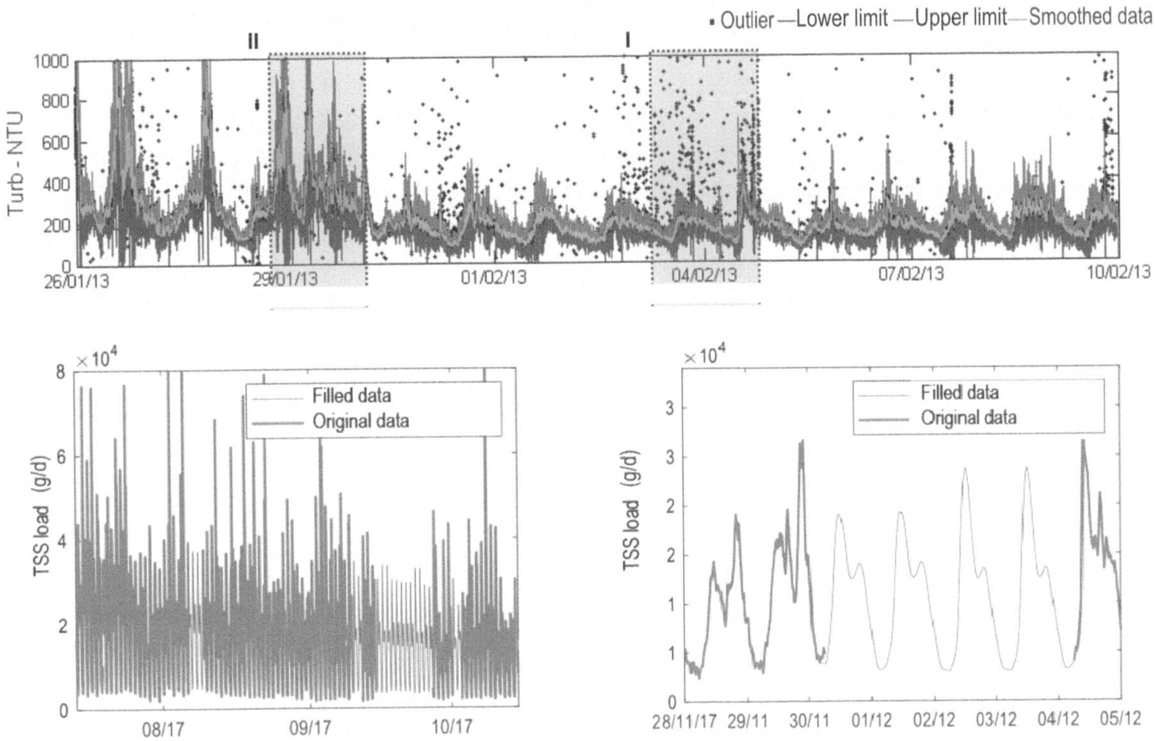

Figure 3 | (a) Raw turbidity sensor data (dots), along with the processed data (green) generated using an exponentially-weighted moving average (EWMA) filter (Alferes *et al.* 2015). (b) Raw total suspended solids (TSS) sensor signal containing gaps (orange), which are filled with average dry-weather daily profiles (blue) (Patry 2020). The full colour version of this figure is available in the online version of this paper, at http://dx.doi.org/10.2166/wst.2020.393.

7. Calibration: The sharp change in sensor output directly following a calibration.
8. Isolated fault: When a single point in a series shows an incorrect value.

Schraa *et al.* (2006) discuss several checks that may be performed on sensor signals to detect faults in the sensors or the process itself, such as whether user-defined bounds on the sensor measurement have been exceeded, or whether a measurement signal is outside the typical 4–20 mA range. However, by far the most common way to detect erroneous data used by analysts is by visual inspection of the time series (Alferes *et al.* 2015). However, given the sheer amount of data collected in any treatment plant, visual fault detection is simply impractical (not to mention, error-prone and requiring lots of expert knowledge). Thus, automatic detection of process and measurement faults is necessary. As it turns out, fault detection (determining whether a fault is present and at which time it occurred), fault isolation (determining which sensor caused the failure) and fault diagnosis (determining what kind of fault has occurred) (Isermann & Ballé 1997) are non-trivial tasks. As such, a lot of research has been carried out over the past decades specifically for faults present in WRRF.

Corominas *et al.* (2011) proposed a practical methodology to compare the accuracy of different univariate algorithms (e.g. exponentially weighted moving average (EWMA)) on artificial faults in time series generated by the Long-Term Control Benchmark Model (BSM_LT) proposed by Rosén *et al.* (2004). Meanwhile, Rosén & Olsson (1998), (Yoo *et al.* 2007) and Alferes *et al.* (2013) describe the use of multivariate techniques such as Principal Component Analysis (PCA) and Partial Least Squares (PLS) to detect anomalies in time series. Russo *et al.* (2019) and Garneau & Vanrolleghem (2018) also demonstrate the use of autoencoder neural networks for the same purpose.

For their part, Ohmura *et al.* (2019) discuss two of the main assumptions made by most fault detection methods; that is, that different sensors used in multivariate fault detection fail at distinct times from each other, and that sensors work perfectly for some identifiable period. They demonstrate that these assumptions do not hold in every situation by simultaneously deploying 8 pH probes in wastewater and observing that all of them started drifting immediately. Consequently, the authors advocate for the development of new fault detection methods that are not based on these assumptions for use in such cases.

On the other hand, instead of focussing on fault detection as such, researchers such as Schneider *et al.* (2019) have investigated whether faulty data itself may be used with reasonable effectiveness despite their flaws. They concluded that unmaintained sensor signals were just as useful as maintained ones to detect specific process features (e.g. the 'knee' in an oxidation-reduction potential curve). Some types of sensors, however (e.g. dissolved oxygen) showed nonlinear disturbances that rendered their signal unusable when unmaintained.

Additionally, Rieger *et al.* (2010) recommend the use of mass balances inside plants to deduce missing information from a given sensor as well as to identify and correct for random and systematic gross errors in the data collection. Applying this procedure in process operation for data evaluation is referred to as data reconciliation. Important to the framework of data reconciliation are the concepts of measurement redundancy and variable observability. A methodology based on (bi)linear mass balances and specifically for WRRF data is proposed by Le *et al.* (2018) and Villez *et al.* 2020, setting up a measurement system to maximize the chance of detecting sensor faults and reconcile the faulty data.

Because of sensor faults and maintenance operations, it should be clear that no sensor boasts 100% uptime. Gap filling of data is a fact of life. Such gaps may be short or very long depending on the dynamics of the phenomenon being monitored and other factors. Different methods have been proposed to fill data gaps, the choice of which is influenced by the duration of the gap as well as the goal being pursued by the analyst. In their open-source wastewater data treatment toolkit, De Mulder *et al.* (2018) suggest five data imputation strategies specifically aimed at data generated by WRRF operation:

1. Interpolate.
2. Use a correlation with other available measurement signals.
3. Replace with a corresponding value in an average daily profile.
4. Repeat the values obtained on the preceding day.
5. Replace with the output of a model.

Though these gap filling methods may be used for reconstructing data series related to the plant itself, this is not their only use. Collecting influent water quality data, for example, is notoriously difficult and time consuming. This is unfortunate, as the state of the plant is directly related to the nature of the influent it is treating, both in terms of quantity and quality. Thus, water professionals may turn to influent generators to fill in the gaps in their influent data. These tools aim to generate complete time series from partial wastewater treatment plant influent data (Martin & Vanrolleghem 2014). As such, they replace expensive and time-consuming data collection campaigns; they allow modellers to infer concentrations of relevant wastewater components from other and correlated components, and they enable one to generate at will any number of instances of similar (and similarly plausible) time series. The generators can either be based on historical data from the influent of the plant (Devisscher *et al.* 2006) or characteristics of the upstream watershed and sewer network (Talebizadeh *et al.* 2016).

Say, in what folder was that already? – data storage and access

There exists a lag between the collection of data and its ultimate use. Hence, data storage and subsequent extraction are integral steps of data-driven systems, though they often go unmentioned in the WRRF literature.

The type of data being collected influences the choice of a storage system. In the case of offline laboratory data, manual record-keeping is still very often used, making data valorization a laborious task. However, laboratories nowadays have the option to use digital Laboratory Information Management Systems (LIMS) to organize their workflows and the results of their analyses in a central database (Skobelev *et al.* 2011). Similarly, data generated by sensors and other assets may be stored using a general-purpose database management system (DBMS) or an application-specific process data historian (Yee & Eren 2012). The difference between the two is that the latter is highly optimised in terms of writing, storing and extracting time series as a result of extensive data filtering, compression and caching. However, as data historians typically compress data to save storage space, their use may result in information loss. Indeed, compression reduces fidelity and increases the granularity of the data. Reconstruction of the original data without losing important statistical features is therefore not guaranteed (Thornhill *et al.* 2004).

In the pursuit of intelligent decision-making for WRRF operation, it is key that stakeholders have access to all relevant available data. Meaning that cross-subject and cross-facility data repositories such as spreadsheet logbooks, programmable logic controllers (PLC), supervisory control and data acquisition systems (SCADA), distributed control system (DCS), historians, lab information management system (LIMS), weather stations, computerized maintenance management systems (CMMS), enterprise asset management systems (EAM), and so on, need to be integrated and centrally accessible. This is where the concept of a data warehouse comes in. Defined by Inmon (1998), a data warehouse is a logically centralized data repository where cleaned data originating from operational data stores are integrated and standardized to support business intelligence. Whereas the operational data stores are typically used to answer short-term questions in real-time, the objective of a data warehouse is to provide decision support for mid to long-term organisational strategies. For WRRFs, the obvious long-term goal stakeholders are trying to get a grip on is to increase operational efficiency, with existing resources, while meeting regulatory compliance. Although the concept of data warehouses has been around for several decades and applied in various contexts, their application in the water and wastewater utility industry remains limited or unreported. Indeed, isolated data stores, without open-data interfaces for enterprise-wide access, are still the norm for most water and wastewater utilities (Sirkiä *et al.* 2017). Recently, software providers are trying to counteract the existence of such data silos by upgrading process historians into full functional data warehouses (Matthews 2017).

Data warehouses are mostly based on the relational database model to store data in a structured format; that is, in a collection of different tables and records, each containing pre-defined fields with data and metadata. Various tables are then linked together by relations between fields, which allows for the cross-referencing of data from separate tables. The database schema prescribes the content of each field of each table and the links between tables. In the case of time series, relationships between the data itself are limited although metadata can provide structural relationships. An example related to the field of water and wastewater quality is the schema developed by Plana *et al.* 2019, which aims to store related values coming from sampling stations within a water network, a sewershed or a WRRF with the relevant metadata (see Figure 3). For their part, OGC Consortium (2011) have suggested a markup language to store relevant metadata along with water data according to a general structure, though without prescribing any specific database configuration for storing such data.

Note that metadata such as sampling location, sensor serial number, sensor manufacturer, measurement unit, etc., are commonly understood as being created and stored early in the process of data collection. However, any step within the data pipeline could potentially generate its own metadata. For example, the use of a fault detection algorithm can generate metadata recording the specific version of the algorithm used, the value of each of its parameters, the subset of data that was used for calibration, etc. Though metadata obtained during collection may provide important context to its associated data, examples of frameworks for automatic generation of metadata further on in the pipeline; that is, post-collection, are rarer for the wastewater operations field. One example, however, is found in De Mulder *et al.* (2018). Though the FAIR guidelines are not explicitly followed in this instance, the authors nonetheless provide a structured framework to also store metadata generated during data pre-processing. Such frameworks are crucial if one is to try to replicate outcomes from experiments or models using raw data that passed through data cleaning and gap-filling procedures. Given a long enough time series, virtually all wastewater-related time series are bound to undergo some degree of processing, hence the crucial necessity of expanding on these frameworks for the entire pipeline.

As the data sources related to the operation of a WRRF multiply, schema-based data storage becomes increasingly strained to its limits. For example, the relationships between the data being collected may become unclear or left undefined as the variety of data increases, or the speed of data creation may well exceed the speed at which the data can be parsed and processed by the database server. One then enters the realm of big data, for which specific large-scale data storage systems are used. Data may be called 'big' according to the degree to which it presents the properties summarized by the four V's (Farley *et al.* 2018):

- *Volume* of data being generated: counting every sensor, actuator, alarm and type of laboratory measurement taking place in WRRFs, typical plants have several thousands, if not tens of thousands of tags being logged. This, of course, generates a lot of digital data; for example, a single measurement with a sampling period of 10 seconds will produce 8,640 records a day, amounting up to 3,153,600 records a year.
- *Velocity* of data creation and collection: on-line sensors and instruments placed throughout a WRRF can collect data at frequencies up to tens of measurements a minute. Often the information content of this data is, however, very low since the dynamics of the process being measured are much slower than the sensor's sampling frequency (Olsson 2012).
- *Veracity* of the data being collected: the harsh environment WRRF sensors find themselves in mean that faults are omnipresent. Pre-processing is therefore critical for the data to be trustworthy.

- *Variety* of collected data: most of the sensor data collected in WRRFs are time-series data. However, a lot of information at WRRFs is nowadays being stored in other unstructured data formats such as photos, videos, spectral measurements, instrument data sheets, standard operating procedures, etc.

Data lakes are a big data alternative to data warehouses that allow for the storage of both structured and unstructured data, without concern for indexing or making sure that the metadata fits a specific schema (Nargesian *et al.* 2018). This proves much more flexible than data warehousing and is, therefore, easier to implement when the variety of data is high. However, data lakes provide no guarantee regarding the integrity of the data they hold, which might result in data swamps rather than lakes. Additionally, since there is no enforcement of metadata upon writing, much of the context in which the data was collected may be lost. Instead, the user reading the data will have the task of reconstructing context from available clues (Liu & Gawlick 2015). The risk of data being overlooked, misused or corrupted is thus increased compared to data warehouses. Though the use of data lakes is increasing in the business intelligence world, no published example of an implementation for a WRRF has been identified by the authors of this review.

Data stored in databases can be accessed either through structured queries, as is the case for relational databases and data warehouses, or using metadata or pattern-recognition with data lakes and other unstructured data sources. To select data to be extracted and used, there needs to be a way for the data analysts to, as it were, know ahead of time what to search for. This is especially difficult to achieve when no schema exists to guide their search. There is thus a need for tools that enable one to preview, select and subset the available data. Query languages together with scripting languages are powerful interfaces, though they require programming literacy. Several closed and open-source software solutions have been developed (Demšar *et al.* 2013; Drucker & Fernandez 2015) over the years to enhance data exploration for a larger user base by creating graphical user interfaces that connect to structured and unstructured data sources alike.

Whereas data storage has traditionally been done on-site, there has been a shift in recent years to the adoption of off-site data storage using cloud computing. With virtually unlimited storage capacities and computing power, cloud solutions remove the burden of server acquisition, maintenance, and eventual upgrade away from their customers. However, such a system requires fast and reliable internet access, which is not available to WRFFs in remote locations. Specialized issues such as cyber security deserve special attention when working in the cloud, as the connection of WRRF data systems to the internet introduces the risk of unauthorized access and tampering (Blumensaat *et al.* 2019).

An aside on data mining

Though it is a separate process, data mining is linked with both data extraction and data analysis. Its increasing relevance, triggered by the ever-growing data sets of WRRFs, means that it warrants a brief discussion. Starting in the 1980s, access to more processing power has allowed for the adoption by businesses of data mining techniques, which are used to discover underlying relationships in data that were originally collected for another purpose. Lovell (1983) and Denton (1985) warned that this repurposing of data might lead researchers to find specious relationships in the mountains of data they analysed instead of detecting actual, meaningful relationships. They, therefore, underlined the necessity for researchers to remember that correlation does not imply causation when attempting data mining. Despite these reserves, data mining has flourished, and the water field has caught on as well.

Data mining should not be conflated with machine learning (ML) however – the former is the process of finding patterns in large amounts of data, while the latter is the process of using algorithms to find those patterns. Both activities are therefore unique but complementary, as both are required to automate the discovery of relevant trends and relationships in WRRF data.

Hadjimichael *et al.* (2016) have explored the potential of data mining for enhancing decision support systems (DSS) for urban water systems. They have found that the literature on data mining for DSS is very sparse compared to literature using data mining to create process models or for process optimization and that several obstacles stand in the way of the adoption of data mining in DSS systems. These challenges are just as applicable to the modelling field. They are that:

- Water professionals tend to lack the expertise in computer engineering required to develop adequate ML-derived models;
- Computer engineers do not possess the field-specific expertise needed to develop adequate ML-derived models for water systems on their own either;

- The tools and interfaces delivered to water professionals to explore their data are too difficult to use and lack adequate long-term support for them to be used in practice.

In light of these issues, it is clear that the wastewater field, though eager to leverage automated data analysis, generally lacks the required expertise to become proactive participants in its adoption and development for wastewater-related purposes.

This mix of eagerness and lack of expertise creates a perfect storm for the creation of data graveyards – large collections of data that never get used. These occur when data is collected without a clear motivation or purpose (Corominas *et al.* 2018). It is thus essential to treat the data pipeline as the critical part of the WRRF ecosystem that it is. Consequently, it must be just as carefully designed and maintained as the rest of the plant, with the collaboration of data governance experts.

A model is worth a thousand datasets – modelling

Modelling encodes, in a circumscribed form, all the knowledge and information relevant to a task. Knowledge may come from prior experience, whereas information must propagate from collected and processed data of the studied system. Though models are often formalized in mathematical language, they are not always. Sometimes, a schematic drawing or a written explanation of a phenomenon is all the modelling you need. However, such models can be considered more conceptual in nature. In the case of water resource recovery, however, mathematical modelling is extremely useful as it allows water professionals to perform virtual simulations of a WRRF system and its subprocesses. Simulations facilitate several tasks, such as the design of new plants, the optimization of existing ones, the prediction of the future behaviour of systems given an initial state, etc. To understand the current modelling trends, it is useful to define the types of models available to modellers developing operational support tools and tuning control systems. Models are typically categorized as follows:

1. White-box models are equivalent to mechanistic and phenomenological models in that their internal structure is legible and interpretable. Hence, they are deduced from first principles and are typically expressed as sets of differential algebraic equations to describe the steady-state or dynamic behaviour of a system (Gernaey *et al.* 2004). These models depend on data through their forcing input variables, but also for the calibration of their parameters, as well as for setting the initial and boundary conditions as required for the numeric calculations. White-box models can be further broken down into:
   a. Mechanistic models: these are based on physical laws; for example, the law of conservation of mass, or the laws of thermodynamics. They rigorously describe the behaviour of the system in an idealized form. A simple example of such a model is the mass transfer of oxygen (Garcia-Ochoa & Gomez 2009). Biofilm models, with their careful characterization of substrate transport through the biofilm (Pizarro *et al.* 2001) or the hydrodynamic flow of the bulk liquid in biofilm channels (Eberl *et al.* 2000) are other excellent examples of such models. Water resource recovery processes such as ammonia stripping and struvite extraction also use mechanistic models to characterize the chemical reactions underpinning these processes (Vaneeckhaute *et al.* 2018).
   b. Phenomenological models: these may be based on physical laws; however, they are not strictly beholden to them. Instead, these models include empirical relationships that describe the patterns of the observed phenomenon without having that description depend on the fundamental processes that generate the behaviour (Martin & Vanrolleghem 2014). For wastewater treatment, the most commonly used models belonging to this category surely are the Activated Sludge Model (ASM) family (Henze *et al.* 2000), which combines mechanistic mass balances of biochemical processes with heuristic relationships. Another widely used phenomenological model in the field of WRRF modelling is the settling model presented by Takács *et al.* (1991), which is based on mass balance equations, but relies on phenomenological descriptions of settling in activated sludge to model the storage and outflows of clarifiers.
2. Black-box models are models that map sets of inputs onto a certain output without any concern for embedding structured knowledge of the real processes that created these outputs. Because they contain no prior knowledge of the process they model, these models are completely dependent on the data being used to build them for their accuracy and applicability. Black-box models have proven to be very effective at finding unsuspected patterns in data generated by various application fields. The techniques employed for black-box modelling are varied, encompassing multivariate statistical models, time-series models (Box & Jenkins 1970), support vector machines (SVM) (Cortes & Vapnik 1995) and artificial neural networks (Werbos 1982). The basic choice of

each usually depends on whether the task at hand falls into the regression, classification or clustering categories. Haimi *et al.* (2013) provide an extensive overview of the various methods being applied for data-derived soft-sensors, specifically for biological wastewater treatment plants.

Because of recent advances in big data analytics and artificial intelligence, data-driven models will become progressively more widespread in the urban water field, hence strengthening the case for well-managed and readily available data (Garrido-Baserba *et al.* 2020). However, data-driven models have also been criticized for their lack of transparency, which could result in mistrust. A new line of explainability research has emerged with the aim of not only comprehending what a model did or might have done but also being able to question and audit the model (Gilpin *et al.* 2018). In this context, it is important to note that not all data-driven models are opaque to the user. For example, Principal Component Analysis (PCA) allows one to readily infer which variable of the input space contributed to the observed response in the output. The problem of explainability is, however, mostly encountered in the context of more sophisticated artificial intelligence (AI) methods based on neural networks.

3. Grey-box models, sometimes referred to as hybrid models, are a mixture of fundamental white-box and empirical black-box components. This category of models deserves special attention, as they combine the explanatory power inherent in first-principles models with the ability to detect subtle patterns in data. This hybrid formula is thus especially useful when white-box models contain parameters or state variables that are not readily evaluated experimentally (Psichogios & Ungar 1992). For example, Shiva Kumar & Venkateswarlu (2012) modelled a fixed bed biofilm reactor using a mechanistic model; however, they used an artificial neural network (ANN) to determine the kinetics of the growth rate, as it was unclear which mechanistic formulation was best suited to the behaviour of their biofilm. Also, Meirlaen (2002) trained a neural network to emulate the biokinetics of ASM2 while maintaining a mechanistic mass balance. Similarly, Lee *et al.* (2005) coupled an ASM1-derived model with different black-box models (namely, a Neural Network PLS scheme) to model the water quality of an industrial plant. They found that this approach not only yielded good performance, it also provided the authors with a readily interpretable signal to detect instances of unstable plant performance. It can thus be seen that the use of data-driven modelling schemes need not come at the cost of interpretability. Because of the growing interest of combining both domain-specific and data-derived knowledge, especially for complex nonlinear systems such as those encountered in WRRFs, new modelling frameworks are constantly being developed. The most notable hybrid modelling methods are currently coming from the emerging field of scientific machine learning (Baker *et al.* 2019), which lays the theoretical framework needed to embed mechanistic differential equations into deep learning models.

Putting on a friendly face – comprehension

As powerful as mathematics are at describing the physical world (possibly, even 'unreasonably effective' as Wigner (1960) famously remarked), the fact remains that very few humans are fluent in the language of maths. Our brains being what they are, most of us process information entirely differently than the machines running our models. There must, therefore, exist in the data pipeline interfaces that translate the information embodied in data, and the outputs of mathematical models, into a form that is more adapted to human cognition such as visual or linguistic models. Graphs are the most ubiquitous of these interfaces; however, all graphs are not created equal. Shah & Hoeffner (2002) explored the impact of graph formatting on the interpretation of data and found, among other things, that several parameters of graph design influence the ability of the user to make sense of the displayed information. More importantly, the authors also emphasize that these effects have different magnitudes based on the level of graph literacy of the user. This means that graphical tools used during model interpretation may need to be vastly different than those necessary to communicate those model results to wider audiences, namely WRRF operators, management and, why not, the general public. The translation of data from mathematical models to human actors must, therefore, be approached with care and with the user's needs in mind.

Knowing this, it must be noted that numerous efforts have been made towards concise but clear visual tools to present data and models (Figure 4). Amerlinck (2015) contributed a tool to quickly identify rate-limiting terms for processes of an ASM model using a colour scale. Similarly, Thürlimann *et al.* (2015) have developed a visual tool, based on colour bars and calendar-based views, for intuitive tracking of plant key performance indicators (KPIs) within a

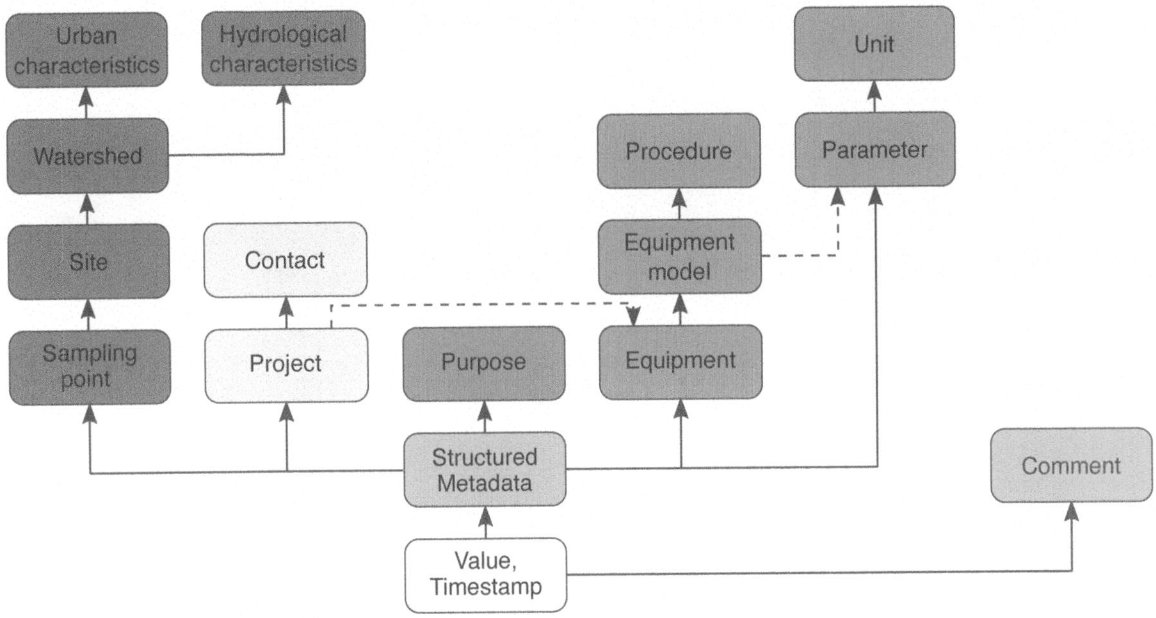

Figure 4 | Water-related metadata considered by Plana *et al.* (2019) (adapted from the original).

process optimisation software for WRRFs. Animation of simulation results may also prove useful. Stepping out of the WRRF context and into the catchment scale, Benedetti *et al.* (2006) proposed using diagrams with arrows of different widths (i.e. Sankey diagrams) to rapidly identify the sources and pathways of pollutants in a watershed. The same idea was also applied to the WRRF system by van der Hoek *et al.* (2018).

Besides presenting model outcomes, it is equally important to make the underlying modelling tools accessible to potential users. Dynamic process simulators for WRRFs, whether built in-house or commercially provided, are eminently fit to this task. By converting the complex mathematical equations of unit operations into drag-and-drop objects with user-friendly interfaces, the threshold to initiate a process modelling exercise is substantially decreased. As a result, simulators are nowadays used to design entire WRRFs, to support process operation, to help in the development of software sensors, and to train process operators. To improve model prediction accuracy, and consequently the perceived trustworthiness of the simulator, measurement data is used for calibration and validation. Although commercial WRRF simulators typically provide intuitive and automated model calibration tools, the implementation of data pre-processing and analysis tools remains somewhat limited. Including more powerful tools for data analysis, fault detection, gap filling, data-driven modelling and the like could be a big leap forward in the path towards even more successful WRRF simulators. With the advent of digital twins that include process simulators with a real-time data feed, the demand for efficient data treatment and hybrid process models will only rise in the predictable future.

Another important instrument used to disseminate information to plant operational staff is the human-machine interface (HMI). In the past, the design of such operator screens was entirely determined by the creativity of the automation engineers programming the system. The result was typically little more than copies of the P&IDs with confusing and distracting graphics. Such poor HMI practices prevent staff from operating plants near their most efficient point, and more importantly, they have been shown to contribute to major accidents in the process industries (HSE 1997). Nowadays, guidelines are available to avoid poor graphical principles during HMI design. Whereas data was previously scattered on the screen with a graphic of the process, the current trend is to add a high degree of context to the data. This way, continuous comparison is made possible, which simplifies the interpretation of complex process operation. Having knowledge directly embedded in the screens of such high-performance HMIs can drastically improve the situational awareness of process operators and thus decrease response time (Hollifield *et al.* 2008). In this context, Rieger & Olsson (2012) recognized the importance of clear visual communication of process control actions. They argue for embedding control actions and controller

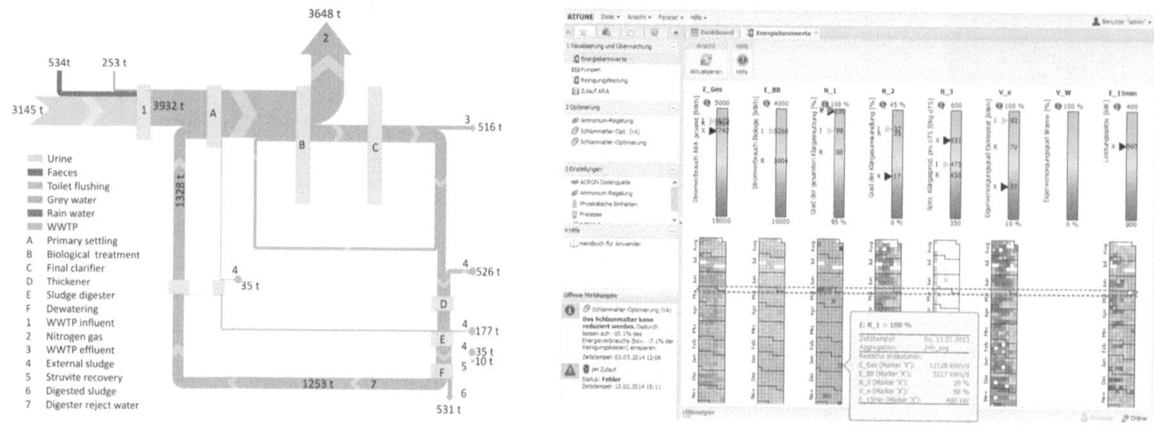

AVERAGE	AN01	AN02	AN03	AN04	DT01	DT02	BT01	BT02	BT03	BT04	BT05	BT06	CA
Monod S_NH	1.00	1.00	1.00	1.00	1.00	0.99	0.99	0.99	0.99	0.98	0.98	0.98	0.98
Monod S_A	0.91	0.91	0.92	0.92	0.63	0.52	0.08	0.06	0.04	0.04	0.03	0.03	0.03
Monod S_F	0.84	0.72	0.57	0.43	0.12	0.12	0.14	0.12	0.11	0.16	0.17	0.20	0.23
Monod S_Alk	0.99	0.99	0.99	0.99	0.99	0.99	0.99	0.99	0.99	0.99	0.99	0.99	0.99
Monod S_PO	1.00	1.00	1.00	1.00	1.00	1.00	0.98	0.94	0.81	0.41	0.30	0.22	0.21
Monod S_NO	0.01	0.00	0.00	0.00	0.09	0.18	0.84	0.84	0.84	0.84	0.85	0.85	0.85
Inhibition S_O	1.00	1.00	1.00	1.00	0.99	0.96	0.12	0.22	0.49	0.05	0.05	0.05	0.03

Figure 5 | Top left: Sankey diagram of nitrogen flow through a WRRF) (van der Hoek *et al.* 2018); Top right: Decision support tool used to visualize WRRF key performance indicators based on energy and process data (Thürlimann *et al.* 2015); Bottom: Systems analysis tool that uses colour coding to identify growth limitations in different sections of an activated sludge tank (Amerlinck 2015).

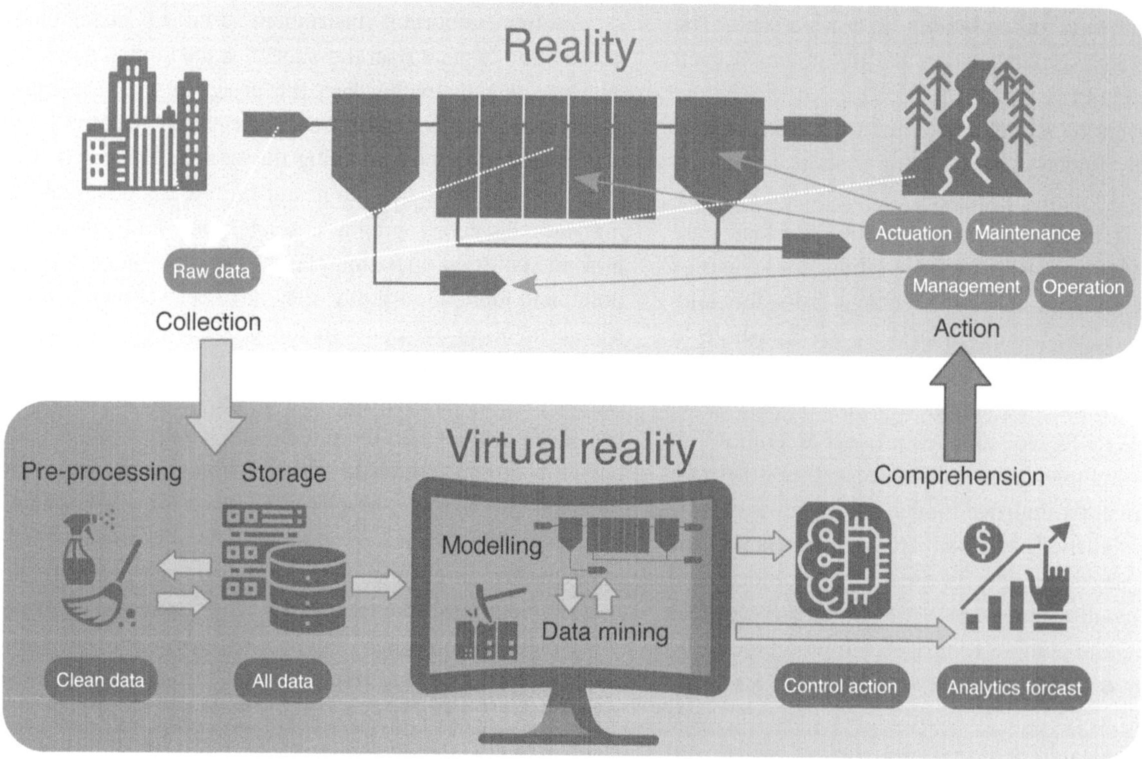

Figure 6 | Components of a digital twin.

settings directly in SCADA HMIs, along with the use of hierarchical parameter displays to guide users to the information most likely to help them accomplish their tasks. It can thus be concluded that, though seemingly simple, visual and intuitive components of the data pipeline are indispensable to ingest the massive amounts of information coming from data and models.

Is there a faucet on this thing? – action

For data to be truly useful, it must be put to work. One way of doing so is through the deployment of automatic process controllers and process models. The role of data in each of those is explored here.

Automation is the process of creating systems that can perform tasks without human intervention. The result is that process operators are alleviated from demanding repetitive tasks required to keep the plant running. In WRRFs, valve opening and shuttering, pump timings and sensor cleaning are among the candidates for automation. The first attempts at automated plant control were undertaken in the 1970s (Olsson 2012). Already at that time, it was clear to some researchers that massive gains in efficiency were possible by using automated controllers that encapsulated some knowledge about the plant. A lot of effort was applied to control the aeration process in particular, as aeration accounts for 45–75% of all wastewater treatment energy expenditures (Rosso *et al.* 2008). Early control strategies consisted of simply maintaining dissolved oxygen (DO) concentration in bioreactors at a fixed low value with the help of conventional feedback controllers. More complex schemes appeared over time, such as DO cascade control (Ingildsen *et al.* 2002), ammonium-based aeration control (Rieger *et al.* 2013) and multiple input multiple output control strategies, which modify the system setpoints to achieve the required oxidation capacity for shifting conditions within the plant (Åmand *et al.* 2013). This increase in complexity marks the shift from regulatory control – automatically manipulating an actuator to reach a setpoint value – to supervisory control – manipulating the set point itself to reach a higher-order objective (Ayesa *et al.* 2006).

As the complexity of control strategies grows, the likelihood that a component critical to that strategy will fail increases. It is therefore important to implement these strategies within a fault-tolerant framework, in which the process can either be automatically transferred to a fallback strategy or a safe parking-point until the fault is identified and corrected (Mhaskar *et al.* 2013). The use of fault-tolerant control is reviewed thoroughly by Blanke *et al.* (1997) and Zhang & Jiang (2008).

In addition to automatic control systems, one must not forget the people who oversee and support the automated systems. If successfully executed, the passage of data through the entire pipeline empowers those people to have a clear view of the state of, and power to act on, the plant through the automated system. The power to act hinges on the final transformation of knowledge into intelligence and intelligent action. Although this transformation is eased by visual interfaces, it ultimately occurs in the minds of water professionals. This means that the tools springing from the data pipeline must help workers synthesize their knowledge of the plant in ways that enable them to act intelligently on it. These tools come in many different shapes and forms. Managers can benefit from the data by having it embedded in a decision support system (Hadjimichael *et al.* 2016). Operators can benefit from dashboards indicating the state of sensors or key performance indicators (KPI) related to the processes they oversee (Thürlimann *et al.* 2015). Even the public can get involved with wastewater data if given the appropriate tools. For example, see the citizen science project of Damman *et al.* (2019), involving a community to sample and analyse the quality of their rivers downstream of their local wastewater treatment plant.

Of course, control for the sake of control achieves nothing: the control goal and whether that goal is reflective of the plant stakeholder's interests is what ultimately makes these systems effective. Weijers (2000) proposes a systematic methodology to derive appropriate control goals and the associated constraints for a given plant goal, as opposed to working from vague intentions such as 'minimizing costs while maximizing water quality'.

For their part, Rieger & Olsson (2012) remark that the stakeholders of wastewater treatment plants may vastly disagree on what constitutes a good control objective. In their view, this is because each of the stakeholders acts under contradictory incentives. Thus, the human aspect of control, including goal setting and the relationships between stakeholders and co-workers, is found to have a tremendous impact on the success of control strategies. Rieger & Olsson (2012) add:

> *'There has to be a qualified team of people who feel a deep sense of ownership of the system and the WWTP, and who are committed to its continuous improvement. It is important that all employees increase their competence through continual education.'*

It is thus crucial that the complex human interactions that form the backbone of wastewater treatment systems stay in view when designing technical solutions seeking to enhance these systems. Nevertheless, there is no denying that technology is a strong ally to the wastewater field in their quest to turn their data into intelligent actions.

DIGITAL TWINS

In recent years, a lot of excitement has been sparked by the perspective of creating so-called digital twins for WRRFs (IWA and Xylem Inc. 2019). In the context of water management, Kolditz *et al.* (2019) define digital twins as virtual systems that 'contain all important characteristics and features of the real system, depending on the specific purpose for an application'. This vague definition does not explain in detail what a digital twin is. It does, however, suggest the following key features:

1. Digital twins are virtual systems that aim to embody and simulate the physical components of a real system. Therefore, not physically existing as such, but made by software to appear to do so.
2. A digital twin sources its data from measurements performed on the physical system itself or from its environment.
3. The digital twin constantly mirrors the current state of the physical plant. Consequently, some of the data the digital twin requires needs to be provided in real-time.
4. The digital twin, having access to the current state of the plant, can make predictions on the future state of the physical system.
5. The predictions that are produced can be fed back into the real physical system in the form of intelligent actions. Depending on the intrinsic characteristics of the system, these actions can be made automatically or manually.

It is easy to see that the concept of a digital twin is reliant on every part of the data pipeline. Indeed, it cannot exist without extensive, fault-free and continuous data from the plant and its environment; it embodies knowledge of the plant through a constantly updated model; it provides insight into the plant via interfaces, visualization and analytics, and it allows for action on the plant via automatic control and the insight it generates for water professionals.

Moreover, the digital twin concept for process industries is often thought of as a close cousin of a model-based control system used to automatically optimise plant performance. To the authors, this point of view limits itself to the continuous operation of the process, meaning that it does not take into account the potential value generated by a digital twin in other aspects of a plant, namely plant maintenance and asset management. Consensus on which components must be present in a digital system for it to be called a digital twin is therefore much needed so as not to misuse this powerful concept. Note that this effort has already been initiated in other industries and other branches of the water industry, including the field of water and wastewater networks (Water Online and SWAN 2019).

LIGHT AT THE END OF THE PIPE?

This review attempted to inventory the steps required to turn raw data into intelligent action for the operation of a WRRF. As these steps were discussed, possible problems were pointed out, as well as possible ways to cope with them. The main sticking points discussed in this review are gathered in Table 1.

For some of these issues (e.g. fault detection or data gaps), there seems to exist convincing technological and scientific tools to alleviate the problems. For others, however, the way forward is simply the continued dedication of water professionals (e.g. sensor maintenance), or their willingness to collaborate with or be trained by experts from other fields (e.g. development of a data warehousing strategy or adapted user interfaces).

The nature of the issues encountered along the pipeline and their potential fixes may be diverse, but they certainly will all require a significant amount of effort and dedication to effectively tackle. This has implications for research, of course, but also for the practical applicability of smart wastewater applications such as the digital twin to WRRFs around the world. Indeed, since the success of such a system requires the creation and the maintenance of an integrated data pipeline, as well as considerable modelling efforts to implement relevant control strategies within the twin, only the most sophisticated WRRFs may be able to attempt to create such a system for many years to come. Nonetheless, progress in the development of robust data pipelines may very well have benefits for smaller WRRFs as well. Indeed, one can easily imagine remote WRRFs benefitting from better fault detection for remote monitoring and from gap-filling to reconstruct faulty time series between maintenance operations, for example. Thus, as WRRFs brace themselves for a smart future, hopefully water professionals will consider patching leaks in the data pipeline to be much more than a chore, but rather see it as participating in the development of a sophisticated system that is just as complex and

Table 1 | Summary of data-related issues encountered in WRRFs

Step of the pipeline	Issue encountered	Paths for improvement
Collection	Harsh WRRF environment negatively affects online sensors	• Careful, regular, sensor maintenance • Fault detection procedures
	Lab data is often recorded manually.	• Use of LIMS software
	Maintenance of sensors can be resource intensive.	• Define a strategy for the entire lifecycle of the sensor before installing.
	Enthusiasm for collection without purpose leads to data graveyards.	• Treat data collection as its own process; let it be engineered by experts, carefully planned and rigorously executed. • Develop a strategy ahead of time for data use.
Pre-processing	Manual fault detection is too time consuming.	• Univariate automatic fault detection algorithms • Multivariate automatic fault detection
	Automatically captured and treated data contain gaps.	• Time-series generators • Model-based gap filling • Interpolation
	Difficult to know which data has been changed by pre-processing algorithms, and what algorithm was used.	• Keep thorough metadata accounting for pre-processing • Data versioning • Version control of pre-processing algorithms
	Fault detection algorithms often assume we can pinpoint a period of 'good data', but sensors may begin to drift immediately after commissioning.	• Sensor redundancy • Mass balances for data reconciliation • Use signal features that are identifiable even if the sensor is unmaintained/faulty
Storage and access	Data historians use destructive compression	• Consider signal properties when choosing a signal resolution
	Simple databases are ill-suited to handle WRRF 'big data'.	• Data warehouses • Data lakes
	Data is often spread in multiple storage sites and is thus difficult to access.	• Improved collaboration between data experts and water professionals.
	IT infrastructure may be difficult to maintain on-premise because of lack of technical know-how.	• Increased use of cloud computing • Training
	Extracting data requires prior knowledge of what data is available, and water professionals do not have access to adequate interfaces to explore their data.	• Access to data exploration software • Collaboration between data workers and water professionals
Data mining	Mining data for relationships increases the odds of finding spurious correlations.	• Stronger statistical tests to determine significance. • Extensive use of domain-specific knowledge.
Modelling	Water professionals don't have the required skills to build data-driven models, while data scientists don't have the WRRF-specific knowledge required to develop adequate wastewater treatment models.	• Strong collaboration between water professionals and data scientists. • Extensive use of domain-specific knowledge. • Training
	Simulation software is mostly aimed at mechanistic modelling and doesn't provide tools to create data-driven models	• Provide an integrated toolchain that supports more modelling methodologies.

(*continued*)

Table 1 | continued

Step of the pipeline	Issue encountered	Paths for improvement
Comprehension	Computers think with math; humans don't.	• Synthesis of information via interactive dashboards, visualizations and reports. • Use of colour, shape, spatial placement. • Provide adapted interfaces for each task.
	Human-machine interfaces are sometimes confusing and distracting	• Make use of high-performance HMI guidelines. • Collaborate with UI designers to create powerful interfaces.
Action	Complex control strategies are prone to fail when data quality drops.	• Online automatic fault detection. • Controller reconfiguration – automatic switch to fallback strategy.
	Control objectives are often ill-defined among different plant stakeholders.	• Systematic objective definition. • Continuous communication and collaboration between all stakeholders.

fascinating as wastewater treatment and resource recovery themselves.

ACKNOWLEDGEMENTS

This work has been carried out at Université Laval, supported financially by the Natural Sciences and Engineering Research Council of Canada (NSERC) through the award of an NSERC Discovery Grant awarded to Peter Vanrolleghem (grant number RGPIN-2016-06522) and an NSERC ES D PhD scholarship for Jean-David Therrien (grant number PGSD3-519336-2018). Peter Vanrolleghem holds the Canada Research Chair on Water Quality Modelling.

DATA AVAILABILITY STATEMENT

All relevant data are included in the paper or its Supplementary Information.

REFERENCES

Alferes, J. & Vanrolleghem, P. A. 2016 Efficient automated quality assessment: dealing with faulty on-line water quality sensors. *AI Communications* **29** (6), 701–709. doi:10.3233/AIC-160713.

Alferes, J., Tik, S., Copp, J. B. & Vanrolleghem, P. A. 2013 Advanced monitoring of water systems using in situ measurement stations: data validation and fault detection. *Water Science and Technology* **68** (5), 1022–1030. doi:10.2166/wst.2013.302.

Alferes, J., Copp, J. B., Weijers, S. & Vanrolleghem, P. A. 2015 Validating data quality for water quality monitoring: Objective comparison of three data quality assessment approaches. In: *Proceedings of the New Developments in IT & Water Conference*, February 8–10, 2015, Rotterdam, The Netherlands.

Al-Omari, A., Wett, B., Nopens, I., De Clippeleir, H., Han, M., Regmi, P., Bott, C. & Murthy, S. 2015 Model-based evaluation of mechanisms and benefits of mainstream shortcut nitrogen removal processes. *Water Science and Technology* **71** (6), 840–847. doi:10.2166/wst.2015.022.

Åmand, L., Olsson, G. & Carlsson, B. 2013 Aeration control – a review. *Water Science and Technology* **67** (11), 2374–2398. doi:10.2166/wst.2013.139.

Amaral, A., Schraa, O., Rieger, L., Gillot, S., Fayolle, Y., Bellandi, G., Amerlinck, Y., Gori, R., Neves, R. & Nopens, I. 2017 Towards advanced aeration modelling: from blower to bubbles to bulk. *Water Science and Technology* **75** (3), 507–517. doi:10.2166/wst.2016.365.

Amerlinck, Y. 2015 *Model Refinements In View of Wastewater Treatment Plant Optimization: Improving The Balance In Sub-Model Detail.* PhD Thesis, Department of Mathematical Modelling, Statistics and Bioinformatics, Ghent University, Ghent, Belgium, pp. 289.

Ayesa, E., De la Sota, A., Grau, P., Sagarna, J. M., Salterain, A. & Suescun, J. 2006 Supervisory control strategies for the new WWTP of Galindo-Bilbao: the long run from the conceptual design to the full-scale experimental validation. *Water Science and Technology* **53** (4–5), 193–201. doi:10.2166/wst.2006.124.

Aymerich, I., Rieger, L., Sobhani, R., Rosso, D. & Corominas, L. 2015 The difference between energy consumption and energy cost: modelling energy tariff structures for water resource recovery facilities. *Water Research* **81** (2015), 113–123. doi:10.1016/j.watres.2015.04.033.

Baker, N., Alexander, F., Bremer, T., Hagberg, A., Kevrekidis, Y., Najm, H., Parashar, M., Patra, A., Sethian, J., Wild, S., Wilcox, K., Karen, L. & Lee, S. 2019 *Workshop Report on Basic Research Needs for Scientific Machine Learning: Core Technologies for Artificial Intelligence*. Richland, WA, USA, p. 96.

Benedetti, L., Dirckx, G., Bixio, D., Thoeye, C. & Vanrolleghem, P. A. 2006 Substance flow analysis of the wastewater collection and treatment system. *Urban Water Journal* **3** (1), 33–42. doi:10.1080/15730620600578694.

Blanke, M., Izadi-Zamanabadi, R., Bøgh, S. A. & Lunau, C. P. 1997 Fault-tolerant control systems – a holistic view. *Control Engineering Practice* **5** (5), 693–702. doi:10.1016/S0967-0661(97)00051-8.

Blumensaat, F., Leitão, J. P., Ort, C., Rieckermann, J., Scheidegger, A., Vanrolleghem, P. A. & Villez, K. 2019 How urban storm- and wastewater management prepares for emerging opportunities and threats: digital transformation, ubiquitous sensing, new data sources, and beyond – a horizon scan. *Environmental Science and Technology* **53** (15), 8488–8498. doi:10.1021/acs.est.8b06481.

Box, G. E. P. & Jenkins, G. M. 1970 *Time Series Analysis, Forecasting, and Control*, 2nd edn. Holden-Day Publishing, San Francisco, CA, USA, pp. 553.

Corominas, L., Villez, K., Aguado, D., Rieger, L., Rosén, C. & Vanrolleghem, P. A. 2011 Performance evaluation of fault detection methods for wastewater treatment processes. *Biotechnology and Bioengineering* **108** (2), 333–344. doi:10.1002/bit.22953.

Corominas, L., Garrido-Baserba, M., Villez, K., Olsson, G., Cortés, U. & Poch, M. 2018 Transforming data into knowledge for improved wastewater treatment operation: a critical review of techniques. *Environmental Modelling and Software* **106** (August), 89–103. doi:10.1016/j.envsoft.2017.11.023.

Cortes, C. & Vapnik, V. 1995 Support-vector networks. *Machine Learning* **20**, 273–297. doi:10.1007/BF00994018.

Damman, S., Helness, H., Grindvoll, I. L. T. & Sun, C. 2019 Citizen science to enhance evaluation of local wastewater treatment – a case study from Oslo. *Water Science and Technology* **79** (10), 1887–1896. doi:10.2166/wst.2019.180.

Demšar, J., Curk, T., Erjavec, A., Gorup, Č., Hočevar, T., Milutinovič, M., Možina, M., Polajnar, M., Toplak, M., Starič, A., Štajdohar, M., Umek, L., Žagar, L., Žbontar, J., Žitnik, M. & Zupan, B. 2013 Orange: data mining toolbox in Python. *Journal of Machine Learning Research* **14**, 2349–2353.

De Mulder, C., Flameling, T., Weijers, S., Amerlinck, Y. & Nopens, I. 2018 An open software package for data reconciliation and gap filling in preparation of water and resource recovery facility modeling. *Environmental Modelling and Software* **107** (September), 186–198. doi:10.1016/j.envsoft.2018.05.015.

Denton, F. T. 1985 Data mining as an industry. *The Review of Economics and Statistics* **67** (1), 124–127. doi:10.2307/1928442.

Devisscher, M., Clacci, G., Fé, L., Benedetti, L., Bixio, D., Thoeye, C., De Gueldre, G., Marsili-Libelli, S. & Vanrolleghem, P. A. 2006 Estimating costs and benefits of advanced control for wastewater treatment plants – the MAgIC methodology. *Water Science and Technology* **53** (4–5), 215–223. doi:10.2166/wst.2006.126.

Drucker, S. M. & Fernandez, R. 2015 *A Unifying Framework for Animated and Interactive Unit Visualizations. Technical Report. Microsoft Research*. p. 9. Available from: http://research.microsoft.com/pubs/252104/sanddance.pdf (accessed 26 May 2020).

Dürrenmatt, D. J. & Gujer, W. 2012 Data-driven modeling approaches to support wastewater treatment plant operation. *Environmental Modelling and Software* **30** (April), 47–56. doi:10.1016/j.envsoft.2011.11.007.

Eberl, H. J., Picioreanu, C., Heijnen, J. J. & Van Loosdrecht, M. C. M. 2000 Three-dimensional numerical study on the correlation of spatial structure, hydrodynamic conditions, and mass transfer and conversion in biofilms. *Chemical Engineering Science* **55** (24), 6209–6222. doi:10.1016/S0009-2509(00)00169-X.

Farley, S. S., Dawson, A., Goring, S. J. & Williams, J. W. 2018 Situating ecology as a big-data science: current advances, challenges, and solutions. *BioScience* **68** (8), 563–576. doi:10.1093/biosci/biy068.

Garcia-Ochoa, F. & Gomez, E. 2009 Bioreactor scale-up and oxygen transfer rate in microbial processes: an overview. *Biotechnology Advances* **27** (2), 153–176. doi:10.1016/j.biotechadv.2008.10.006.

Garneau, C. & Vanrolleghem, P. A. 2018 Neural network for tuning-friendly automatic outlier detection in water quality time series. In *Proceedings of the 9th International Congress on Environmental Modelling and Software, Fort Collins, CO, USA*.

Garrido-Baserba, M., Corominas, L., Cortés, U., Rosso, D. & Poch, M. 2020 The fourth-revolution in the water sector encounters the digital revolution. *Environmental Science and Technology* **54** (8), 4698–4705. doi:10.1021/acs.est.9b04251.

Gernaey, K. V., Van Loosdrecht, M. C. M., Henze, M., Lind, M. & Jørgensen, S. B. 2004 Activated sludge wastewater treatment plant modelling and simulation: state of the art. *Environmental Modelling and Software* **19** (9), 763–783. doi:10.1016/j.envsoft.2003.03.005.

Gilpin, L. H., Bau, D., Yuan, B. Z., Bajwa, A., Specter, M. & Kagal, L. 2018 Explaining explanations: an overview of interpretability of machine learning. In: *Proceedings of the 2018 IEEE 5th International Conference on Data Science and Advanced Analytics (DSAA)*, October 1–4 2018, *Turin, Italy*, pp. 80–89.

Hadjimichael, A., Comas, J. & Corominas, L. 2016 Do machine learning methods used in data mining enhance the potential of decision support systems? A review for the urban water sector. *AI Communications* **29** (6), 747–756. doi:10.3233/AIC-160714.

Haimi, H., Mulas, M., Corona, F. & Vahala, R. 2013 Data-derived soft-sensors for biological wastewater treatment plants: an overview. *Environmental Modelling and Software* **47**, 88–107. doi:10.1016/j.envsoft.2013.05.009.

Henze, M., Gujer, W., Mino, T. & van Loosdrecht, M. C. M. 2000 *Activated Sludge Models. Reprint. Technical Report.*

Scientific and Technical Report Series. IWA Publishing, London, UK, p. 130.

Hernández-del-Olmo, F., Gaudioso, E., Dormido, R. & Duro, N. 2016 Energy and environmental efficiency for the N-ammonia removal process in wastewater treatment plants by means of reinforcement learning. *Energies* **9** (9), 755. doi:10.3390/en9090755.

Hernández-del-Olmo, F., Gaudioso, E., Duro, N. & Dormido, R. 2019 Machine learning weather soft-sensor for advanced control of wastewater treatment plants. *Sensors (Switzerland)* **19** (14), 1–12. doi:10.3390/s19143139.

Hollifield, B. R., Oliver, D., Nimmo, I. & Habibi, E. 2008 *The High Performance HMI Handbook: A Comprehensive Guide to Designing, Implementing and Maintaining Effective HMIs for Industrial Plant Operations*, 1st edn. PAS, Inc., Houston, TX, USA, p. 206.

HSE GB 1997 *The Explosion and Fires at the Texaco Refinery, Milford Haven, 24 July 1994: A Report of the Investigation by the Health and Safety Executive Into the Explosion and Fires on the Pembroke Cracking Company Plant at the Texaco Refinery, Milford Haven on 24 J. Incident Report Series*. HSE Books, London, UK, p. 66.

Ingildsen, P. & Olsson, G. 2016 *Smart Water Utilities: Complexity Made Simple*. IWA Publishing, London, UK, p. 304.

Ingildsen, P., Jeppsson, U. & Olsson, G. 2002 Dissolved oxygen controller based on on-line measurements of ammonium combining feed-forward and feedback. *Water Science and Technology* **45** (4–5), 453–460. doi:10.2166/wst.2002.0649.

Inmon, W. H. 1998 *Building the Data Warehouse*, 3rd edn. Wiley Computer Publishing, New York, NY, USA, p. 412.

Isermann, R. & Ballé, P. 1997 Trends in the application of model-based fault detection and diagnosis of technical processes. *Control Engineering Practice* **5** (5), 709–719. doi:10.1016/S0967-0661(97)00053-1.

IWA and Xylem Inc. 2019 *Digital Water Report. Technical Report*. London, UK, p. 44.

Jimenez, J., Latrille, E., Harmand, J., Robles, A., Ferrer, J., Gaida, D., Wolf, C., Mairet, F., Bernard, O., Alcaraz-Gonzalez, V., Mendez-Acosta, H., Zitomer, D., Totzke, D., Spanjers, H., Jacobi, F., Guwy, A., Dinsdale, R., Premier, G., Mazhegrane, S., Ruiz-Filippi, G., Seco, A., Ribeiro, T., Pauss, A. & Steyer, J. P. 2015 Instrumentation and control of anaerobic digestion processes: a review and some research challenges. *Reviews in Environmental Science and Biotechnology* **14** (4), 615–648. doi:10.1007/s11157-015-9382-6.

Kolditz, O., Rink, K., Nixdorf, E., Fischer, T., Bilke, L., Naumov, D., Liao, Z. & Yue, T. 2019 Environmental information systems: paving the path for digitally facilitated water management (Water 4.0). *Engineering* **5** (5), 828–832. doi:10.1016/j.eng.2019.08.002.

Korodi, A., Radu, M. A. & Crisan, R. 2018 Non-invasive control solution inside higher-level OPC UA based wrapper for optimizing groups of wastewater systems. In: *Proceedings of the IEEE International Conference on Emerging Technologies and Factory Automation (ETFA)*, September 4–7 2018, *Turin, Italy*, pp. 597–604.

Le, Q. H., Verheijen, P. J. T., van Loosdrecht, M. C. M. & Volcke, E. I. P. 2018 Experimental design for evaluating WWTP data by linear mass balances. *Water Research* **142**, 415–425. doi:10.1016/j.watres.2018.05.026.

Lee, D. S., Vanrolleghem, P. A. & Jong, M. P. 2005 Parallel hybrid modeling methods for a full-scale cokes wastewater treatment plant. *Journal of Biotechnology* **115** (3), 317–328. doi:10.1016/j.jbiotec.2004.09.001.

Liu, Z. H. & Gawlick, D. 2015 Management of flexible schema data in RDBMSs – opportunities and limitations for NoSQL-. In: *Proceedings of the 7th Biennial Conference on Innovative Data Systems Research (CIDR)*, January 4–7 2015, *Asilomar, CA, USA*.

Lovell, M. C. 1983 Data mining. *The Review of Economics and Statistics* **65** (1), 1–12. doi:10.2307/1924403.

Makropoulos and Savić 2019 Urban hydroinformatics: past, present and future. *Water (Switzerland)* **11** (10), 1959. doi:10.3390/w11101959.

Mannina, G., Ekama, G., Caniani, D., Cosenza, A., Esposito, G., Gori, R., Garrido-Baserba, M., Rosso, D. & Olsson, G. 2016 Greenhouse gases from wastewater treatment – a review of modelling tools. *Science of the Total Environment* **551–552** (1 May 2016), 254–270. doi:10.1016/j.scitotenv.2016.01.163.

Martin, C. & Vanrolleghem, P. A. 2014 Analysing, completing, and generating influent data for WWTP modelling: a critical review. *Environmental Modelling & Software* **60** (October 2014), 188–201. doi:10.1016/j.envsoft.2014.05.008.

Matthews, W. 2017 *New Roles for Process Historians. InTech Magazine* Nov–Dec, https://ww2.isa.org/intech/20171202/.

Meirlaen, J. 2002 *Immission Based Real-Time Control of the Integrated Urban Wastewater System*. PhD Thesis, Department of Mathematical Modelling, Biometrics and Process Control, Gent University, Ghent, Belgium, p. 246.

Mhaskar, P., Liu, J. & Christofides, P. D. 2013 *Fault-Tolerant Process Control: Methods and Applications*. Springer London, London, UK, p. 261.

Nargesian, F., Zhu, E., Miller, R. J., Pu, K. Q. & Arocena, P. C. 2018 Data lake management: challenges and opportunities. *Proceedings of the VLDB Endowment* **12** (12), 1986–1989. doi:10.14778/3352063.3352116.

Newhart, K. B., Holloway, R. W., Hering, A. S. & Cath, T. Y. 2019 Data-driven performance analyses of wastewater treatment plants: a review. *Water Research* **157**, 498–513. doi:10.1016/j.watres.2019.03.030.

OGC Consortium 2011 OGC WaterML 2.0: Part 1- Timeseries. Measurement (1–151). http://www.opengeospatial.org/.

Ohmura, K., Thürlimann, C. M., Kipf, M., Carbajal, J. P. & Villez, K. 2019 Characterizing long-term wear and tear of ion-selective pH sensors. *Water Science and Technology* **80** (3), 541–550. doi:10.2166/wst.2019.301.

Olsson, G. 2012 ICA and me – a subjective review. *Water Research* **46** (6), 1585–1624. doi:10.1016/j.watres.2011.12.054.

Patry, B. 2020 *Suivi, Compréhension et Modélisation d'une Technologie à Biofilm Pour l'augmentation de La Capacité Des Étangs Aérés*. PhD Thesis, Département de génie civil et de génie des eaux, Université Laval, Québec, Canada, p. 190.

Pizarro, G., Griffeath, D. & Noguera, D. R. 2001 Quantitative cellular automaton model for biofilms. *Journal of Environmental Engineering* **127** (9), 782–789. doi:10.1061/(ASCE)0733-9372(2001)127:9(782).

Plana, Q., Alferes, J., Fuks, K., Kraft, T., Maruéjouls, T., Torfs, E. & Vanrolleghem, P. A. 2019 Towards a water quality database for raw and validated data with emphasis on structured metadata. *Water Quality Research Journal* **54** (1), 1–9. doi:10.2166/wqrj.2018.013.

Psichogios, D. C. & Ungar, L. H. 1992 A hybrid neural network-first principles approach to process modeling. *AIChE Journal* **38** (10), 1499–1511. doi:10.1002/aic.690381003.

Regmi, P., Holgate, B., Fredericks, D., Miller, M. W., Wett, B., Murthy, S. & Bott, C. B. 2015 Optimization of a mainstream nitritation-denitritation process and anammox polishing. *Water Science and Technology* **72** (4), 632–642. doi:10.2166/wst.2015.261.

Regmi, P., Stewart, H., Amerlinck, Y., Arnell, M., García, P. J., Johnson, B., Maere, T., Miletić, I., Miller, M., Rieger, L., Samstag, R., Santoro, D., Schraa, O., Snowling, S., Takács, I., Torfs, E., van Loosdrecht, M. C. M., Vanrolleghem, P. A., Villez, K., Volcke, E. I. P., Weijers, S., Grau, P., Jimenez, J. & Rosso, D. 2019 The future of WRRF modelling – outlook and challenges. *Water Science and Technology* **79** (1), 3–14. doi:10.2166/wst.2018.498.

Revollar, S., Vilanova, R., Vega, P., Francisco, M. & Meneses, M. 2020 Wastewater treatment plant operation: simple control schemes with a holistic perspective. *Sustainability (Switzerland)* **12** (3), 1–28. doi:10.3390/su12030768.

Rieger, L. & Olsson, G. 2012 Why many control systems fail. *Water Environment and Technology* **2012** (June), 42–45. doi:10.2175/193864711802764779.

Rieger, L. & Vanrolleghem, P. A. 2008 monEAU: a platform for water quality monitoring networks. *Water Science and Technology* **57** (7), 1079–1086. doi:10.2166/wst.2008.135.

Rieger, L., Alex, J., Winkler, S., Boehler, M., Thomann, M. & Siegrist, H. 2003 Progress in sensor technology – progress in process control? Part I: sensor property investigation and classification. *Water Science and Technology* **47** (2), 103–112. doi:10.2166/wst.2003.0096.

Rieger, L., Takács, I., Villez, K., Siegrist, H., Lessard, P., Vanrolleghem, P. A. & Comeau, Y. 2010 Data reconciliation for wastewater treatment plant simulation studies – planning for high-quality data and typical sources of errors. *Water Environment Research* **82** (5), 426–433. doi:10.2175/106143009×12529484815511.

Rieger, L., Jones, R. M., Dold, P. L. & Bott, C. B. 2013 Ammonia-based feedforward and feedback aeration control in activated sludge processes. *Water Environment Research* **86** (1), 63–73. doi:10.2175/106143013×13596524516987.

Rittmann, B. E., Boltz, J. P., Brockmann, D., Daigger, G. T., Morgenroth, E., Sørensen, K. H., Takács, I., van Loosdrecht, M. C. M. & Vanrolleghem, P. A. 2018 A framework for good biofilm reactor modeling practice (GBRMP). *Water Science and Technology* **77** (5), 1149–1164. doi:10.2166/wst.2018.021.

Rosén, C. & Olsson, G. 1998 Disturbance detection in wastewater treatment plants. *Water Science and Technology* **37** (12), 197–205. doi:10.2166/wst.1998.0542.

Rosén, C., Jeppsson, U. & Vanrolleghem, P. A. 2004 Towards a common benchmark for long-term process control and monitoring performance evaluation. *Water Science and Technology* **50** (11), 41–49. doi:10.2166/wst.2004.0669.

Rosén, C., Rieger, L., Jeppsson, U. & Vanrolleghem, P. A. 2008 Adding realism to simulated sensors and actuators. *Water Science and Technology* **57** (3), 337–344. doi:10.2166/wst.2008.130.

Rosso, D., Larson, L. E. & Stenstrom, M. K. 2008 Aeration of large-scale municipal wastewater treatment plants: state of the art. *Water Science and Technology* **57** (7), 973–978. doi:10.2166/wst.2008.218.

Russo, S., Disch, A., Blumensaat, F. & Villez, K. 2019 Anomaly detection using deep autoencoders for in-situ wastewater systems monitoring data. In: *Proceedings of the 10th IWA Symposium on Systems Analysis and Integrated Assessment (Watermatex2019)*, September 1–4 2019, Copenhagen, Denmark.

Schneider, M. Y., Carbajal, J. P., Furrer, V., Sterkele, B., Maurer, M. & Villez, K. 2019 Beyond signal quality: the value of unmaintained pH, dissolved oxygen, and oxidation-reduction potential sensors for remote performance monitoring of on-site sequencing batch reactors. *Water Research* **161**, 639–651. doi:10.1016/j.watres.2019.06.007.

Schraa, O., Tolc, B. & Copp, J. B. 2006 Fault detection for control of wastewater treatment plants. *Water Science and Technology* **53** (4–5), 375–382. doi:10.2166/wst.2006.143.

Shah, P. & Hoeffner, J. 2002 Review of graph comprehension research: implications for instruction. *Educational Psychology Review* **14** (1), 47–69. doi:10.1023/a:1013180410169.

Shiva Kumar, B. & Venkateswarlu, C. 2012 Estimating biofilm reaction kinetics using hybrid mechanistic-neural network rate function model. *Bioresource Technology* **103** (1), 300–308. doi:10.1016/j.biortech.2011.10.006.

Shyamal, S. & Swartz, C. L. E. 2019 Real-time energy management for electric arc furnace operation. *Journal of Process Control* **74** (February 2019), 50–62. doi:10.1016/j.jprocont.2018.03.002.

Sirkiä, J., Laakso, T., Ahopelto, S., Ylijoki, O., Porras, J. & Vahala, R. 2017 Data utilization at Finnish water and wastewater utilities: current practices vs. state of the art. *Utilities Policy* **45** (April 2017), 69–75. doi:10.1016/j.jup.2017.02.002.

Skobelev, D. O., Zaytseva, T. M., Kozlov, A. D., Perepelitsa, V. L. & Makarova, A. S. 2011 Laboratory information management systems in the work of the analytic laboratory. *Measurement Techniques* **53** (10), 1182–1189. doi:10.1007/s11018-011-9638-7.

Solon, K., Flores-Alsina, X., Kazadi Mbamba, C., Ikumi, D., Volcke, E. I. P., Vaneeckhaute, C., Ekama, G., Vanrolleghem, P. A., Batstone, D. J., Gernaey K, V. & Jeppsson, U. 2017 Plant-wide modelling of phosphorus transformations in wastewater treatment systems: impacts of control and operational strategies. *Water Research* **113**, 97–110. doi:10.1016/j.watres.2017.02.007.

Spérandio, M., Pocquet, M., Guo, L., Ni, B. J., Vanrolleghem, P. A. & Yuan, Z. 2016 Evaluation of different nitrous oxide production models with four continuous long-term wastewater treatment process data series. *Bioprocess and Biosystems Engineering* **39** (3), 493–510. doi:10.1007/s00449-015-1532-2.

Takács, I., Patry, G. G. & Nolasco, D. 1991 A dynamic model of the clarification-thickening process. *Water Research* **25** (10), 1263–1271. https://doi.org/10.1016/0043-1354(91)90066-Y.

Talebizadeh, M., Belia, E. & Vanrolleghem, P. A. 2016 Influent generator for probabilistic modeling of nutrient removal wastewater treatment plants. *Environmental Modelling and Software* **77**, 32–49. doi:10.1016/j.envsoft.2015.11.005.

Thornhill, N. F., Shoukat Choudhury, M. A. A. & Shah, S. L. 2004 The impact of compression on data-driven process analyses. *Journal of Process Control* **14** (4), 389–398. doi:10.1016/j.jprocont.2003.06.003.

Thürlimann, C. M., Dürrenmatt, D. J. & Villez, K. 2015 Energy and process data processing and visualisation for optimising wastewater treatment plants. *Water Practice and Technology* **10** (1), 10–18. doi:10.2166/wpt.2015.002.

van der Hoek, J., Duijff, R. & Reinstra, O. 2018 Nitrogen recovery from wastewater: possibilities, competition with other resources, and adaptation pathways. *Sustainability (Switzerland)* **10** (12), 4605. doi:10.3390/su10124605.

Vaneeckhaute, C., Claeys, F., Tack, F., Meers, E., Belia, E. & Vanrolleghem, P. A. 2018 Development, implementation, and validation of a generic nutrient recovery model (NRM) library. *Environmental Modelling & Software* **99**, 170–209.

Vanrolleghem, P. A. 2014 Bits and bytes and bugs – On monitoring and control in WRRF's (aka WWTPs). In *Smart Wastewater Virginia WEA Seminar*, April 30-May 1 2014, Richmond, VA, USA (Invited Keynote Lecture).

Vanrolleghem, P. A. & Lee, D. S. 2003 On-line monitoring equipment for wastewater treatment processes: state of the art. *Water Science and Technology* **47** (2), 1–34. doi:10.2166/wst.2003.0074.

Varga, E., Hauduc, H., Barnard, J., Dunlap, P., Jimenez, J., Menniti, A., Schauer, P., Lopez Vazquez, C. M., Gu, A. Z., Sperandio, M. & Takács, I. 2018 Recent advances in bio-P modelling – a new approach verified by full-scale observations. *Water Science and Technology* **78** (10), 2119–2130. doi:10.2166/wst.2018.490.

Vezzaro, L., Pedersen, J. W., Larsen, L. H., Thirsing, C., Duus, L. B. & Mikkelsen, P. S. 2020 Evaluating the performance of a simple phenomenological model for online forecasting of ammonium concentrations at WWTP inlets. *Water Science and Technology* **81** (1), 109–120. doi:10.2166/wst.2020.085.

Villez, K., Vanrolleghem, P. A. & Corominas, L. 2020 A general-purpose method for pareto optimal placement of flow rate and concentration sensors in networked systems – with application to wastewater treatment plants. *Computers and Chemical Engineering* **139**, 106880. doi:10.1016/j.compchemeng.2020.106880.

Water Environment Federation 2014 *Moving Toward Resource Recovery Facilities. Technical Report.* Alexandria, VA, USA, pp. 16.

Water Online and SWAN 2019 *Smart Water Report: Navigating The Smart Water Journey: From Leadership to Results.* Horsham, PA, USA. Water Online, p. 58.

Weijers, S. 2000 *Modelling, Identification and Control of Activated Sludge Plants for Nitrogen Removal.* PhD Thesis, Eindhoven University of Technology, Eindoven, The Netherlands, p. 235.

Werbos, P. J. 1982 Applications of advances in nonlinear sensitivity analysis. In: *System Modeling and Optimization* (R. F. Drenick & F. Kozin, eds). Springer-Verlag, Berlin/Heidelberg, Germany, pp. 762–770.

Whitt, M. D. 2012 *Successful Instrumentation and Control Systems Design*, 2nd edn. International Society of Automation, Durham, NC, USA, p. 531.

Wigner, E. P. 1960 The unreasonable effectiveness of mathematics in the natural sciences. *Communications on Pure and Applied Mathematics* **13** (1), 1–14. doi:10.1002/cpa.3160130102.

Wilkinson, M. D., Dumontier, M., IjJ, A., Appleton, G., Axton, M., Baak, A., Blomberg, N., Boiten, J. W., da Silva Santos, L. B., Bourne, P. E., Bouwman, J., Brookes, A. J., Clark, T., Crosas, M., Dillo, I., Dumon, O., Edmunds, S., Evelo, C. T., Finkers, R., Gonzalez-Beltran, A., Gray, A. J. G., Groth, P., Goble, C., Grethe, J. S., Heringa, J., 't Hoen, P. A., Hooft, R., Kuhn, T., Kok, R., Kok, J., Lusher, S. J., Martone, M. E., Mons, A., Packer, A. L., Persson, B., Rocca-Serra, P., Roos, M., van Schaik, R., Sansone, S. A., Schultes, E., Sengstag, T., Slater, T., Strawn, G., Swertz, M. A., Thompson, M., van der Lei, J., van Mulligen, E., Velterop, J., Waagmeester, A., Wittenburg, P., Wolstencroft, K., Zhao, J. & Mons, B. 2016 The FAIR guiding principles for scientific data management and stewardship. *Scientific Data* **3** (1), 160018. doi:10.1038/sdata.2016.18.

Yee, I. & Eren, H. 2012 'Data historian.' In: *Instrument Engineers' Handbook, Volume Three: Process Software and Digital Networks* (B. G. Liptak & H. Eren, eds). CRC Press, Boca Raton, FL, USA, pp. 465–470.

Yoo, C. K., Villez, K., Lee, I. B., Rosén, C. & Vanrolleghem, P. A. 2007 Multi-model statistical process monitoring and diagnosis of a sequencing batch reactor. *Biotechnology and Bioengineering* **96** (4), 687–701. doi:10.1002/bit.21220.

Yuan, Z., Olsson, G., Cardell-Oliver, R., van Schagen, K., Marchi, A., Deletic, A., Urich, C., Rauch, W., Liu, Y. & Jiang, G. 2019 Sweating the assets – the role of instrumentation, control and automation in urban water systems. *Water Research* **155**, 381–402. doi:10.1016/j.watres.2019.02.034.

Zegers, E., Atlin, S., Nyman, K. & Berenyi, A. 2019 Toronto's new data management platform and how we plan to manage the quality of data that feeds it. In *Proceedings of WEFTEC 2019*, September 21–25 2019, Chicago, IL, USA.

Zhang, Y. & Jiang, J. 2008 Bibliographical review on reconfigurable fault-tolerant control systems. *Annual Reviews in Control* **32** (2), 229–252. doi:10.1016/j.arcontrol.2008.03.008.

First received 29 May 2020; accepted in revised form 5 August 2020. Available online 18 August 2020

© IWA Publishing 2020

A comprehensive review of deep learning applications in hydrology and water resources

Muhammed Sit, Bekir Z. Demiray, Zhongrun Xiang, Gregory J. Ewing, Yusuf Sermet and Ibrahim Demir IWA

ABSTRACT

The global volume of digital data is expected to reach 175 zettabytes by 2025. The volume, variety and velocity of water-related data are increasing due to large-scale sensor networks and increased attention to topics such as disaster response, water resources management, and climate change. Combined with the growing availability of computational resources and popularity of deep learning, these data are transformed into actionable and practical knowledge, revolutionizing the water industry. In this article, a systematic review of literature is conducted to identify existing research that incorporates deep learning methods in the water sector, with regard to monitoring, management, governance and communication of water resources. The study provides a comprehensive review of state-of-the-art deep learning approaches used in the water industry for generation, prediction, enhancement, and classification tasks, and serves as a guide for how to utilize available deep learning methods for future water resources challenges. Key issues and challenges in the application of these techniques in the water domain are discussed, including the ethics of these technologies for decision-making in water resources management and governance. Finally, we provide recommendations and future directions for the application of deep learning models in hydrology and water resources.

Key words | artificial intelligence, deep learning, hydroscience, machine learning, review, water

Muhammed Sit (corresponding author)
Interdisciplinary Graduate Program in Informatics,
University of Iowa, Iowa City,
USA
and
IIHR – Hydroscience & Engineering,
University of Iowa,
100 C. Maxwell Stanley Hydraulics Laboratory,
Iowa City,
Iowa 52242-1585,
USA
E-mail: *muhammed-sit@uiowa.edu*

Bekir Z. Demiray
Department of Computer Science,
University of Iowa, Iowa City,
USA

Zhongrun Xiang
Gregory J. Ewing
Ibrahim Demir IWA
Department of Civil and Environmental Engineering,
University of Iowa, Iowa City,
USA

Yusuf Sermet
Department of Electrical and Computer Engineering,
University of Iowa, Iowa City,
USA

HIGHLIGHTS

- A comprehensive review of deep learning applications in hydrology and water resources fields is conducted.
- Summaries of papers that employ deep learning for hydrologic modeling are presented, along with a concise deep learning synopsis.
- Key issues within the water-related deep learning application literature are reported.
- Future directions for research in the intersection of deep learning and hydroscience is presented.

LIST OF ACRONYMS

AE	Autoencoder	CNN, ConvNet	Convolutional Neural Network
AI	Artificial Intelligence	DBM	Deep Boltzman Machines
ANN	Artificial Neural Networks	DBN	Deep Belief Network
BMP	Best Management Practices	DEM	Digital Elevation Model
BP	BackPropagation	DNN	Deep Neural Network
ConvLSTM	Convolutional LSTM	DQN	Deep Q-Network

doi: 10.2166/wst.2020.369

EC	Electrical Conductivity
EEMD	Empirical Mode Decomposition
ELM	Extreme Learning Machines
ENN	Elman Neural Network
FCN	Fully Convolutional Network
FFNN	Feed-Forward Neural Network
GAN	Generative Adversarial Networks
GIS	Geographic Information System
GPU	Graphics Processing Unit
GRACE	Gravity Recovery and Climate Experiment
GRU	Gated Recurrent Unit
IMF	Intrinsic Mode Function
LSTM	Long Short-Term Memory
LiDAR	Light Detection and Ranging
MAP	Mean Areal Precipitation
MAPLE	McGill Algorithm for Precipitation nowcasting by Lagrangian Extrapolation
MLP	Multilayer Perceptron
MSE	Mean Squared Error
NAR	Nonlinear Autoregressive
NARX	Nonlinear Autoregressive eXogenous
NDVI	Normalized Difference Vegetation Index
NOAH	National Oceanic and Atmospheric Administration
PDE	Partial Differential Equation
PLSR	Partial Least Square Regression
RBM	Restricted Boltzman Machine
RGB	Red Green Blue
RNN	Recurrent Neural Network
ReLU	Rectified Linear Unit
SAE	Stack AutoEncoder
SSIM	Structural Similarity
SST	Sea Surface Temperature
SVM	Support Vector Machines
SVR	Support Vector Regression
VAE	Variational AutoEncoders
VMD	Variational Mode Decomposition

INTRODUCTION

The global volume of digital data is expected to reach 175 zettabytes by 2025 (Reinsel *et al.* 2018). Large-scale sensor networks, as well as the increased awareness of climate change, water resources management and the monitoring of water-related hazards, led to the substantial growth of the volume, variety, and velocity of water-related data (Weber *et al.* 2018; Sit *et al.* 2019a, 2019b). Modern data collection techniques, including satellite hydrology, Internet of Things for on-site measurements (Kruger *et al.* 2016), and crowdsourcing tools (Sermet *et al.* 2020b), has revolutionized the water science and industry as approached by the government, academia, and private sector (Krajewski *et al.* 2017). The effective utilization of vast water data holds the key for long-term sustainability and resilience, and presents opportunities to transform water governance for the upcoming decades (Grossman *et al.* 2015). In the hydrological domain, multivariate analysis relying on extensive and semantically connected data resources is required to generate actionable knowledge and produce realistic and beneficial solutions to water challenges facing communities (Carson *et al.* 2018; Jadidoleslam *et al.* 2019). However, the inaccessible, unstructured, nonstandardized, and incompatible nature of the data makes optimized data models (Demir & Szczepanek 2017) and smarter analytics approaches a necessity (Sermet & Demir 2018a).

Computerized methods to create an understanding of hydrological phenomena are based on various modeling strategies, which simplify a hydrological system to simulate its behavior (Antonetti & Zappa 2018). Physical models aim to achieve this goal by specifically designing complex simulations that are powered by mathematical and numeric specifications of conceptualized physical characteristics (Jaiswal *et al.* 2020). However, hydrological systems, as is the case with other natural systems, are inherently heterogeneous (Marçais & De Dreuzy 2017) as opposed to less complex human-made systems with defined rules. Therefore, physical models, although deterministic and reliable, do not always perform and scale well due to their intrinsic limitations (Islam 2011). As an alternative, statistical models have been employed to make use of the comprehensive set of available hydrological, environmental, and geophysical data (Evora & Coulibaly 2009). These approaches assume minimum awareness of the underlying mechanism and receive their strength by eliciting useful information and patterns from the available data through statistical analysis (McCuen 2016). However, they have displayed shortcomings in terms of accuracy and certainty, and also require excessive computational power (Ardabili *et al.* 2019; Agliamzanov *et al.* 2020).

Recent developments in artificial intelligence (AI) and graphics processor units (GPU) have paved the way for deep learning, a pioneering approach that is fueled by multilayer artificial neural networks (ANN) (LeCun *et al.* 2015). Deep learning provides a black-box method to learn from complex and high-dimensional data to infer robust and scalable insights while minimizing the degree to which manual labor is needed (Sengupta *et al.* 2020). One feature that separates deep learning from its superset machine learning is the use of multilayer models, which leads to a higher level representation of the underlying data sources (Saba *et al.* 2019). Furthermore, deep learning is capable of extracting substantial features without being explicitly instructed, and thus, is more immune to raw and noisy data (Sahiner *et al.* 2019). Successful implementations of deep learning permeate numerous domains and industries, including medical imaging (Haskins *et al.* 2020), healthcare (Esteva *et al.* 2019), finance (Heaton *et al.* 2017), geophysical sciences (Shen 2018), remote sensing (Ma *et al.* 2019), and hydrology. Due to its significant adoption rate and potential to be applicable to any domain that encompasses problems that can be expressed as control systems, numerous open-source and for-profit software tools, educational resources and generalized algorithms have been made available for use, opening up countless paths to advance hydrological studies.

This paper presents a systematic review of applications of deep learning within the scope of the hydrological domain. The literature has been thoroughly examined to identify the use cases of deep learning in the subfields of the water sector, including flooding, weather, land use and soil, water quality, surface water, water resources, and groundwater. Each study has been evaluated to extract information that is scientifically relevant to assess the study's contribution and reproducibility, including the hydrological tasks that were taken on to be approached by deep learning along with the utilized network architectures, datasets, software tools and frameworks, licenses, and deep learning practices and algorithms. The paper explores modern deep learning networks from the lens of the hydrological domain to investigate the shortcomings and challenges of the status quo and to highlight the opportunities and future directions to serve as a guide to researchers, professionals, and governmental organizations in the water sector.

The major contributions of this paper can be summarized as follows. Although there are various configurations of ANNs optimized for various data types and use cases, it is challenging to reduce a real-life hydrological task to a certain predefined approach given the depth and complexity of the tasks as well as the diversity of networks. The methodologies that need to be employed while developing deep learning-powered solutions in hydrology are not standardized in terms of data quality and preparation, execution, validation, and documentation. Furthermore, the strength, usability, and reliability of a model lie on clearly set descriptions and procedures for deterministic reproducibility, given the variety of development frameworks, as well as the application areas. To the best of our knowledge, there has not been a thorough investigation of systematically approaching water challenges with deep learning. Thus, this paper serves as a meticulous guide for the stakeholders of the hydrology domain to advance the water industry with intelligent systems that are revolutionized by multifaceted data (Sermet *et al.* 2020a).

The remainder of this article is organized as follows. The Literature Review section provides the review methodology followed by a comprehensive literature review of deep learning applications in the water domain. Descriptions of deep learning concepts, tasks, and architectures are described to summarize the available methodology for use by the hydrological community. The Results section presents a detailed summary and analysis of reviewed papers grouped by their application area. The Key Issues and Challenges section highlights the key issues and challenges facing the stakeholders utilizing deep learning in the water domain with respect to technical limitations as well as ethical considerations. The Recommendations and Conclusions section outlines a vision entailing the adoption of prominent and deep learning-powered technologies to solve the water challenges of the future, and then concludes the paper with a concise summary of findings.

LITERATURE REVIEW

This section starts with a detailed description of the literature search methodology in the first subsection Review Methodology and then presents the information extracted from each reviewed manuscript. The subsection Deep Learning gives a brief overview of deep learning history, describes various neural network architectures, and elaborates on different machine learning task types. At the end of this section we share the summary of the literature as figures to provide an understanding of this review and a table of all the papers reviewed.

Review methodology

A systematic literature search on water domain was employed for this review. Web of Science, Scopus, Springer Link, Wiley Online Library, and The International Water Association Publishing Online were used as the databases and the keywords included 'deep neural network', 'deep neural networks', 'deep learning', 'lstm', 'long short term memory', 'cnn', 'convolutional', 'gan', 'generative adversarial', 'rnn', 'recurrent neural', 'gru', and 'gated recurrent'. Hydroscience and water resources specific keywords were not included in the search queries because of the cardinality of domain-specific keywords. The cardinality is a result of many application fields within these domains and also many interchangeable and synonymous words used to refer the same terms within different scientific communities around the globe. Hydrology and water resources related papers are filtered at the journal level which provided a comprehensive coverage compared to keyword level filtering. After limiting the search with these keywords in the publication title, abstract, or keywords, an additional exclusion criterion was applied through each database's categorization system if applicable only to include the publications within the environmental fields. Also, searches were limited to only include journal publications. All articles published in 2018, 2019, and 2020 up until the end of March containing these keywords in their titles, keyword fields, or abstracts were included in the first list of articles gathered. This time interval is primarily chosen based on our initial literature search and availability of deep learning application papers to create a comprehensive review and curate insights within the water domain between 2018 and 2020. There were also other review articles partially covering the water domain and timeline (Shen 2018). After gathering the initial list totaling 1,515 publications, each of them was briefly reviewed to determine whether they were in alignment with the scope of this study. All publications that were not research papers were excluded, namely vision papers, editorials, and review papers. From the initial list of 1,515 publications, 315 remained after this filtering step. These publications were filtered further to keep publications that met certain technical criteria. This step eliminated all publications that did not involve some form of deep ANN in their pipeline of work.

After this step, 129 publications remained and were included in our comprehensive review. The comprehensive review process consisted of manually reviewing the papers one by one to extract specific publication features, including Architecture, Framework/Library/Programming Language, Dataset, Source Code Sharing, Reproducibility, Subfield: Deep Learning, Subfield: Environment, Summary. Each of the feature categories are described below:

- **Architecture** – The type of deep neural network (DNN) architecture(s) employed in the study. This could be simply ANN or more complex architectures like Generative Adversarial Networks (GAN) or Variational Auto Encoders (VAE).
- **Framework/Library** – This column serves as a survey within the field to understand the programming language and numeric computation library choices of researchers.
- **Dataset** – Whether the dataset(s) used in the study were collected specifically for the study, acquired from an authority resource, or previously existing standalone.
- **Source Code Sharing** – A boolean field indicating if the code of the study is open-sourced and accessible by the public.
- **Reproducibility** – A boolean field indicating if the results of the study could be reproduced just by using the information provided in the manuscript.
- **Subfield: Deep Learning** – A classification of the machine learning task tackled in the paper. This field uses one of the following values: Regression, Classification, Sequence Prediction, Matrix Prediction, Unsupervised Learning, and Reinforcement Learning. Details of these are given in the next subsection where we describe deep learning practices.
- **Subfield: Environment** – A classification of the task carried out in an environmental field. This field uses one of the following values: Flood, Groundwater, Land Usage and Soil, Surface Water, Water Quality, Water Resources Management, Weather, and Others. Others include papers within the environmental field that do not exactly fit with the other subfields included here.
- **Summary** – The brief summary of the study.

The data for each publication reviewed herein are shared later in this section with figures and a table. Technical summaries of the papers reviewed are given in the Results section. Conclusions drawn from the acquired data are shared in the Key Issues and Challenges section.

Deep learning

Deep learning is a subfield of machine learning where a long-known algorithm, a neural network, an ANN, a DNN, or a Feed-Forward Neural Network (FFNN), referring to the same algorithm, is used to map features into an output or a set of outputs. Formed by an input layer, intermediate hidden layers, and an output layer, ANNs present an efficient way to learn linear and nonlinear relationships between input and output pairs (Figure 1). Neural networks, when formed by many stacked layers, can represent complex features in later layers by using simpler representations formed by earlier layers in the network

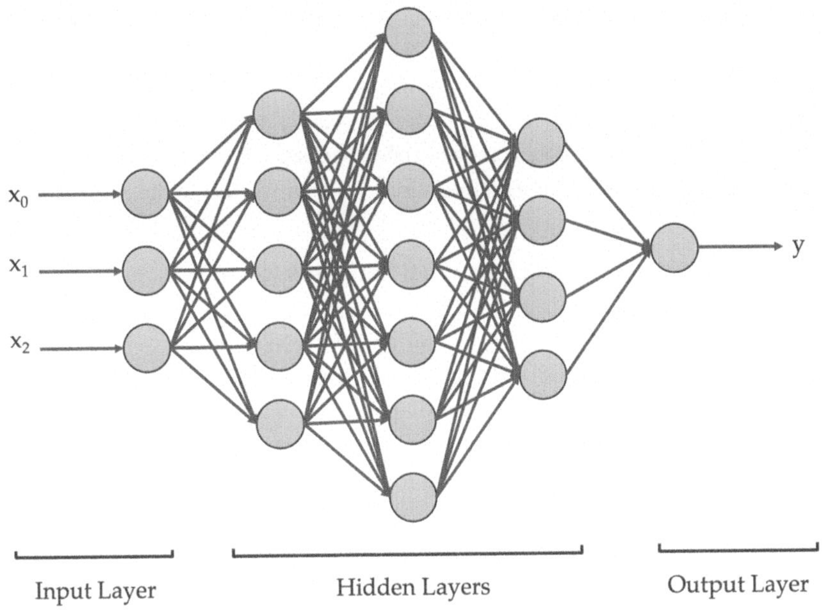

Figure 1 | A densely connected ANN architecture.

(Goodfellow *et al.* 2016). Each layer within an ANN comprises at least one neuron. ANN is a network of these neurons connected to each other with some weights and these neurons run specific functions, namely activation functions, mapping it's input to an output. Stacked on top of each other, the series of functions runs over the input of the network and translates the input to the output in the output layer. Typically, each neuron within a layer runs the same activation function and the type of the layer is determined by this activation function. Network type is determined by the combination of layers used and how neurons are connected to each other within and between layers. The quintessential form of an ANN is the Multilayer Perceptron (MLP). An MLP contains at least a single hidden layer while each neuron within the network is connected to every neuron within the next layer. This architecture forms a fully connected neural network.

An activation function in a typical MLP multiplies the input by a weight and outputs it to the next neuron in line. In modern neural networks the common (Goodfellow *et al.* 2016) and recommended (Nair & Hinton 2010) activation function is the Rectified Linear Unit (ReLU) function, which introduces nonlinearity to the network. A hidden layer that applies ReLU to the input could be referred to as an ReLU layer or an activation layer using ReLU. ANN are generally trained using the BackPropagation (BP) algorithm (Rumelhart *et al.* 1986) to fit the network to the set of input–output pairs using a loss function. A loss function or a cost function is used to determine how successful the mapping is done by a model. An easy and go-to loss function is Mean Squared Error (MSE), which computes the difference between an output of the ANN and the ground truth for each input–output sample, squares the difference to avoid negative values and computes the mean error of all the samples. BP trains an ANN using the loss function by computing the gradient of the loss function with respect to the weights between neurons and updates the weights to better fit the data samples in order to minimize the loss function.

ANNs are not a new concept, considering primitive versions were around in the 1940s (Goodfellow *et al.* 2016). Instead, ANNs attracted attention of researchers from various scientific disciplines when it became clear that they were extremely powerful in capturing representations in the data and with advances in GPUs, which enable extremely fast matrix operations (Goodfellow *et al.* 2016). In this way, a neural network architecture that previously was infeasible to utilize due to time complexity as the number of hidden layers increased, could be used in training on complex datasets learning representations.

For each of the reviewed papers we identify the machine learning subfield, which corresponds to task types. Task types are defined by the output form of a network. We consider the following as distinct task types: Regression, Classification, Sequence Prediction, Matrix Prediction,

Unsupervised Learning, and Reinforcement Learning. These task types are briefly discussed below:

- **Regression** – A machine learning task that predicts a continuous value. Forecasting the next measurement from a series of measurements would be an example of a regression task.
- **Classification** – When the aim of a model is to predict whether the set of inputs can be categorized into some classes, the task is a classification task. The number of classes are not limited. Predicting whether the next day will be a dry or wet one is a classification task.
- **Sequence Prediction** – Regression of a sequence of numeric values or a vector. Forecasting the next 24 h of measurements for a stream sensor is a sequence prediction task.
- **Matrix Prediction** – Regression of a matrix of numeric values. Forecasting the next precipitation measurements for a rectangular region would be a matrix prediction task. Each of the values in the predicted matrix would be the precipitation value of a subregion within the actual region.
- **Unsupervised Learning** – A learning task applying competitive learning instead of error correction learning like previous task types. Decreasing dimension of a high-dimensional hydrological input data to lower dimension visualizable data would be an example.
- **Reinforcement Learning** – A learning task where the output is unspecified. In a reinforcement learning task, the algorithm tries to find the optimal solution for any given input using a reward–penalty policy and a try–error mechanism. An example of a reinforcement learning task is an AI model that learns when to release water from a dam.

ARCHITECTURES

This section summarizes cornerstone neural network architectures used by papers reviewed in this study. Also, some ANN concepts are briefly discussed. For further understanding of these architectures, we refer readers to the cited works.

Convolutional neural networks (CNN)

A CNN (LeCun 1989) or a ConvNet comprises at least one convolutional layer, which methodologically expects a 3D tensor as the input (Figure 2). A convolutional layer applies multiple cascaded convolution kernels to extract intricate knowledge from the input. For example, a CNN for a red, green, blue (RGB) image tensor with a shape of image width × image height × 3 would have a convolutional layer that applies three different convolution operations with three separate kernels to each of the color channel matrices. Using a convolution kernel matrix, a convolutional layer that processes an image as such can extract 2D positional information from images, such as understanding objects that are close to each other.

To make a neural network with a convolutional layer cognize nonlinear correlations between the input and output along with linear correlations, one needs to introduce nonlinearity to the network via an activation layer. Typically, that is done using the ReLU as the activation function following the convolution layer. Another common layer used within a CNN is a pooling layer. A pooling layer is used to reduce the size of the input while keeping the positional knowledge intact. A frequently used pooling method within CNN literature is Max Pooling (Zhou & Chellappa 1988). This sample-based discretization moves the most

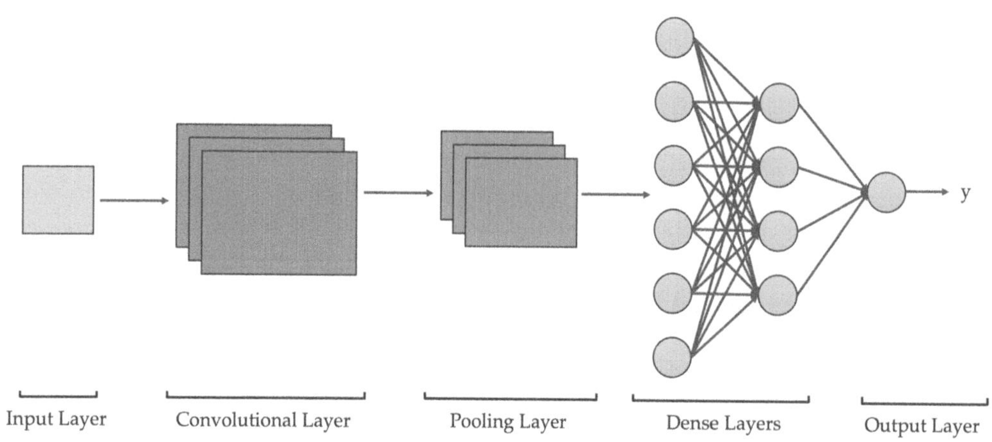

Figure 2 | A basic CNN structure for image classification.

important learned features to subsequent layers while reducing the size. Consequently, CNNs make good architectures for deep learning tasks with images or image-like objects as inputs. This ability of CNNs makes way for various breakthroughs in the fields of object detection, super-resolution, image classification, and computer vision.

Generative adversarial networks (GAN)

GANs (Goodfellow *et al.* 2014) consist of two seemingly separate CNNs working in unison and competing in a min–max game (Figure 3). One of these CNN, the generator, aims to generate fake examples out of a dataset while the other, discriminator, aims to reveal whether its input is fake or not. Since they try to beat each other, it causes them to improve over time in both generating fake outputs and discriminating fake from real.

GANs are initially used as generative models, as in randomly generating new samples from a dataset to appear as if they are from the originating dataset when visualized (Gautam *et al.* 2020). They achieve this goal by mapping random noise to real samples from the given dataset, and then they generate new instances from new random noise tensors. Despite their success at generation, GANs are also capable of learning translation tasks such as super-resolution (Demiray *et al.* 2020) or image-to-image translation (Isola *et al.* 2017).

Recurrent neural networks (RNN)

RNNs (Pollack 1990) are a type of ANN that includes a recurrent layer. The difference of a recurrent layer from a regular fully connected hidden layer is that neurons within a recurrent layer could also be connected to each other (Figure 4). In other words, the output of a neuron is conveyed both to the neuron(s) within the next layer and to the next neuron within the same layer. Using this mechanism, RNNs can carry information learned within a neuron to the next neuron in the same layer.

This procedure becomes convenient when the data to be fed to the network is in sequential nature, such as a time-series data or a text. When training a model over a

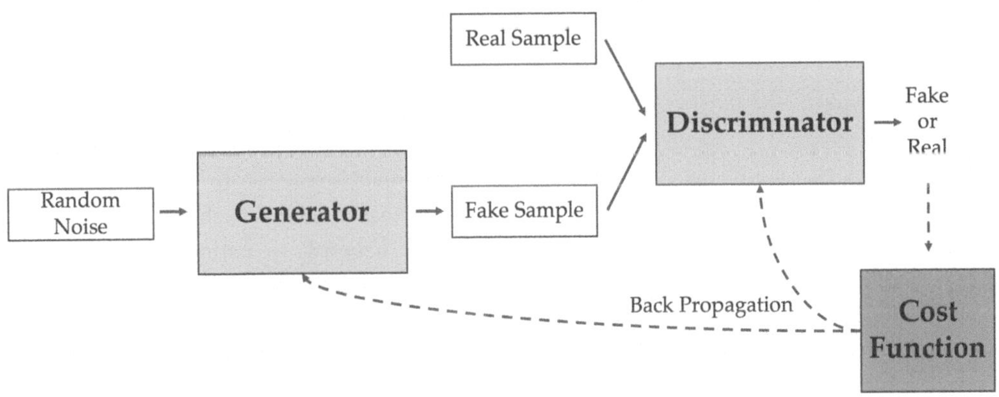

Figure 3 | Overview of GAN.

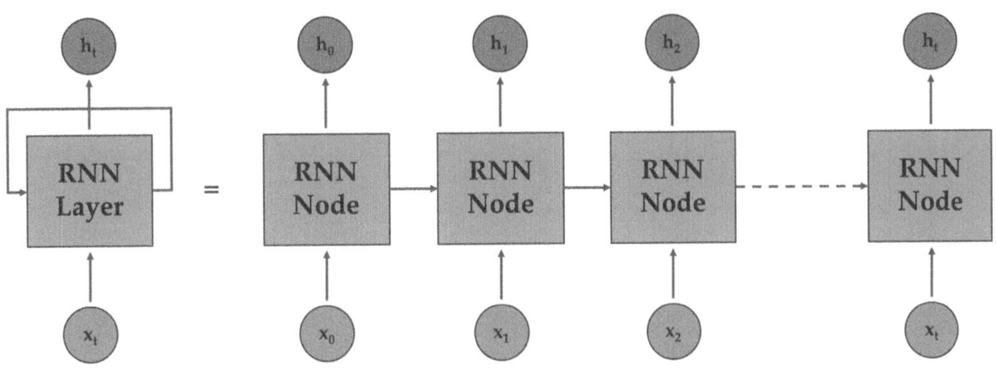

Figure 4 | Connections of a recurrent layer.

data sample like a text to extract the meaning, most of the time the beginning of the text could change the meaning that is to be extracted from the end of the text. RNNs aim to keep the information gained from earlier parts of a data sample in the memory and move it to the later parts of the same data sample to ensure better knowledge discovery (Goodfellow *et al.* 2016). A simple RNN implementation lacks the practicality in long sequences, such as long paragraphs, as it is common to encounter the vanishing gradient problem while training (Bengio *et al.* 1993). With the vanishing gradient problem, the gradients of the loss function become extremely high in some cases during training and consequently make the training process and the trained network deficient (Goodfellow *et al.* 2016). More complex RNN implementations like Long Short-Term Memory (LSTM) (Hochreiter & Schmidhuber 1997) networks or Gated Recurrent Unit (GRU) networks (Cho *et al.* 2014) solve this problem, but have greater computational complexity. Various RNN structures can be used for tasks that somewhat rely on sequential understanding of datasets such as language modeling (Sundermeyer *et al.* 2012), text classification (Yin *et al.* 2017; Sit *et al.* 2019a), and time-series forecasting (Sit & Demir 2019; Xiang & Demir 2020).

Long short-term memory (LSTM) networks

LSTM (Hochreiter & Schmidhuber 1997) Networks are developed for longer short-term memory life over the input, paving the way for more efficient but more resource-intensive training over datasets consisting of sequential samples. Instead of an activation function producing one output and carrying the output to immediate neurons both in the next layer and the same layer, LSTM neurons produce two different values yielded by a series of activations and operations. Although both outputs are kept within the LSTM layer to keep track of things learned over the past part of the sequence, one of the outputs is transferred to the next layer (Figure 5).

Gated recurrent unit (GRU) networks

Although LSTM networks, most of the time, solved the vanishing gradients problem and helped many breakthroughs within the fields of Natural Language Processing and time-series prediction, their time complexity emerges as a downside. GRU (Cho *et al.* 2014) networks reduce the complexity while keeping the efficacy intact. Similar to a simple RNN neuron, a GRU neuron produces only one output after a series of computations and uses the same output to convey important features learned to both the next layer and the next neuron within the same layer.

Nonlinear autoregressive (NAR) models

A NAR model is not necessarily a neural network model, but a model that is used for time-series prediction, taking into account both current and previous samples from a time-series to map input sequence to outputs. A NAR model needs a nonlinear function such as a polynomial function or a neural network to perform training. If a neural network is used, a NAR network would classify as

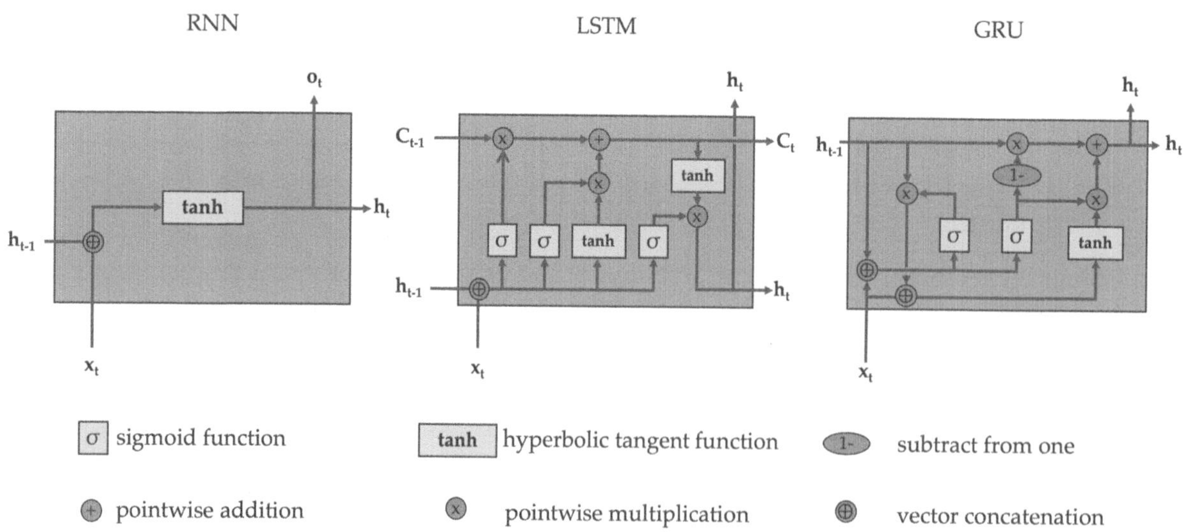

Figure 5 | Computation-wise comparison of RNN, LSTM, and GRU nodes.

an RNN based on the fact that it utilizes sequential complexion of the given input. Papers reviewed within this study that employ NAR thus implement a neural network as the function in their proposed models. There are many NAR variations and one that deserves a mention, due to cardinality of papers reviewed in this study that employ it, is the Nonlinear Autoregressive Exogenous Model (NARX) (Lin *et al.* 1996). NARX is an RNN implementation that takes advantage of exogenous inputs, hence the name.

Elman network (ENN)

An Elman Network (Elman 1990) is yet another RNN implementation that has three layers, only one being a hidden layer (Figure 6). The hidden layer of the ENN is connected to a set of neurons called context units. In each iteration after the first one over the network, the state of the hidden layer is copied into the context units. Thus, the state of the network for the previous sample in the data stream is kept in the network each time to be used in the next iterations. An ENN can train over sequential datasets better than a regular ANN due to this mechanism acting like a memory.

Autoencoders (AE)

Autoencoders (Rumelhart *et al.* 1985) are neural networks that are used to reduce the dimensionality of datasets (Figure 7). They are implemented in an unsupervised fashion to generate only a representation of the dataset within their hidden layer neurons, also called the latent vector. Taking the same set of values for both input and output of the network, an AE learns to reduce a dataset into a representation state and additionally learns how to reconstruct the data sample to its original form from the learned representation.

Restricted boltzmann machines (RBM) and deep belief networks (DBN)

RBMs (Hinton 2002) present two-layer stochastic generative models that are in the shape of bipartite graphs. RBMs form the building block for DBN (Hinton 2009), but they can also be used as standalone models. RBMs were initially used in unsupervised tasks; they also enable the user to tackle classification and regression tasks by implementing them within other networks. The paper reviewed in this study that employs RBMs uses it in a setting where RBMs are followed by a set of fully connected layers in order to perform a classification task. Formed by stacked RBMs, DBNs, similarly to RBMs, could also be used to tackle many types of tasks by training unsupervised beforehand. The bipartite connections of RBMs are altered when they are stacked onto each other to form a DBN (Figure 8). Only the top two layers of a DBN have bipartite connections, while the rest of the layers have one-way connections with each other. Although RBMs are useful for many tasks, their use in research is decreasing in the machine learning field as researchers adopt newer architectures that could be utilized in the same fashion such as AE, DBNs, and GANs.

Extreme learning machines (ELM)

ELMs (Huang *et al.* 2006) are three-layer neural networks in which the weights connecting to the second layer from the input layer are randomized, and the weights connecting to

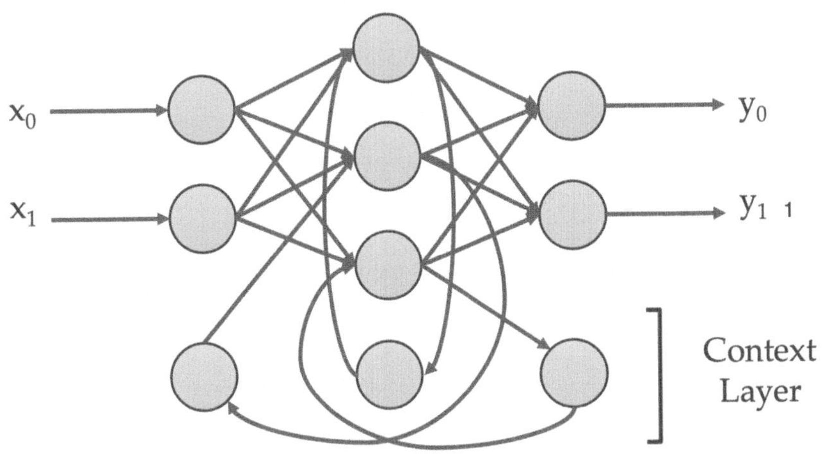

Figure 6 | An ENN architecture with two input and two output neurons.

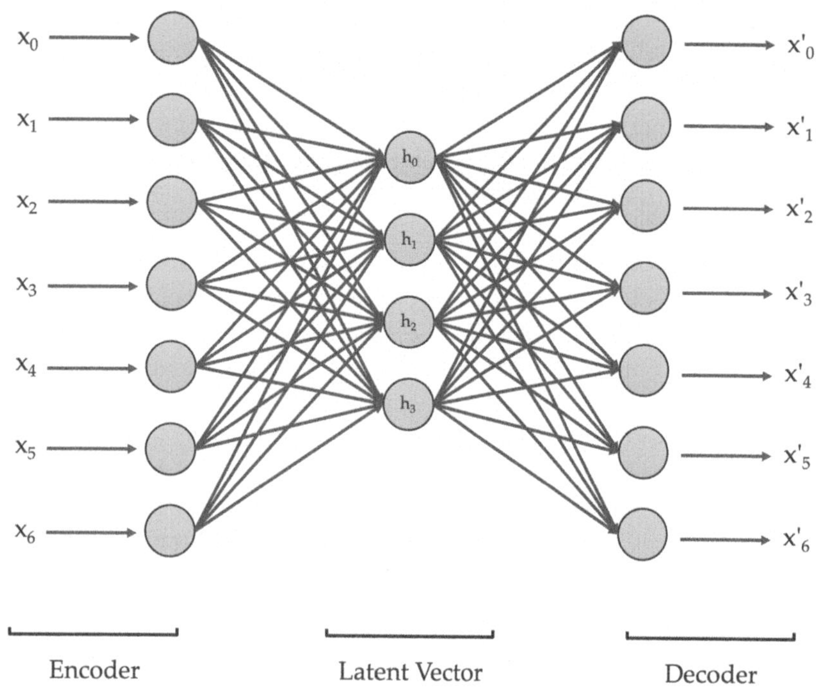

Figure 7 | Visualization of an AE network.

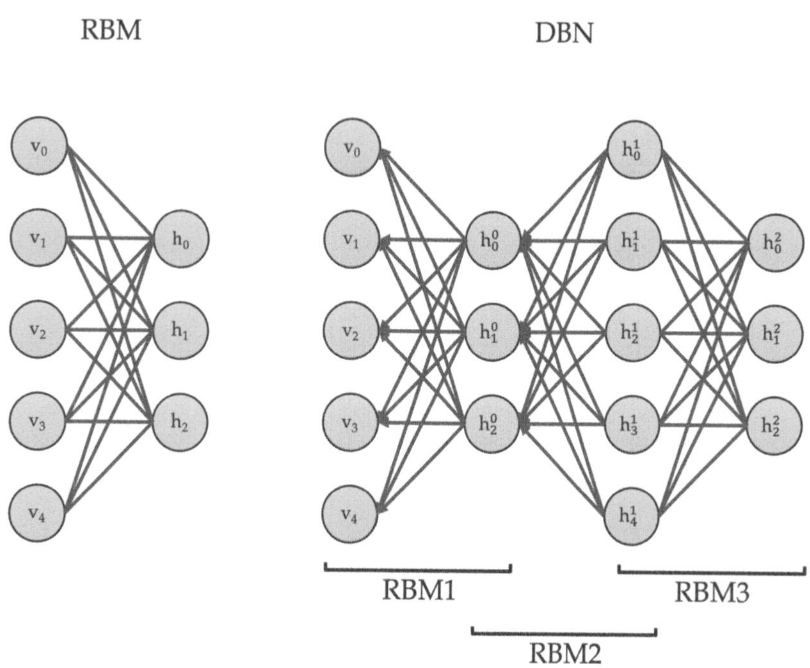

Figure 8 | RBM and DBN examples.

the third are trained. ELMs are designed to produce better generalization performance while training faster than BP-based ANNs consisting of one hidden layer on large datasets thanks to its randomization mechanism. ELM networks have been criticized for being unoriginal (Wang & Wan 2008).

Deep Q networks (DQN)

DQN (Mnih *et al.* 2013) is a reinforcement learning algorithm in which there is a predetermined rewarding policy. The DQN works similarly to the Q-Learning algorithm, which works by providing a medium for the agent and the environment and tries to maximize the rewards the agent gets through its actions in the environment. A DQN differentiates from the conventional Q-Learning algorithm with how it generalizes. In other words, the agent in the Q-Learning algorithm cannot estimate the outcome from action–environment pairs it did not see before, while in DQN the agent is able to produce a reinforced action.

SUMMARY OF ARTICLES

This section presents visual summaries of the reviewed papers, as well as a summary table (Table 1) showing extracted information from each paper that received comprehensive review. As previously stated, information from a total of 129 papers are given; however, data points in shared figures do not always equal to this value depending on the data column, this is because some entries in the table may have multiple values in some data columns instead of one.

Among the reviewed papers, as in the deep learning literature, most used architectures are CNNs and LSTMs (Figure 9). We explain this aspect with their respective success in matrix prediction and sequence prediction, tasks that have high importance in hydrologic modeling. One confounding thing is that even though LSTM networks were widely employed, GRU networks, an architecture that yields similar performance, did not receive significant usage in the field. Additionally, we found it surprising that even though most of the studies reviewed here tackle tasks involving sequential data, Transformers were not employed by any of the studies we reviewed. It should be noted that a Transformer is a neural network architecture that is widely used in the field of natural language processing, which is another field that focuses on sequential data.

As the utilization of deep learning in the water field in a broad sense increases over time (Figure 10), annual usage of the DNN architectures increases. Figure 11 shows the change in usage up to March 2020 and also presents simple projections for the rest of the year built on top of the number of publications after March 2020. We expect to see growth in the usage of neural network architectures that have been widely used in other disciplines but not in the water field, such as DQNs and GANs.

The datasets used in publications reviewed are typically datasets acquired from authorities or governmental agencies (Figure 12). Even though in deep learning literature, the dataset acquisition is primarily done by using previously existing datasets, in the water field, on the contrary, this does not seem to be the case. Code accessibility of the papers is another aspect of the studies published in the water field that differentiates it from the deep learning field in general. Although open-sourced models are widely expected from deep learning researchers, open-sourcing the software built for a study is unusual for a publication in the water field, if not rare, as can be seen in Table 1. Cumulating from both the data acquisition type and code accessibility, reproducing the outcomes of a paper does not seem to be an easy task for the authorities and other researchers in the field.

Due to the fact that the field of hydrology relies on sequential data, most studies seem to work on sequence prediction and regression tasks (Figure 13). This phenomenon might also be linked with the fact that most of the studies reviewed were classified in the Flood subdomain (Figure 14). Figure 15 summarizes the usage of numerical computing frameworks. Although TensorFlow seems to be the first choice among water domain researchers, it should be noted that most of the usage comes from Keras, the second most-used framework, which typically works on top of TensorFlow by providing a higher level interface. Thus, our inference is that Keras is the most-used deep learning framework within the water field. In contrast, libraries like PyTorch, which is highly endorsed in deep learning literature, find a smaller place to themselves. The reason for this might be the ability of fast ANN prototyping and modeling that Keras provides to programmers without expecting many implementation-relevant details.

RESULTS

In this section we provide brief summaries of papers presented in the previous section (Table 1). Papers are grouped and interpreted by their use.

Streamflow and flood

Runoff prediction and flood forecasting are major tasks in rainfall-runoff modeling. Toward this end, many researchers have applied cutting-edge deep learning architectures to the runoff prediction and flood forecasting tasks. Since rainfall and runoff are both time-series data, the common networks for the streamflow prediction and flood forecast are RNN, LSTM, NAR, and ENN. Kratzert *et al.* (2018) applied an

Table 1 | Reviewed papers with curated data points

Paper	Network type	Framework	Dataset	Open-source	Reproducible	Deep learning task	Water field
Yuan *et al.* (2018)	LSTM	MATLAB	Acquired	No	No	Sequence Prediction	Flood
Zhang *et al.* (2018e)	LSTM	–	Acquired	No	No	Sequence Prediction	Flood
Kratzert *et al.* (2018)	LSTM	Keras, TensorFlow	Existing	No	Yes	Sequence Prediction	Flood
He *et al.* (2019b)	ANN, DBN	–	Acquired	No	Yes	Sequence Prediction	Flood
Hu *et al.* (2019)	LSTM	–	Acquired	No	No	Regression	Flood
Wang *et al.* (2019b)	CNN	–	Acquired	No	No	Sequence Prediction	Flood
Yang *et al.* (2019a)	NAR, LSTM	Keras, TensorFlow	Acquired	No	No	Regression	Flood
Sankaranarayanan *et al.* (2019)	ANN	Keras	Acquired	No	No	Classification	Flood
Ni *et al.* (2019)	LSTM, CNN	–	Acquired	No	No	Sequence Prediction	Flood, Weather
Bai *et al.* (2019)	LSTM, AE	–	Acquired	No	No	Sequence Prediction	Flood
Yang *et al.* (2019b)	LSTM	–	Acquired	No	No	Regression	Flood
Bhola *et al.* (2019)	CNN	–	Collected	No	No	Segmentation	Flood
Damavandi *et al.* (2019)	LSTM	Keras, TensorFlow	Acquired	No	Yes	Sequence Prediction	Flood
Moy de Vitry *et al.* (2019)	CNN	TensorFlow	Collected	Yes	Yes	Matrix Prediction	Flood
Worland *et al.* (2019)	ANN	Keras, TensorFlow	Acquired	No	No	Regression	Flood
Kratzert *et al.* (2019a)	LSTM	PyTorch	Acquired	Yes	Yes	Sequence Prediction	Flood
Kumar *et al.* (2019)	RNN, LSTM	Keras	Acquired	No	Yes	Sequence Prediction	Flood
Wan *et al.* (2019)	ENN	–	Acquired	No	No	Sequence Prediction	Flood
Qin *et al.* (2019)	LSTM	TensorFlow	Acquired	No	No	Sequence Prediction	Flood
Kratzert *et al.* (2019b)	LSTM	PyTorch	Existing	Yes	Yes	Regression	Flood
Bui *et al.* (2020a)	ANN	MATLAB	Acquired	No	No	Classification	Flood
Nguyen & Bae (2020)	LSTM	TensorFlow	Acquired	No	No	Sequence Prediction	Flood
Bui *et al.* (2020b)	ANN	–	Acquired	No	No	Classification	Flood
Kabir *et al.* (2020)	DBN	TensorFlow	Acquired	No	No	Sequence Prediction	Flood
Kao *et al.* (2020)	LSTM	Keras	Acquired	No	No	Sequence Prediction	Flood
Xiang *et al.* (2020)	LSTM	Keras, TensorFlow	Acquired	No	No	Sequence Prediction	Flood
Zuo *et al.* (2020)	LSTM	Matlab, TensorFlow	Acquired	Yes	Yes	Sequence Prediction	Flood
Ren *et al.* (2020)	ANN, LSTM, GRU	–	Acquired	No	No	Sequence Prediction	Flood
Zhu *et al.* (2020b)	LSTM	Keras	Acquired	No	Yes	Sequence Prediction	Flood
Wang *et al.* (2020d)	CNN	Keras	Acquired	No	No	Classification	Flood
Laloy *et al.* (2018)	GAN	Lasagne, Theano	Existing	No	No	Matrix Prediction	Groundwater
Wang *et al.* (2020b)	ANN	–	Collected	Yes	No	Matrix Prediction	Groundwater
Santos *et al.* (2020)	CNN	Keras	Existing	Yes	Yes	Matrix Prediction	Groundwater
Mo *et al.* (2019a)	CNN	PyTorch	Collected	Yes	Yes	Matrix Prediction	Groundwater
Mo *et al.* (2019b)	CNN	PyTorch	Collected	Yes	Yes	Matrix Prediction	Groundwater

Sun *et al.* (2019a)	CNN	Keras, TensorFlow	Acquired	No	No	Sequence Prediction	Groundwater
Jeong & Park (2019)	NARX, LSTM, GRU	TensorFlow	Acquired	No	No	Sequence Prediction	Groundwater
Zhou *et al.* (2020)	CNN	PyTorch	Acquired	No	No	Regression	Groundwater
Jeong *et al.* (2020)	LSTM	TensorFlow	Acquired	No	No	Sequence Prediction	Groundwater
Zhang *et al.* (2018d)	CNN	Keras, TensorFlow	Collected	No	No	Segmentation	Land Use and Soil
Zhang *et al.* (2018 g)	LSTM	TensorFlow	Acquired	No	No	Regression	Land Use and Soil
Cao *et al.* (2018)	CNN, ELM	–	Existing	No	No	Segmentation	Land Use and Soil
Zeng *et al.* (2018)	ANN	–	Acquired	No	No	Regression	Land Use and Soil
Reddy & Prasad (2018)	LSTM	–	Acquired	No	Yes	Sequence Prediction	Land Use and Soil
Fu *et al.* (2018)	CNN	–	Acquired	No	No	Classification	Land Use and Soil
Jiang (2018)	AE	–	Acquired	No	No	Classification	Land Use and Soil
Shen *et al.* (2019)	ANN	R/H_2O	Acquired	No	No	Regression	Land Use and Soil
Jin *et al.* (2019)	CNN	TensorFlow	Acquired	No	No	Segmentation	Land Use and Soil
Persello *et al.* (2019)	CNN	–	Acquired	No	No	Segmentation	Land Use and Soil
Kroupi *et al.* (2019)	CNN	TensorFlow	Existing	No	No	Classification	Land Use and Soil
Sun *et al.* (2019b)	CNN	PyTorch	Acquired	No	No	Classification	Land Use and Soil
Meng *et al.* (2019)	CNN	–	Acquired	No	No	Classification	Land Use and Soil
Kopp *et al.* (2019)	CNN	–	Collected	No	No	Segmentation	Land Use and Soil
Jiang *et al.* (2019)	CNN	–	Acquired	No	Yes	Matrix Prediction	Land Use and Soil
Bhosle & Musande (2019)	CNN	TensorFlow	Existing	No	No	Classification	Land Use and Soil
Zhang et al. (2019a)	ANN, CNN	–	Existing	No	No	Matrix Prediction	Land Use and Soil
Wang *et al.* (2020a)	DBN	–	Acquired	No	No	Matrix Prediction, Regression	Land Use and Soil
Nam & Wang (2020)	AE	R/H_2O	Acquired	No	No	Classification	Land Use and Soil
O'Neil *et al.* (2020)	CNN	TensorFlow	Acquired	Yes	Yes	Classification	Land Use and Soil
Yang *et al.* (2020)	CNN	Keras	Acquired	No	Yes	Segmentation	Land Use and Soil
Abdi *et al.* (2018)	CNN	–	Existing	No	Yes	Classification	Others
Li *et al.* (2018)	CNN, AE	PyTorch	Acquired	No	Yes	Segmentation	Others
Rohmat *et al.* (2019)	ANN	MATLAB	Acquired	No	No	Regression	Others
Amirkolaee & Arefi (2019)	CNN	MATLAB	Existing	No	Yes	Matrix Prediction	Others
Wang *et al.* (2019a)	CNN	Keras	Existing	No	No	Classification	Others
Kylili *et al.* (2019)	CNN	–	Collected	No	No	Classification	Others
Kang (2019)	CNN	Keras, TensorFlow	Acquired	No	No	Classification	Others
Haklidir & Haklidir (2019)	ANN	–	Acquired	No	Yes	Regression	Others
Kim *et al.* (2020)	GAN, AE	–	Acquired	No	No	Unsupervised Learning	Others
Zhang *et al.* (2018f)	LSTM	Theano	Acquired	No	No	Regression	Surface Water
Liu *et al.* (2019c)	ANN	Keras, TensorFlow	Acquired	No	Yes	Sequence Prediction	Surface Water

(continued)

Table 1 | continued

Paper	Network type	Framework	Dataset	Open-source	Reproducible	Deep learning task	Water field
Xiao *et al.* (2019a)	CNN, LSTM	Keras, TensorFlow	Acquired	No	No	Matrix Prediction	Surface Water
Read *et al.* (2019)	LSTM	–	Acquired	No	No	Regression	Surface Water
Liu *et al.* (2019b)	CNN	Caffe	Existing	No	No	Segmentation	Surface Water
Xiao *et al.* (2019b)	CNN	PyTorch	Existing	No	No	Classification	Surface Water
Mei *et al.* (2019)	CNN	PyTorch	Existing	No	No	Regression	Surface Water
Ling *et al.* (2019)	CNN	MATLAB	Acquired	No	No	Segmentation	Surface Water
Song *et al.* (2019)	CNN	Caffe	Existing	No	No	Classification	Surface Water
Hrnjica & Bonacci (2019)	LSTM	–	Acquired	No	No	Sequence Prediction	Surface Water
Qi *et al.* (2019)	LSTM	–	Acquired	No	No	Sequence Prediction	Surface Water
Zhu *et al.* (2020a)	LSTM	ANNdotNET, CNTK	Acquired	No	No	Sequence Prediction	Surface Water
Lee & Lee (2018)	RNN, LSTM	–	Acquired	No	No	Regression	Water Quality
Hamshaw *et al.* (2018)	RBM	MATLAB	Acquired	Yes	Yes	Classification	Water Quality
Liu *et al.* (2019a)	LSTM	Keras, TensorFlow	Acquired	No	No	Sequence Prediction	Water Quality
Yurtsever & Yurtsever (2019)	CNN	Caffe	Collected	No	No	Classification	Water Quality
Li *et al.* (2019)	LSTM, GRU, ENN	–	Acquired	No	No	Sequence Prediction	Water Quality
Shin *et al.* (2019)	LSTM	–	Acquired	No	No	Sequence Prediction	Water Quality
Banerjee *et al.* (2019)	ANN	R/H_2O	Collected	No	No	Regression	Water Quality
Wang *et al.* (2019c)	LSTM	–	Acquired	No	No	Sequence Prediction	Water Quality
Yim *et al.* (2020)	ANN	MATLAB	Collected	No	No	Regression	Water Quality
Zou *et al.* (2020)	LSTM	Keras	Acquired	No	No	Sequence Prediction	Water Quality
Liang *et al.* (2020)	LSTM	Keras	Acquired	No	No	Sequence Prediction	Water Quality
Barzegar *et al.* (2020)	CNN, LSTM	–	Collected	No	No	Sequence Prediction	Water Quality
Yu *et al.* (2020)	LSTM	–	Acquired	No	No	Sequence Prediction	Water Quality
Zhang *et al.* (2018a)	LSTM	Keras, TensorFlow	Acquired	No	No	Sequence Prediction	Water Resources Management
Zhang *et al.* (2018b)	LSTM, GRU	Keras, TensorFlow	Acquired	No	No	Sequence Prediction	Water Resources Management
Zhang *et al.* (2018c)	LSTM, NARX, ENN	Keras, TensorFlow	Collected	Yes	Yes	Regression	Water Resources Management
Harrou *et al.* (2018)	DBN	–	Collected	No	No	Sequence Prediction	Water Resources Management
Shi & Xu (2018)	AE	–	Collected	No	No	Regression	Water Resources Management
Zhou *et al.* (2019)	ANN, CNN	PyTorch	Existing	Yes	Yes	Classification	Water Resources Management
Fang *et al.* (2019)	CNN	–	Acquired	No	No	Classification	Water Resources Management
Sun *et al.* (2020)	ANN	–	Acquired	No	No	Regression	Water Resources Management
Karimi *et al.* (2019)	ANN, LSTM	MATLAB	Collected	No	No	Sequence Prediction	Water Resources Management
Xu *et al.* (2020)	LSTM	–	Acquired	No	No	Regression	Water Resources Management
Nam *et al.* (2020)	DQN	–	Acquired	No	No	Reinforcement Learning	Water Resources Management
Mamandipoor *et al.* (2020)	LSTM	Keras, TensorFlow	Acquired	No	No	Classification	Water Resources Management

Tang *et al.* (2018)	ANN	–	Acquired	No	No	Regression	Weather
Klampanos *et al.* (2018)	AE, CNN	–	Acquired	No	No	Unsupervised Learning	Weather
Scher & Messori (2018)	CNN	Keras, TensorFlow	Acquired	No	No	Matrix Prediction	Weather
Ukkonen & Mäkelä (2019)	ANN	Keras	Acquired	No	Yes	Classification	Weather
He *et al.* (2019a)	LSTM	–	Acquired	No	No	Classification	Weather
Jeppesen *et al.* (2019)	CNN	Keras, TensorFlow	Existing	Yes	Yes	Segmentation	Weather
Chen *et al.* (2019)	GAN	PyTorch	Acquired	No	No	Matrix Prediction	Weather
Wieland *et al.* (2019)	CNN	Keras, TensorFlow	Existing	No	Yes	Segmentation	Weather
Weyn *et al.* (2019)	CNN, LSTM	Keras, TensorFlow	Acquired	No	No	Matrix Prediction	Weather
Wei & Cheng (2020)	ANN	Weka	Acquired	No	No	Sequence Prediction	Weather
Zhang *et al.* (2019b)	CNN	–	Acquired	No	Yes	Classification	Weather
Kim *et al.* (2019)	ANN	–	Acquired	No	No	Regression	Weather
Pan *et al.* (2019)	CNN	–	Acquired	No	Yes	Matrix Prediction	Weather
Tran & Song (2019)	LSTM, GRU, CNN	TensorFlow	Existing	No	Yes	Matrix Prediction	Weather
Poornima & Pushpalatha (2019)	LSTM, ELM	Keras, TensorFlow	Acquired	No	Yes	Sequence Prediction	Weather
Chai *et al.* (2019)	CNN	–	Acquired	No	Yes	Segmentation	Weather
Wu *et al.* (2020)	LSTM, CNN	–	Acquired	No	Yes	Matrix Prediction	Weather
Zhang *et al.* (2020)	LSTM	–	Acquired	No	No	Regression	Weather
Su *et al.* (2020)	LSTM, CNN	–	Acquired	No	No	Sequence Prediction	Weather
Chen *et al.* (2020)	CNN, LSTM	PyTorch	Existing	No	Yes	Matrix Prediction	Weather
Weber *et al.* (2020)	CNN	TensorFlow	Acquired	Yes	Yes	Matrix Prediction	Weather
Yan *et al.* (2020)	CNN	TensorFlow	Acquired	No	Yes	Sequence Prediction	Weather
Wang *et al.* (2020c)	RNN	–	Acquired	No	No	Matrix Prediction	Weather

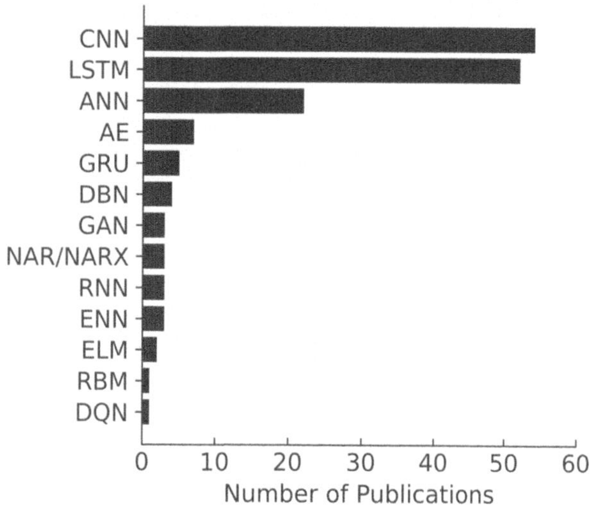

Figure 9 | Architectures described in the study and their usage in reviewed papers.

LSTM model on daily runoff prediction for the first time, which included meteorological observations, and obtained results better than a well-established physical model – the Sacramento Soil Moisture Accounting Model SAC-SMA + Snow-17. In 2019, Kratzert *et al.* (2019a) further applied the LSTM model to 531 watersheds in the USA with k-fold cross-validation and showed that LSTM can be applied on ungauged watersheds with better results than physical models, such as calibrated SAC-SMA and the National Water Model. Other researchers applied RNN to the runoff forecast and compared their outputs to other machine learning models. Damavandi *et al.* (2019) proposed an LSTM model on a Texas watershed predicting the next day's daily streamflow using climate data and the current day's

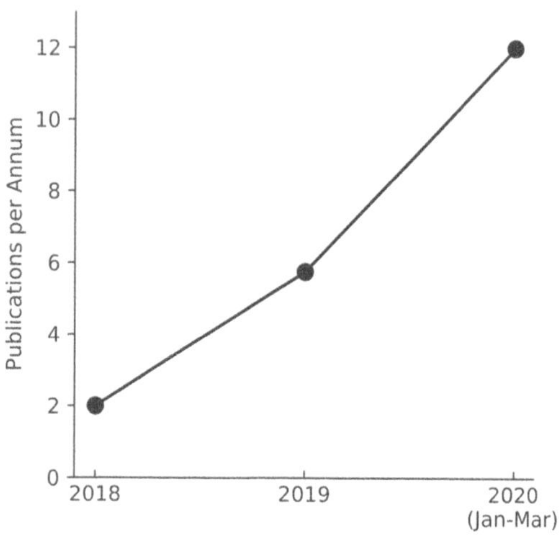

Figure 10 | Average number of papers published each month during 2018, 2019, and 2020 from January to March.

streamflow. Their results showed that LSTM performs better than the physical model CaMa-Flood. Zhang *et al.* (2018e) applied an LSTM model on monthly reservoir inflow and outflow predictions in hourly, daily, and monthly basis and achieved better results than Support Vector Machines (SVM) and BackPropagation Neural Networks. Kumar *et al.* (2019) applied RNN and LSTM models for the monthly rainfall prediction in India and showed that LSTM provided better results. Qin *et al.* (2019) applied the LSTM on the streamflow prediction and compared it with the Autoregression model. Yang *et al.* (2019a) applied an LSTM model to the daily reservoir overflow prediction. NAR with external input and LSTM are used for three reservoirs with geographic information, daily precipitation, air temperature, wind speeds, relative humidity, and sunshine duration. Wan *et al.* (2019) also successfully applied ENN to a real-time 3-h ahead flood forecasting.

Since hyper-parameter optimization is a problem in deep learning, some studies focused on applying additional optimization algorithms on deep learning models. Yuan *et al.* (2018) proposed two models that use the Ant Lion Optimization (ALO) and Particle Swarm Optimization (PSO) to optimize the parameters and hidden layers of an LSTM model, which are named LSTM-ALO and LSTM-PSO. Their results show that the LSTM-PSO outperformed the LSTM and LSTM-ALO. Yang *et al.* (2019a) proposed a genetic algorithm-based NARX, which outperforms the NARX and LSTM. Ni *et al.* (2019) proposed two LSTM based models, which are wavelet-LSTM and CNN + LSTM for the rainfall and streamflow forecasting, and both models have better results than LSTM. Kabir *et al.* (2020) proposed a wavelet-ANN to make the hourly streamflow predictions. Results show that wavelet-ANN can make acceptable predictions for, at most, 6 h ahead, which outperforms ANN, DBN, and Support Vector Regression (SVR).

Some studies focused on the data pretreatment such as the decomposition of rainfall and runoff prior to deep learning models. He *et al.* (2019b) proposed a DNN model for daily runoff prediction where the inputs are the runoff series that were decomposed into multiple Intrinsic Mode Functions (IMF) with Variational Mode Decomposition (VMD). Hu *et al.* (2019) proposed an LSTM model for flood forecast with the preprocessing of Proper Orthogonal Decomposition and singular value decomposition. Zuo *et al.* (2020) proposed three LSTM models with different pretreatments, which are VMD, Ensemble Empirical Mode Decomposition (EEMD), and Discrete Wavelet Transform for daily streamflow up to 7 days ahead. Zhu *et al.* (2020b) proposed a probabilistic LSTM model coupled with the

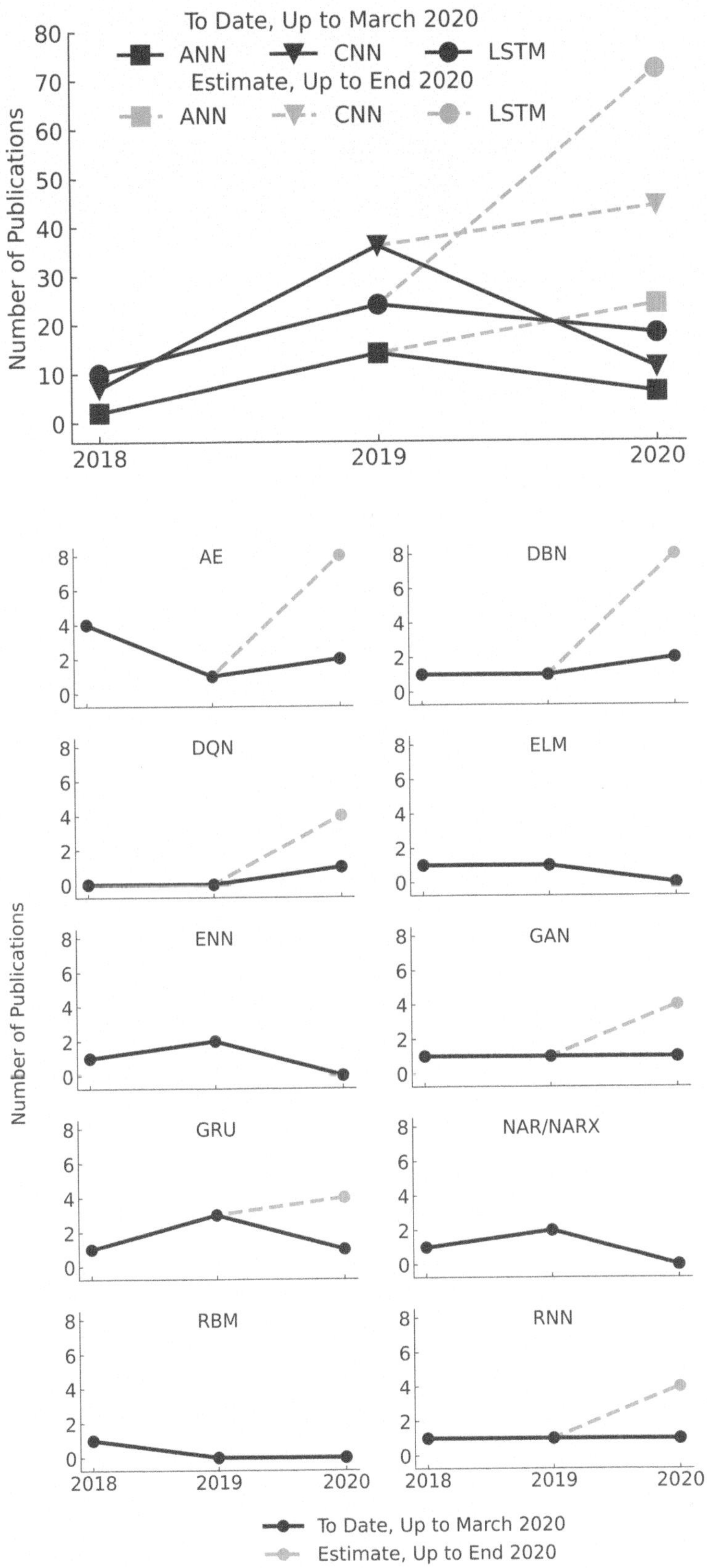

Figure 11 | Architectures described and their annual usage in reviewed papers.

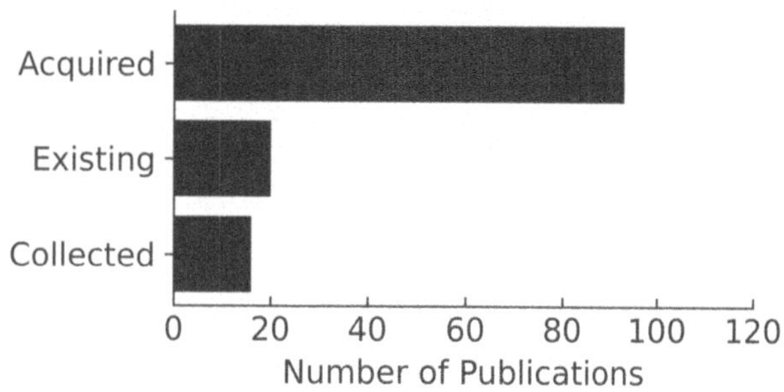

Figure 12 | Distribution of dataset acquisition types in reviewed papers.

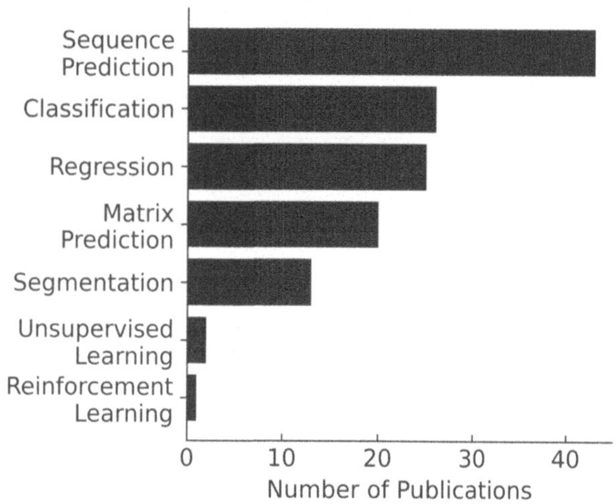

Figure 13 | Histogram of machine learning task types studied in reviewed papers.

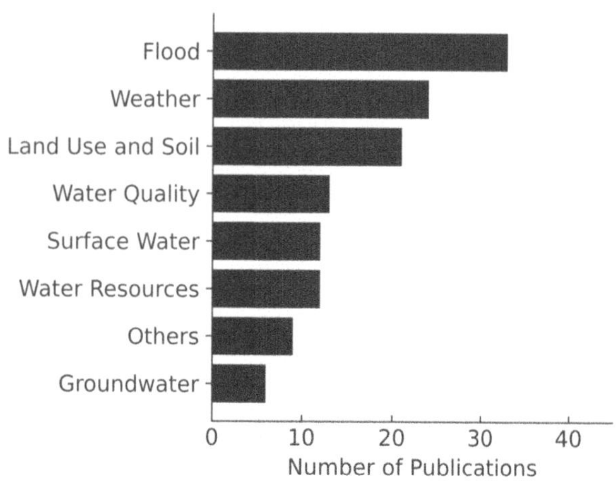

Figure 14 | Histogram of Water domain subfields studied in reviewed papers.

Gaussian process to deal with the probabilistic daily streamflow forecasting. These studies show that the pretreatment of input may help to improve the deep learning model accuracy.

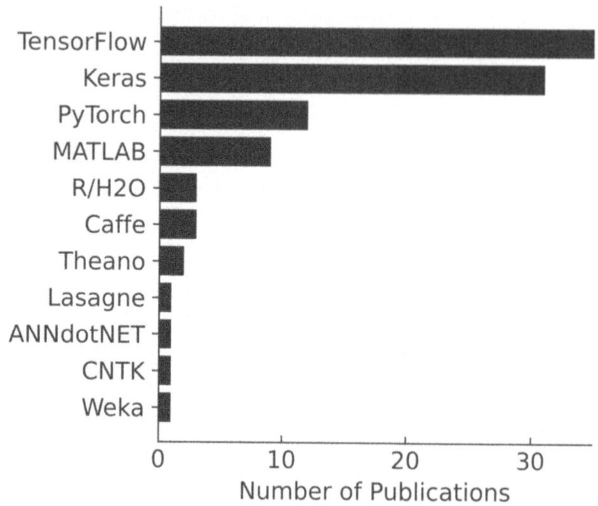

Figure 15 | Machine learning frameworks described in the study and their usage in reviewed papers.

Other studies focused on constructing more complex deep learning architectures such as autoencoder, encoder–decoder, and customized layer based on the LSTM. Bai *et al.* (2019) proposed an LSTM model with Stack Autoencoder (SAE) to predict daily discharge values based on one-week discharge values, and the results of SAE-LSTM outperform LSTM alone. For multiple time-step flood forecast tasks, an encoder–decoder LSTM is proposed for the runoff prediction by Kao *et al.* (2020) and Xiang *et al.* (2020). Kao *et al.* (2020) proposed an encoder–decoder LSTM model that can be used on multi-timestep output predictions for up to 6 h. Xiang *et al.* (2020) proposed an encoder–decoder LSTM model that can be used to predict for up to 24 h ahead. Both studies showed the encoder–decoder LSTM is better than LSTM. In particular, Kratzert *et al.* (2019b) proposed the Entity-Aware-LSTM network, which designed a specific layer for the rainfall-runoff modeling based on LSTM. This network allows for learning catchment similarities as a feature layer, and data from multiple watersheds can be used to provide runoff for a watershed. These studies applied some high-level designs to the LSTM network and can perform much better than normal LSTM on the long-term or multiple watersheds.

Several studies coupled physical models with deep learning networks. Yang *et al.* (2019b) proposed a model using LSTM to improve the performance of flood simulations of a physical model. The watershed-averaged daily precipitation, wind, temperature, and model-simulated discharge from the Global Hydrological Model GHMs + CaMa-Flood between 1971 and 2020 were used to model the discharge. This indicates the deep learning models can be used to improve the streamflow forecast accuracy of physical models. Nguyen & Bae (2020) proposed an LSTM network using the Quantitative Precipitation Forecasts of the McGill Algorithm for Precipitation nowcasting by Lagrangian Extrapolation (MAPLE) system to reproduce 3-h Mean Areal Precipitation (MAP) forecasts. Corrected MAPs are used as input to a coupled 1D/2D urban inundation model to predict water levels and relevant inundation areas (1D conduit network model and 2D overland flow model). In this model, LSTM takes a forecast from MAPLE and then reproduced MAPs are used by the coupled 1D/2D urban inundation model. Worland *et al.* (2019) proposed a DNN model to predict the flow-duration curves using United States Geological Survey streamflow data by constructing 15 output values representing 15 quantiles of the curve. These indicate that deep learning models can be used as surrogate models of physical models or curves.

Some studies focused on flood susceptibility and flood prediction capabilities. Bui *et al.* (2020b) applied a DNN network with 11 variables such as Digital Elevation Model (DEM), aspect, slope, etc. to predict the flood susceptibility, a value between 0 and 1, for an area. Wang *et al.* (2020d) proposed a CNN network with 13 flood-triggering factors in the study area to map the flood susceptibility for the study area. In this project, each pixel may have different flood susceptibility values. Sankaranarayanan *et al.* (2019) applied a DNN network to classify if flooding is possible in an area for the month given total precipitation and temperature. This study simplified the project into a binary prediction task of if there will be a flood event or not. The same simplification is done by Bui *et al.* (2020a), who applied a DNN network to classify if a region is flash flood prone or not using the elevation, slope, aspect, curvature, stream density, Normalized Difference Vegetation Index (NDVI), soil type, lithology, and rainfall data as input.

With CNNs, some innovative flood monitor and forecast projects are now possible. Wang *et al.* (2019b) applied a CNN model that predicted the real-time hourly water levels for warning systems during typhoons using satellite images. Bhola *et al.* (2019) proposed a model that estimates the water level from CCTV camera images using a CNN model. The model performed the edge detection to find the water body in the photo, and the water level is then calculated from physical measurements. Moy de Vitry *et al.* (2019) proposed a model that calculated the flooded area and the surface observed flooding index with CCTV cameras using CNN. U-net CNN is used to segment the stream shape from a CCTV camera and another CNN model is used to calculate the flooded areas in pixels from the segmented photo.

Subsurface and groundwater

There are several different types of studies in subsurface and groundwater. Different deep learning models can be used in each of different types of studies. One type of groundwater study is the estimation of water table level or flow rate. With the groundwater monitoring wells data, this is a 1D regression task similar to the surface water predictions using recurrent models. Jeong & Park (2019) applied the NARX, LSTM, and GRU on the water table level estimations with observed data from monitoring wells. It found that the estimations from the NARX and LSTM models are superior to those of the other models in terms of prediction accuracy. Jeong *et al.* (2020) further applied LSTM with multiple loss functions, which shows that the proposed LSTM model with cost function of MSE with Whittaker smoother, least trimmed squares, and asymmetric weighting has the best performance on groundwater level prediction with data corrupted by outliers and noise. For groundwater flow prediction, Wang *et al.* (2020b) proposed a theory-guided DNN, which not only makes the predictions of groundwater flow, but also estimates the parameters of the Partial Differential Equation (PDE) as well as the initial condition and boundary condition of the PDE. By constructing the loss function, the theory-guided neural network can provide the prediction results with reasonable parameters of the physical model PDE.

More groundwater studies take a cross-section in time-series and focus on a 2D map. Examples include the groundwater water balance map (Sun *et al.* 2019a), hydraulic conductivity field map (Zhou *et al.* 2020), pressure, and CO_2 saturation field map (Mo *et al.* 2019a). CNN models can be used in these studies. Sun *et al.* (2019a) proposed a CNN model using the physical model National Oceanic and Atmospheric Administration (NOAH) simulation results as input to predict the groundwater water balance. Thus, the CNN model is used to correct the physical model results by learning the spatial and temporal patterns of residuals between Gravity Recovery and Climate Experiment (GRACE) observations and NOAH simulations. Results show that the CNN can significantly improve the physical model simulation accuracy. Zhou *et al.* (2020) proposed a CNN of eight layers to learn a map between stochastic conductivity field and longitudinal macro-dispersivity based on synthetic 2D conductivity fields. The estimations are in acceptable accuracy with moderate heterogeneity. Mo *et al.* (2019a) proposed an Encoder-Decoder CNN to approximate the pressure and CO_2 saturation field map in different time-steps as a surrogate model. Furthermore, some studies in groundwater focus on 3D mapping. An example is the flow-rate estimation in a 3D rock. Santos *et al.* (2020) proposed a 3D CNN to predict the estimated state solution of the Navier–Stokes equation for laminar flow, which is the flow rate, with 3D rock images. This 3D CNN is a surrogate model of the Navier–Stokes equation, and takes less than a second.

In groundwater studies, one of the other deep learning applications is inversion, such as to identify the contaminate source in groundwater. Some complex traditional algorithms can be used to solve this problem. However, it will be not efficient when the data is in high dimensions. Mo *et al.* (2019b) developed a deep autoregressive neural network which was used as the surrogate model of this high-dimensional inverse problem and provides more accurate inversion results and predictive uncertainty estimations than physical models. Laloy *et al.* (2018) proposed another approach for the inversion by using GAN. With the GAN models, the inversion rapidly explores the posterior model distribution of the 2D steady state flow and recovers model realizations that fit the data close to the true model results.

Finally, Sun *et al.* (2020) investigated three learning-based models, namely DNN, Multiple Linear Regression, and SARIMAX, to find missing monthly data in total water from GRACE data (Tapley *et al.* 2004). Based on the results, the performance of DNN is slightly better than SARIMAX, and significantly better than MLP in most of the basins. The three learning-based models are reliable for the reconstruction of GRACE data in areas with humid and no/low human interventions. Dynamic multiphase flow in heterogeneous media is a problem in groundwater studies. These studies show that deep learning models can work as surrogate models by improving computational efficiency in groundwater dynamic predictions and inversions.

Land and soil

The segmentation and classification of land use and land cover are important for water and soil resources management. Many studies have applied the deep learning networks to create a more accurate land cover map from satellite or radar imagery. Abdi *et al.* (2018) applied SAE and CNN on land cover classification of urban remote sensing data. Tests on nine datasets show that SAE and CNN are better than machine learning models like logistic regression, Naive Bayes, K-Nearest Neighbors, and SVM. Cao *et al.* (2018) applied ELM and CNNs to classify land cover categories on satellite imagery and the results show that the combined CNN–ELM method has the highest accuracy. Shen *et al.* (2019) applied a DNN network to predict the land drought index using the precipitation data, meteorological drought index data, and soil relative moisture data. Bhosle & Musande (2019) applied a CNN model on the land cover land-use classification on Indian Pines dataset. Some studies applied more complex coupled models, including encoder–decoder, AE, 3D networks, and coupled machine learning models to achieve a higher model accuracy. Zhang *et al.* (2019a) applied 3D-CNN and 3D-DenseNet models on the land cover land-use classification on Indian Pines and Pavia University datasets. All these studies showed a high accuracy of the land cover identification task. Nam & Wang (2020) applied the AE prior to the random forest, SVM, and other machine learning models to the landslide susceptibility prediction. Results show that random forest with an AE gives more accurate results than other machine learning methods. In the study of wetland-type identification by Meng *et al.* (2019), results show that the ensemble model SVM-CNN performs better than CNN and SVM. O'Neil *et al.* (2020) applied an Encoder-Decoder CNN model to classify the wetland types using Light Detection and Ranging (LiDAR) radar dataset and NDVI index data. This study used the LiDAR DEM and remote sensing images to generate physically informed input, including Slope, NDVI, DTW, and Topographic Wetness Index. In particular, Kroupi *et al.* (2019) applied a CNN-based model to land-cover classification that trained on the European satellite dataset, EuroSAT, and tested in a region outside Europe. It still provided promising results despite differences in tested and trained regions. These results indicate the CNN models have a high model accuracy as well as high robustness in the land-use land-type classification task.

In addition to applying known networks, some researchers developed objective-based CNN rather than traditional pixel-based CNN to better identify the land use and land types. Zhang *et al.* (2018d) proposed an object-based CNN to label very fine spatial resolution remotely sensed images to do object-wise segmentation rather than pixel-wise segmentation in urban areas. Fu *et al.* (2018) proposed a blocks-based object-based CNN for classification of land use and land-cover types and achieved end-to-end classification. This model works well on irregular segmentation objects, which is a common in-land use classification. Jiang (2018) proposed an object-based CNN with an AE for extracting high-level features automatically. Results show the AE-object-CNN is better than three manual design feature systems. Jin *et al.* (2019)

proposed an object-oriented CNN that used a typical rule set of feature objects to construct the object-oriented segmentation results before using a CNN to make the classification. These studies show object-based studies have a better performance than simple CNN.

Some studies focus on more specific tasks based on different study purposes such as the identification of agricultural fields, impervious surfaces, wetland types, water body types, and crop types. Persello *et al.* (2019) applied two CNN models, SegNet (Badrinarayanan *et al.* 2017) and VGG16 (Simonyan & Zisserman 2014), for the segmentation and classification of agricultural fields. Sun *et al.* (2019b) applied a CNN model to classify if the land is impervious surfaces, vegetation, or bare soil from satellite imagery or both the satellite and LiDAR data. Results show that the model with LiDAR data provides better results than the model with satellite imagery only. Meng *et al.* (2019) identified the wetland types in one lake using the Chinese remote sensing imagery GF-2. Bhosle & Musande (2019) applied a CNN model to identify crop types on EO-1 Hyperion sensor hyperspectral imagery. Yang *et al.* (2020) applied CNNs to identify the water body types from remote sensing images. Mask R-CNN is used to segment the water body, and the ResNets (He *et al.* 2015), including ResNet-50 and ResNet-101, are used to identify the water body types. Results show a high accuracy on regular-shaped water bodies.

Similar networks can be used to generate DEMs, which is another type of land use study. Jiang *et al.* (2019) applied a CNN model to predict the paleo-valley DEM using the original DEM data and electrical conductivity (EC). The EC data are collected from the field study, and the CNN models used in this study distinguish the valley and non-valley pixels, which can find the spatial connectivity of the paleo valley. The CNN can efficiently constrain 3D paleo-valley geometry DEM.

Snow cover, a special study in land-cover studies, can be measured by deep learning models in different approaches. Kopp *et al.* (2019) proposed a model to predict the snow depth using OpenCV and Mask R-CNN on the surveillance camera photos. The Mask R-CNN is used to segment the detectable measuring rod, and then the OpenCV library can be used to identify the snow depth by measuring how much of the measuring rod is covered by snow. Wang *et al.* (2020a) applied a DBN and CNN to estimate snow depth from the geographical data. DBN takes multiple inputs, including latitude, longitude, elevation, forest cover fraction, time, passive microwave horizontal, and vertical polarization brightness temperatures of 19 and 37 GHz. Results show that the DBN outperforms CNN in this study.

Additional land and soil related studies include the land surface temperature, soil salinity, and vegetation dynamics over time, and these can be done with time-series related models such as LSTM. Zhang *et al.* (2018g) applied the ensemble composition-based LSTM models on the daily land surface temperature simulation. The original daily land surface temperature data series were decomposed into many IMFs and a single residue item. The Partial Autocorrelation Function is used to obtain the number of input data sample points for LSTM models. Zeng *et al.* (2018) applied Partial Least Square Regression (PLSR), SVM, and DNN for predicting soil salinity from images. Surprisingly, DNN performs worse than PLSR for this task. Reddy & Prasad (2018) applied a LSTM model to predict the vegetation dynamics using NDVI images from 2000 to 2016 in a 7-day gap. This study shows that the single feature NDVI can be used to provide accurate vegetation prediction over time.

Surface water

The prediction of water level is crucial for water resources management and protecting the natural environment. Many studies have applied deep learning methods, such as LSTM and ANN, to forecast the water level from one day to one year. Hrnjica & Bonacci (2019) investigated two different ANNs, namely LSTM and FFNN, to forecast the lake water level. Monthly measurements for the last 38 years of Vrana Lake, Croatia, is utilized for training the models to predict 6 or 12 months ahead of the lake water level. The set of sequences with different lengths created from the obtained data is used in the networks, instead of using classical lagged data. The results of LSTM and FFNN are compared with classical time forecasting methods and ANN. According to the results, the performance of LSTM is the best among the models in all scenarios, while FFNN provides better accuracy than the compared methods in both 6 and 12 months prediction. Zhu *et al.* (2020a) also investigates LSTM and FFNN in their work with data from a different region. However, the models in this paper are designed to predict one month ahead of the lake water level for 69 temperate lakes in Poland. Their results indicate that LSTM and FFNN perform similarly most of the time, unlike in the previous paper. The reasons for this situation can be the differences between the datasets, prediction intervals, or model designs.

In addition to lake water-level prediction, Liu *et al.* (2019c) developed a Bayesian neural network, which is based on ANN with posterior inference, to forecast the water level in a reservoir to derive operation rules. According to the paper, the current reservoir status and future inflows are the primary factors that affect operational decisions. The influence of inflow uncertainties on reservoir operations is also more than model parameter uncertainty. Their findings show the impact of the input data alongside the promising results of the Bayesian neural network. Qi *et al.* (2019) forecasted daily reservoir inflow by ensembling the different results from the LSTM models with different decomposed inflow data as inputs for more accurate assumptions. Zhang *et al.* (2018f) proposed an LSTM model to predict water table depth in agricultural areas. The LSTM model takes monthly water diversion, evaporation, precipitation, temperature, and time as inputs for prediction of the water table depth. The results of the LSTM model are compared with the FFNN model, and the LSTM model performs better than the FFNN model. It also highlights that the dropout method increases the LSTM model's accuracy in this task.

Alongside the water level or flow, the prediction of water temperature has received much attention in the scholarship. Xiao *et al.* (2019a) proposed a convolutional long short-term memory (ConvLSTM) model to predict the sea surface temperature (SST). In the paper, 36 years of satellite-derived SST data are used. Based on the results, the ConvLSTM model produces more accurate results than the linear SVR model and two different LSTM models for short and mid-term temperature prediction. Read *et al.* (2019) aimed to predict lake water temperatures based on depth. A hybrid LSTM and theory-based feedbacks (model penalties) model was developed. According to the results, the hybrid model produces the best results among the tested methods in the paper. The results from the paper can be seen as an example of improving predictions by integrating scientific rules with deep learning methods.

In addition to the aforementioned subtopics in surface water, various tasks are also investigated by scholars such as segmentation, change detection, or super-resolution. Despite extensive usage of LSTM or ANN in previously aforementioned papers, CNN models are generally used in these tasks. Liu *et al.* (2019c) labeled objects in aerial images as water, tree, etc. with the help of a fully convolutional neural network (FCN) model and multinomial logistic regression. FCN takes the aerial image and returns a probability. At the same time, LiDAR data passes into multinomial logistic regression and returns another probability. These two probabilities are combined with higher-order conditional random field formulation to produce a final probability. Based on the final probability, objects are labeled with the corresponding group. Mei *et al.* (2019) developed a CNN-based model to measure the sea-ice thickness in Antarctica from LiDAR data. The input is a windowed LiDAR scan (snow freeboard), and the mean ice thickness is the output of the model. In addition to the LiDAR scan, the paper investigated the effects of different inputs such as surface roughness and snow depth on the task. Ling *et al.* (2019) used the CNN model to generate a finer resolution of the image to measure the wetted river width. Then, the output of CNN is used to measure the width of the river. Song *et al.* (2019) aimed to detect the change in surface water in remote sensing images. An FCN model is proposed and used for the same regions at different times. The FCN model returns the surface water regions for each time. The outputs for the same regions are compared, and a change map created to show the difference. In addition to these publications, Xiao et al. (2019b) introduced a dataset for the classification of ice crystals. The dataset contains 7,282 images in ten different categories. The performance of numerous pre-trained models, such as AlexNet (Krizhevsky *et al.* 2012) and VGGNet(s) (Simonyan & Zisserman 2014), are also provided in the paper.

Water quality

Water quality monitoring and prediction are vital operations for many fields, such as water resources management and water treatment. Water quality and safety depend on numerous parameters with complex biological, chemical, and physical interactions. As such, deterministic water quality models are a realistic option in only the simplest and idealized scenarios. However, data-driven models are increasingly used in a variety of water quality applications. One such application area is predicting surface water quality. Li *et al.* (2019) proposed an ensemble approach that combines three RNN models with Dempster–Shafer (D–S) evidence theory (Shafer 1976) to predict the quality of water. The results of three RNN models, namely LSTM, GRU, and ENN, are combined by D–S evidence theory for the final output. The combined model predicts, at most, 50 h in advance, and the results show that the model accuracy reduces significantly over 25 h. Liu *et al.* (2019a) used an LSTM model to forecast drinking water quality for up to 6 months. Zou *et al.* (2020) developed an ensemble approach to predict water quality data, such as pH, dissolved oxygen (DO), Permangate Index (CODMn), and NH_3–N. The approach is based on using three LSTM models with different size interval data feeding each of them. The final prediction is a combination of the results of three LSTM models. Banerjee *et al.* (2019) choose the indicators, namely DO and zooplankton abundance, to reflect the water quality level of a reservoir.

An ANN model is proposed to model the selected indicators to represent the water quality level. Yu *et al.* (2020) combined the LSTM model with Wavelet Mean Fusion and Wavelet Domain Threshold Denoising to simulate the change of chlorophyll-a concentration in Dianchi Lake, China, and used 15 water quality parameters as inputs, such as pH and DO. Liang *et al.* (2020) also worked with the prediction of the chlorophyll-a concentration level. Fabricated data are created by the Environmental Fluid Dynamics Code to train an LSTM model. The LSTM model can forecast the chlorophyll-a concentration level for up to one month. Chlorophyll-a, water temperature, and total phosphorus are identified as critical inputs that affect the performance of the LSTM model. Barzegar *et al.* (2020) investigated multiple models to predict the level of DO and chlorophyll-a in Small Prespa Lake, Greece. Three different neural network models, namely CNN, LSTM, and CNN-LSTM, were developed to forecast the DO and chlorophyll-a concentrations and pH, oxidation–reduction potential, water temperature, and EC were used as inputs for the models. In addition to the ANN models, SVM and decision tree models are used for the performance comparison. According to results, the hybrid CNN-LSTM model provides the best accuracy to predict both DO and chlorophyll-a.

Yim *et al.* (2020) developed a stacked autoencoder-deep neural network (SAE-DNN) to predict phycocyanin concentrations in inland waters from in-situ hyperspectral data. The proposed architecture's ability for the prediction from airborne hyperspectral imagery is examined. Shin *et al.* (2019) introduced an LSTM model to forecast the occurrence of harmful algal blooms in the South Sea of Korea. SST and photosynthetically available radiation are extracted from satellite data to be used as inputs for the LSTM model to minimize damage. Lee & Lee (2018) aimed to predict the occurrence and number of harmful algal blooms with an LSTM model by providing weekly water quality and quantity data. The LSTM model's performance is compared with RNN and MLP models, and the LSTM provides better results among all the investigated methods based on the results. Hamshaw *et al.* (2018) used the RBM to classify the sediment-discharge curve in 14 categories from the 2D image of the suspended-sediment discharge plots from more than 600 storm events. Finally, an LSTM-based system was proposed to identify the characteristics of the water pollutants and trace its sources in the work of Wang *et al.* (2019c). In the system, a water quality cross-validation map is generated to identify pollutants and, based on defined rules, track the pollutants to common industries.

Water resources management

Urban water systems are essential to modern cities. Efficient operation of water treatment plants, wastewater treatment plants, and conveyance networks require accurate modeling of these interconnected systems. Zhang *et al.* (2018b) investigated the multiple models to simulate and predict the water level of Combined Sewer Overflow structure. In the study, the collected data from Internet of Things is used separately with four different neural networks, namely MLP, ANN with Wavelet, LSTM, and GRU, to compare the networks with each other. According to the results, LSTM and GRU have good performances with respect to others, but GRU has a quicker learning curve, fewer parameters, and simpler architecture. Despite these advantages, the accuracy of GRU is slightly lower than the LSTM. Zhang *et al.* (2018a) predicted the next hour's wastewater flow to avoid sewer overflow using LSTM with the traditional hydraulic model on the sewer system. This 1-h prediction is tested in several scenarios, which are: 1-time step prediction at the 1-h sampling frequency, 2-time steps prediction at the 30-min sampling frequency, 4-time steps prediction at 15-min sampling frequency, and 6-time steps prediction at 10-min sampling frequency. The performance of the LSTM-based model is compared with SVR and FFNN, and the LSTM-based model has the best accuracy in all scenarios. Zhang *et al.* (2018c) also predicted the next hour's wastewater inflow for the wastewater treatment plant. The paper aims to identify which parts of the sewer system have more free space and take action based on the outcome. LSTM, NARX, and ENN are compared, and the LSTM model provides better results than other methods based on the results. Karimi *et al.* (2019) proposed an LSTM model to forecast flow in sanitary sewer systems. It claimed that accepting the groundwater as an additional input for the LSTM model increases the overall accuracy of the task.

In addition to the prediction of the wastewater level or flow, some studies aimed to detect conveyance network conditions. Xu *et al.* (2020) aimed to detect abnormal working conditions in the water supply network. In addition to the detection of abnormal conditions, pressure in the water supply network is predicted. An LSTM model is developed to achieve these goals. The performance of the LSTM model outperforms the traditional prediction models, such as SVM. Zhou *et al.* (2019) used an ANN model to identify burst locations in a water distribution network. The model takes the pressure data as input and returns one or several possible pipes, which can be the location of the burst. Fang *et al.* (2019) focused on detecting multiple leakage points in a water distribution network with a CNN-based model. The model accepts the pressure measurements of the distributed water system as input and returns the possible locations of leakage points.

Various studies explored the strength of deep learning-powered modeling in water and wastewater treatment procedures. Shi & Xu (2018) developed a Stacked Denoising Autoencoders to predict a two-stage biofilm system's performance. Nam *et al.* (2020) proposed a DQN-based system to operate a membrane bioreactor more efficiently. It aims to maximize the system's energy efficiency while meeting stringent discharge qualities. GoogLeNet (Szegedy *et al.* 2015) architecture is used to identify and classify microbeads in urban wastewater into five categories based on microscopic images in the work of Yurtsever & Yurtsever (2019). Harrou *et al.* (2018) provided a case study using a DBM-SVM model to identify abnormal signals from the water features such as pH, conductivity, etc. in wastewater treatment plants. The results show that it is possible to detect the abnormal conditions in order to alert the system early based on the outcome of the DBM-SVM model. Mamandipoor *et al.* (2020) developed an LSTM model to monitor a wastewater treatment plant for detecting faults during the oxidation and nitrification processes.

Weather

Rainfall forecasting is one of the significant tasks in the domain of meteorology. Several techniques have been proposed to forecast rainfall with the help of statistics, machine learning, and deep learning. Zhang *et al.* (2020) introduced an ensemble approach to forecast rainfall. In the first step, eight major meteorological factors are selected via correlation analysis between control forecast meteorological factors and real-time rainfall. Then, samples are divided into four categories by K-means clustering. The LSTM-based model is fed by each cluster, and outputs are combined to reach the final prediction. Weber *et al.* (2020) developed a CNN-based surrogate model for one of the global climate model CanESM2 (Arora *et al.* 2011). The CNN model is fed by 97 years of monthly precipitation output from CanESM2, and the model preserves its performance even when the forecast length is expanded to 120 months. According to the paper, the accuracy of the model can be increased by deeper networks. Poornima & Pushpalatha (2019) investigated an LSTM model to forecast the rainfall using 34 years of rainfall data. In the paper, the LSTM model results are compared with multiple methods, such as Holt–Winters and AutoRegressive Integrated Moving Average (ARIMA). Tang *et al.* (2018) introduced a DNN model to predict rain and snow rates at high altitudes. In the paper, passive microwave, infrared and environmental data are trained to the reference precipitation datasets, which are obtained by two space-borne radars for the estimations. The results of the DNN model are compared with many methods, such as the Goddard Profiling Algorithm. The experiment results show that the DNN model is capable of predicting snow and rain rate more accurately than other tested methods at high altitudes.

Some studies specifically focused on the improvement of quantitive precipitation estimation accuracy, alongside precipitation nowcasting. Since the prediction of precipitation generally is a time-series problem, the usage of LSTM architecture is common for this task. Wu *et al.* (2020) designed a fusion model, which is a combination of CNN and LSTM, to improve quantitative precipitation estimation accuracy. The proposed model uses satellite data, rain gauge data, and thermal infrared images. The CNN part of the model extracts the spatial characteristics of the satellite, rain gauge, and thermal infrared data, where the LSTM part of the model handles the time dependencies of the provided data. The performance of the CNN-LSTM model is better than the comparative models, such as CNN, LSTM, and MLP. Yan *et al.* (2020) used a CNN model to forecast short-term precipitation with the help of radar reflectance images for a local area in China. As a dataset, the radar reflection images and the corresponding precipitation values for 1 h are collected. The model takes the images as inputs and returns the forecast value for 1-h precipitation. The CNN model contains residual links between the layers, which increase the efficiency of the model. Chen *et al.* (2020) focused on precipitation nowcasting using a ConvLSTM model. The model accepts the radar echo data to forecast 30 or 60 min of precipitation value. According to results, using the customized multisigmoid loss function and group normalization provides better performance than ConvLSTM with classical loss functions, such as cross-entropy loss function, and conventional extrapolation methods.

Statistical downscaling methods often provide more accurate precipitation estimation than using raw precipitation values in the models.

Deep learning methods can be used as statistical downscaling methods to improve the accuracy of the tasks. Pan *et al.* (2019) proposed a CNN model as a statistical downscaling method for daily precipitation prediction. The method is tested with 14 geogrid points in the USA, and statistical downscaling results from the CNN model outperform other tested methods, including linear regression, nearest neighbor, random forest, and DNN. Wang *et al.* (2020c) developed an RNN model to perform statistical downscaling on temperature and precipitation in order to improve the accuracy of hydrological models. The RNN model provides better accuracy than the compared methods, such as ANN, for extreme temperature and precipitation downscaling based on the evaluation of downscaled data on the Soil and Water Assessment Tool model.

In meteorology and remote sensing, cloud or cloud shadow detection has received attention in the recent literature. Because U-net (Ronneberger *et al.* 2015), a CNN model for biomedical image segmentation, and SegNet (Badrinarayanan *et al.* 2017), a convolutional encoder–decoder architecture for image segmentation, provide successful results in their domain, many researchers use those models as a base model for their works. Jeppesen *et al.* (2019) used U-net to detect clouds in satellite imagery. The model is trained and tested with Landsat 8 Biome and SPARCS datasets. Wieland *et al.* (2019) aimed to segment cloud, shadow, snow/ice, water, and land in multispectral satellite images. The U-net based model is proposed and trained with Landsat datasets to segment images into five categories. According to results, contrast and brightness augmentations of the training data improve the segmentation accuracy, alongside adding shortwave-infrared bands. Zhang *et al.* (2019b) applied the U-net based model on the RGB and infrared waveband images from the Landsat-8 dataset for cloud detection. LeGall-5/3 wavelet transform is used on the dataset to accelerate the model and make it feasible to implement on-board satellite. Chai *et al.* (2019) proposed a CNN model based on SegNet to detect clouds and cloud shadow in Landsat imagery.

Thunderstorms and typhoons are one of the extreme natural disasters that can cause massive damages. Some studies focused on the prediction of those hazardous natural events to take early actions to minimize the damage. Ukkonen & Mäkelä (2019) aimed to predict the occurrence of thunderstorms from parameters related to instability, inhibition, and moisture. A DNN model is trained with lightning data and a high-resolution global reanalysis. Various regions, such as Sri Lanka and Europe, are used as test areas for the model. Many valuable findings are provided related to the correlation between thunderstorm occurrence and parameters specific to regions. Kim *et al.* (2019) aimed to find similarities between a typhoon and a previous typhoon for helping to mitigate the effect of the typhoon. In the study, a DNN model is used to encode the typhoon event by typhoon parameters, such as route, pressure, and moving speed. The model returns the typhoon events' similarities to the historical ones, which provide insights for officials to take early action. During the work, a database was created for 189 typhoons that occurred between 1950 and 2017. Wei & Cheng (2020) aimed to predict wind speed and wave height of a typhoon with the help of an RNN model, namely the Two-step Wind-Wave Prediction (TSWP). The TSWP model predicts wind speed first, followed by wind height for 1–6 h in the future. The results of the TSWP model outperform the comparative methods, such as MLP, DNN, and logistic regression.

In addition to the aforementioned studies, various works have been published for different meteorological tasks. Su *et al.* (2020) used the pyramid delaminating technique to generate the global optical flow field. A ConvLSTM model takes the RGB image and generated flow field in order to improve the forecast accuracy of echo position and intensity. The generation, dissipation, and merging of convective cells were also better identified comparison to other classical methods. Weyn *et al.* (2019) forecast weather at 500-hPa geopotential height using a CNN model. The paper uses historical gridded reanalysis data in the model without explicit knowledge about the physical process. The CNN model produces promising results for capturing the climatology and annual variability of 500-hPa heights and predicts realistic atmospheric states for 14 days. However, the CNN model still did not perform as well as an operational weather model. He *et al.* (2019a) proposed an LSTM model to classify periods as rainy or dry using microwave links. Scher & Messori (2018) aimed to predict the uncertainty of a forecast based on past forecasts and their occurrence rate with a CNN model. The CNN model takes the atmospheric fields at three different heights as inputs and returns the scalar value representing the predictability (or the uncertainty) of precipitation forecast. Klampanos *et al.* (2018) studied the relationship between nuclear events and weather. A model with AE and CNN is used for rapid source estimation during radiological releases. The model clusters weather events first and looks over their correlations with nuclear events. Chen *et al.* (2019) designed a GAN model based on Super-Resolution with Generative Adversarial Network (Ledig *et al.* 2017), which is a GAN architecture for single image super-resolution, to improve the resolution of radar echo images of weather radar systems in order to increase the accuracy of the tasks that accepts echo images as inputs. Tran & Song (2019) designed a Convolutional RNN model for the radar echo extrapolation task. The model uses multiple (five) satellite images and predicts ten steps ahead at the pixel level. During the training process, Structural Similarity (SSIM) and multiscale SSIM are used to obtain better results.

Unclassified studies

Some literature included in this review did not fit within any of the defined categories in this review. Summaries of these papers are provided in this section. Many of these papers apply deep learning methods to ocean processes. In coastal hydraulics, Kang (2019) used an improved CNN on images to classify and monitor waves. Also in coastal hydraulics, Kim *et al.* (2020) used a number of DNN with coastal video imagery to develop a framework to track nearshore waves. Kylili *et al.* (2019) used a CNN to identify floating plastic debris in ocean images. Wang *et al.* (2019a) used a CNN architecture with a satellite radar product to classify ocean surface roughness into ten geophysical phenomena. Li *et al.* (2018) also used a CNN to classify hurricane damage from post-event aerial imagery.

The following three papers applied deep learning in a variety of disciplines. Haklidir & Haklidir (2019) used a DNN to predict the temperature of geothermal springs and wells using hydrogeochemical data, including chemical concentrations. Amirkolaee & Arefi (2019) implemented a very deep CNN structure to estimate a digital elevation model from single airborne or spaceborne images. Finally, Rohmat *et al.* (2019) developed and embedded a DNN into a GIS-based basin-scale decision support system to assess the impacts of best management practices (BMP). Based on geographic information and potential BMP, the DNN returns multiple outputs that describe net flow separation between groundwater recharge, groundwater return flow, and overland flow. This project demonstrates the potential of deep learning to integrate into decision-making on large-scale water resources projects.

KEY ISSUES AND CHALLENGES

Deep learning captures the nonlinear complex representations with datasets, making it a viable and efficient option in predictive model creation. In machine learning literature, capturing the representation is known as representation learning (Goodfellow *et al.* 2016). Representation learning relies on ANN's ability in acting as a universal approximator (Cybenko 1989; Hornik *et al.* 1989; Leshno *et al.* 1993), meaning that ANNs with only one hidden layer, theoretically, could represent any function. The drawback regarding this ability is that the task in hand might need a hidden layer that is too large to be feasible to be trained and executed. Adding new hidden layers to neural networks comes into play in order to cover this negation. Consequently, one important principle of representation learning is that, with minimal data engineering, the datasets should be fed to the neural network and the neural network should decide which features within the dataset are important for the goal of representing that dataset. Although this should be the case, the literature in the water domain does not widely apply this principle. We attribute this to the fact that datasets are neither extensive enough in terms of the number of given data, nor have vast spatial and/or temporal coverage.

As opposed to fields like computer vision and natural language processing, the water field lacks high quality, collected, curated, labeled, and published datasets that are used for benchmarking and method development. We identify this lack of benchmarking datasets as a key challenge (Ebert-Uphoff *et al.* 2017) that slows the state-of-the-science deep learning applications in the water domain. Most studies reviewed herein acquire datasets from governmental agencies, depending on their needs. Although potentially convenient for a contained effort of a single work, each one-off dataset dampens the speed of improvement of the state-of-the-science in this field. If researchers had the opportunity to build on a widely accepted dataset in their fields, they would be able to improve the accuracy of their models by taking advantage of previous models created using the same dataset. This collaboration around common datasets and models would open opportunities in the field, such as paving the way for their real-time usage. However, as we do not have many benchmarking datasets, many research groups around the globe run similar networks on custom datasets that they acquired with limited scientific interaction.

One other problem with the data provided by authorities is that they are dispersed among different agencies and they occasionally have mismatches in temporal and/or spatial coverage. Although the data provided extends to many years before and covers decades, one might need to access various databases created by several different government agencies in order to build a dataset. The period of record also commonly differs across agencies and areas, which causes the data acquisition process to be time-consuming and sometimes inconclusive. Further complicating data acquisition, water data are suppressed due to their military significance in some strategic instances, which contributes to limiting progress in the field. We support growing access to governmental and agency-collected water data and its use in deep learning.

We consider the fundamental understanding of deep learning within the literature as another problematic issue emerging from the reviews made in this study. A common mistake is what deep learning is and what deep learning is not. Most papers seem to interpret deep learning as a specific technique. However, deep learning is simply a broad term for various machine learning algorithms centered upon ANNs. Various studies reviewed here claim that they employ deep learning, yet they only take advantage of traditional ANN approaches. Most of the time this utilization does not surpass the extent of a study that employs conventional statistical modeling. This phenomenon raises questions about whether these studies attempt to exploit the keywords, using the term 'deep learning' to take advantage of the current scientific zeitgeist. We observe that these motives rarely result in work that forwards the deep learning literature. We particularly note the poor practice of ascribing work 'deep learning' while not employing representation learning principles.

This comprehensive review identified that literature at the intersection of deep learning and water do not cover the used methods in detail. Most of the studies appear reluctant to give model/architecture details that are vital to reproduce the proposed training pipeline. Combined with the aforementioned dataset problems, this causes consistent barriers to reproducibility of work. This overshadows the reported accuracy of studies and slows advancement in the deep learning-powered water field.

In contrast, some studies discuss unnecessary details over and over again in their manuscript. Discussing how the most efficient number of hidden layers or hyper-parameters are found by trial and error extensively without a hint of intuition, or discussing the optimal batch size beyond a few sentences do not appear because the authors use pertinent deep learning theory. Not sharing intuition regarding technical choices also prevents the papers reaching their goal. There are many papers utilizing LSTM networks, but few discuss the rationale behind their decision. One would at least expect a paper to describe why they choose LSTM networks over GRU networks in their setting. Although this needs a comparison of two similar RNN implementations, it should also be noted that networks like GANs and DQNs find a very limited place for themselves in the literature. Furthermore, researchers use networks like ELM, ENN, and NAR in various tasks without sharing their relative success to more complex architectures widely used in state-of-the-art deep learning models.

Ethics in deep learning applications

Ethics are broadly concerned with describing the appropriateness of an act; whether something is 'right' or 'wrong' to do. Meanwhile, deep learning is a tool that can be used to produce algorithms that automate prediction and decision-making. Thus, most ethical concerns of deep learning derive from the central question: what will the deep learning application do? The ethical considerations of deep learning in water are no exception. Take, for example, a deep learning application that predicts streamflow within an urban catchment. By appearances, the act of streamflow prediction alone lacks any ethical character. However, if the same tool is then used to make decisions in disaster mitigation (Demir *et al.* 2018) or public planning, many questions with ethical dimensions arise (Ewing & Demir 2020). What data were, and were not, used to develop the model? What biases exist in the dataset? How do these biases affect decision-making and human lives, and do the decisions reveal any discriminating behaviors? From this example, it is clear that the primary ethical considerations for the application of deep learning in water should be concerned with how entities – people, the environment, communities – will be affected by deep learning in decision-making workflows.

Although powerful, these deep learning tools simultaneously expand the reach and speed of decision-making, while also stripping away layers of context that would possibly be relevant to a humans' decision-making process. However, many of the ethical decisions in the water field remain unchanged and are primarily distributional: who receives water services, and their level of quality, and what level, and for whom, of risk is acceptable. These persistent ethical water dilemmas must be resolved within the new paradigm of deep learning in water.

As revealed by this literature review, few deep learning applications in the water literature include decision-making components. One paper, Rohmat *et al.* (2019), that was reviewed in this paper, explicitly states that their deep learning tool is integrated into a decision support system. This lack of attention to deep learning in service of decision-making in water academia presents an opportunity for a new line of research. It is also an opportunity to incorporate the work of other fields early, such as the well-documented ethical concerns stemming from deep learning decision and recommendation engines in social and civil applications (such as policing, criminal justice, and self-driving cars (Angwin *et al.* 2016; O'Neil 2016).) Furthermore, a proliferation of AI ethics frameworks have been developed and reviewed in Hagendorff (2019) and Jobin *et al.* (2019), as well as numerous guidelines for applying and assessing algorithms in social and civil domains (O'Reilly 2013; Rahwan 2018; Reisman Schultz Crawford & Whittaker 2018). These guidelines stress a deep understanding of the task the algorithm is in service of, feedback on algorithm performance, and rigorous assessment procedures.

The discussions of ethics in the context of deep learning are part of a continued conversation of ethics generally. Questions of how to treat those in our communities, how to treat the environment, what is the 'right' priority, or priorities, are not questions exclusive to the AI/deep learning domain; rather, the new scale and speed provided by deep learning require the inspection of age-old ethical questions in a new light.

RECOMMENDATIONS AND CONCLUSIONS

This paper provides a comprehensive review of the recent application of DNN as novel solutions to tackle hydrological tasks and challenges. A total of 129 publications were systematically selected for rigorous review and analysis as grouped by their application area. Based on the statistical meta-analysis of journal publications dating between January 2018 and March 2020, it was empirically observed that the average number of deep learning applications per month in the water sector steadily increased in an exponential fashion. Furthermore, the rapidly increasing body of work from these publications show deep learning's potential in a diversity of water sector applications. Key issues and challenges that may constitute setbacks and hindrances of deep learning adoption in the water industry have been identified and reported, accompanied by recommendations to persevere in spite of the logistical, computational, expertise-related, and data-related challenges to its principled adoption. Based on the extent of this review and the broad spectrum of application areas, we anticipate the water sector will continue to incorporate deep learning at an accelerating rate and deep learning will play a key role in the future of water.

Deep learning-powered technologies opened up a plethora of application and research opportunities to revolutionize hydrological science and workflow. In a bird's eye view fashion, key areas of innovation for future research include:

1. Automated forecasting: As this review outlined, the majority of current deep learning applications in the hydrological domain focus on forecasting of numerous variables (e.g. water level, discharge) given the problem's suitability for machine learning. In the future, efforts can be coordinated between agencies and research organizations to collaboratively develop complementary models that will yield actionable and reliable information. These models can be maintained and powered by a stream of real-time data influx to constitute the future of decision-making systems and geographical information systems.
2. Published datasets: Lack of deep learning-ready datasets within the water field was stressed in the previous section. The main problem caused by this absence of many datasets is that the research community does not build upon previous work in terms of constructing better neural network architectures and moving the state of art to the next iteration. This inference is supported by the fact that among the 30 papers related to flooding reviewed in this study, there are only a few that use a previously curated, labeled dataset. The result is that many papers are published that achieve the same task with almost identical methods but different data. This absence implicitly causes redundancy in the field. We believe if more studies focus on creating benchmark datasets that are open to researchers, both cumulativeness of the science would be satisfied and deep learning-powered modeling in water resources research would go further in terms of generic applicability.
3. AI as a service: As the popularity and usefulness of AI tools increase, a new research area came to prominence in the computer science field. This area is focused on developing generalized and centralized web frameworks that can readily provide the means to develop custom AI solutions through Platform-as-a-Service systems. These systems hold great potential for the hydrological community because they allow developers to focus on designing and managing intelligent applications without the burden of infrastructure maintenance and adjustment of computing resources. They can provide intuitive graphical user interfaces to allow the development of hydrological deep learning applications by connecting predefined and custom models with provided datasets.
4. Edge computing: The main propeller in the creation of smart applications is the consistent and diverse data flux. However, as the frequency and type of data resources expand, a centralized approach to collect and analyze data simply may not be viable for multivariate tasks. Furthermore, the costs associated with the transfer of large data from distributed sensors are not trivial and may discourage stakeholders to increase sensor coverage and data reporting interval. As a solution, edge computing offers a new perspective to process the data on the sensor. There is extensive research on utilizing deep learning for the

Internet of Things through edge computing, which can allow the stakeholders of the water domain to innovate novel applications with existing or low-cost sensor networks. As a tangible example, a camera-equipped river monitoring sensor can employ deep learning to analyze pictures on the edge to detect any foreign objects (e.g. tree, human) in the river, and transmit only the useful information to a centralized system.

5. Intelligent assistants: The massive amount of environmental and hydrological data makes it challenging to efficiently and effectively extract required knowledge in a timely manner. Manual efforts are often needed to analyze comprehensive raw data for decision-making purposes. As a solution, intelligent assistants serve as voice-enabled knowledge engines that can interpret a natural language question, apply human-like reasoning, and extract the desired factual response out of curated data. These assistants can be integrated into various communication channels for easy access, including web systems, messaging applications (e.g. Skype, Messenger, SMS), personal assistants (e.g. Google Assistant, Apple Siri), home automation devices (e.g. Google Home, Amazon Alexa), augmented and virtual reality systems (e.g. HoloLens, Magic Leap, Oculus Quest), and automated workflow systems. The voice-enabled communication and immediate access to knowledge can facilitate hydrological research as well as natural disaster preparedness and response (Sermet & Demir 2018a).
6. Virtual and augmented reality: Incorporation of deep learning with virtual and augmented reality environments provides a prominent research area due to its immersive nature that allows effective analysis of complex environmental phenomena that is not feasible to be orchestrated in real-life. For example, deep learning can power realistic flood simulations while mimicking human behavior to train first-responders, aid decision-makers, and educate the public (Sermet & Demir 2018b). Another use may be guiding on-site personnel that may require expertise, such as sensor maintenance, structural renovation, and field experiments, through heads-up displays and deep learning-powered recognition and decision support applications (Sermet & Demir 2020).

DATA AVAILABILITY STATEMENT

All relevant data are included in the paper or its Supplementary Information.

REFERENCES

Abdi, G., Samadzadegan, F. & Reinartz, P. 2018 Deep learning decision fusion for the classification of urban remote sensing data. *Journal of Applied Remote Sensing* **12** (1), 016038.

Agliamzanov, R., Sit, M. & Demir, I. 2020 Hydrology@ home: a distributed volunteer computing framework for hydrological research and applications. *Journal of Hydroinformatics* **22** (2), 235–248.

Amirkolaee, H. A. & Arefi, H. 2019 Convolutional neural network architecture for digital surface model estimation from single remote sensing image. *Journal of Applied Remote Sensing* **13** (1), 016522.

Angwin, J., Larson, J., Mattu, S. & Kirchner, L. 2016 *Machine Bias. ProPublica, May, 23*, p. 2016.

Antonetti, M. & Zappa, M. 2018 How can expert knowledge increase the realism of conceptual hydrological models? A case study based on the concept of dominant runoff process in the Swiss Pre-Alps. *Hydrology & Earth System Sciences* **22** (8), 4425–4447.

Ardabili, S., Mosavi, A., Dehghani, M. & Várkonyi-Kóczy, A. R. 2019 Deep learning and machine learning in hydrological processes climate change and earth systems a systematic review. In *International Conference on Global Research and Education*. Springer, Cham, pp. 52–62.

Arora, V. K., Scinocca, J. F., Boer, G. J., Christian, J. R., Denman, K. L., Flato, G. M., Kharin, V. V., Lee, W. G. & Merryfield, W. J. 2011 Carbon emission limits required to satisfy future representative concentration pathways of greenhouse gases. *Geophysical Research Letters* **38** (5).

Badrinarayanan, V., Kendall, A. & Cipolla, R. 2017 Segnet: a deep convolutional encoder-decoder architecture for image segmentation. *IEEE Transactions on Pattern Analysis and Machine Intelligence* **39** (12), 2481–2495.

Bai, Y., Bezak, N., Sapač, K., Klun, M. & Zhang, J. 2019 Short-term streamflow forecasting using the feature-enhanced regression model. *Water Resources Management* **33** (14), 4783–4797.

Banerjee, A., Chakrabarty, M., Rakshit, N., Bhowmick, A. R. & Ray, S. 2019 Environmental factors as indicators of dissolved oxygen concentration and zooplankton abundance: deep learning versus traditional regression approach. *Ecological Indicators* **100**, 99–117.

Barzegar, R., Aalami, M. T. & Adamowski, J. 2020 Short-term water quality variable prediction using a hybrid CNN–LSTM deep learning model. *Stochastic Environmental Research and Risk Assessment* **34**, 415–433.

Bengio, Y., Frasconi, P. & Simard, P. 1993 The problem of learning long-term dependencies in recurrent networks. In: *IEEE International Conference on Neural Networks*. IEEE, pp. 1183–1188.

Bhola, P. K., Nair, B. B., Leandro, J., Rao, S. N. & Disse, M. 2019 Flood inundation forecasts using validation data generated with the assistance of computer vision. *Journal of Hydroinformatics* **21** (2), 240–256.

Bhosle, K. & Musande, V. 2019 Evaluation of deep learning CNN model for land use land cover classification and crop identification using hyperspectral remote sensing images.

Journal of the Indian Society of Remote Sensing **47** (11), 1949–1958.

Bui, D. T., Hoang, N. D., Martínez-Álvarez, F., Ngo, P. T. T., Hoa, P. V., Pham, T. D., Samui, P. & Costache, R. 2020a A novel deep learning neural network approach for predicting flash flood susceptibility: a case study at a high frequency tropical storm area. *Science of The Total Environment* **701**, 134413.

Bui, Q. T., Nguyen, Q. H., Nguyen, X. L., Pham, V. D., Nguyen, H. D. & Pham, V. M. 2020b Verification of novel integrations of swarm intelligence algorithms into deep learning neural network for flood susceptibility mapping. *Journal of Hydrology* **581**, 124379.

Cao, F., Yang, Z., Ren, J. & Ling, B. W. K. 2018 Convolutional neural network extreme learning machine for effective classification of hyperspectral images. *Journal of Applied Remote Sensing* **12** (3), 035003.

Carson, A., Windsor, M., Hill, H., Haigh, T., Wall, N., Smith, J., Olsen, R., Bathke, D., Demir, I. & Muste, M. 2018 Serious gaming for participatory planning of multi-hazard mitigation. *International Journal of River Basin Management* **16** (3), 379–391.

Chai, D., Newsam, S., Zhang, H. K., Qiu, Y. & Huang, J. 2019 Cloud and cloud shadow detection in Landsat imagery based on deep convolutional neural networks. *Remote Sensing of Environment* **225**, 307–316.

Chen, H., Zhang, X., Liu, Y. & Zeng, Q. 2019 Generative adversarial networks capabilities for super-resolution reconstruction of weather radar echo images. *Atmosphere* **10** (9), 555.

Chen, L., Cao, Y., Ma, L. & Zhang, J. 2020 A deep learning-based methodology for precipitation nowcasting with radar. *Earth and Space Science* **7** (2), e2019EA000812.

Cho, K., Van Merriënboer, B., Bahdanau, D. & Bengio, Y. 2014 *On the Properties of Neural Machine Translation: Encoder-Decoder Approaches. arXiv preprint arXiv:1409.1259.*

Cybenko, G. 1989 Approximation by superpositions of a sigmoidal function. *Mathematics of Control, Signals and Systems* **2** (4), 303–314.

Damavandi, H. G., Shah, R., Stampoulis, D., Wei, Y., Boscovic, D. & Sabo, J. 2019 Accurate prediction of streamflow using long short-term memory network: a case study in the Brazos River Basin in Texas. *International Journal of Environmental Science and Development* **10** (10), 294–300.

Demir, I. & Szczepanek, R. 2017 Optimization of river network representation data models for web-based systems. *Earth and Space Science* **4** (6), 336–347.

Demir, I., Yildirim, E., Sermet, Y. & Sit, M. A. 2018 FLOODSS: Iowa flood information system as a generalized flood cyberinfrastructure. *International Journal of River Basin Management* **16** (3), 393–400.

Demiray, B. Z., Sit, M. & Demir, I. 2020 *D-SRGAN: DEM Super-Resolution with Generative Adversarial Network. arXiv preprint arXiv:2004.04788.*

Ebert-Uphoff, I., Thompson, D. R., Demir, I., Gel, Y. R., Karpatne, A., Guereque, M., Kumar, V., Cabral-Cano, E. & Smyth, P. 2017 A vision for the development of benchmarks to bridge geoscience and data science. In: *17th International Workshop on Climate Informatics*. National Center for Atmospheric Research, Boulder, CO, USA.

Elman, J. L. 1990 Finding structure in time. *Cognitive Science* **14** (2), 179–211.

Esteva, A., Robicquet, A., Ramsundar, B., Kuleshov, V., DePristo, M., Chou, K., Cui, C., Corrado, G., Thrun, S. & Dean, J. 2019 A guide to deep learning in healthcare. *Nature Medicine* **25** (1), 24–29.

Evora, N. D. & Coulibaly, P. 2009 Recent advances in data-driven modeling of remote sensing applications in hydrology. *Journal of Hydroinformatics* **11** (3–4), 194–201.

Ewing, G. & Demir, I. 2020 An ethical decision-making framework with serious gaming: smart water case study on flooding. *EarthArXiv*. doi:10.31223/osf.io/y873 g.

Fang, Q., Zhang, J., Xie, C. & Yang, Y. 2019 Detection of multiple leakage points in water distribution networks based on convolutional neural networks. *Water Supply* **19** (8), 2231–2239.

Fu, T., Ma, L., Li, M. & Johnson, B. A. 2018 Using convolutional neural network to identify irregular segmentation objects from very high-resolution remote sensing imagery. *Journal of Applied Remote Sensing* **12** (2), 025010.

Gautam, A., Sit, M. & Demir, I. 2020 *Realistic River Image Synthesis Using Deep Generative Adversarial Networks. arXiv preprint arXiv:2003.00826.*

Goodfellow, I., Pouget-Abadie, J., Mirza, M., Xu, B., Warde-Farley, D., Ozair, S., Courville, A. & Bengio, Y. 2014 Generative adversarial nets. In: *Advances in Neural Information Processing Systems*. pp. 2672–2680.

Goodfellow, I., Bengio, Y. & Courville, A. 2016 *Deep Learning*. MIT Press, Cambridge, MA.

Grossman, D., Buckley, N. & Doyle, M. 2015 *Data Intelligence for 21st Century Water Management: A Report From the 2015 Aspen-Nicholas Water Forum*. Aspen-Nicholas Water Forum. Available from https://www.aspeninstitute.org/publications/data-intelligence-21st-century-water-management-report-2015-aspen-nicholas-water-forum/ (accessed 31 May 2020).

Hagendorff, T. 2019 *The Ethics of AI Ethics–An Evaluation of Guidelines. arXiv preprint arXiv:1903.03425.*

Haklidir, F. S. T. & Haklidir, M. 2019 Prediction of reservoir temperatures using hydrogeochemical data, Western Anatolia geothermal systems (Turkey): a machine learning approach. *Natural Resources Research* **29**, 2333–2346.

Hamshaw, S. D., Dewoolkar, M. M., Schroth, A. W., Wemple, B. C. & Rizzo, D. M. 2018 A new machine-learning approach for classifying hysteresis in suspended-sediment discharge relationships using high-frequency monitoring data. *Water Resources Research* **54** (6), 4040–4058.

Harrou, F., Dairi, A., Sun, Y. & Senouci, M. 2018 Statistical monitoring of a wastewater treatment plant: a case study. *Journal of Environmental Management* **223**, 807–814.

Haskins, G., Kruger, U. & Yan, P. 2020 Deep learning in medical image registration: a survey. *Machine Vision and Applications* **31** (1), 8.

He, K., Zhang, X., Ren, S. & Sun, J. 2015 *Deep Residual Learning for Image Recognition. arXiv preprint arXiv:1512.03385.*

He, B., Liu, X., Hu, S., Song, K. & Gao, T. 2019a Use of the C-band microwave link to distinguish between rainy and dry periods. *Advances in Meteorology* **2019**. doi:10.1155/2019/3428786.

He, X., Luo, J., Zuo, G. & Xie, J. 2019b Daily runoff forecasting using a hybrid model based on variational mode decomposition and deep neural networks. *Water Resources Management* **33** (4), 1571–1590.

Heaton, J. B., Polson, N. G. & Witte, J. H. 2017 Deep learning for finance: deep portfolios. *Applied Stochastic Models in Business and Industry* **33** (1), 3–12.

Hinton, G. E. 2002 Training products of experts by minimizing contrastive divergence. *Neural Computation* **14** (8), 1771–1800.

Hinton, G. E. 2009 Deep belief networks. *Scholarpedia* **4** (5), 5947.

Hochreiter, S. & Schmidhuber, J. 1997 Long short-term memory. *Neural Computation* **9** (8), 1735–1780.

Hornik, K., Stinchcombe, M. & White, H. 1989 Multilayer feedforward networks are universal approximators. *Neural Networks* **2** (5), 359–366.

Hrnjica, B. & Bonacci, O. 2019 Lake level prediction using feed forward and recurrent neural networks. *Water Resources Management* **33** (7), 2471–2484.

Hu, R., Fang, F., Pain, C. C. & Navon, I. M. 2019 Rapid spatio-temporal flood prediction and uncertainty quantification using a deep learning method. *Journal of Hydrology* **575**, 911–920.

Huang, G. B., Zhu, Q. Y. & Siew, C. K. 2006 Extreme learning machine: theory and applications. *Neurocomputing* **70** (1–3), 489–501.

Islam, Z. 2011 *A Review on Physically Based Hydrologic Modeling*. University of Alberta, Edmonton, AB, Canada. doi:10.13140/2.1.4544.5924.

Isola, P., Zhu, J. Y., Zhou, T. & Efros, A. A. 2017 Image-to-image translation with conditional adversarial networks. In *Proceedings of the IEEE Conference on Computer Vision and Pattern Recognition*. pp. 1125–1134.

Jadidoleslam, N., Mantilla, R., Krajewski, W. F. & Cosh, M. H. 2019 Data-driven stochastic model for basin and sub-grid variability of SMAP satellite soil moisture. *Journal of Hydrology* **576**, 85–97.

Jaiswal, R. K., Ali, S. & Bharti, B. 2020 Comparative evaluation of conceptual and physical rainfall–runoff models. *Applied Water Science* **10** (1), 1–14.

Jeong, J. & Park, E. 2019 Comparative applications of data-driven models representing water table fluctuations. *Journal of Hydrology* **572**, 261–273.

Jeong, J., Park, E., Chen, H., Kim, K. Y., Han, W. S. & Suk, H. 2020 Estimation of groundwater level based on the robust training of recurrent neural networks using corrupted data. *Journal of Hydrology* **582**, 124512.

Jeppesen, J. H., Jacobsen, R. H., Inceoglu, F. & Toftegaard, T. S. 2019 A cloud detection algorithm for satellite imagery based on deep learning. *Remote Sensing of Environment* **229**, 247–259.

Jiang, W. 2018 Object-based deep convolutional autoencoders for high-resolution remote sensing image classification. *Journal of Applied Remote Sensing* **12** (3), 035002.

Jiang, Z., Mallants, D., Peeters, L., Gao, L., Soerensen, C. & Mariethoz, G. 2019 High-resolution paleovalley classification from airborne electromagnetic imaging and deep neural network training using digital elevation model data. *Hydrology and Earth System Sciences* **23** (6), 2561–2580.

Jin, B., Ye, P., Zhang, X., Song, W. & Li, S. 2019 Object-oriented method combined with deep convolutional neural networks for land-use-type classification of remote sensing images. *Journal of the Indian Society of Remote Sensing* **47** (6), 951–965.

Jobin, A., Ienca, M. & Vayena, E. 2019 *Artificial Intelligence: The Global Landscape of Ethics Guidelines. arXiv preprint arXiv:1906.11668.*

Kabir, S., Patidar, S. & Pender, G. 2020 Investigating capabilities of machine learning techniques in forecasting stream flow. In: *Proceedings of the Institution of Civil Engineers-Water Management*. Thomas Telford Ltd, Vol. 173, No. 2, pp. 69–86.

Kang, L. 2019 Wave monitoring based on improved convolution neural network. *Journal of Coastal Research* **94** (sp1), 186–190.

Kao, I. F., Zhou, Y., Chang, L. C. & Chang, F. J. 2020 Exploring a long short-term memory based encoder-decoder framework for multi-step-ahead flood forecasting. *Journal of Hydrology* **583**, 124631.

Karimi, H. S., Natarajan, B., Ramsey, C. L., Henson, J., Tedder, J. L. & Kemper, E. 2019 Comparison of learning-based wastewater flow prediction methodologies for smart sewer management. *Journal of Hydrology* **577**, 123977.

Kim, Y. J., Yura, E., Kim, T. W. & Yoon, J. S. 2019 Development of disaster prevention system based on deep neural network using deep learning with dropout. *Journal of Coastal Research* **91** (sp1), 186–190.

Kim, J., Kim, J., Kim, T., Huh, D. & Caires, S. 2020 Wave-tracking in the surf zone using coastal video imagery with deep neural networks. *Atmosphere* **11** (3), 304.

Klampanos, I. A., Davvetas, A., Andronopoulos, S., Pappas, C., Ikonomopoulos, A. & Karkaletsis, V. 2018 Autoencoder-driven weather clustering for source estimation during nuclear events. *Environmental Modelling & Software* **102**, 84–93.

Kopp, M., Tuo, Y. & Disse, M. 2019 Fully automated snow depth measurements from time-lapse images applying a convolutional neural network. *Science of The Total Environment* **697**, 134213.

Krajewski, W. F., Ceynar, D., Demir, I., Goska, R., Kruger, A., Langel, C., Mantilla, R., Niemeier, J., Quintero, F., Seo, B. C. & Small, S. J. 2017 Real-time flood forecasting and information system for the state of Iowa. *Bulletin of the American Meteorological Society* **98** (3), 539–554.

Kratzert, F., Klotz, D., Brenner, C., Schulz, K. & Herrnegger, M. 2018 Rainfall–runoff modelling using long short-term memory (LSTM) networks. *Hydrology and Earth System Sciences* **22** (11), 6005–6022.

Kratzert, F., Klotz, D., Herrnegger, M., Sampson, A. K., Hochreiter, S. & Nearing, G. S. 2019a Toward improved predictions in ungauged basins: exploiting the power of machine learning. *Water Resources Research* **55**, 11344–11354.

Kratzert, F., Klotz, D., Shalev, G., Klambauer, G., Hochreiter, S. & Nearing, G. 2019b Towards learning universal, regional, and local hydrological behaviors via machine learning applied to large-sample datasets. *Hydrology & Earth System Sciences* **23** (12), 5089–5110.

Krizhevsky, A., Sutskever, I. & Hinton, G. E. 2012 Imagenet classification with deep convolutional neural networks. In: *Advances in Neural Information Processing Systems*. pp. 1097–1105.

Kroupi, E., Kesa, M., Navarro-Sánchez, V. D., Saeed, S., Pelloquin, C., Alhaddad, B., Moreno, L., Soria-Frisch, A. & Ruffini, G. 2019 Deep convolutional neural networks for land-cover classification with Sentinel-2 images. *Journal of Applied Remote Sensing* **13** (2), 024525.

Kruger, A., Krajewski, W. F., Niemeier, J. J., Ceynar, D. L. & Goska, R. 2016 Bridge-mounted river stage sensors (BMRSS). *IEEE Access* **4**, 8948–8966.

Kumar, D., Singh, A., Samui, P. & Jha, R. K. 2019 Forecasting monthly precipitation using sequential modelling. *Hydrological Sciences Journal* **64** (6), 690–700.

Kylili, K., Kyriakides, I., Artusi, A. & Hadjistassou, C. 2019 Identifying floating plastic marine debris using a deep learning approach. *Environmental Science and Pollution Research* **26** (17), 17091–17099.

Laloy, E., Hérault, R., Jacques, D. & Linde, N. 2018 Training-image based geostatistical inversion using a spatial generative adversarial neural network. *Water Resources Research* **54** (1), 381–406.

LeCun, Y. 1989 Generalization and network design strategies. *Connectionism in Perspective* **19**, 143–155.

LeCun, Y., Bengio, Y. & Hinton, G. 2015 Deep learning. *Nature* **521** (7553), 436–444.

Ledig, C., Theis, L., Huszár, F., Caballero, J., Cunningham, A., Acosta, A., Aitken, A., Tejani, A., Totz, J., Wang, Z. & Shi, W. 2017 Photo-realistic single image super-resolution using a generative adversarial network. In: *Proceedings of the IEEE Conference on Computer Vision and Pattern Recognition*. pp. 4681–4690.

Lee, S. & Lee, D. 2018 Improved prediction of harmful algal blooms in four Major South Korea's rivers using deep learning models. *International Journal of Environmental Research and Public Health* **15** (7), 1322.

Leshno, M., Lin, V. Y., Pinkus, A. & Schocken, S. 1993 Multilayer feedforward networks with a nonpolynomial activation function can approximate any function. *Neural Networks* **6** (6), 861–867.

Li, Y., Ye, S. & Bartoli, I. 2018 Semisupervised classification of hurricane damage from postevent aerial imagery using deep learning. *Journal of Applied Remote Sensing* **12** (4), 045008.

Li, L., Jiang, P., Xu, H., Lin, G., Guo, D. & Wu, H. 2019 Water quality prediction based on recurrent neural network and improved evidence theory: a case study of Qiantang River. *China. Environmental Science and Pollution Research* **26** (19), 19879–19896.

Liang, Z., Zou, R., Chen, X., Ren, T., Su, H. & Liu, Y. 2020 Simulate the forecast capacity of a complicated water quality model using the long short-term memory approach. *Journal of Hydrology* **581**, 124432.

Lin, T., Horne, B. G., Tino, P. & Giles, C. L. 1996 Learning long-term dependencies in NARX recurrent neural networks. *IEEE Transactions on Neural Networks* **7** (6), 1329–1338.

Ling, F., Boyd, D., Ge, Y., Foody, G. M., Li, X., Wang, L., Zhang, Y., Shi, L., Shang, C., Li, X. & Du, Y. 2019 Measuring river wetted width from remotely sensed imagery at the subpixel scale with a deep convolutional neural network. *Water Resources Research* **55** (7), 5631–5649.

Liu, P., Wang, J., Sangaiah, A. K., Xie, Y. & Yin, X. 2019a Analysis and prediction of water quality using LSTM deep neural networks in IoT environment. *Sustainability* **11** (7), 2058.

Liu, Y., Piramanayagam, S., Monteiro, S. T. & Saber, E. 2019b Semantic segmentation of multisensor remote sensing imagery with deep ConvNets and higher-order conditional random fields. *Journal of Applied Remote Sensing* **13** (1), 016501.

Liu, Y., Qin, H., Zhang, Z., Yao, L., Wang, Y., Li, J., Liu, G. & Zhou, J. 2019c Deriving reservoir operation rule based on Bayesian deep learning method considering multiple uncertainties. *Journal of Hydrology* **579**, 124207.

Ma, L., Liu, Y., Zhang, X., Ye, Y., Yin, G. & Johnson, B. A. 2019 Deep learning in remote sensing applications: a meta-analysis and review. *ISPRS Journal of Photogrammetry and Remote Sensing* **152**, 166–177.

Mamandipoor, B., Majd, M., Sheikhalishahi, S., Modena, C. & Osmani, V. 2020 Monitoring and detecting faults in wastewater treatment plants using deep learning. *Environmental Monitoring and Assessment* **192** (2), 148.

Marçais, J. & de Dreuzy, J. R. 2017 Prospective interest of deep learning for hydrological inference. *Groundwater* **55** (5), 688–692.

McCuen, R. H. 2016 *Modeling Hydrologic Change: Statistical Methods*. CRC Press, Boca Raton, FL.

Mei, M. J., Maksym, T., Weissling, B. & Singh, H. 2019 Estimating early-winter Antarctic sea ice thickness from deformed ice morphology. *The Cryosphere* **13** (11), 2915–2934.

Meng, X., Zhang, S. & Zang, S. 2019 Lake wetland classification based on an SVM-CNN composite classifier and high-resolution images using wudalianchi as an example. *Journal of Coastal Research* **93** (sp1), 153–162.

Mnih, V., Kavukcuoglu, K., Silver, D., Graves, A., Antonoglou, I., Wierstra, D. & Riedmiller, M. 2013 *Playing Atari with Deep Reinforcement Learning. arXiv preprint arXiv:1312.5602*.

Mo, S., Zabaras, N., Shi, X. & Wu, J. 2019a Deep autoregressive neural networks for high-dimensional inverse problems in groundwater contaminant source identification. *Water Resources Research* **55** (5), 3856–3881.

Mo, S., Zhu, Y., Zabaras, N., Shi, X. & Wu, J. 2019b Deep convolutional encoder-decoder networks for uncertainty quantification of dynamic multiphase flow in heterogeneous media. *Water Resources Research* **55** (1), 703–728.

Moy de Vitry, M., Kramer, S., Wegner, J. D. & Leitão, J. P. 2019 Scalable flood level trend monitoring with surveillance cameras using a deep convolutional neural network. *Hydrology and Earth System Sciences* **23** (11), 4621–4634.

Nair, V. & Hinton, G. E. 2010 Rectified linear units improve restricted boltzmann machines. In: *Proceedings of the 27th*

International Conference on Machine Learning (ICML-10). pp. 807–814.

Nam, K. & Wang, F. 2020 An extreme rainfall-induced landslide susceptibility assessment using autoencoder combined with random forest in Shimane Prefecture, Japan. *Geoenvironmental Disasters* **7** (1), 6.

Nam, K., Heo, S., Loy-Benitez, J., Ifaei, P., Lee, S., Kang, S. & Yoo, C. 2020 An autonomous operational trajectory searching system for an economic and environmental membrane bioreactor plant using deep reinforcement learning. *Water Science and Technology* **81**, 1578–1587.

Nguyen, D. H. & Bae, D. H. 2020 Correcting mean areal precipitation forecasts to improve urban flooding predictions by using long short-term memory network. *Journal of Hydrology* **584**, 124710.

Ni, L., Wang, D., Singh, V. P., Wu, J., Wang, Y., Tao, Y. & Zhang, J. 2019 Streamflow and rainfall forecasting by two long short-term memory-based models. *Journal of Hydrology* **583**, 124296.

O'Neil, C. 2016 *Weapons of Math Destruction: How Big Data Increases Inequality and Threatens Democracy*. Broadway Books, New York, NY.

O'Neil, G. L., Goodall, J. L., Behl, M. & Saby, L. 2020 Deep learning using physically-informed input data for wetland identification. *Environmental Modelling & Software* **126**, 104665.

O'Reilly, T. 2013 Open data and algorithmic regulation. In: *Beyond Transparency: Open Data and the Future of Civic Innovation* (B. Goldstein & L. Dyson, eds). Code for America Press, San Francisco, CA. pp. 289–300.

Pan, B., Hsu, K., AghaKouchak, A. & Sorooshian, S. 2019 Improving precipitation estimation using convolutional neural network. *Water Resources Research* **55** (3), 2301–2321.

Persello, C., Tolpekin, V. A., Bergado, J. R. & de By, R. A. 2019 Delineation of agricultural fields in smallholder farms from satellite images using fully convolutional networks and combinatorial grouping. *Remote Sensing of Environment* **231**, 111253.

Pollack, J. B. 1990 Recursive distributed representations. *Artificial Intelligence* **46** (1–2), 77–105.

Poornima, S. & Pushpalatha, M. 2019 Prediction of rainfall using intensified LSTM based recurrent neural network with weighted linear units. *Atmosphere* **10** (11), 668.

Qi, Y., Zhou, Z., Yang, L., Quan, Y. & Miao, Q. 2019 A decomposition-ensemble learning model based on LSTM neural network for daily reservoir inflow forecasting. *Water Resources Management* **33** (12), 4123–4139.

Qin, J., Liang, J., Chen, T., Lei, X. & Kang, A. 2019 Simulating and predicting of hydrological time series based on tensorFlow deep learning. *Polish Journal of Environmental Studies* **28** (2), 796–802.

Rahwan, I. 2018 Society-in-the-loop: programming the algorithmic social contract. *Ethics and Information Technology* **20** (1), 5–14.

Read, J. S., Jia, X., Willard, J., Appling, A. P., Zwart, J. A., Oliver, S. K., Karpatne, A., Hansen, G. J., Hanson, P. C., Watkins, W. & Steinbach, M. 2019 Process-guided deep learning predictions of lake water temperature. *Water Resources Research* **55**, 9173–9190.

Reddy, D. S. & Prasad, P. R. C. 2018 Prediction of vegetation dynamics using NDVI time series data and LSTM. *Modeling Earth Systems and Environment* **4** (1), 409–419.

Reinsel, D., Gantz, J. & Rydning, J. 2018 *The Digitization of the World From Edge to Core. IDC White Paper.*

Reisman, D., Schultz, J., Crawford, K. & Whittaker, M. 2018 *Algorithmic Impact Assessments: A Practical Framework for Public Agency Accountability*. AI Now Institute, pp. 1–22.

Ren, T., Liu, X., Niu, J., Lei, X. & Zhang, Z. 2020 Real-time water level prediction of cascaded channels based on multilayer perception and recurrent neural network. *Journal of Hydrology* **585**, 124783.

Rohmat, F. I., Labadie, J. W. & Gates, T. K. 2019 Deep learning for compute-efficient modeling of BMP impacts on stream-aquifer exchange and water law compliance in an irrigated river basin. *Environmental Modelling & Software* **122**, 104529.

Ronneberger, O., Fischer, P. & Brox, T. 2015 U-net: convolutional networks for biomedical image segmentation. In: *International Conference on Medical Image Computing and Computer-Assisted Intervention*. Springer, Cham, pp. 234–241.

Rumelhart, D. E., Hinton, G. E. & Williams, R. J. 1985 *Learning Internal Representations by Error Propagation (No. ICS-8506)*. University of California San Diego La Jolla Institute for Cognitive Science.

Rumelhart, D. E., Hinton, G. E. & Williams, R. J. 1986 Learning representations by back-propagating errors. *Nature* **323** (6088), 533–536.

Saba, L., Biswas, M., Kuppili, V., Godia, E. C., Suri, H. S., Edla, D. R., Omerzu, T., Laird, J. R., Khanna, N. N., Mavrogeni, S. & Protogerou, A. 2019 The present and future of deep learning in radiology. *European Journal of Radiology* **114**, 14–24.

Sahiner, B., Pezeshk, A., Hadjiiski, L. M., Wang, X., Drukker, K., Cha, K. H., Summers, R. M. & Giger, M. L. 2019 Deep learning in medical imaging and radiation therapy. *Medical Physics* **46** (1), e1–e36.

Sankaranarayanan, S., Prabhakar, M., Satish, S., Jain, P., Ramprasad, A. & Krishnan, A. 2019 Flood prediction based on weather parameters using deep learning. *Journal of Water and Climate Change.*

Santos, J. E., Xu, D., Jo, H., Landry, C. J., Prodanović, M. & Pyrcz, M. J. 2020 PoreFlow-Net: A 3D convolutional neural network to predict fluid flow through porous media. *Advances in Water Resources* **138**, 103539.

Scher, S. & Messori, G. 2018 Predicting weather forecast uncertainty with machine learning. *Quarterly Journal of the Royal Meteorological Society* **144** (717), 2830–2841.

Sengupta, S., Basak, S., Saikia, P., Paul, S., Tsalavoutis, V., Atiah, F., Ravi, V. & Peters, A. 2020 A review of deep learning with special emphasis on architectures, applications and recent trends. *Knowledge-Based Systems* **194**, 105596.

Sermet, Y. & Demir, I. 2018a An intelligent system on knowledge generation and communication about flooding. *Environmental Modelling & Software* **108**, 51–60.

Sermet, Y. & Demir, I. 2018b Flood action VR: a virtual reality framework for disaster awareness and emergency response training. In: *Proceedings of the International Conference on Modeling, Simulation and Visualization Methods (MSV)*. pp. 65–68.

Sermet, Y. & Demir, I. 2020 Virtual and augmented reality applications for environmental science education and training. In: *New Perspectives on Virtual and Augmented Reality: Finding New Ways to Teach in A Transformed Learning Environment*. (L. Daniela, ed.). Routledge, London, UK, p. 261–275.

Sermet, Y., Demir, I. & Muste, M. 2020a A serious gaming framework for decision support on hydrological hazards. *Science of The Total Environment* **728**, 138895.

Sermet, Y., Villanueva, P., Sit, M. A. & Demir, I. 2020b Crowdsourced approaches for stage measurements at ungauged locations using smartphones. *Hydrological Sciences Journal* **65** (5), 813–822.

Shafer, G. 1976 *A Mathematical Theory of Evidence*, Vol. 42. Princeton University Press, Princeton, NJ.

Shen, C. 2018 A transdisciplinary review of deep learning research and its relevance for water resources scientists. *Water Resources Research* **54** (11), 8558–8593.

Shen, R., Huang, A., Li, B. & Guo, J. 2019 Construction of a drought monitoring model using deep learning based on multi-source remote sensing data. *International Journal of Applied Earth Observation and Geoinformation* **79**, 48–57.

Shi, S. & Xu, G. 2018 Novel performance prediction model of a biofilm system treating domestic wastewater based on stacked denoising auto-encoders deep learning network. *Chemical Engineering Journal* **347**, 280–290.

Shin, J., Kim, S. M., Son, Y. B., Kim, K. & Ryu, J. H. 2019 Early prediction of Margalefidinium polykrikoides bloom using a LSTM neural network model in the South Sea of Korea. *Journal of Coastal Research* **90** (sp1), 236–242.

Simonyan, K. & Zisserman, A. 2014 *Very Deep Convolutional Networks for Large-Scale Image Recognition. arXiv preprint arXiv:1409.1556*.

Sit, M. & Demir, I. 2019 *Decentralized Flood Forecasting Using Deep Neural Networks. arXiv preprint arXiv:1902.02308*.

Sit, M., Sermet, Y. & Demir, I. 2019a Optimized watershed delineation library for server-side and client-side web applications. *Open Geospatial Data, Software and Standards* **4** (1), 8.

Sit, M. A., Koylu, C. & Demir, I. 2019b Identifying disaster-related tweets and their semantic, spatial and temporal context using deep learning, natural language processing and spatial analysis: a case study of Hurricane Irma. *International Journal of Digital Earth* **12** (11), 1205–1229.

Song, A., Kim, Y. & Kim, Y. 2019 Change detection of surface water in remote sensing images based on fully convolutional network. *Journal of Coastal Research* **91** (sp1), 426–430.

Su, A., Li, H., Cui, L. & Chen, Y. 2020 A convection nowcasting method based on machine learning. *Advances in Meteorology* **2020**, 5124274. https://doi.org/10.1155/2020/5124274.

Sun, A. Y., Scanlon, B. R., Zhang, Z., Walling, D., Bhanja, S. N., Mukherjee, A. & Zhong, Z. 2019a Combining physically based modeling and deep learning for fusing GRACE satellite data: can we learn from mismatch? *Water Resources Research* **55** (2), 1179–1195.

Sun, Z., Zhao, X., Wu, M. & Wang, C. 2019b Extracting urban impervious surface from worldView-2 and airborne LiDAR data using 3D convolutional neural networks. *Journal of the Indian Society of Remote Sensing* **47** (3), 401–412.

Sun, Z., Long, D., Yang, W., Li, X. & Pan, Y. 2020 Reconstruction of GRACE data on changes in total water storage over the global land surface and 60 basins. *Water Resources Research* **56** (4), e2019WR026250.

Sundermeyer, M., Schlüter, R. & Ney, H. 2012 LSTM neural networks for language modeling. In: *Thirteenth Annual Conference of the International Speech Communication Association*. International Speech Communication Association, Portland, OR, USA.

Szegedy, C., Liu, W., Jia, Y., Sermanet, P., Reed, S., Anguelov, D., Erhan, D., Vanhoucke, V. & Rabinovich, A. 2015 Going deeper with convolutions. In *Proceedings of the IEEE Conference on Computer Vision and Pattern Recognition*. The Computer Vision Foundation & IEEE, Boston, MA, USA, pp. 1–9.

Tang, G., Long, D., Behrangi, A., Wang, C. & Hong, Y. 2018 Exploring deep neural networks to retrieve rain and snow in high latitudes using multisensor and reanalysis data. *Water Resources Research* **54** (10), 8253–8278.

Tapley, B. D., Bettadpur, S., Watkins, M. & Reigber, C. 2004 The gravity recovery and climate experiment: mission overview and early results. *Geophysical Research Letters* **31** (9).

Tran, Q. K. & Song, S. K. 2019 Computer vision in precipitation nowcasting: applying image quality assessment metrics for training deep neural networks. *Atmosphere* **10** (5), 244.

Ukkonen, P. & Mäkelä, A. 2019 Evaluation of machine learning classifiers for predicting deep convection. *Journal of Advances in Modeling Earth Systems* **11** (6), 1784–1802.

Wan, X., Yang, Q., Jiang, P. & Zhong, P. A. 2019 A hybrid model for real-time probabilistic flood forecasting using elman neural network with heterogeneity of error distributions. *Water Resources Management* **33** (11), 4027–4050.

Wang, L. P. & Wan, C. R. 2008 Comments on' The extreme learning machine. *IEEE Transactions on Neural Networks* **19** (8), 1494–1495.

Wang, C., Tandeo, P., Mouche, A., Stopa, J. E., Gressani, V., Longepe, N., Vandemark, D., Foster, R. C. & Chapron, B. 2019a Classification of the global Sentinel-1 SAR vignettes for ocean surface process studies. *Remote Sensing of Environment* **234**, 111457.

Wang, J. H., Lin, G. F., Chang, M. J., Huang, I. H. & Chen, Y. R. 2019b Real-time water-level forecasting using dilated causal convolutional neural networks. *Water Resources Management* **33** (11), 3759–3780.

Wang, P., Yao, J., Wang, G., Hao, F., Shrestha, S., Xue, B., Xie, G. & Peng, Y. 2019c Exploring the application of artificial intelligence technology for identification of water pollution characteristics and tracing the source of water quality pollutants. *Science of The Total Environment* **693**, 133440.

Wang, J., Yuan, Q., Shen, H., Liu, T., Li, T., Yue, L., Shi, X. & Zhang, L. 2020a Estimating snow depth by combining satellite data and ground-based observations over Alaska: a deep learning approach. *Journal of Hydrology* **585**, 124828.

Wang, N., Zhang, D., Chang, H. & Li, H. 2020b Deep learning of subsurface flow via theory-guided neural network. *Journal of Hydrology* **584**, 124700.

Wang, Q., Huang, J., Liu, R., Men, C., Guo, L., Miao, Y., Jiao, L., Wang, Y., Shoaib, M. & Xia, X. 2020c Sequence-based statistical downscaling and its application to hydrologic simulations based on machine learning and big data. *Journal of Hydrology* **586**, 124875.

Wang, Y., Fang, Z., Hong, H. & Peng, L. 2020d Flood susceptibility mapping using convolutional neural network frameworks. *Journal of Hydrology* **582**, 124482.

Weber, L. J., Muste, M., Bradley, A. A., Amado, A. A., Demir, I., Drake, C. W., Krajewski, W. F., Loeser, T. J., Politano, M. S., Shea, B. R. & Thomas, N. W. 2018 The Iowa watersheds project: Iowa's prototype for engaging communities and professionals in watershed hazard mitigation. *International Journal of River Basin Management* **16** (3), 315–328.

Weber, T., Corotan, A., Hutchinson, B., Kravitz, B. & Link, R. 2020 Technical note: Deep learning for creating surrogate models of precipitation in Earth system models. *Atmospheric Chemistry and Physics* **20**, 2303–2317. doi:10.5194/acp-20-2303-2020.

Wei, C. C. & Cheng, J. Y. 2020 Nearshore two-step typhoon wind-wave prediction using deep recurrent neural networks. *Journal of Hydroinformatics* **22** (2), 346–367.

Weyn, J. A., Durran, D. R. & Caruana, R. 2019 Can machines learn to predict weather? Using deep learning to predict gridded 500-hPa geopotential height from historical weather data. *Journal of Advances in Modeling Earth Systems* **11** (8), 2680–2693.

Wieland, M., Li, Y. & Martinis, S. 2019 Multi-sensor cloud and cloud shadow segmentation with a convolutional neural network. *Remote Sensing of Environment* **230**, 111203.

Worland, S. C., Steinschneider, S., Asquith, W., Knight, R. & Wieczorek, M. 2019 Prediction and inference of flow duration curves using multioutput neural networks. *Water Resources Research* **55** (8), 6850–6868.

Wu, H., Yang, Q., Liu, J. & Wang, G. 2020 A spatiotemporal deep fusion model for merging satellite and gauge precipitation in China. *Journal of Hydrology* **584**, 124664.

Xiang, Z. & Demir, I. 2020 Distributed long-term hourly streamflow predictions using deep learning–a case study for state of Iowa. *Environmental Modeling & Software* **131**, 104761.

Xiang, Z., Yan, J. & Demir, I. 2020 A rainfall-runoff model with LSTM-based sequence-to-sequence learning. *Water Resources Research* **56** (1).

Xiao, C., Chen, N., Hu, C., Wang, K., Xu, Z., Cai, Y., Xu, L., Chen, Z. & Gong, J. 2019a A spatiotemporal deep learning model for sea surface temperature field prediction using time-series satellite data. *Environmental Modelling & Software* **120**, 104502.

Xiao, H., Zhang, F., He, Q., Liu, P., Yan, F., Miao, L. & Yang, Z. 2019b Classification of ice crystal habits observed from airborne Cloud Particle Imager by deep transfer learning. *Earth and Space Science* **6** (10), 1877–1886.

Xu, Z., Ying, Z., Li, Y., He, B. & Chen, Y. 2020 Pressure prediction and abnormal working conditions detection of water supply network based on LSTM. *Water Supply* **20**, 963–974.

Yang, S., Yang, D., Chen, J. & Zhao, B. 2019a Real-time reservoir operation using recurrent neural networks and inflow forecast from a distributed hydrological model. *Journal of Hydrology* **579**, 124229.

Yang, T., Sun, F., Gentine, P., Liu, W., Wang, H., Yin, J., Du, M. & Liu, C. 2019b Evaluation and machine learning improvement of global hydrological model-based flood simulations. *Environmental Research Letters* **14** (11), 114027.

Yan, Q., Ji, F., Miao, K., Wu, Q., Xia, Y. & Li, T. 2020 Convolutional residual-attention: a deep learning approach for precipitation nowcasting. *Advances in Meteorology* **2020**. https://doi.org/10.1155/2020/6484812

Yang, F., Feng, T., Xu, G. & Chen, Y. 2020 Applied method for water-body segmentation based on mask R-CNN. *Journal of Applied Remote Sensing* **14** (1), 014502.

Yim, I., Shin, J., Lee, H., Park, S., Nam, G., Kang, T., Cho, K. H. & Cha, Y. 2020 Deep learning-based retrieval of cyanobacteria pigment in inland water for in-situ and airborne hyperspectral data. *Ecological Indicators* **110**, 105879.

Yin, W., Kann, K., Yu, M. & Schütze, H. 2017 *Comparative Study of cnn and rnn for Natural Language Processing. arXiv preprint arXiv:1702.01923.*

Yu, Z., Yang, K., Luo, Y. & Shang, C. 2020 Spatial-temporal process simulation and prediction of chlorophyll-a concentration in Dianchi Lake based on wavelet analysis and long-short term memory network. *Journal of Hydrology* **582**, 124488.

Yuan, X., Chen, C., Lei, X., Yuan, Y. & Adnan, R. M. 2018 Monthly runoff forecasting based on LSTM–ALO model. *Stochastic Environmental Research and Risk Assessment* **32** (8), 2199–2212.

Yurtsever, M. & Yurtsever, U. 2019 Use of a convolutional neural network for the classification of microbeads in urban wastewater. *Chemosphere* **216**, 271–280. https://doi.org/10.1016/j.chemosphere.2018.10.084.

Zeng, W., Zhang, D., Fang, Y., Wu, J. & Huang, J. 2018 Comparison of partial least square regression, support vector machine, and deep-learning techniques for estimating soil salinity from hyperspectral data. *Journal of Applied Remote Sensing* **12** (2), 022204. https://doi.org/10.1117/1.JRS.12.022204.

Zhang, D., Hølland, E. S., Lindholm, G. & Ratnaweera, H. 2018a Hydraulic modeling and deep learning based flow forecasting for optimizing inter catchment wastewater transfer. *Journal of Hydrology* **567**, 792–802. https://doi.org/10.1016/j.jhydrol.2017.11.029.

Zhang, D., Lindholm, G. & Ratnaweera, H. 2018b Use long short-term memory to enhance internet of things for combined sewer overflow monitoring. *Journal of Hydrology* **556**, 409–418. https://doi.org/10.1016/j.jhydrol.2017.11.018.

Zhang, D., Martinez, N., Lindholm, G. & Ratnaweera, H. 2018c Manage sewer in-line storage control using hydraulic model and recurrent neural network. *Water Resources Management* **32** (6), 2079–2098. https://doi.org/10.1007/s11269-018-1919-3.

Zhang, C., Sargent, I., Pan, X., Li, H., Gardiner, A., Hare, J. & Atkinson, P. M. 2018d An object-based convolutional neural network (OCNN) for urban land use classification. *Remote Sensing of Environment* **216**, 57–70. https://doi.org/10.1016/j.rse.2018.06.034.

Zhang, D., Lin, J., Peng, Q., Wang, D., Yang, T., Sorooshian, S., Liu, X. & Zhuang, J. 2018e Modeling and simulating of reservoir operation using the artificial neural network, support vector regression, deep learning algorithm. *Journal of Hydrology* **565**, 720–736. https://doi.org/10.1016/j.jhydrol.2018.08.050.

Zhang, J., Zhu, Y., Zhang, X., Ye, M. & Yang, J. 2018f Developing a long short-term memory (LSTM) based model for predicting water table depth in agricultural areas. *Journal of Hydrology* **561**, 918–929. https://doi.org/10.1016/j.jhydrol.2018.04.065.

Zhang, X., Zhang, Q., Zhang, G., Nie, Z., Gui, Z. & Que, H. 2018g A novel hybrid data-driven model for daily land surface temperature forecasting using long short-term memory neural network based on ensemble empirical mode decomposition. *International Journal of Environmental Research and Public Health* **15** (5), 1032. https://doi.org/10.3390/ijerph15051032.

Zhang, C., Li, G., Du, S., Tan, W. & Gao, F. 2019a Three-dimensional densely connected convolutional network for hyperspectral remote sensing image classification. *Journal of Applied Remote Sensing* **13** (1), 016519. https://doi.org/10.1117/1.JRS.13.016519.

Zhang, Z., Iwasaki, A., Xu, G. & Song, J. 2019b Cloud detection on small satellites based on lightweight U-net and image compression. *Journal of Applied Remote Sensing* **13** (2), 026502. https://doi.org/10.1117/1.JRS.13.026502.

Zhang, C. J., Zeng, J., Wang, H. Y., Ma, L. M. & Chu, H. 2020 Correction model for rainfall forecasts using the LSTM with multiple meteorological factors. *Meteorological Applications* **27** (1), e1852. https://doi.org/10.1002/met.1852.

Zhou, Y. T. & Chellappa, R. 1988 Computation of optical flow using a neural network. In: *IEEE International Conference on Neural Networks*. Vol. 1998, IEEE, Anchorage, AK, USA, pp. 71–78.

Zhou, X., Tang, Z., Xu, W., Meng, F., Chu, X., Xin, K. & Fu, G. 2019 Deep learning identifies accurate burst locations in water distribution networks. *Water Research* **166**, 115058.

Zhou, Z., Shi, L. & Zha, Y. 2020 Seeing macro-dispersivity from hydraulic conductivity field with convolutional neural network. *Advances in Water Resources* **138**, 103545.

Zhu, S., Hrnjica, B., Ptak, M., Choiński, A. & Sivakumar, B. 2020a Forecasting of water level in multiple temperate lakes using machine learning models. *Journal of Hydrology* **585**, 124819.

Zhu, S., Luo, X., Yuan, X. & Xu, Z. 2020b An improved long short-term memory network for streamflow forecasting in the upper Yangtze River. *Stochastic Environmental Research and Risk Assessment* 1–17.

Zou, Q., Xiong, Q., Li, Q., Yi, H., Yu, Y. & Wu, C. 2020 A water quality prediction method based on the multi-time scale bidirectional long short-term memory network. *Environmental Science and Pollution Research* **27**, 1–12.

Zuo, G., Luo, J., Wang, N., Lian, Y. & He, X. 2020 Decomposition ensemble model based on variational mode decomposition and long short-term memory for streamflow forecasting. *Journal of Hydrology* **585**, 124776.

First received 16 June 2020; accepted in revised form 22 July 2020. Available online 5 August 2020

© IWA Publishing 2020

Performance improvement of wastewater treatment processes by application of machine learning

O. Icke, D. M. van Es, M. F. de Koning, J. J. G. Wuister, J. Ng, K. M. Phua, Y. K. K. Koh, W. J. Chan and G. Tao

ABSTRACT

Improving wastewater treatment processes is becoming increasingly important, due to more stringent effluent quality requirements, the need to reduce energy consumption and chemical dosing. This can be achieved by applying artificial intelligence. Machine learning is implemented in two domains: (1) predictive control and (2) advanced analytics. This is currently being piloted at the integrated validation plant of PUB, Singapore's National Water Agency. (1) Primarily, predictive control is applied for optimised nutrient removal. This is obtained by application of a self-learning feedforward algorithm, which uses load prediction and machine learning, fine–tuned with feedback on ammonium effluent. Operational results with predictive control show that the load prediction has an accuracy of ≈88%. It is also shown that an up to ≈15% reduction of aeration amount is achieved compared to conventional control. It is proven that this load prediction-based control leads to stable operation and meeting effluent quality requirements as an autopilot system. (2) Additionally, advanced analytics are being developed for operational support. This is obtained by application of quantile regression neural network modelling for anomaly detection. Preliminary results illustrate the ability to autodetect process and instrument anomalies. These can be used as early warnings to deliver data-driven operational support to process operators.

Key words | anomaly detection, autopilot, machine learning, operational support, predictive control, wastewater treatment

O. Icke (corresponding author)
D. M. van Es
M. F. de Koning
J. J. G. Wuister
Aquasuite, Royal HaskoningDHV,
Laan 1914 35, 3818 EX Amersfoort,
The Netherlands
E-mail: *otto.icke@rhdhv.com*

J. Ng
K. M. Phua
Y. K. K. Koh
W. J. Chan
G. Tao
PUB, Singapore's National Water Agency,
40 Scotts Rd,
Singapore 228231,
Singapore

HIGHLIGHTS

- Autopilot delivers stable operation, meeting requirements and reducing energy.
- Self-learning predictive control optimises wastewater treatment performance.
- Autodetection of process and instrument anomalies offers operational support.

INTRODUCTION

Improving wastewater treatment processes is becoming increasingly important. This is caused by more stringent effluent quality requirements and the need to reduce both energy consumption and chemical dosing. At the same time, decreasing availability of skilled operators can demand an autopilot solution. Such challenges also apply to PUB, Singapore's National Water Agency. Growing water demands and rising operational costs require innovative technologies to guarantee water safety and security.

To assess the feasibility of deploying innovative process technologies in water reclamation plants (WRPs), an integrated validation plant (IVP) was commissioned in 2017. IVP is a demonstration plant at Ulu Pandan WRP which has a capacity of ≈12,500 m^3/day. It has a membrane bioreactor (MBR) system, where the bioreactors consist of non-aerated and aerated plug flow step-feed activated sludge basins for organics and nutrient removal, and MBR membranes for sludge water separation. A schematic overview of the IVP is shown in Figure 1.

doi: 10.2166/wst.2020.382

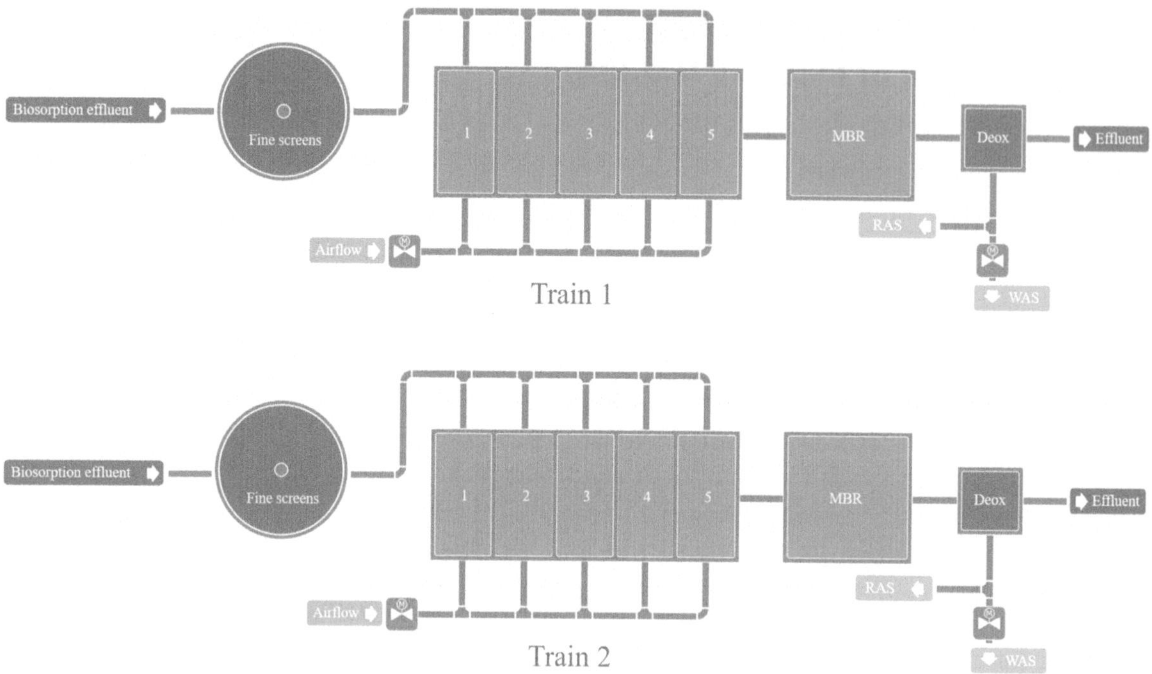

Figure 1 | Schematic overview of the integrated validation plant (IVP).

The IVP consists of two trains namely Train 1 and Train 2. Each train has an aerated plug flow step-feed reactor with five basins. Each basin has two zones with adjustable valves for the air flow and online dissolved oxygen (DO) measurements. Basins 1, 3 and 5 are equipped with online measurements for ammonium (NH_4), and additional measurements of nitrite (NO_2) and nitrate (NO_3) for Basin 5. In addition, the ammonium concentrations and flow rates of the effluent of the fine screens to the reactors are measured online. This makes it possible to determine the ammonium influent load to each bioreactor.

An example of these innovative technologies applied to the IVP is the development and deployment of an autopilot solution with advanced analytics and predictive control for the biological processes of the WRP. The aim of this autopilot solution is to achieve stable operation while meeting product water quality as feed for the downstream reverse osmosis (RO) process. Other objectives include an accurate influent load prediction for optimised nutrient removal which in turn substantially reduces aeration energy usage. Early warnings for the process operators for process and instrument anomalies are also achieved with the solution.

To achieve these goals, application of self-learning predictive control, instead of reactive control, is preferred. Conventional control solely using proportional-integral-derivative (PID) loops can cause instabilities as it can only be reactive in the shorter term and cannot be anticipatory in the longer term. In wastewater treatment, the key aberrations are primarily caused by the daily pattern of the influent load and rain events and should, therefore, be regarded for optimal operation (de Koning *et al.* 2013). Dry-weather flow (DWF) can be predicted, and rain-weather flow (RWF) detected. Utilising these in predictive control enables stable processes as it accurately forecasts the requested output and subtly adjusts for target variations. Additionally, predictive analytics provides the opportunity for early detection of anomalies before measured and predicted values diverge sharply.

In this study, the methods and results of the application of predictive control and advanced analytics in the IVP are explained and discussed.

METHODS

The autopilot solution consists of an advanced analytics and predictive control system for the IVP. A brief description of the methods and techniques used is subsequently presented for each system.

Predictive control

The predictive control applies self-learning feedforward control using load prediction and machine learning fine-tuned with feedback control on effluent ammonium levels. The determination of the aeration setpoints consists of several steps which are described below.

Firstly, an influent ammonium load prediction is determined with a self-learning algorithm (Bakker *et al.* 2013a, 2013b). It captures the day-patterns in hourly (or quarterly) values, obtaining the regular day-patterns for each day of the week. It has a 48 (or 72) h prediction horizon. Days with large deviations are rejected automatically. This occurs if any of the array elements of the dimensionless daily curve factors exceeds the maximum allowable error, which has an adjustable default value (½). The day-patterns can also be automatically rejected due to the signal status indicating an invalid value or manually rejected by the operator. Each daily data is classified with a reason for approval or rejection. Days with deviations not exceeding the tolerance factor are approved but labelled with a warning.

This specific technique was originally developed as a fully adaptive model for the prediction of the short-term drinking water demand. Subsequently, this technique was also successfully applied for dry-weather flow (DWF) prediction for aeration control (de Koning *et al.* 2013) and also combined with rainfall forecast data, to carry out flow control (Icke *et al.* 2017). For the IVP, it is used to predict the influent flow and diurnal pattern of ammonium concentration.

Secondly, the actual influent load prediction is used in a self-learning regression model which estimates the required amount of aeration for each train. It aggregates the online measured values of aeration flow and influent load to averages and assigns weighting factors. As recent aggregates are more significant for the current process, the weighting factors start at one for the most recent ones (over a range of few days) and decreases linearly to zero (over a range of few weeks). Invalid values, due to an erroneous signal or being outside the fit constraints, are automatically rejected. The aggregates and fit are updated multiple times a day. This makes the regression model with the self-learning prediction robust: it adapts quickly, and it hardly requires maintenance.

Thirdly, the resulting estimated amount of required aeration flow is subsequently fine-tuned with feedback control. This is obtained with a proportional-integral (PI) controller on ammonium for each separated bioreactor. Additionally, for each aerator valve, the amount of aeration is adjusted with a PI controller on oxygen, fine-tuning the aeration flow per aeration zone (Yuan *et al.* 2019). In this way, the amount of aeration is adapted to the needs of each train and specific zone to achieve optimised control on the dynamic setpoints for ammonium and DO.

Finally, the amount of aeration per train can be dynamically suppressed with an adjustable limit for nitrate. In case nitrate concentrations exceed the ammonium concentrations, aeration is suppressed to prevent over aeration, balancing ammonium versus nitrate concentrations, thereby saving energy.

The above-mentioned predictive control philosophies were installed within an on-premise configuration for real time aeration control of each of the IVP bioreactors.

Advanced analytics

The advanced analytics applies neural network modelling for early detection of different anomalies. The detection of the anomalies consists of several steps which are described sequentially.

Firstly, pre-processing of the data is executed. All data are normalized with respect to their mean and standard deviation (i.e. z-scored). Missing values are inserted using the arithmetic mean value for the respective feature. All data are resampled to the median frequency of the target variable using mean aggregation. Additionally, measurements that fall outside an expected ratio are qualified as not a number (NaN), which are consecutively filled with the mean. Valid ranges are defined for each signal separately.

Secondly, to detect anomalies in signal time series, an artificial neural network (ANN) is applied: quantile regression in combination with multi-layer perceptron (QR-MLP) (Koenker 2005; Hastie *et al.* 2009). This approach created predicted intervals within which expected data values should lie. The main benefit of using a neural network is that different and tilted loss-functions can be implemented for each output unit (Cannon 2018, Rodrigues & Pereira 2018). In this way, the network can output a range of quantiles for every timepoint in a single architecture.

The neural networks all contain two hidden layers with 64 neurons each. Each hidden layer is followed by a rectified linear unit (Re LU) activation function (Hinton 2010) and then followed by a dropout layer (Srivastava *et al.* 2014) with a rate of 0.2. The networks are trained using a batch size of 64 and a learning rate of 0.0005. Training lasts for a maximum of 200 epochs but stops earlier if validation loss does not improve for five consecutive epochs.

Finally, post-processing of the results is executed. Resampling of data is applied, and data resolution can range from 5 min refresh rate up until 24 h. The fine resolution can highlight individual datapoint outliers, whereas the coarse resolution can highlight trend outliers, or drifts. The QR-MLP modelling including pre- and postprocessing is applied to the IVP for the detection of process and instrument anomalies. Several anomaly models are subsequently developed.

One of the developed models predicts the aeration flow needed for each train of the IVP from the influent load. It is plausible that higher influent load results in increased oxygen demand from the process and, therefore, resulting in a higher aeration flow. However, if measured aeration flow increases and cannot be explained from the measured influent load, it can be detected as a process anomaly as it exceeds the predicted interval for aeration flow. As input to the model, the ammonium influent load was summarised as the current value in addition to the value of ¼, ½, 1, 2, 4, 6, 8, 16 and 32 hours ago, as well as the mean of the past of ¼, ½, 1, 2, 4, 6, 8, 16 and 32 h. In addition, one-hot-encoded features that represented day of the week and cyclically encoded features that represented second-of-day, were setup. The latter two feature categories serve to capture daily patterns, that can differ across each weekday. In total, this resulted in 28 features.

Another model predicts the ammonium concentration in the effluent of each IVP bioreactor, based on the influent ammonium load and the ammonium concentration measured in the basins of the other bioreactor. If the measured ammonium effluent concentration deviates from the expected range, based on ammonium measurements from the influent and from the other basins, it shall be designated as an instrument anomaly. In this model, the current values of all measurements were populated. In addition, one-hot-encoded features that represented day of the week and cyclically encoded features that represented second-of-day were setup again. This resulted in 13 features.

A similar approach was applied for nitrate instruments. The nitrate concentration in the effluent of each bioreactor of the IVP is predicted based on the current value of ammonium load in the influent and the online ammonium measurements in the basins of all bioreactors. Again, one-hot-encoded features that represented day of the week and cyclically encoded features that represented second-of-day were added. In total, this resulted in 16 features.

Before implementation and integration in the online advanced analytics environment in the cloud, these models were trained on data from 15 May 2019 to 31 July 2019 and were evaluated on the period between 1 and 16 August 2019. Due to the equatorial location (climate) of IVP (Ulu Pandan Water Reclamation Plant in Singapore), process temperature remained quite constant throughout the year. Therefore, a regular (e.g. monthly) retraining schedule should cover any remaining changes after implementation.

RESULTS AND DISCUSSION

The autopilot solution consists of an advanced analytics and predictive control system for the IVP. Results of each system are presented and discussed below.

Predictive control

After implementation, Train 1 became operational on 11 April 2019, followed by Train 2 on 13 May 2019. The predicted and measured influent ammonium load values are shown in Figure 2. After an initial training period (until 21 June 2019), the prediction algorithm has learned the daily patterns of the influent load. The influent load is then (from 21 June 2019) predicted with a usable forecast horizon of 48 h.

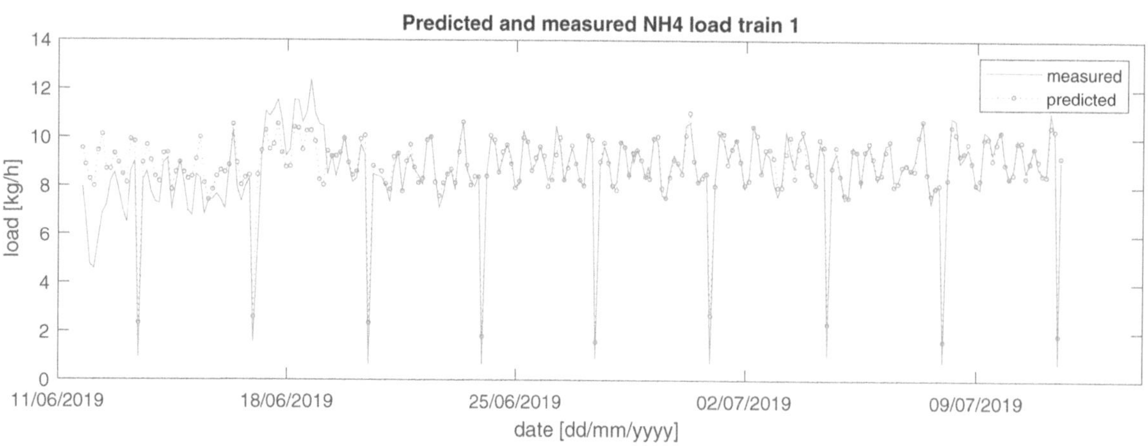

Figure 2 | Trend of the predicted values (green -o-) and measured values (red -) of the influent load. Please refer to the online version of this paper to see this figure in colour: http://dx.doi.org/10.2166/wst.2020.382.

In Figure 3, an XY plot of predicted and measured values of the influent ammonium load is shown. Operational results with predictive control show that the load prediction has an accuracy (R^2) of 88%. This is verified for the period 21 June–21 August 2019, and rainfall influence has not been filtered out. The prediction of the load pattern is self-learning without further manual intervention required. Using the predicted instead of the measured value of the influent load as an input for feedforward has the advantage of considering the daily pattern rather than short-term variations.

In Figure 4, an XY plot of aeration flow and influent load is showing a fit line through the aggregates. The figure shows a clear relationship between aeration flow and load which confirms that the amount of aeration needed for feedforward control can be estimated very accurately based on fit and prediction. The linear regression is refit periodically without manual tuning required.

Results show that up to 15% reduction of aeration amount can be achieved as compared to conventional control with ammonium level near setpoints. This is determined by comparison of the aeration of one train operating with predictive control (Train 1: avg. of 833 Nm^3/h), and another train operating with conventional control (Train 2: avg. of 981 Nm^3/h) for the period 11 April–11 May 2019. The 15% reduction of aeration amount also yields a corresponding reduction of aeration energy.

Advanced analytics

For illustration of the data science models, an event at IVP was selected. Figure 5 shows the data trend for ammonium load in the influent and concentrations in the effluent of the Trains 1 and 2 bioreactors.

It is apparent, that the ammonium concentrations in Trains 1 and 2 diverge in the period between 18 and

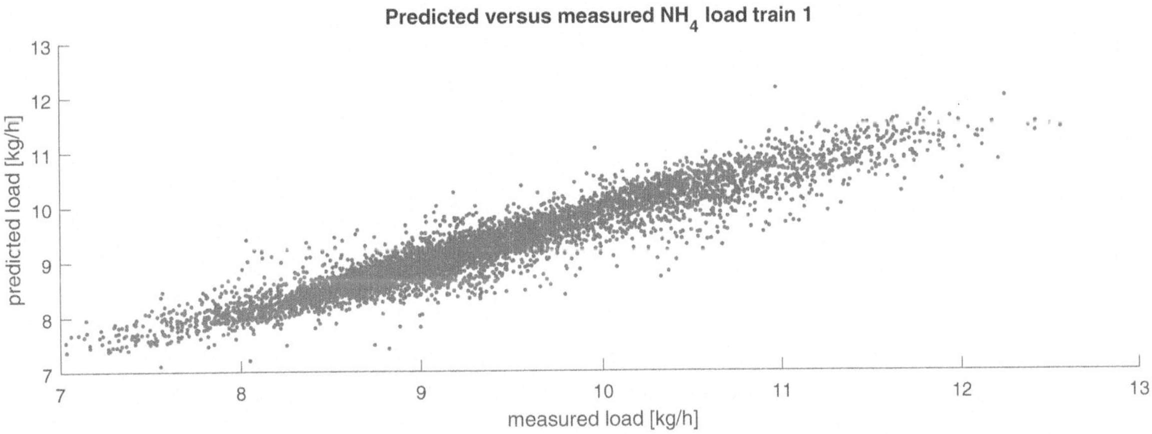

Figure 3 | Trend of predicted values (y) versus measured values (x) of the influent load with $R^2 = 0.88$.

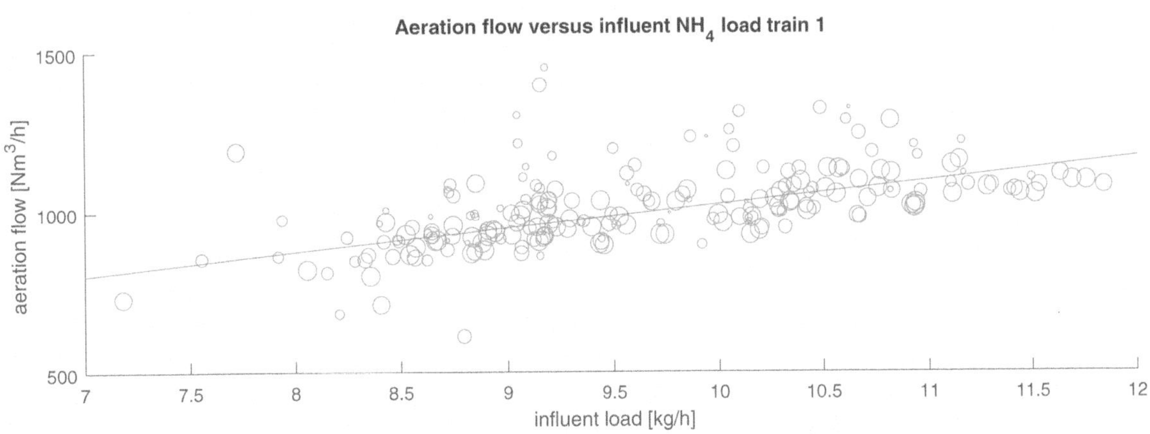

Figure 4 | Plot of estimated relation between aeration flow (y) and influent load (x). The size of the marker shows the weighted value declining longer ago used in the fit.

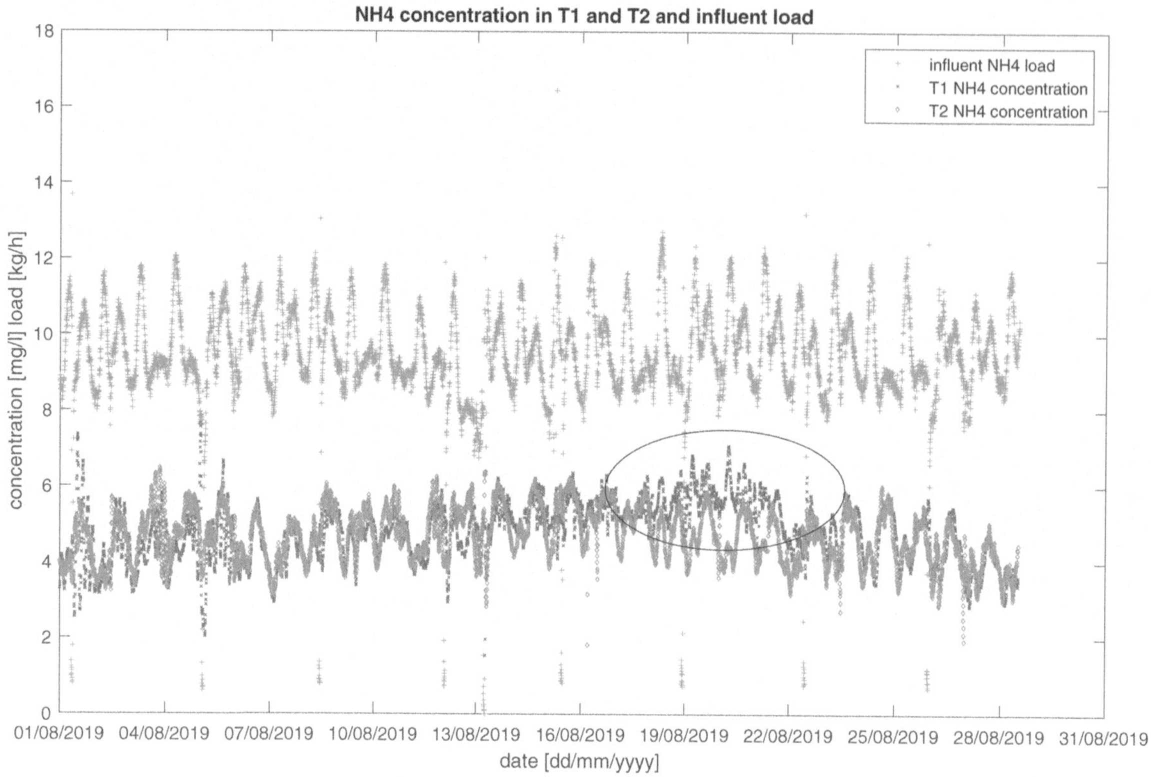

Figure 5 | Plot of the measured influent load (green+) and ammonium concentration for Trains 1 and 2 (blue × vs. red ◇). Divergence between trains is indicated by the black oval. Please refer to the online version of this paper to see this figure in colour: http://dx.doi.org/10.2166/wst.2020.382.

23 August 2019. The ammonium in Train 1 increases, which cannot be explained from the ammonium influent load, whereas the ammonium in Train 2 follows the influent load. Ammonium setpoints were identical for both trains.

This event is ideally suited to analyse the QR-MLP models and check whether they are able to detect this anomaly. Also, it enables the combination of outcomes from different models. The higher ammonium concentration could either be caused by a lower aeration amount compared to the influent load or caused by instrument drift that can lead to changes in nitrate levels.

Detecting aeration anomalies

The trained model which predicts the aeration amount based on influent load was evaluated for both trains. For Train 1, the 5-min resolution model shows various extreme outliers (i.e. values that fell outside the predicted 99.7% confidence interval (CI); see Figure 6). However, the figure also shows that the observed values are on average higher than the predicted values.

To quantify this observation, the 5-min results were resampled to daily (24 h) averages (see Figure 7). Then, outliers where daily averages fell outside the 68.3% CI were quantified. This shows that the slow drift becomes visible on 13 August and that it exceeds the CI on 19 August. Further optimization of the resampling time could offer the best result by balancing between higher resolution of single events and early detection of drifts.

Detecting ammonium anomalies

The aeration controller uses the online measured ammonium concentration of the effluent of the bioreactor as input. A higher ammonium concentration results in a higher amount of aeration. The degree of the increase depends on the setting of the proportional gain of the aeration controller.

To uncover whether the increased amount of aeration was due to increase in measured ammonium, a second model was set up to predict the ammonium concentration based on the influent load and concentrations within the other basins. The same 5-min and daily-average approach was used to visualize the modelling results for Train 1 (Figures 8 and 9, respectively), which showed and corroborated the increase in ammonium within the same period.

For this model, it is apparent that the measured ammonium concentration was higher than predicted

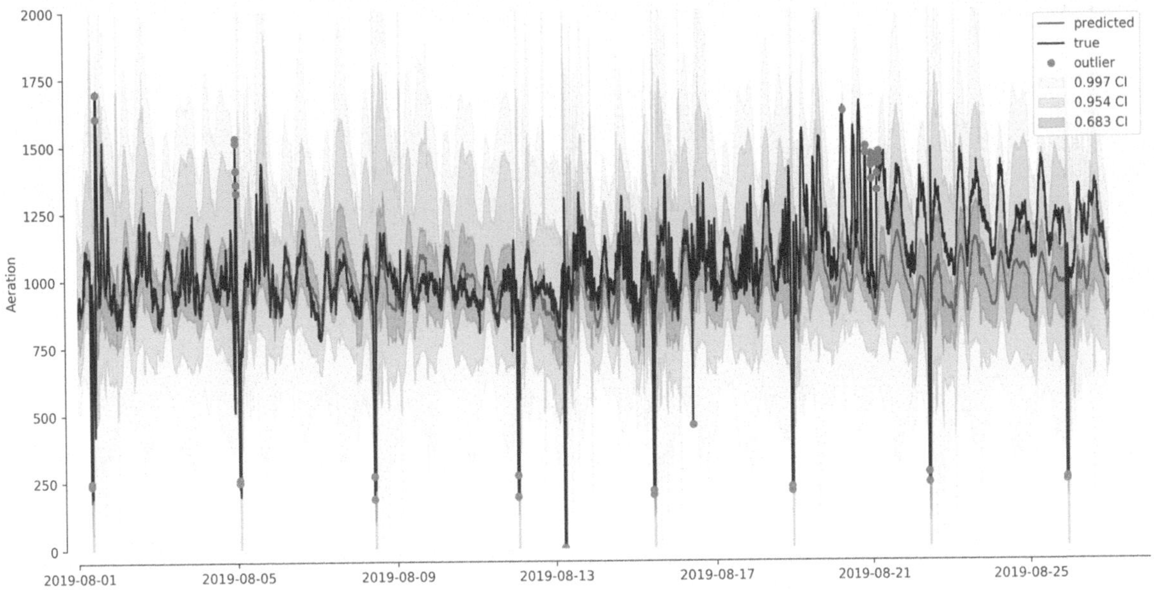

Figure 6 | Plot of predicted aeration values (bandwidths) versus measured values (black line) for Train 1 of IVP. The resolution is 5 min. The outliers (red dots) detected at prediction >0.997 CI.

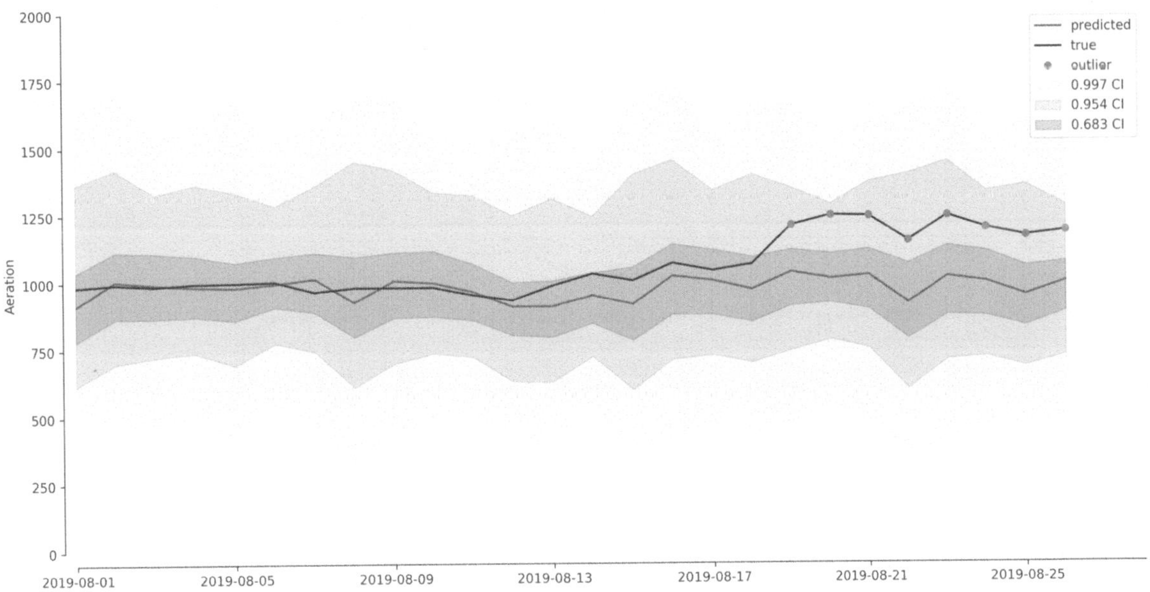

Figure 7 | Plot of predicted aeration values (bandwidths) versus measured values (black line) for Train 1 of IVP. The resolution is 1 day. The outliers (red dots) detected at prediction >0.683 CI.

based on the ammonium influent load and concentrations of the other basins during the same event. Also, the resampling again resulted in an earlier (13 August) and more consistent anomaly detection. This model points out that it is plausible that the sensor value of the ammonium sensor installed in the effluent of Train 1 bioreactor has drifted.

However, the measured ammonium concentration of this model trends back to predicted values earlier than the previous model for aeration anomalies. After 22 August, the ammonium concentration is within the predicted range as seen in Figures 8 and 9, whereas the aeration is still outside the predicted range as seen in Figures 6 and 7. This can be explained by the mechanism

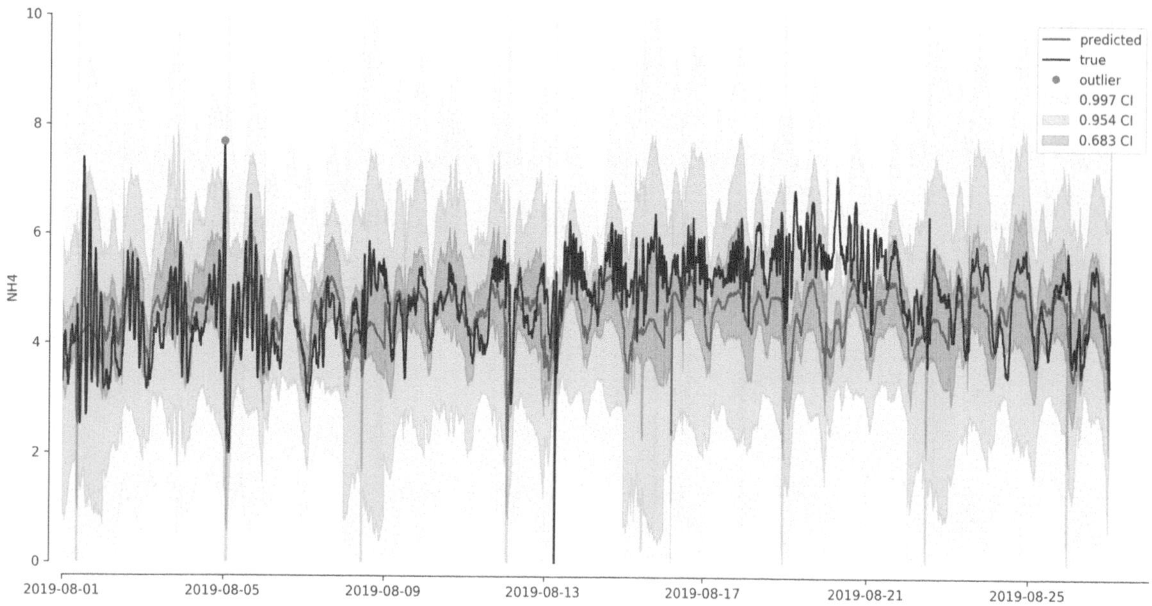

Figure 8 | Plot of predicted ammonium values (bandwidths) versus measured values (black line) for Train 1 of IVP. The resolution is 5 minutes. The outliers (red dots) detected at prediction >0.997 CI.

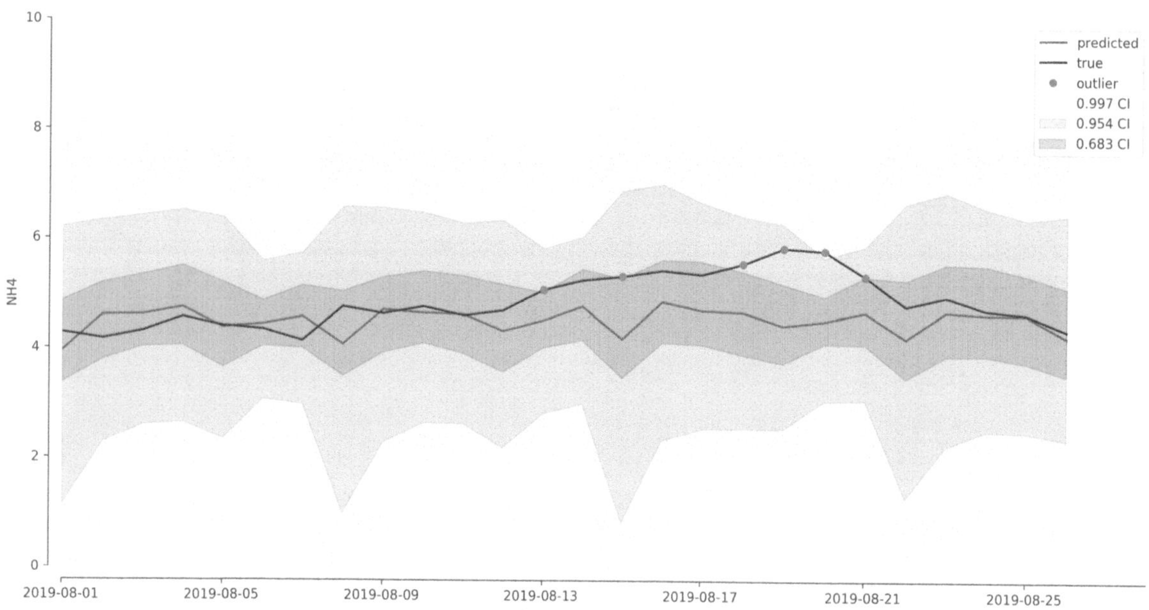

Figure 9 | Plot of predicted ammonium values (bandwidths) versus measured values (black line) for Train 1 of IVP. The resolution is 1 day. The outliers (red dots) detected at prediction >0.683 CI.

of the aeration controller. After some days, the machine learning algorithm of the aeration controller picked up that for the same influent load, higher aeration is needed. It also must retrain some days, once the ammonium sensor drift is resolved. This also emphasized the importance of detecting instrument anomalies early.

Detecting nitrate anomalies

The increased aeration that resulted from abnormally high measures of ammonium should also result in increased nitrate concentrations, that is relative to the influent load. It can be expected that with increased aeration, more nitrate is produced through nitrification, while the amount of denitrification gets suppressed due to less anoxic capacity. Both changes will increase the nitrate concentration in the effluent of Train 1 bioreactor.

To investigate this, a model was setup to predict nitrate concentration based on influent load and ammonium concentrations of all basins. Displaying these results in 5-min and daily-average resolution for Train 1 (Figures 10 and 11 respectively) exhibited a similar trend of increasing nitrate levels.

This model produced similar results as the aeration model that is based on influent load. Nitrate is higher than expected based on the influent load, thus resulting in a

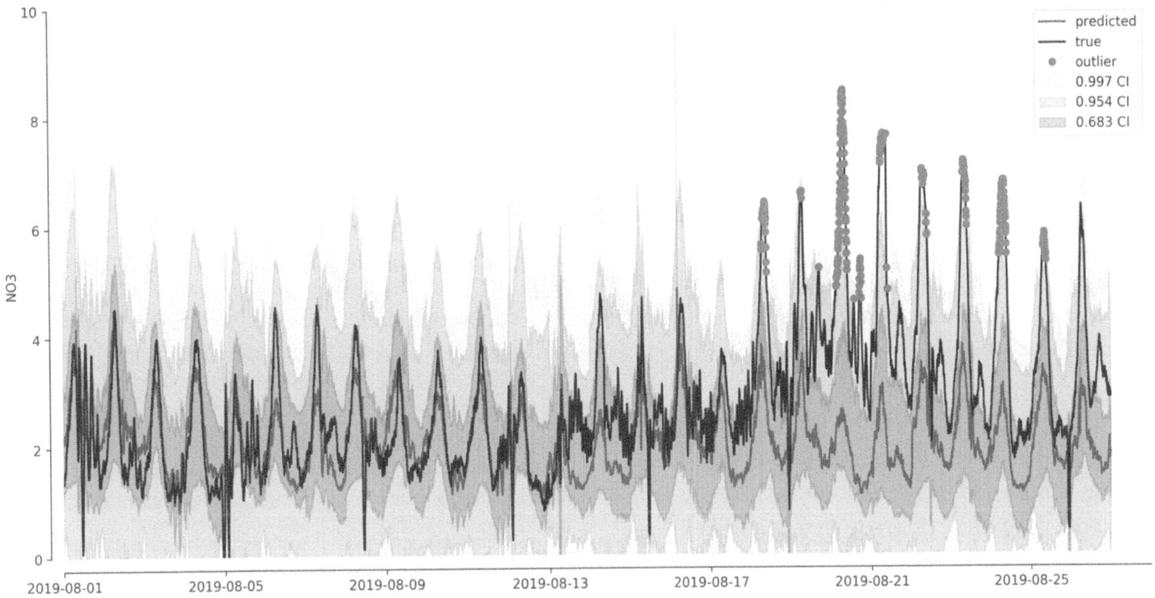

Figure 10 | Plot of predicted nitrate values (bandwidths) versus measured values (black line) for Train 1 of IVP. The resolution is 5 min. The outliers (red dots) detected at prediction >0.997 CI.

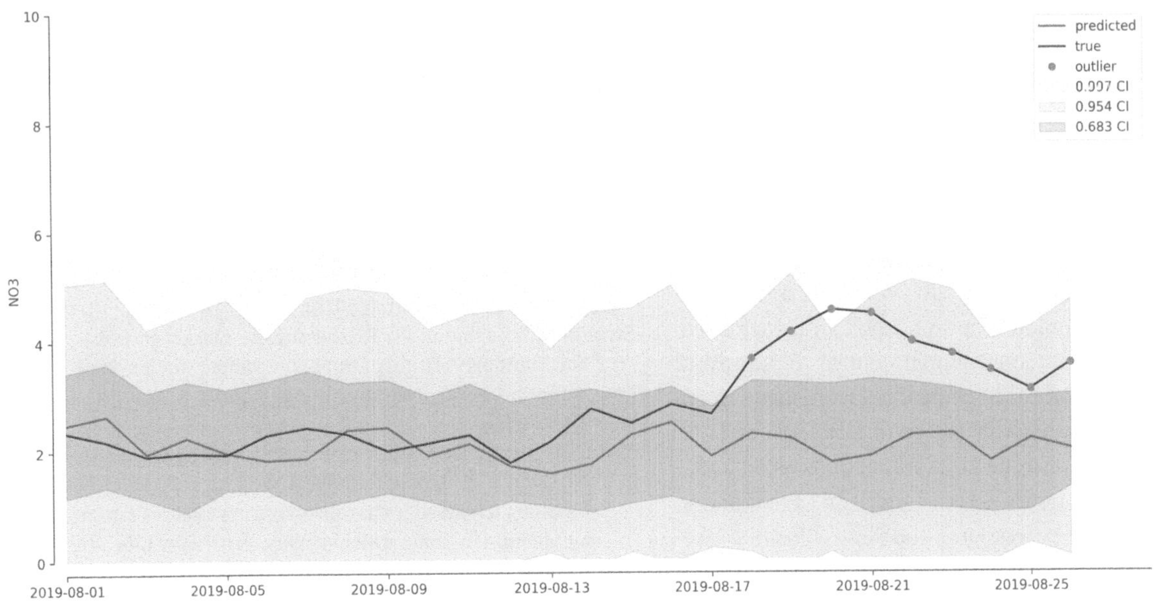

Figure 11 | Plot of predicted nitrate values (bandwidths) versus measured values (black line) for Train 1 of IVP. The resolution is 1 day. The outliers (red dots) detected at prediction >0.683 CI.

higher amount of aeration supply. After 22 August, the ammonium concentration is within its range, while nitrate concentration is still high, largely because there is a time lag inherent from the machine learning of the aeration controller based on the weight factors.

Evaluation

These three quantile regression models illustrate that anomaly detection during early stage is possible. Based on the outcomes of these different models, it is also possible to determine in which direction to search for the cause of anomaly so that it can be resolved early. This delivers direct value to the process operator. Therefore, these three models are currently being implemented and integrated in an online advanced analytics environment that is hosted in the cloud for real-time operational support.

Discussion

Resampling resolution and quantiles are important optimisation parameters for further research. The 4-h resolution as an alternative for the 5-min and daily resolution is currently being researched upon. Also, the criteria to detect outliers, where measured values exceed CI thresholds of the model repeatedly and consecutively, are being investigated to improve the robustness of the analytics module.

For other geographical locations that display strong seasonal patterns, the currently developed models must be incorporated with additional data parameters, such as process temperature and sludge solids content. Also, several years of data needs to be collected and utilized for model training to achieve optimal performance.

CONCLUSIONS

Operational results with predictive control show that the load prediction has an accuracy (R^2) of ≈88% and that a reduction of aeration amount up to ≈15% can be achieved at the IVP compared to conventional control. It is proven that this load prediction-based control leads to stable operation and effluent quality with autopilot functions at IVP. Additionally, it is demonstrated that it functions as an autopilot system with 48 hours forecast horizon. Preliminary results with quantile regression modelling illustrate the ability of automated early detection of process and instrument anomalies. These can be used as an advanced process control and decision support system for operations.

DATA AVAILABILITY STATEMENT

All relevant data are included in the paper or its Supplementary Information.

REFERENCES

Bakker, M., Vreeburg, J. H. G., van Schagen, K. M. & Rietveld, L. C. 2013a A fully adaptive forecasting model for short-term drinking water demand. *Environ. Modell. Software* **48**, 141–151.

Bakker, M., Vreeburg, J. H. G., Palmen, L. J., Sperber, V., Bakker, G. & Rietveld, L. C. 2013b Better water quality and higher energy efficiency by using model predictive flow control at water supply systems. *J. Water Supply Res. Technol.* **62**, 1–13.

Cannon, A. J. 2018 Non-crossing nonlinear regression quantiles by monotone composite quantile regression neural network, with application to rainfall extremes. *Stochastic Environ. Res. Risk Assess* **32**, 3207–3225.

de Koning, M., van Schagen, K. M., Agarwalla, B., Trolio, R. & Domuard, M. 2013 Predictive control in wastewater treatment. In *OZ Water '13 Conference, Perth, Australia.*

Hastie, T., Tibshirani, R. & Friedman, J. 2009 *The Elements of Statistical Learning: Data Mining, Inference, and Prediction.* Springer, New York, NY, USA.

Hinton, G. 2010 Rectified linear units improve restricted boltzmann machines vinod nair. *Proceedings of ICML* **27**, 807–814.

Icke, O., van Schagen, K., Huising, C., Wuister, J., van Dijk, E. & Budding, A. 2017 Flow intake control using dry-weather forecast. *Drinking Water Eng. Sci.* **10** (2), 69–74.

Koenker, R. 2005 *Quantile Regression.* Cambridge University Press, Cambridge, UK, pp. 146–147.

Rodrigues, F. & Pereira, F. C. 2018 *Beyond Expectation: Deep Joint Mean and Quantile Regression for Spatial-Temporal Problems.* Available from: https://arxiv.org/pdf/1808.08798.pdf (accessed 16 March 2020).

Srivastava, N., Hinton, G., Krizhevsky, A., Sutskever, I. & Salakhutdinov, R. 2014 Dropout: a simple way to prevent neural networks from overfitting. *J. Mach. Learn. Res.* **15**, 1929–1958.

Yuan, Z., Olsson, G., Cardell-Oliver, R., van Schagen, K., Marchi, A., Deletic, A., Urich, C., Rauch, W., Liu, Y. & Jiang, G. 2019 Sweating the assets – the role of instrumentation, control and automation in urban water systems. *Water Res.* **155**, 381–402.

First received 30 April 2020; accepted in revised form 30 July 2020. Available online 12 August 2020

© IWA Publishing 2020

Data analytics in control and operation of municipal wastewater treatment plants: qualitative analysis of needs and barriers

S. Eerikäinen, H. Haimi IWA, A. Mikola IWA and R. Vahala IWA

ABSTRACT

This study aims to identify barriers and needs for the application of data analytics in municipal wastewater treatment. The study was conducted through a series of interviews with stakeholders involved in instrumentation, control, and automation of wastewater treatment plants. Opportunities and limitations observed by different stakeholders were assessed with a thematic analysis. Thematic analysis enabled a broader consideration of social and organizational aspects related to process control, operation, and maintenance. Identified key barriers for applying data analytics included laborious instrumentation maintenance, unstable control loops, and deficient customization of digital tools for users at wastewater treatment plants. Development needs include easier data processing tools, quality assurance of instrumentation, and controller tuning. Results indicate that the perceived potential of data analytics is highly dependent on the performance of underlying physical and digital systems, as well as the control strategies and operating environment of the plant. Despite the barriers, users and developers see many potential applications for data analytics and expect them to have a central role in the control and operation of wastewater treatment plants in the future.

Key words | data analytics, municipal wastewater treatment, operation and maintenance, process control, stakeholders, thematic analysis

S. Eerikäinen (corresponding author)
A. Mikola IWA
R. Vahala IWA
Aalto University,
Finland
E-mail: *sanni.eerikainen@aalto.fi*

H. Haimi IWA
FCG Finnish Consulting Group Ltd,
Finland

HIGHLIGHTS

- Barriers and needs for advanced data processing and analytics at wastewater treatment plants (WWTPs) seen by practitioners were uncovered through interviews.
- Views of different stakeholders were systematically assessed using thematic analysis.
- Many barriers arise from lack of quality assurance for instrumentation and control loops, resulting in poor control performance and data quality.
- Current software for data processing, analysis, and visualization are seldom well customized for user needs at WWTPs.
- Centralized operation of WWTPs will push the development towards more systematic data management, predictive control, and a higher level of automation.

doi: 10.2166/wst.2020.311

GRAPHICAL ABSTRACT

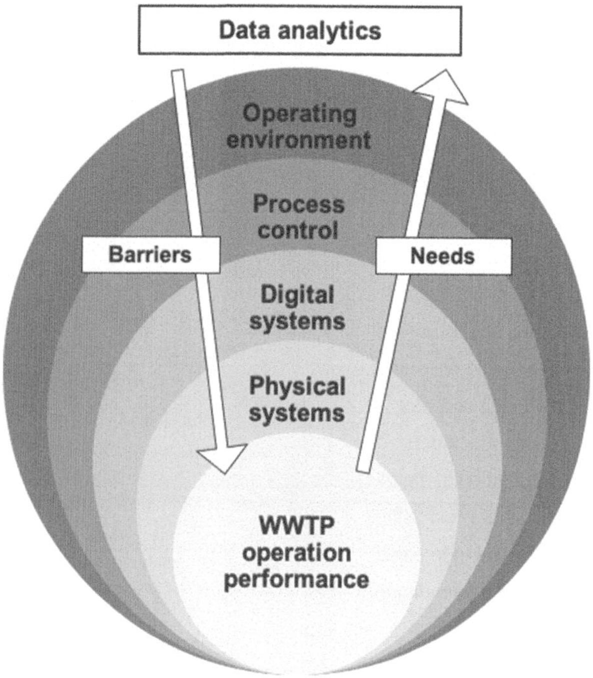

INTRODUCTION

Process instrumentation, control and automation (ICA) are an integral part of the design and operation of today's municipal wastewater treatment plants (WWTPs). ICA systems allow efficient operation of the plant to meet the effluent criteria while minimizing the operational and capital expenses (Olsson 2012). Jeppsson *et al.* (2002) discuss that in addition to efficiency requirements, increased complexity of the plant will push the development of ICA systems. Future demands for resource recovery from wastewater will further increase the importance of understanding and controlling plant-wide and even system-wide interactions (Solon *et al.* 2019).

Olsson *et al.* (2005) state that increasing complexity of WWTPs has led to a more comprehensive and systematic monitoring of the treatment process and its environment. As a result, the availability and quality of monitoring data at WWTPs have significantly improved. However, collected data is not analyzed further to increase process knowledge and select optimal operation strategies. Ingildsen & Olsson (2016) describe measuring as only the first step of effective ICA systems, followed by the analysis and integration of information to decision-making.

Researchers have proposed several new ICA tools for WWTPs based on data analytics, i.e. deriving meaningful patterns from the raw process data. Solutions such as model-predictive control (MPC), process fault detection, and condition monitoring of hardware have been widely studied (see e.g. Åmand *et al.* 2013; Haimi *et al.* 2013; Newhart *et al.* 2019). Despite the promising research and piloting results, a review by Corominas *et al.* (2018) revealed that only 9% of studied computer-based data analysis solutions had been validated and used in full-scale applications, indicating a gap between research and practice.

Meanwhile, techniques studied for wastewater treatment are already applied in different fields of process industries. Jämsä-Jounela (2007) reviewed the future of process automation in the process industry, stating that MPC already has thousands of applications. Most of the applications are in the chemical and oil and gas industries, but MPC is increasingly adopted in the pulp and paper and food processing sectors. Online prediction of process disturbances and end-product quality are said to be successfully applied in the same sectors. Bauer *et al.* (2016) demonstrated that various tools for control loop performance monitoring have been adopted in the chemical, oil and gas, minerals processing, and pulp and paper sectors.

Examples from other industries seem to indicate that the barriers in applying data analytics and new ICA tools in general, are not purely technical, but rather contextual. For example, Smuts & Hussey (2011) studied why the power generation industry in the USA had not adopted advanced control applications at the same pace as other industries. Identified key barriers included perceived complexity of the technologies, lack of skilled personnel, uncertainty of cost-benefits, and lack of regulatory incentives.

Dybå *et al.* (2011) state that in software engineering, qualitative methods are necessary for assessing complex human behavior and organizational aspects. Whereas quantitative research aims to measure and quantify the problem, qualitative research aims to interpret and understand the underlying behaviors, beliefs, and motivation.

This study aims to review the barriers for adoption of new ICA tools and data analytics in municipal WWTPs in Finland. To gather practical views from both user and developer perspectives, the study was conducted through stakeholder interviews. To conclude general views from individual interviews, the transcripts were assessed using thematic analysis. The study complements and updates the results of state-of-the-art survey by Haimi *et al.* (2009) on ICA practices in Finnish WWTPs, while also discussing broader aspects of data analysis and management related to ICA.

Qualitative research methods have occasionally been applied in the water sector (see e.g. Bamberger 2000; Naman & Gibson 2015). Thematic analysis of ICA systems and advanced data analytics provides a new perspective on non-technical aspects that guide, drive, and hinder their technological development. The systematic approach enables the translation of personal views and experiences into generalized findings. We believe that these results will be of use in the design of future research and contribute to reducing the gap between research and practice.

METHODS

Interviews

The study was implemented through a range of stakeholder interviews. Interviewed stakeholders included personnel of WWTPs, consulting companies, automation companies and technology companies. In total, 19 interviews of approximately 2–3 hours length were held during spring 2020. From these, ten interviews were with staff and management of Finnish WWTPs. There are over 500 WWTPs in Finland, from which about 90 WWTP's population equivalent is over 10,000 (Talvitie 2018). Exact treatment requirements vary and are defined in the environmental permit. Average removal rate in 2017 for BOD_7 (biological oxygen demand) was 97%, total phosphorus 96%, and total nitrogen 56%. Various types of WWTPs organizations, with average flowrates from 2,000 to 250,000 m^3/d were interviewed to obtain a representative sample. Interviewed organizations are presented in Table 1.

Nine professionals from Finnish consulting, automation, and technology companies were interviewed. The selection of companies for the interview was based on their references and experience of municipal WWTPs in Finland. Based on the share of population and market covered by these WWTPs and companies, their perceptions were presumed to provide a good overview of the municipal wastewater sector in Finland.

Interviews were held as semi-structured, meaning that the key themes and questions were defined beforehand, but responded and discussed further in a conversational manner. Questions for WWTPs dealt with process control, monitoring, data management, and maintenance, including the challenges, plans, and future views regarding the use of data analytics. Questions for companies were focused on the development opportunities and barriers of ICA and data analytics they have faced at WWTPs, including comparison to other sectors they work with. Some of the interviewees sent additional material, such as process descriptions, which were included in the analysis materials. Interviews were recorded and transcribed afterwards.

Thematic analysis

After the interviews, all collected materials were processed with thematic analysis. Thematic analysis is a transparent

Table 1 | Interviewed organizations

Type of organization	No. of interviews	Additional information
WWTP	10	Eight CAS, one MBBR, and one MBR plant
Consulting	3	Senior process and automation designers
Automation	3	Engineers or managers of water business unit
Technology	3	Instrumentation and process equipment providers

CAS = conventional activated sludge, MBBR = moving bed biofilm reactor, MBR = membrane bioreactor.

method of identifying, analyzing, and reporting patterns in qualitative data (Braun & Clarke 2006). Quotations from the research data, e.g. interview transcripts, are systematically coded with their meanings. A code can be considered to be a class of several quotations indicating the same idea, view, or opinion. From these codes, patterns are identified to construct generalized findings, i.e. themes.

The phases of the analysis process are presented in Figure 1. First, the overall material is reviewed to revise potential gaps or inconsistencies. After that, the initial coding framework is developed based on the research questions. When the coding process is started, all quotations in the material, considered as relevant for the research topic, are coded with their general meaning. While more codes are created, other codes can be reviewed and adjusted to a logical set of findings. Codes can be merged or split to find a balance between a desired level of detail and groundedness, i.e. the number of quotations linked to a single code.

Once all the material has been reviewed and coded, themes are explored. Codes related to each other are grouped and organized to form a complete picture of the theme and views that it encases. Individual codes and their linked quotations are reviewed again to verify their linkage to the theme. Throughout the process, the material can still be revised, and codes created and adjusted as new findings arise. Finally, structured themes are written out into results.

The analysis process was carried out with ATLAS.ti software (Scientific Software Development GmbH, v.8.4.4.). Further instructions on doing a thematic analysis with ATLAS.ti have been given by Friese *et al.* (2018).

RESULTS AND DISCUSSION

Both users and providers of ICA systems were interviewed. Based on the interviews, 198 codes were created with an average groundedness of 2.6, indicating from how many quotations the code has been concluded. Further on, codes identified as significant for the research question, were organized into themes consisting of 65 codes with an average of 5.3 quotations linked to each code. This indicates that the selected codes are well grounded on the interview findings. For code lists with their respective number of quotations, the reader is referred to the Supplementary Material. Main themes and sub-themes used in thematic analysis are presented in Table 2.

Perceived barriers and future development were categorized by the source, i.e. the type of interviewee presenting these views. Topics per type of interviewee are presented in Figure 2. For WWTP staff, a variety of barriers for optimizing the process control, operation, and maintenance with data analytics originate from the limitations in the physical systems. Interviewed companies perceived most significant barriers in the digital systems, such as software integration.

Consultants shared a similar view with WWTPs but emphasized more the barriers in maintenance work and

Table 2 | Analysis themes

Main theme	Sub-theme	Definition
Physical systems	Actuators	Process equipment, e.g. pumps, compressors, valves, motors, centrifuges
	Instrumentation	Sensors and analyzers in the process and its surroundings
	Maintenance	Maintenance and condition monitoring of process equipment and sensors
Digital systems	Data analysis	Tasks related to data processing, analysis, and visualization
	Digital tools	Tools used for data collection, processing, analysis, and visualization
Process control		Process control logics and strategies, including use of controllers
Operating environment		Non-technical issues, e.g. personnel, management and purchasing, cooperation with contractors and providers etc.

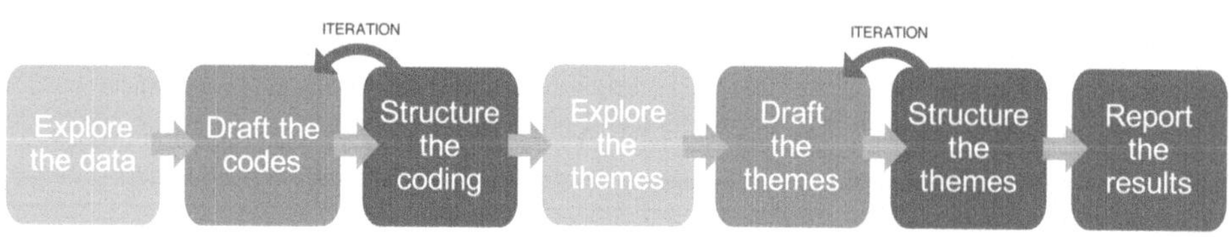

Figure 1 | Thematic analysis process.

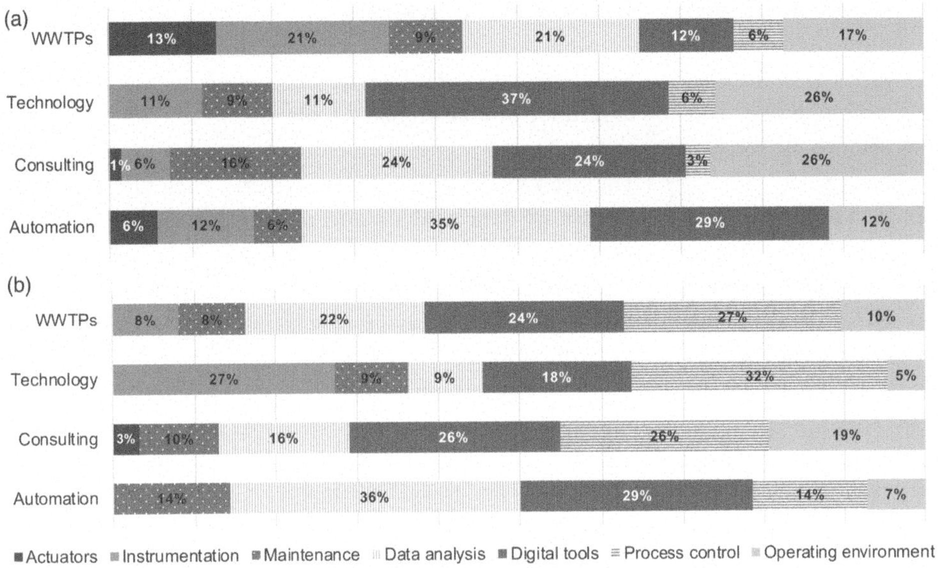

Figure 2 | Division of (a) barriers and (b) future development areas considered by different interviewees.

less the equipment and instrumentation. Technology providers emphasized barriers in digital tools, such as integration of software, while automation contractors highlighted the barriers in actual analysis phase, such as maintaining and updating process models. It is worth noting that the background and professional experience of interviewees clearly guided their answers. This suggests that more active interaction between different professions might fill some gaps and allow more interdisciplinary solutions.

When discussing future development, very few interviewees considered the development of equipment and instrumentation. Most expectations were focused on process control strategies developing towards more predictive and automated control. Maintenance practices were expected to develop mostly along with new digital tools.

Overall themes and their key content were compiled into figures and underlying views described in further detail. Theme figures conclude frequently mentioned views from all interviewees related to the respective theme: descriptions of current practices (now), limitations for more optimal use (barriers), identified development needs (needs), and how they expect the practices in this theme to develop (future). The first theme of physical systems, consisting of the sub-themes of instrumentation, process equipment, and maintenance, is presented in Figure 3.

Theme 1: Physical systems

Key conclusion: limited resources are prioritized for basic operation and maintenance tasks

Maintenance need of instrumentation appeared as the key barrier for several control applications, contributing also to a lack of personnel resources, data quality issues, and unstable control loops. Consequently, the potential benefit from data analytics is reduced. Condition monitoring of sensors and equipment is often based on alarm limits and human senses with some exceptions. For example, sludge

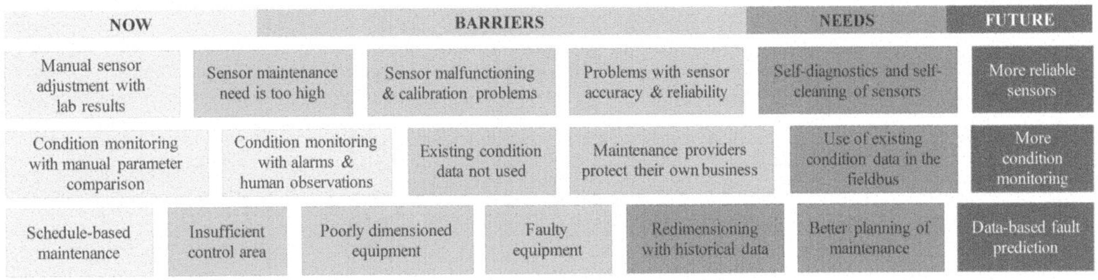

Figure 3 | Theme 1: Physical systems.

dewatering centrifuges commonly have vibration and temperature measurements and some utilities monitor the temperature of critical bearings. Although it is recognized that sensor signals often contain relevant condition data, extracting the information from the fieldbus network is deemed difficult and laborious by the personnel.

Functionalities of instrumentation, such as self-cleaning and self-calibration, have largely developed and, according to the providers, will continue to do so. Intelligent sensors, i.e. sensors with built-in data processing capabilities, have gained some ground at WWTPs, but providers still consider that wider application is hindered by the lack of requirements in the purchasing phase.

Optimizing of control is sometimes also prevented by the physical limits of equipment, e.g. the minimum frequency of aeration compressor is too high for night-time wastewater flow or capacity of polymer dosing equipment is not sufficient for large flows. Poorly dimensioned equipment was found to be more common in smaller WWTPs, in which the actual flowrates is easily derived from the original design values. Some companies considered that re-dimensioning of the equipment based on historical data could often be economically feasible. Calculating the payback time for replacing the equipment, e.g. changing one of the three aeration compressors to a smaller one, could justify the investment.

Currently, some data-driven fault detection and fault prediction for process equipment is offered directly by the technology providers. In these cases, condition data is collected and analyzed by the individual provider. For now, this service has been utilized only for the most critical equipment, e.g. aeration compressors. Fault prediction is expected to further develop in the future, but there is no clear indication yet whether WWTPs will expand the outsourcing of equipment condition data collection and analysis to several different providers, in a distributed manner, or aim to centralize the data collection and analysis to either the automation or the maintenance software.

Theme 2: Digital systems

Key conclusion: data analysis is limited by lack of quality assurance and customized tools

The second theme of digital systems, including the sub-themes of data analysis and digital tools, is presented in Figure 4. There are few tools for automatically monitoring or verifying the data quality, or for easy marking of faulty data. Lack of quality assurance makes data processing excessively laborious, further hindering its daily use for analysis. The automation contractor often designs and updates the automation and reporting dashboards and functions based directly on the client's requests. Involving more process expertise in the design of digital tools for WWTPs was considered beneficial by several interviewees. For now, the involvement of consultants as the source of process expertise is not seen as economically feasible.

Almost all interviewed utilities planned to analyze their process data more in the future, but for most utilities clear targets were not yet defined. The most frequently mentioned goal was monitoring the energy efficiency, for which the current reporting and analysis tools in the automation system were considered as insufficient. Automation companies saw that a lot more analysis could already be done with the current software, but utilities have no time or expertise to adjust the system to their needs. Definition of common key performance indicators (KPIs) for energy efficiency to enable utility benchmarking was considered beneficial by some WWTPs and companies.

An increasing interest was mentioned for identifying gradual changes in equipment performance, such as degradation of pumps or fouling of sensors. There are also development plans for basic statistical tools to distinguish trends in alarm sources and types. Some potential was seen in both predictive and descriptive analytics for overall supervision of the process state and verifying of impacts from operator actions.

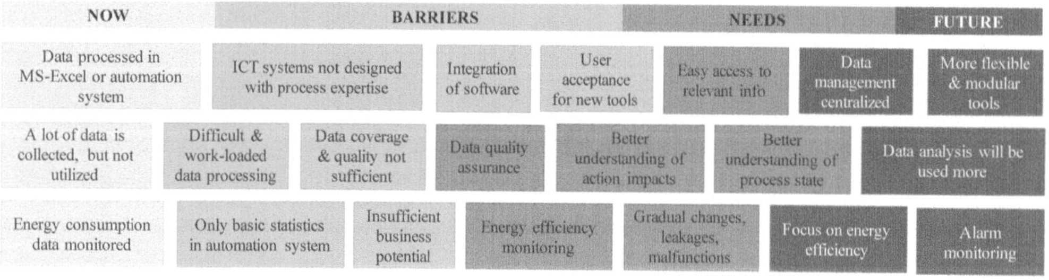

Figure 4 | Theme 2: Digital systems.

In addition to current tools used in the automation software, some utilities are planning to purchase or have already purchased additional third-party software for further data analysis and visualization, not only for use at the plant but also within all units of the company. It is not yet clear how the overall data infrastructure will develop and who will provide the next generation of tools used in overall process monitoring. Automation providers probably have an advantage over competitors to provide simple add-ins to the already familiar software. However, some interviewees deem the viewpoint of automation companies to utility operations as too narrow for tools that would require a thorough understanding of process behavior.

Theme 3: Process control

Key conclusion: process control suffers from instability and requires predictability

Key findings related to process control are presented in Figure 5. Several optimized control strategies that the interviewed WWTPs had tested, e.g. ammonium-based feedforward control of aeration, were eventually disregarded due to the increased need for sensor maintenance or instability of the control loop in varying plant conditions. In some cases, instability could also be partly due to poor controller tuning, an issue that various companies considered as a major barrier for process optimization. Tuning of controllers was considered to be often inadequate in the start-up phase and decayed over the years if the plant had no dedicated staff with sufficient expertise for the task.

Most of the barriers mentioned for applying more optimal control strategies are directly or indirectly influenced by the large variation of wastewater flowrate and quality. Integration of plant operation with sewage network operation was rarely mentioned, with a few exceptions of individual development projects to improve access of plant personnel to network data. Several companies mentioned the detached network operation as a major development barrier. Most WWTPs considered that stabilizing the influent flowrate is prevented by the lack of a dedicated equalization basin, but a few had also considered equalization of the influent flowrate with predictive data analytics.

Despite the strong focus on energy efficiency, optimization of chemicals was also deemed relevant both now and in the future. Many utilities already had some development ideas and plans for reducing chemical consumption by using new parameters for dosage control, e.g. alkalinity, and were either purchasing a new instrument for the purpose or using a linear correlation value from an already measured parameter, i.e. simple soft sensor.

Highly complex or black-box control systems did not seem to attract a lot of interest among practitioners. Instead, operator support systems to help the operator to identify and prevent disturbances or decide suitable control actions, especially in the complex parts of the process like aeration, would be adopted more easily.

A lot of future expectations involved more predictive control. This could indicate some interest in integrating the network and plant operation and moving towards a system-wide control. The level of automation was expected to increase, which was seen to result in operator duties changing in the future.

Theme 4: Operating environment

Key conclusion: new generation and operation model will push demand

Key findings related to the operating environment are presented in Figure 6. Lack of utility personnel time and resources were considered major barriers by all

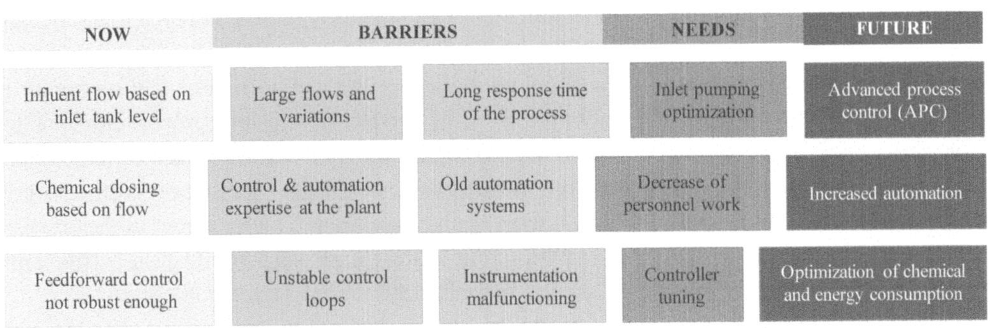

Figure 5 | Theme 3: Process control.

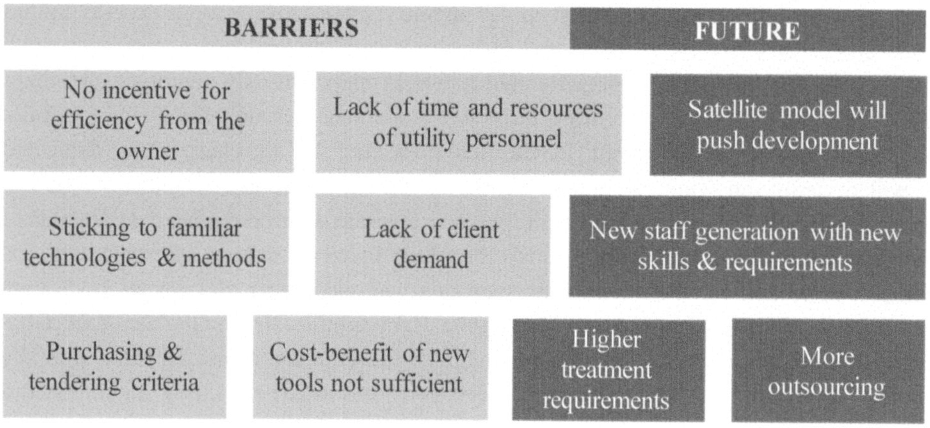

Figure 6 | Theme 4: Operating environment.

interviewees. Interviewed WWTP staff mostly felt that apart from their obligatory daily tasks, they simply do not have enough time for additional process optimization or major development tasks.

Several companies and some WWTP personnel also mentioned the lack of incentives from the owner, i.e. municipality, to improve the cost-efficiency of the plant. Consequently, interviewed technology and service providers saw little business potential in expert services and tools for the sole purpose of process optimization. Low quality criteria in tenders, e.g. in the purchasing of instrumentation, was considered to hinder the application of more reliable solutions. Insufficient emphasis on life-cycle costs, in terms of duration and maintenance need, in the purchasing criteria was also highlighted by the providers.

In addition to cost-efficiency, greenhouse gas emissions and overall environmental impact of WWTPs are receiving increased attention especially in larger utilities. Some WWTPs mentioned that small cost savings of a certain control strategy, which resulted in better performance but required more operator work, would not in all cases be a sufficient motive for implementation. However, the reduction of negative environmental impacts, in addition to the cost savings, was seen to justify the investment of personnel time and resources. These examples were mentioned only by WWTPs with concrete development targets and incentives defined by the company board and management.

All interviewee types shared the view that utility operations will be more centralized and outsourced in the future. A satellite model, i.e. centralizing operation of several small plants to one larger one, was expected to gradually take place and push the development towards more centralized data management, increased process automation, and higher level of outsourcing. These changes were seen to provide further motivation for using computer-based data analysis to quickly extract relevant information from the vast amount of process data, reducing the need for site visits and manual inspections. Another highlighted factor was the new generation of personnel, who are expected to bring in new skills and demand better performance and usability from the digital tools.

Comparison to process industries

In addition to the themes and general views, perceptions of technology and service providers about the differences between municipal WWTPs and process industries as operating environment were analyzed. These are only preliminary findings for the planned future work of assessing and identifying potential data analytics and ICA approaches currently used in other industrial processes. These findings will provide additional perspective for development needs and opportunities of WWTPs. Key considerations are concluded in Figure 7.

Many differences in daily process control, operation, and maintenance of industrial plants compared to municipal WWTPs were seen to originate from fundamental differences between private and public sector as the operating environment. The private purchasing process appears to enable negotiation for a better outcome, such as functionalities or control logics without additional cost. Quality criteria were often considered to be set higher in the process industries with more emphasis on life-cycle costs and reliability, instead of the initial acquisition cost.

In the process industries, incentives for resource efficiency were considered to be better implemented

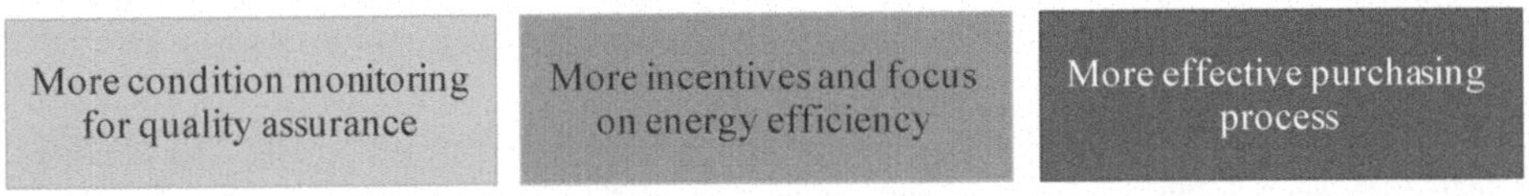

Figure 7 | Key conclusions from comparison of process industries with WWTPs.

throughout the organization. Straight-forward rewarding of each employee, e.g. for energy efficiency, was seen to push the performance towards the set targets. Performance monitoring was more systematic and continuous, e.g. real-time dashboards for energy efficiency KPIs are commonly customized and used for process supervision in the control room.

In addition to resource efficiency, many examples indicated that more effort is made to prevent process failures and the costly production downtime resulting from them. This is reflected in the daily operation in terms of comprehensive quality assurance of sensors, equipment, control loops, and the process in general. The condition of instrumentation is often monitored, e.g. with automatic mass balance checks and sensor self-diagnostics. Control loop performance is automatically monitored and condition-based maintenance, e.g. vibration and temperature monitoring, is widely applied. Advanced process control tools, e.g. fault detection with multivariate statistics, are also applied to identify process disturbances and prevent potential losses of production.

CONCLUSIONS

In this study, several types of stakeholders were interviewed, and a thematic analysis was conducted to assess the barriers and needs of WWTPs to apply data analytics for optimizing the process control, operation and maintenance. Many barriers discussed here are not unique to the wastewater industry, and future work will focus on reviewing solutions applied in process industries for similar challenges.

Results in this study indicate that the perceived potential of data analytics is highly dependent on the performance of underlying physical and digital systems, as well as the control strategies and operating environment of the WWTP. Currently, deficits in data collection and processing negatively impact the perceived cost-benefit of actual data analytics. Identified key barriers include laborious instrumentation maintenance, lack of process expertise in the design of current software, instability of control loops, and insufficient incentives for resource efficiency. Needs can be summarized to higher attention for self-diagnostics of instrumentation, data quality assurance, tuning of controllers, and clear development incentives on both individual and organizational levels. Despite the barriers, many potential applications for data analytics are identified, e.g. easy monitoring of energy efficiency, process disturbances, and process supervision in general. In the future, it is expected that centralization and outsourcing of operations, in addition to the next generation of staff personnel, will push the development towards a higher level of automation and thereby increase the importance of data analytics for WWTPs.

This study was conducted across a group of wastewater industry professionals in Finland. Interviews with a range of stakeholders were seen as an effective method of data collection. Expanding the interviews to a larger variety of stakeholders, e.g. policymakers, education and professional training institutes, or simply a larger variety of employees within the same plants and companies, could provide additional insights to barriers and needs in practice.

Thematic analysis provided a systematic framework for analyzing and summarizing a variety of sector expertise, experiences, and personal opinions. The method could be further utilized to its full potential by studying the user acceptance of a specific software or technology, revealing more factors affecting the perceived usefulness and ease-of-use of ICA and data analytics in practice.

ACKNOWLEDGEMENT

We would like to acknowledge the valuable support of six Finnish water utilities for participating in the steering committee and the funding of this project. In addition, we would like to thank the Water Utilities Development Fund of Finland for funding this project.

DATA AVAILABILITY STATEMENT

All relevant data are included in the paper or its Supplementary Information.

REFERENCES

Åmand, L., Olsson, G. & Carlsson, B. 2013 Aeration control–a review. *Water Science and Technology* **67** (11), 2374–2398.

Bamberger, M. 2000 *Integrating Quantitative and Qualitative Research in Development Projects: Lessons From the Field.* The World Bank, Washington DC, USA.

Bauer, M., Horch, A., Xie, L., Jelali, M. & Thornhill, N. 2016 The current state of control loop performance monitoring–A survey of application in industry. *Journal of Process Control* **38**, 1–10.

Braun, V. & Clarke, V. 2006 Using thematic analysis in psychology. *Qualitative Research in Psychology* **3** (2), 77–101.

Corominas, L., Garrido-Baserba, M., Villez, K., Olsson, G., Cortés, U. & Poch, M. 2018 Transforming data into knowledge for improved wastewater treatment operation: a critical review of techniques. *Environmental Modelling & Software* **106**, 89–103.

Dybå, T., Prikladnicki, R., Rönkkö, K., Seaman, C. & Sillito, J. 2011 Qualitative research in software engineering. *Empirical Software Engineering* **16** (4), 425–429.

Friese, S., Soratto, J. & Pires, D. 2018 Carrying out a computer-aided thematic content analysis with ATLAS.ti. https://pure.mpg.de/rest/items/item_2582914/component/file_2582912/content (accessed 28 April 2020).

Haimi, H., Mulas, M., Sahlstedt, K. & Vahala, R. 2009 *Advanced Operation and Control Methods of Municipal Wastewater Treatment Processes in Finland.* Water and Wastewater Engineering, Helsinki University of Technology, Helsinki, Finland.

Haimi, H., Mulas, M., Corona, F. & Vahala, R. 2013 Data-derived soft-sensors for biological wastewater treatment plants: an overview. *Environmental Modelling & Software* **47**, 88–107.

Ingildsen, P. & Olsson, G. 2016 *Smart Water Utilities: Complexity Made Simple.* IWA Publishing, London, UK.

Jämsä-Jounela, S. 2007 Future trends in process automation. *Annual Reviews in Control* **31** (2), 211–220.

Jeppsson, U., Alex, J., Pons, M., Spanjers, H. & Vanrolleghem, P. A. 2002 Status and future trends of ICA in wastewater treatment–a European perspective. *Water Science and Technology* **45** (4–5), 485–494.

Naman, J. M. & Gibson, J. M. 2015 Disparities in water and sewer services in North Carolina: an analysis of the decision-making process. *American Journal of Public Health* **105** (10), e20–e26.

Newhart, K. B., Holloway, R. W., Hering, A. S. & Cath, T. Y. 2019 Data-driven performance analyses of wastewater treatment plants: a review. *Water Research* **157**, 498–513.

Olsson, G. 2012 ICA and me–a subjective review. *Water Research* **46** (6), 1585–1624.

Olsson, G., Nielsen, M., Yuan, Z., Lynggaard-Jenssen, A. & Steyer, J. 2005 *Instrumentation, Control and Automation in Wastewater Systems.* IWA Publishing, London, UK.

Smuts, J. F. & Hussey, A. 2011 Requirements for successfully implementing and sustaining advanced control applications. In: *Proceedings of the 54th ISA POWID Symposium 2011*, p. 105.

Solon, K., Volcke, E. I., Spérandio, M. & Van Loosdrecht, M. C. 2019 Resource recovery and wastewater treatment modelling. *Environmental Science: Water Research & Technology* **5** (4), 631–642.

Talvitie, J. 2018 *Wastewater Treatment Plants as Pathways of Microlitter to the Aquatic Environment.* PhD thesis, School of Engineering, Aalto University, Finland.

First received 30 April 2020; accepted in revised form 19 June 2020. Available online 6 July 2020

© 2020 The Authors

Making urban water smart: the SMART-WATER solution

Gerasimos Antzoulatos, Christos Mourtzios, Panagiota Stournara, Ioannis-Omiros Kouloglou, Nikolaos Papadimitriou, Dimitrios Spyrou, Alexandros Mentes, Efstathios Nikolaidis, Anastasios Karakostas, Dimitrios Kourtesis, Stefanos Vrochidis and Ioannis Kompatsiaris

ABSTRACT

The rise of Internet of Things (IoT), coupled with the advances in Artificial Intelligence technologies and cloud-based applications, have caused fundamental changes in the way societies behave. Enhanced connectivity and interactions between physical and cyber worlds create 'smart' solutions and applications to serve society's needs. Water is a vital resource and its management is a critical issue. ICT achievements gradually deployed within the water industry provide an alternative, smart and novel way to improve water management efficiently. Contributing to this direction, we propose a unified framework for urban water management, exploiting state-of-the-art IoT solutions for remote telemetry and control of water consumption in combination with machine learning-based processes. The SMART-WATER platform aims to foster water utility companies by enhancing water management and decision-making processes, providing innovative solutions to consumers for smart water utilisation.

Key words | Internet of Things, LPWAN, predictive big data analytics, smart metering, telemetry, water management

Gerasimos Antzoulatos (corresponding author)
Ioannis-Omiros Kouloglou
Efstathios Nikolaidis
Anastasios Karakostas
Stefanos Vrochidis
Ioannis Kompatsiaris
Information Technologies Institute (ITI), Centre for Research and Technology Hellas (CERTH),
Thessaloniki,
Greece
E-mail: *gantzoulatos@iti.gr*

Christos Mourtzios
Nikolaos Papadimitriou
Dimitrios Kourtesis
Research and Development Department,
APIFON S.A. Telecommunications,
Thessaloniki,
Greece

Panagiota Stournara
Dimitrios Spyrou
Alexandros Mentes
Thessaloniki Water Supply and Sewerage Company S.A. (EYATH S.A.),
Tsimiski 98, 54622 Thessaloniki,
Greece

HIGHLIGHTS

- Smart Automated Metering and Remote Control – design and deploy a novel integrated infrastructure that encompasses telemetry and remote-control technologies for the efficient management of the water supply network.
- Device Connectivity and Data Management – develop alternative technologies for the implementation of a fixed wireless network that will support real-time telemetry and remote-control services. The fixed network's performance and reliability are tested and evaluated in real conditions in Thessaloniki's urban environment.
- Data Processing and Visualisation – develop data analysis and visualization tools for complex event processing, identification of patterns behind periodic peak demands, forecasting water consumption and demand, assisting the water utility company to efficient water management and decision-making processes.
- Water Management – create new products for the consumers and the water utility company, assisting the former to build consumption patterns and raise their water consciousness and awareness, and the latter to sustainably manage water, including issues concerning consumption, demand, resources as well as to detect leaks etc.

This is an Open Access article distributed under the terms of the Creative Commons Attribution Licence (CC BY-NC-ND 4.0), which permits copying and redistribution for non-commercial purposes with no derivatives, provided the original work is properly cited (http://creativecommons.org/licenses/by-nc-nd/4.0/).

doi: 10.2166/wst.2020.391

- Feedback of water usage – evaluate the overall infrastructure based on operational efficiency, usability and reliability, and consumer satisfaction.

INTRODUCTION

Factors such as climate change, escalating urban water demand, customer requirements and the integration of new technologies in customers' service have all been increasing the onus on water providers to adopt more effective and sustainable approaches for urban water management. The combination of smart metering and Internet of Things (IoT) allows the development of sophisticated systems to better manage resources and develop new applications to facilitate various parts of society; that is, citizens/consumers, public and private organizations, businesses etc. In the water industry, this combination is the key to develop smart water management systems serving both consumers and water utility companies and fostering sustainability, through water leak detection (Saraswathi *et al.* 2018), river water quality monitoring in real-time (Chowdury *et al.* 2019), water flow monitoring (Kusuma & Anil 2018), short-term water consumption and water demand forecasting (Antunes *et al.* 2018; Benítez *et al.* 2019). Smart water systems are inherently connected with the concept of smart cities, since they comprise part of the city's infrastructure. Future smart cities will face challenges of privacy and cybersecurity that should be taken under consideration to minimize potential risks and strengthen the important role of smart water in urban areas (Moy de Vitry *et al.* 2019).

The smart metering industry is rapidly evolving. The term '*smart*' defines any new systems employing the latest in communication capabilities and enhanced functionalities (Boyle *et al.* 2013). There is a plethora of innovative smart water meters that frequently record and transmit metering data. The absence of an open and widely accepted communication standard in the smart water metering industry led most vendors to introduce proprietary solutions at protocol and platform level (Diehl Stiftung & Co. KG 2020; Sensus 2020) adding deployment complexity and resulting in vendor and protocol 'lock-in' issues (Alvisi *et al.* 2019).

Water management infrastructures typically make use of traditional 'walk-by' and 'drive by' methods, which do not allow frequent data collection and cannot support bidirectional communication (Cole *et al.* 2012). Real time telemetry can only be feasible via a fixed wireless network consisting of data collectors at fixed installation points. Designing and implementing such a network in urban areas is challenging. Short-range technologies like BLE (Gomez *et al.* 2012) and ZigBee (Gislason 2008) can no longer meet the growing demands of modern applications. Alternatives based on cellular networks (2G/3G/4G) provide wider coverage with the expense of cost. Till now, the majority of Measuring Instruments Directive (M.I.D.) certified water meters by top-notch vendors adopt short range protocols, i.e. wM-Bus (EN 13757-4) (Carratù *et al.* 2017), which hinder large scale deployments due to the large number of required gateways. Low Power Wide Area networks (LPWAN) technologies (Sanchez & Cano 2016) provide increased connectivity range (>5 km), low power consumption and low infrastructural costs. The main LPWAN technologies used are: Sigfox (Sigfox 2016), LoRa (Semtech 2015) and NarrowBand-IoT (NB-IoT) (3GPP 2016).

Intelligent water metering effectively generates a big volume of data streams that can be stored into the water utility providers' information systems. However, the need for real-time monitoring as well as for advanced automated reporting tools and predictive analytical processes have emerged. The reliable and accurate short-term water consumption and demand predictions can significantly improve water management, thus bringing significant financial savings to a water utility company. Hence, efficient demand forecasting facilitates on one hand identification of water network failure or water loss and on the other effective scheduling of the utilisation of the energy-related operations for treatment and pumping during the low-cost periods, etc. (Anele *et al.* 2017; Antunes *et al.* 2018; Guancheng & Shuming 2018; Xenochristou *et al.* 2018; Benítez *et al.* 2019; Oyebode & Ighravwe 2019). In the last decades, different linear and nonlinear approaches for short-term water demand forecasting have been proposed in the literature. Two well-known representatives of the first category are the Auto-Regressive Moving Average (ARMA) and Auto-Regressive Integrated Moving Average (ARIMA) approaches, which have been used in water demand time series modeling (Donkor *et al.* 2014; Kofinas *et al.* 2014; Antunes *et al.* 2018; Anele *et al.* 2017). The nonlinear ones aim to fit machine learning techniques to model the intrinsic nonlinearity into the water demand data and predict the time series behaviour, increasing the forecasting accuracy. A variety of studies have explored the application of machine learning methods such as Artificial Neural Networks, Support Vector Machines for Regression (SVR), Random Forests, Deep Neural Networks etc. in short-term water demand forecasting or even hybrid models that combine one or more methods (Kofinas *et al.* 2014; Walker *et al.* 2015; Anele *et al.* 2017; Antunes *et al.* 2018; Guancheng & Shuming 2018; Vijai & Sivakumar 2018; Xenochristou *et al.* 2018; Oyebode & Ighravwe 2019).

In the framework of the SMART-WATER (Mourtzios *et al.* 2019) research project, a unified system consisting of smart water meters coupled with smart water valves and data analysis processes were utilized in the urban area of Thessaloniki, Greece. The proposed innovative combination makes its application an additional challenge.

In the following sections, a brief reference to related projects is exhibited and is followed with a short description of the SMART-WATER project. Then, the methodological framework, as well as the main methods and approaches, are mentioned in detail. Finally, the preliminary results of the project will be discussed in the last section of this work.

RELATED WORK

During the last decade, realizing the need to improve water management resources along with the achievements of evolving IoT technology, significant progress has been made and various projects have come up in the field of water, aiming to address the arising challenges. In the study of Tsavdaridou *et al.* (2012), Automated Meter Reading (AMR) was utilized with positive displacement water meters to detect leakages experimentally in Thessaloniki, in cooperation with the city's water utility company (EYATH S.A.).

In the *IceWater* project, which is presented in Fantozzi *et al.* (2014), focusing on improving energy efficiency and leak detection, ICT solutions were utilized. The possibilities offered by using AMR systems to collect data at the consumers' level were taken into account in Candelieri & Archetti (2014). In this work, an approach for water demand forecasting on an hourly basis was developed, aiming at optimizing operations of the water network in order to reduce energy costs.

Another project, the *iWIDGET* project (Makropoulos *et al.* 2014) proposed two web-based platforms developed mainly for consumers: (a) a web-based platform providing consumers with information for domestic water consumption accessible through the respective platform (Kossieris *et al.* 2014a). This web tool made use of time series charts helping consumers gain deeper insight into their consumption profile and past water consumption data, (b) an educational e-Learning platform, which was presented in Kossieris *et al.* (2014b), providing various applications, aiming to motivate and help consumers improve their water consumption profile. The iWIDGET project aimed at developing a collaboration between consumers and water utilities, exploring huge data sets from smart water meters (Ribeiro *et al.* 2015).

In the *WATERNOMICS* project, a system infrastructure and platform were developed using ICT technologies for water resource management aiming at increasing water efficiency. It proposed three (3) water consumption levels, namely domestic, corporate and municipal. The platform was the main tool to achieve improvements in water consumption behaviour (Perfido *et al.* 2015; Kouroupetroglou *et al.* 2015). Another smart water metering approach is presented in Mudumbe & Abu-Mahfouz 2015, including a web-based visualization tool for real-time and historical water consumption data. In Singapore, a Smart Water Grid is intended by its Public Utility Board for real-time monitoring of pressure and water quality throughout the network as well as for real-time water consumption with the use of Automated Meter Reading (AMR) (PUB 2016). Furthermore, the *SmartH2O* platform was developed (Novak *et al.* 2018; Rizzoli *et al.* 2018) aiming at water consumption reduction through behavioural change in consumers' water consumption. It was based on the combination of smart metering data and visualization tools contributing to creating feedback between the consumers and the water utility sector.

SMART-WATER PROJECT

SMART-WATER is an ongoing pilot project that puts forward a novel design of an infrastructure that utilizes modern telemetry and remote-control technologies to provide innovative services to consumers and water utility companies. Custom-made gateways constitute the backbone of a fixed wireless and multiprotocol network that makes real time telemetry and remote control viable. The gateway functions overcome 'lock-in' restrictions since they can interface with any wM-Bus and LoRa sensors. SMART-WATER's platform can decrypt and manage end device data from water valves and a wide range of different vendor water meters. State-of-the-art data analytics tools and techniques from the fields of statistics and Machine Learning are applied to predict consumption and identify patterns behind periodic peak demands. Finally, two web-based platform applications are developed to provide end users, consumers and water utility companies with real time telemetry and remote-control services. Specifically, the goals of the SMART-WATER project can be summarised according to the five directions, as illustrated in the Figure 1.

These goals have been contextualized in a five-layer unified generic architecture (Figure 2). The proposed platform supports the pipeline process from automatic metering of water consumption to the water management and decision support process by the water utility company as well as by the consumers.

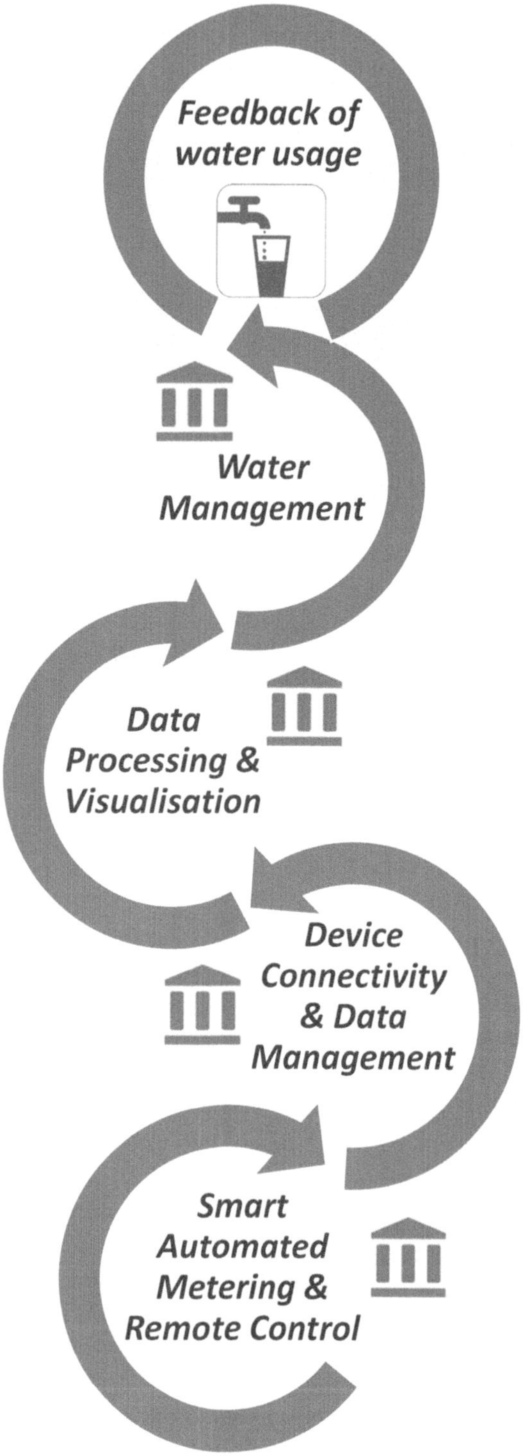

Figure 1 | SMART-WATER five main goals.

End Device Layer: The primary goals of SMART-WATER project are the frequent recording and wireless transmission of water consumption (smart automated telemetry) data combined with remote control of water supply in almost real time.

Device Connectivity Layer: This layer is in charge of the bidirectional connection between end devices and the central server infrastructure using multiple wireless telecommunication protocols.

Device and Data Management Layer: At the 3rd level of the network architecture, the reception, decoding, decryption and storage of the data transmitted by the water meters, as well as the management of the actuation commands to the valves, take place.

Data Processing and Analytics Layer: In this level, the storage and processing of the metering data is realised. This layer employs state-of-the-art machine learning techniques to perform predictive analysis on water consumption and demand, identify patterns and trends in time series data, and detect abnormal water consumption behaviour of consumers.

User Interaction Layer: The system's end-users are water utility company authorized personnel and consumers. Two different User Interfaces (U.I.) were designed and developed covering each group's needs for real time water consumption monitoring and water supply control.

The main contribution of this work is to focus on the backbone aspects of the SMART-WATER project, by illustrating the proposed approaches, in each one of the above layers, which are able to tackle the aforementioned limitations and challenges. Specifically, in this work, a holistic framework is proposed that encapsulates innovative smart metering and remote-control services in real time, designs and deploys a novel fixed wireless network to connect the smart devices with the central infrastructure and develops state-of-the-art methods to manage and analyse the obtained data, aiming to assist the water utility company to achieve efficient water management and provide high quality services to consumers.

METHODS

In this section, the methods, tools and techniques that were designed, implemented and applied in the field through the whole life span of the project are exhibited thoroughly, by following the ordering of the architecture layers.

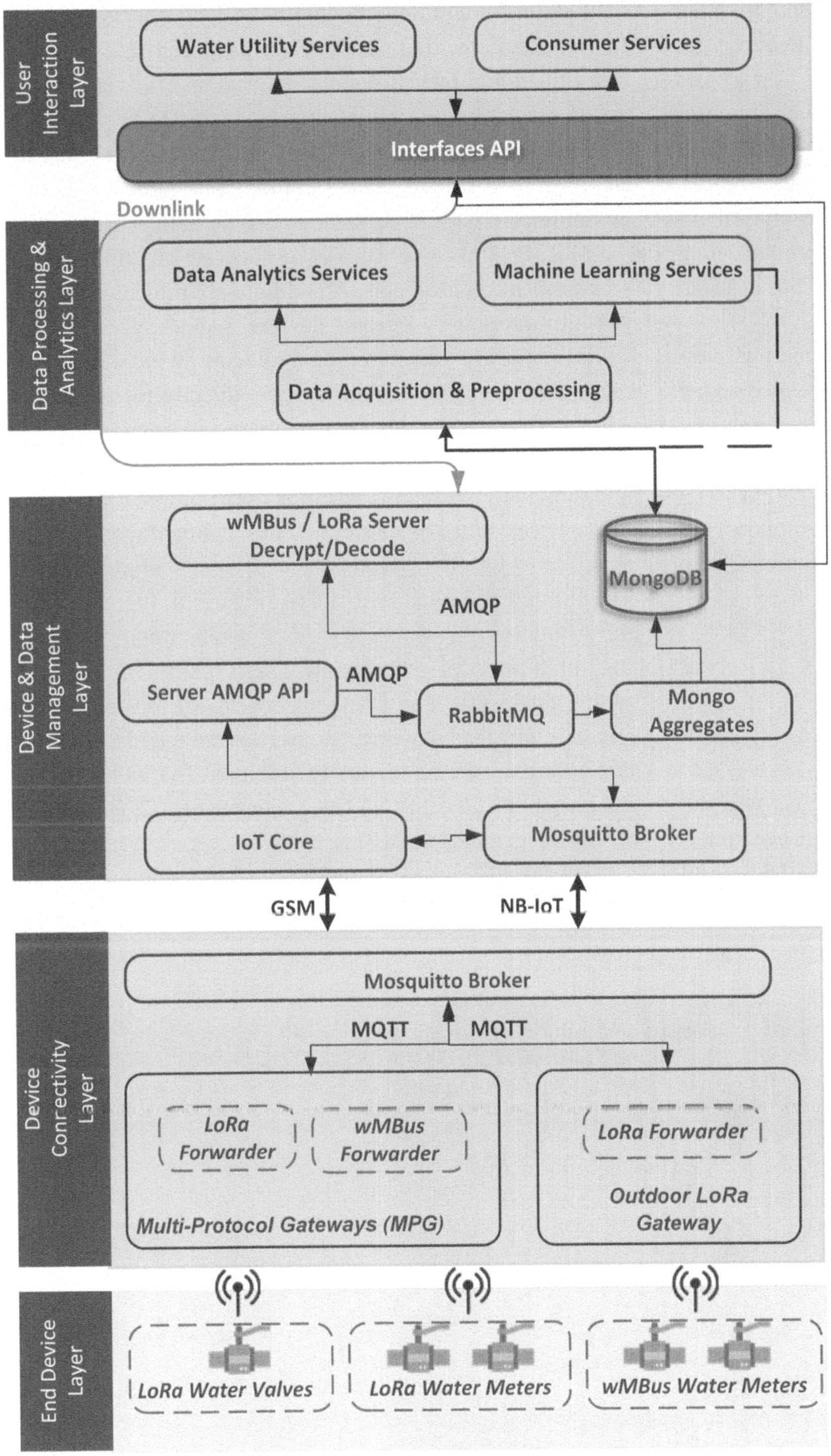

Figure 2 | SMART-WATER five layered architecture.

End device layer

The Sensors layer is the source of data to the SMART-WATER system infrastructure. Smart water meters consist a novel metering technology. Built-in communication systems enable efficient remote reading while sophisticated software alerts effectively for leak detection, bursts or other irregularities such as tampering attempts and reverse flows. The prerequisite features they should demonstrate to constitute the building blocks of a water management infrastructure are M.I.D. certification, high metering accuracy (ultrasound or electromagnetic), wirelessly transmitting metering data at dense regular intervals and ultra-low consumption (10+ years autonomy). Based on the aforementioned features, water meters from five different vendors were chosen for the pilot testing in order to assess their metering and telecommunication performance. Water meters use wM-Bus or LoRa telecommunication protocols.

The second category of end devices are the smart wireless water valves, which make it possible to control water supply and comprise one of the main innovative features of the proposed system. The wireless valves should exhibit certain features: remotely controlled via a wireless and bidirectional telecommunication protocol, ultra-low consumption (10 years autonomy) and tamper alert. Telecommunications-wise, water valves use LoRaWAN protocol.

Device connectivity layer

Multi-Protocol gateways (MPG)

An important challenge tackled during the project was the lack of any commercially available gateway that can simultaneously communicate bidirectionally with wM-Bus and LoRa end devices and at the same time establish bidirectional communication with the central infrastructure. In the literature, there have been proposed IoT gateways that utilize multiple protocols. Amiruddin *et al.* (2018) proposed a gateway that could interface with Zigbee *και* BLE sensors while Kaur & Singh 2013 and Gunasagarana *et al.* 2015 proposed similar telecommunication nodes that could communicate with RF, Zigbee and Bluetooth sensors.

Water meters and water valves operate on different telecommunication protocols. An additional problem is that the two protocols used by end devices, namely wM-Bus and LoRaWAN, are not internet protocols, which does not allow end devices to communicate directly to the server. All the above issues led us to design and manufacture an electronic device that: (a) communicates bidirectionally and simultaneously with wM-Bus and LoRaWAN technology devices and (b) communicates bidirectionally via at least one internet protocol with the central server. The Printed Circuit Board (PCB) (Figure 3), together with-the electronic components (modules/boards), comprise the multi-protocol gateway, which is placed in an IP55 plastic enclosure. The breakout boards are placed on the female pinheads, allowing the user to easily add-drop them depending on the connectivity scenario. The gateway uses Raspberry Pi Zero W (1 GHz, single-core CPU, 512MB RAM) as its main processing unit.

The Multi-Protocol Gateway performs numerous functions simultaneously. Firstly, it collects wM-Bus telegrams

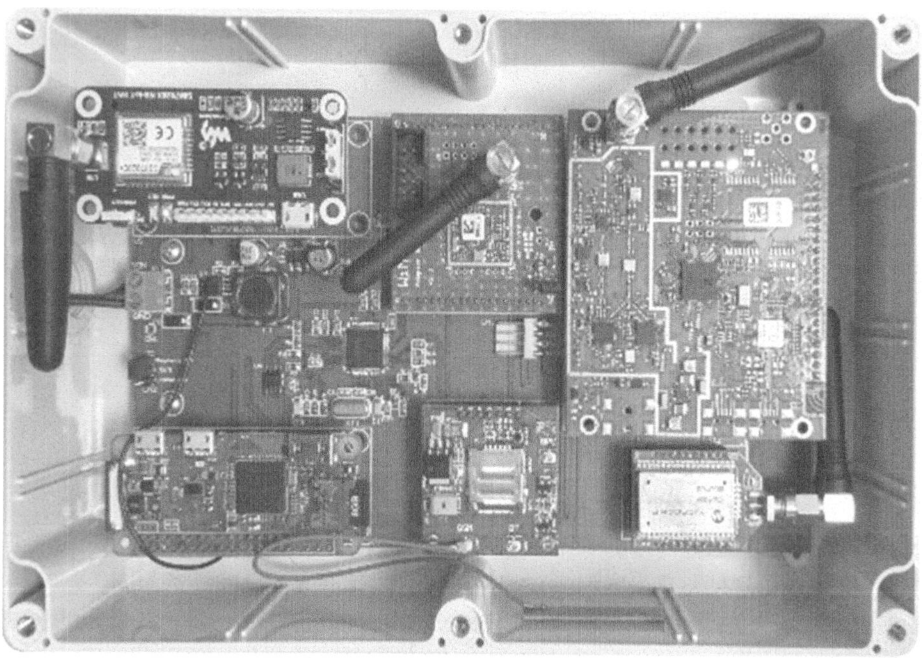

Figure 3 | PCB with all modules attached on female pinheads. Dimensions: 22 cm(L) × 14 cm (W) × 5 cm(H).

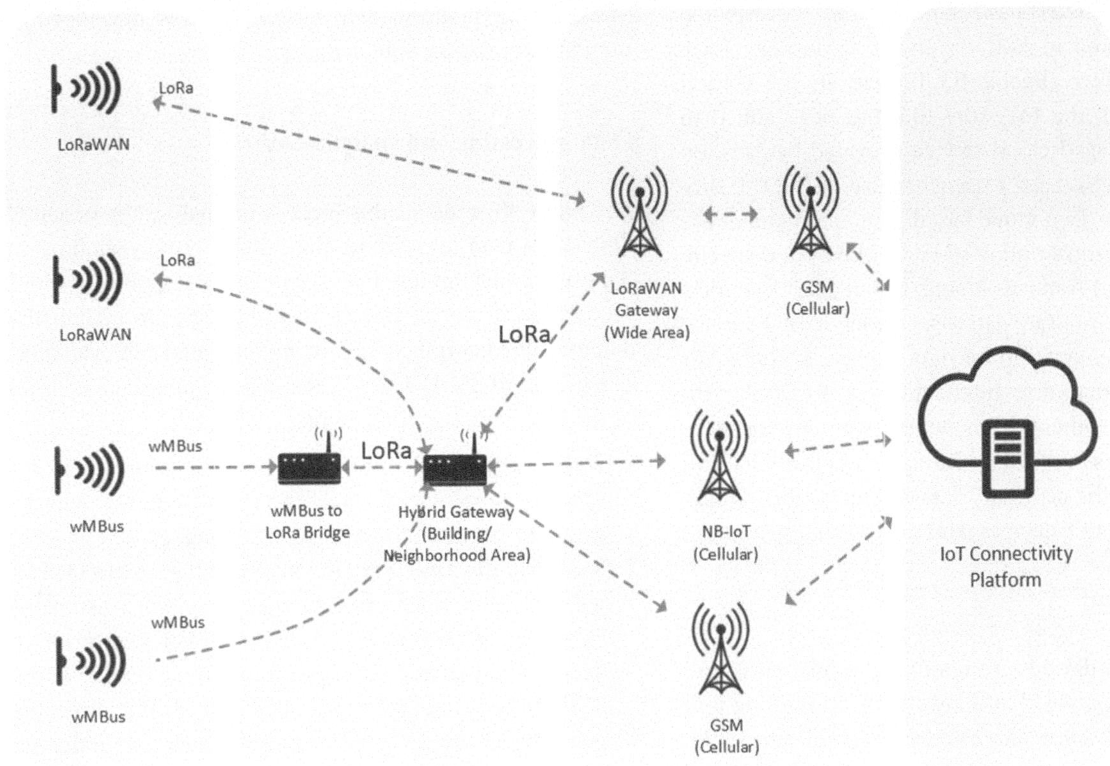

Figure 4 | Fixed network's connectivity scenarios and gateway types.

(T, S, C modes) transmitted by wM-Bus water meters within range. Secondly, it communicates bidirectionally with LoRaWAN end devices. To overcome the issue that the telecommunications protocols used by end devices are not IP protocols, a GSM module was integrated in each gateway to allow two-way communication with the server. The gateways use NB-IoT network as a secondary two-way protocol in premises where there is limited GSM coverage or as a fail-over solution in cases where GSM outage take place.

Hybrid fixed network

One of the most challenging tasks of the SMART-WATER project was the design and implementation of the bidirectional and wireless fixed network, which is necessary for real time telemetry and remote-control services. The dense urban environment and reinforced structural materials in Greek cities, combined with the limitations of RF technologies, made network design and deployment more complicated. Figure 4 depicts the architecture of the fixed wireless network, realized through different connectivity scenarios. SMART-WATER's hybrid network makes use of multiple telecommunication protocols, both legacy (wM-Bus, GSM) and LPWAN (LoRa, NB-IoT), employing different gateway types; for example, off-the-shelf outdoor LoRaWAN Gateways, wM-Bus-to-LoRa Bridges, LoRa repeaters (the last two are realized through MPG), to ensure the optimal connectivity conditions and minimum packet loss.

The gateway's installation points need to be carefully selected in terms of connectivity requirements and cost constraints. In this direction, a custom 'tester' device was designed and assembled in order to perform on site connectivity tests, which defined the optimal gateway installation point inside consumer's premises and verified the connectivity scenarios described in Figure 4.

Device and data management layer

At the 3rd level of the proposed system architecture, the reception, decoding, decryption and storage of metering data, as well as the management of the downlink commands, take place. The server unit can simultaneously process and decode/decrypt payload data from water meters and water valves. This level hosts three main services: (a) end device management, (b) gateway monitoring and (c) LoRa and wM-Bus servers.

In the proposed architecture (Figure 2), the distributed layers from end devices to UIs are depicted. An MQTT

Broker (Mosquito Broker) has been used in each distributed unit of the system and an AMQP (Advanced Message Queuing Protocol) Broker (RabbitMQ Broker) in the central server. In addition, the IoT Core tool has been added to the system, a managed cloud service provided by Amazon Web Services (AWS), used to synchronize all MQTT Brokers in the system. Two cloud-based Linux computers are used through Amazon's online services (AWS) and specifically through the Elastic Compute Cloud (EC2) service. The server uses a noSQL database (MongoDB) both for end device management and for data storage. Server's API (Application Programming Interface) is a RESTful API, with token-based authentication which was developed to support consumer's and administrator's applications and provide them with the necessary data and metadata originating from the network's devices (end devices and gateways).

Device management

Given that the end devices are remotely installed, it is crucial to develop a system that allows their monitoring and registration. The *'Device Management System'* allows the network administrator to register new end devices with all necessary information (e.g. 'Device id', 'Customer id', 'decryption keys', 'installation address', 'coordinates').

Gateway monitoring

MPGs are remotely controlled and monitored. Virtual Private Network (VPN) technology is used for remote management and monitoring of all gateways. The user can perform routine debugging actions, software updates, system monitoring etc. The gateways additionally transmit frequent 'heartbeat' uplink messages to the server through the packet forwarder services.

Lora server/ WM-Bus server

When an uplink message from a LoRa end device reaches the central server, it is directed to the *'LoRaWAN server'* as shown in Figure 2. The LoRa server is an open source application (ChirpStack 2017) developed according to the LoRaWAN protocol in the Go programming language. *'wMbus Server'* is a custom application developed for SMART-WATER. wM-Bus messages are directly processed by the AMQP API service and then transferred to the wM-Bus server in order to be decrypted and decoded. Each wM-Bus end device has a unique decryption key stored in a database upon subscription through the Device Management platform. wM-Bus server was developed in the Python programming language.

Data processing and analytics layer

The main objective of this layer is to analyse the acquired time-series data in order to make short-term predictions of the water consumption in a specific urban region of interest. To achieve this goal, various methodologies from statistics and machine learning fields are implemented and evaluated. The proposed SMART-WATER analytical approach consists of the following seven (7) steps:

1. **Data acquisition**: the water consumption data for specific time period and region of interest will be acquired from the SMART-WATER data storage.
2. **Visualise the Time Series**: visualising the insight characteristics in the time series, such as the trend, seasonality, existence of outliers etc., is an essential step prior to deeper analysis and building a more refined model. The SMART-WATER dashboard aims towards this direction, providing filtering, searching and zooming functionalities to easily explore time-series data.
3. **Preprocessing:** Depending on the dataset, the steps of preprocessing will be defined, including the creation of timestamps, the detection and elimination of outliers, the stationary test, the normalisation process etc.
   i ***Outlier detection and removal***: Outliers are observations that are very different from the majority of the observations in the time series (Hyndman & Athanasopoulos 2018). They may be errors, due to hardware issues, sometimes extremely high values of water consumption appear, or they may simply be unusual. These values should be removed because the majority of analytical techniques are not able to deal with them efficiently. A common and simple way to detect outliers is using the z-score method.
   ii ***Test for Stationary Time Series***: A well-known test for stationarity is the *Augmented Dickey Fuller Test (ADF)* which is a form of Hypothesis Testing. The null hypothesis of the ADF test is that the time series is non-stationary. If the *ADF Test Statistic* is less than the *Critical Value*, the null hypothesis should be rejected, concluding that the series is stationary and it is expected with a very high probability that its previous behaviour will be continued in the future
   iii ***Normalization***: Data normalization involves converting time series data into a given range. It is needed for the application of machine learning models,

Neural Networks. A common technique is the *Min-Max Normalization*, in which the original data are transformed linearly.

4. **Choose the input data features**: a crucial step for the quality of the forecasting process is the right choice of input features, especially during the creation of a model. Each feature represents an input variable that highly affects the outcome of the forecast (Benitez *et al.* 2019). In the SMART-WATER framework, it is considered that previous water consumption observations affect the short-term water demand forecast. Although all the range of data has to be utilised during the training phase, only a small portion has a direct influence on each step of the forecasting process. Thus, the predicted water consumption for the timestamp T + 1, *WCF(T + 1)*, (e.g. day, hour etc.) can be predicted by employing the past observations $O_{T-n}, \ldots, O_T$ in the timestamps (*T-n, …, T*), namely $WCF(T+1) = M(O_T, O_{T-1}, \ldots, O_{T-n})$, where M is the model.
5. **Training phase**: The main goal of this step is the creation of a forecasting model that will enable correct prediction of the future water consumption. Thus, the appropriate method should be chosen along with splitting the time-series dataset for training and testing purposes.
   i *Develop base-line approach*: create/train ARIMA and SARIMA models; namely, the appropriate models' parameters (p, d, q, P, D, Q, m) should be determined.
   ii *Develop machine learning approaches*: create/train machine learning models such as Support Vector Machines (SVMs), Artificial Neural Networks (ANNs), Long Short-Term Memory (LSTM) Neural Networks.
6. **Testing phase**: The trained model is evaluated in terms of its ability to correctly predict future water consumption. The testing set is employed and its performance is estimated using suitable metrics such as Mean Absolute Percentage Error (MAPE), Root Mean Square Error (RMSE) and Coefficient of determination (R^2).
7. **Compare the results and choose the most accurate model**: the best model is chosen in terms of the performance indicators over the testing set. It should be noted that better forecasting models are extracted as the *MAPE* and *RMSE* values tend to zero (0) and the R^2 values approach unity (1).

The proposed schema is easily customised and self-learning as it has been designed and developed to be applicable both at the individual consumers' level as well as for clusters of consumers with similar water consumption patterns. Furthermore, it is completely data-driven and independent of the machine learning model employed for forecasting. In this work, the ARIMA and LSTM methods are employed and their forecasting performance evaluated. A brief mathematical formalization of these methods is presented as follows.

Auto-Regressive Integrated with Moving Average Model (***ARIMA***) is an efficient approach for short-term forecasting (Hyndman & Athanasopoulos 2018). Usually, it denotes as ***ARIMA(p, d, q)*** model, where *p* is the order of Autoregression (AR) terms (number of time-step lags to be used), which represents the relation between past and current observations, *d* is the order of differencing for attaining stationarity (integrated) and *q* is the order of Moving Average (MA) terms, which means the size of the MA window. In the ARIMA model, the forecast value is a linear combination of past observations (linear combination lags of *y* up to *p*) and past errors (linear combination of lagged forecast errors, up to *q*), expressed as follows:

$$y_t = c + \varphi_1 y_{t-1} + \ldots + \varphi_p y_{t-p} + \theta_1 \varepsilon_{t-1} + \ldots + \theta_q \varepsilon_{t-q} + \varepsilon_t$$

For the estimation of the *p*, *d* and *q* parameters, an iterative strategy is applied consisting of the identification step, where a sub-class of models that might better represent the data is chosen, the estimation step, where the parameters of the model are trained using the historical data and finally the diagnostic checking, where the trained model is validated using a disposal dataset.

LSTM stands for **Long Short-Term Memory**; Neural Network is an enhancement of the Recurrent Neural Network (RNN) in relation to their ability to capture the long-term dependencies (Hochreiter & Schmidhuber 1997). Specifically, RNNs use previous information in the present task allowing historical information to be stored in the network's internal state and mapping them to the final output. RNNs suffer from short term memory due to the vanishing gradient problem (Masum *et al.* 2018; Hua *et al.* 2019). Alternatively, the LSTMs overcome this limitation as they are equipped with unique 'Gates' enabled to avoid the long terms for vanishing (Figure 5).

Particularly, an LSTM memory block is a core component that encompasses a memory cell and gates, namely an Input Gate, an Output Gate and a Forget Gate. It computes the mapping from an input sequence $\boldsymbol{x} = (x_1, \ldots, x_t)$ to an output sequence $\boldsymbol{y} = (y_1, \ldots, y_t)$. The following steps are executed inside the memory cell:

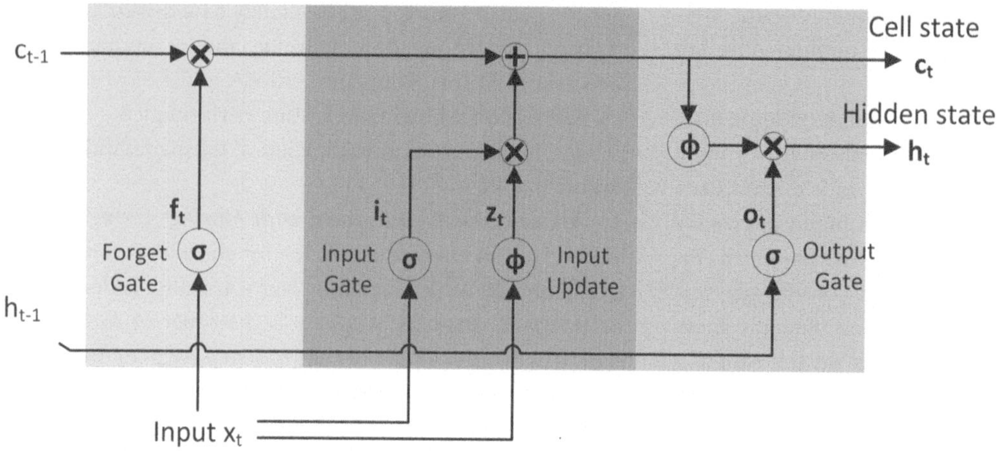

Figure 5 | An illustration of LSTM memory block.

1. In the *Forget Gate*, the input sequence from the current input x_t, is combined with the previous hidden state h_{t-1}, through the sigmoid function $\sigma(x) = \frac{1}{1+e^{-x}}$, which returns values in the range between 0 and 1. Hence,

$$f_t = \sigma(W_f x_t + U_f h_{t-1} + b_f)$$

The closer to 0 means to forget, and the closer to 1 means to keep. In this way, the Forget Gate assists in removing information from the cell state.

2. The *Input Gate* aims to store new information into the memory cell state. Initially, a sigmoid layer determines what will be updated via the formula:

$$i_t = \sigma(W_i x_t + U_i h_{t-1} + b_i)$$

Secondly, the input will be updated by employing the tanh layer, by creating a vector of new candidate values z_t that can be added to the memory cell state, as stated in the following equation:

$$z_t = \varphi(W_z x_t + U_z h_{t-1} + b_z)$$

where $\varphi(\cdot)$ denotes the tanh function, $\varphi(x) = 2\sigma(2x) - 1$. The values fluctuate between -1 and 1.

3. The new state of the memory cell, c_t, should be updated. The estimated values, f_t, i_t, z_t, are fused along with the old memory cell state c_{t-1}. Specifically:

$$c_t = i_t \times z_t + f_t \times c_{t-1}$$

4. The output of the memory cell is realised by element-wise multiplication between the value obtained from a tanh function of c_t and the output of a sigmoid layer, i.e., the Output Gate activation vector o_t, as follows:

$$\begin{cases} o_t = \sigma(W_o x_t + U_o h_{t-1} + V_o c_t + b_o) \\ h_t = o_t \times tanh\,(c_t) \end{cases}$$

The weight matrices W_f, W_i, W_z, W_o, U_f, U_i, U_z, U_o and V_o along with the bias vectors b_f, b_i, b_z, b_o can be learnt in the training stage estimated. Through the cooperation between the memory cell and the gates, LSTM is endowed with a powerful ability to predict time series with long-term dependences (Hochreiter & Schmidhuber 1997; Masum *et al.* 2018; Hua *et al.* 2019).

User interaction layer

Water utility company (administrator) Web application

The ***Main Dashboard*** (Figure 6) is the cornerstone component of the Water Utility Company Web Application, as it encompasses crucial functionalities enabling the Water Utility organisation to improve its daily operations. Therefore, end devices are mapped based on their geographical location. Furthermore, it enables the operators to gain a general glance and overview of the water consumption by region of interest as well as monitor hourly water consumption in near 'real-time'.

Figure 6 | Water Utility Company Web Application dashboards: Main Dashboard.

The ***Single Consumer Dashboard*** (Figure 7) enables the practitioners to monitor the water consumption of a specific consumer in a particular time horizon, to compare his/her consumption with personal past or average consumptions of the consumer's region. The interactive visualisations fuel the operators with functionalities to navigate in specific time periods, aggregate and illustrate time series data hourly, every six hours, daily and monthly.

The ***Alert Dashboard*** (Figure 8) notifies in a concise way the alerts that are received from the end devices. The alerts are categorized into 14 classes, each one indicating a specific malfunction of the end device.

Consumer's Web application

The consumer's application is a web based and responsive application. After logging in (Figure 9), the consumer is

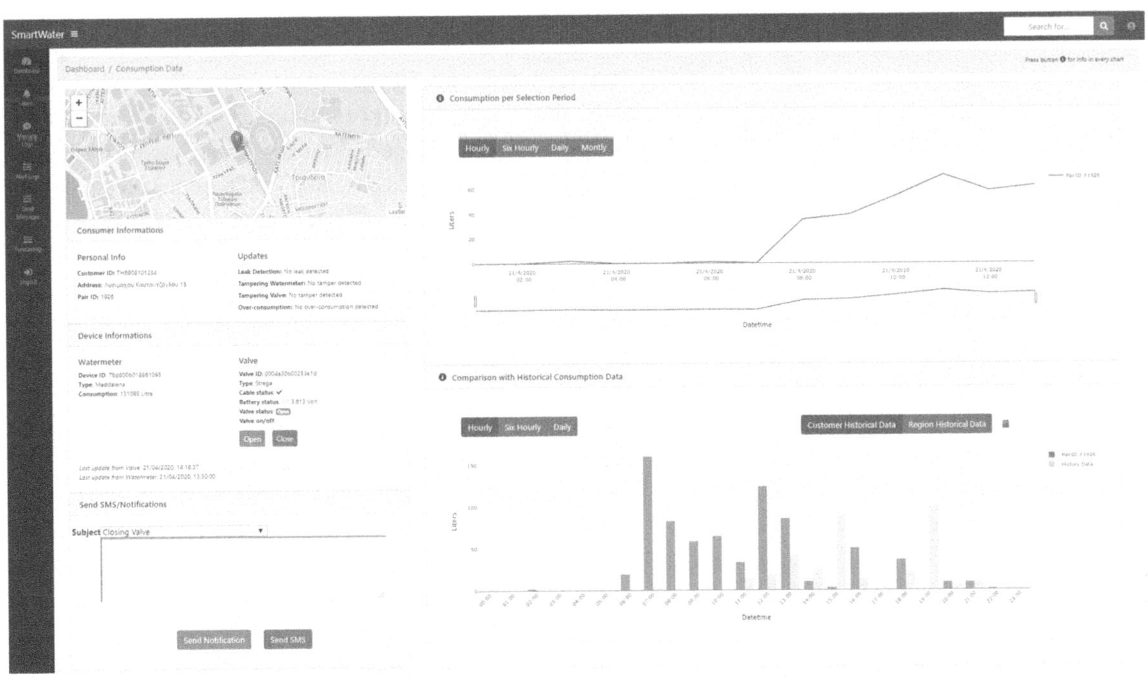

Figure 7 | Water Utility Company Web Application dashboards: Single Consumer Dashboard.

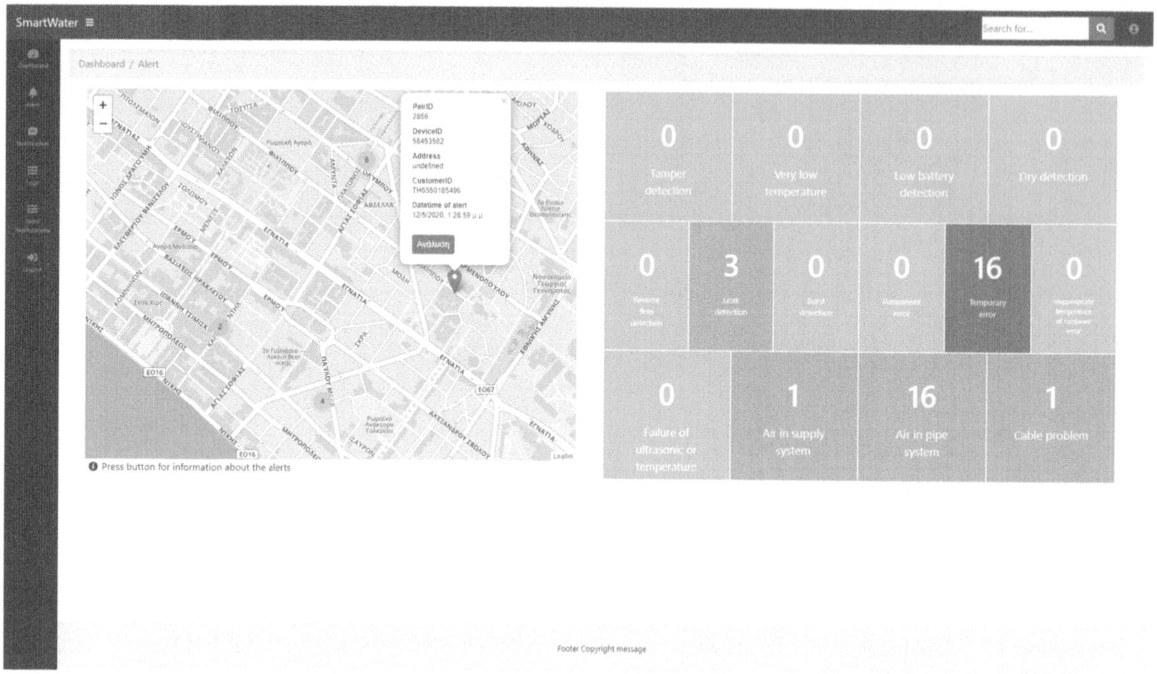

Figure 8 | Water Utility Company Web Application dashboards: Alert Dashboard functions and utilities.

directed to the consumer ***Main Homepage*** (Figure 10) in case he/she owns more than one pair of end devices. Otherwise, he/she is directed to the ***Main Dashboard*** (Figure 11) where he/she can track water consumption by hour/day/week/month and actuate on water valves. Information about former bills is also available and a prediction based on current consumption, for the next billing is also depicted (Figure 10).

RESULTS

This section summarizes findings based on pilot tests in real conditions. Having installed the majority of end devices (75 water meters and valves out of 100) and MPGs, useful conclusions were drawn regarding SMART-WATER system's infrastructure. The system underwent rigorous stress tests to validate its stability across all architecture layers and the ability to serve multiple users simultaneously. Furthermore, the data analysis processes for short-term forecasting of water consumption in the region of interest were assessed in terms of their accuracy.

Time performance

The server infrastructure was initially assessed in terms of its API performance and secondly for the latency of data acquisition. In the SMART-WATER case, users are consumers and water utility company's administrators who access the respective web applications. It is of prime importance to design and develop a reliable and robust server system which can respond to multiple user calls simultaneously. In that direction, the server's API was tested using performance tool Apache Jmeter (https://jmeter.apache.org/). Several critical scenarios were simulated in order to extract *Response Time* (ms) (amount of time needed to complete a call, Wescott 2013) and *Throughput* (k*B*/s) (number of calls per second, Wescott 2013) as key performance metrics for the API. The API's performance was assessed in different cases realizing 12 scenarios. The scenarios entail both GET and POST requests spanning from simple 'Login' requests to more demanding ones. For example, Scenario 10 describes the case of 200 (virtual) users doing consecutive calls (6 GET calls) for 200 seconds in the Consumers' application home page. The description of the scenarios, including their specific features as well as the response times and throughput simulated results per scenario, are listed in Table 1.

It is evident that as the number of users increase, a decrease in throughput is observed as expected due to the additional load created. Scenarios 1, 2 and 3 exhibit the lowest response time (<100 ms) while Scenario 11 the highest one exceeding 1.8 sec. Particularly, it represents the most challenging test case, which is also reflected in the lowest throughput (53 kB/s). Scenario 1 exhibits the highest

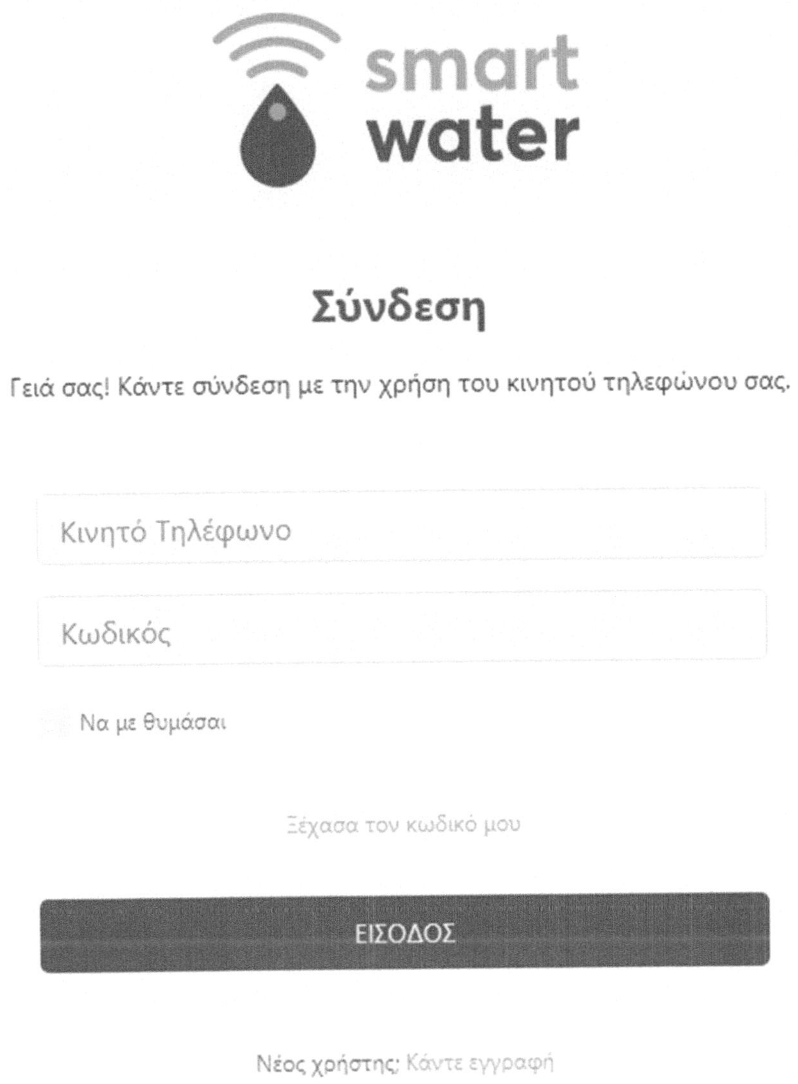

Figure 9 | SMART-WATER Consumer's web application: Log-in page.

simulated throughput reaching 103 kB/s (Figure 12). The results are more than acceptable, since scenarios describe challenging cases hardly possible to occur with real users which means that the API server is designed to serve numerous users at the same time without any significant delay.

Secondly, data acquisition times were recorded to evaluate the system's processing speed and traffic load management. The data acquisition pipeline was divided into three fundamental stages. Each stage is assigned to a respective time slot, namely '***Gateway to Server***' stage (time for a message to reach the server upon reception at the gateway), '***Processing***' stage (time for decoding, decryption, authentication processes after the message reaches the server) and '***Persist***' stage (time for data aggregation). The total statistical sample was 3,000 received uplink messages from water meters installed in consumer's premises. The total simulation time was 2.8 days. The time data points of the three stages are presented in respective box plots (Figure 13). For the '*Gateway to server*' plot, the median time (50% percentile) value is 0.87 s (orange line in the box) while the inter-quartile range (IQR) of the distribution extends from 0.66 s (percentile 25%) to 0.99 s (percentile 75%). For the second stage of the data acquisition pipeline, '*Processing*', the respective values are 0.00312 s (percentile 25%), 0.00319 s (median), and 0.00329 s (percentile 75%). Finally, the statistics for the '*Persist*' time samples are 0.0334 s (percentile 25%), 0.034 s (percentile 50%), and 0.036 s (percentile 75%).

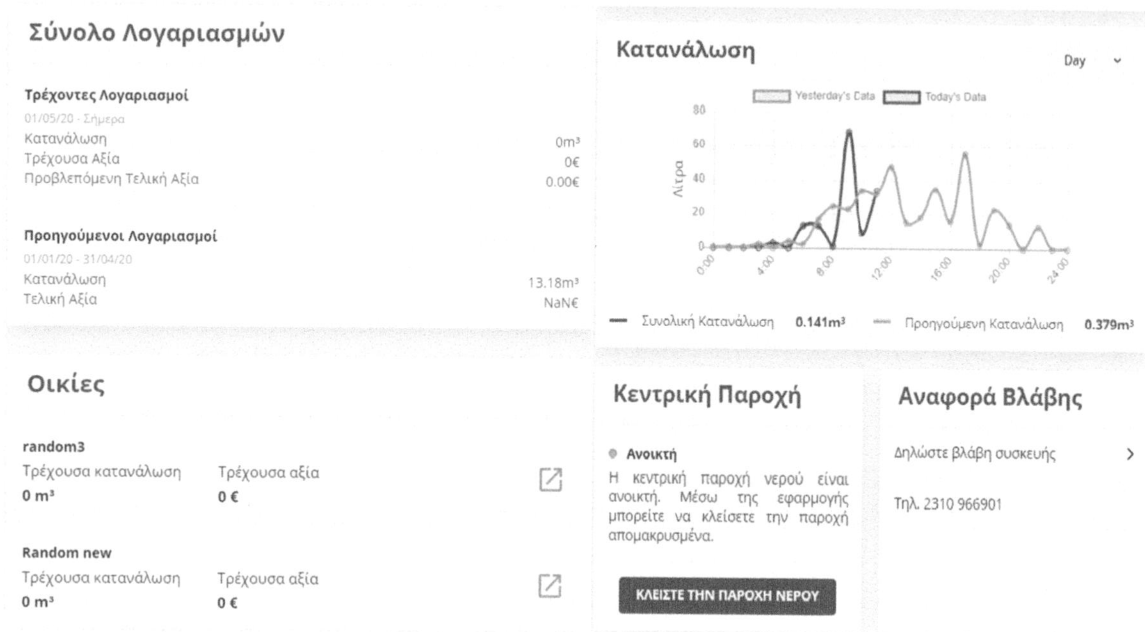

Figure 10 | SMART-WATER Consumer's web application: Main Homepage.

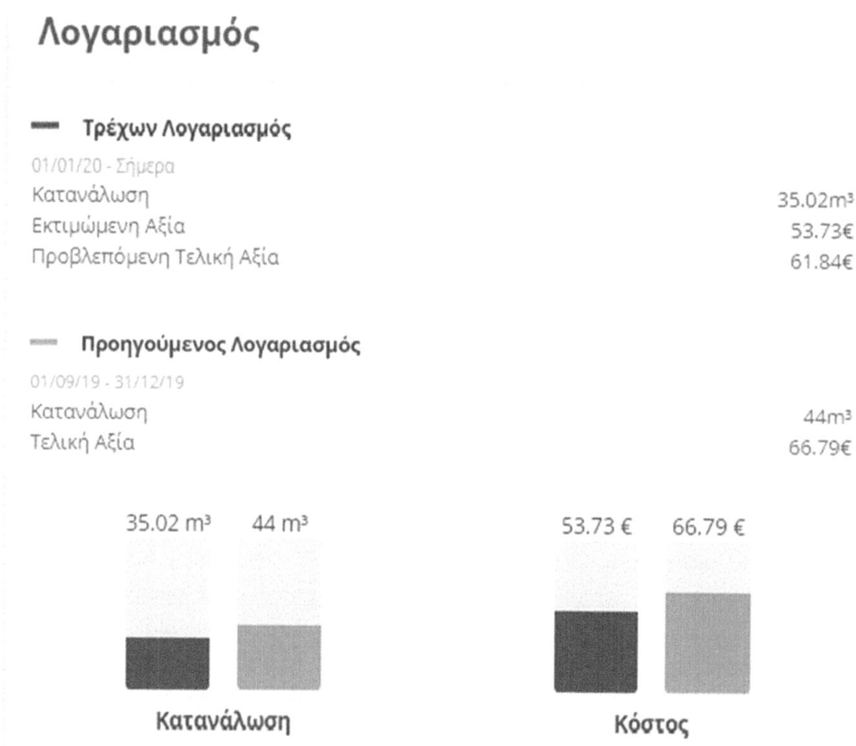

Figure 11 | SMART-WATER Consumer's web application: Main Dashboard.

The longest timeslot among the three stages is the '*Gateway to server*' case, which also exhibits the greatest deviation. Over-the-internet transmission introduces the longest time delay, due to the limited connectivity conditions at the gateways' installation points. The remarkable fluctuation in time values is related to RF

Table 1 | Description of the critical scenarios and API's performance in terms of Response time (ms) and Throughput (kB/s) using the Apache Jmeter performance tool

Scenario No.	Scenario Description	GET/POST	No. Users	Duration (sec.)	Response time (ms)	Throughput (kB/s)
Scenario 1	Consumer App home page	6 GET calls	10	60	96	103
Scenario 2	Login request	1 POST call	10	60	98	102
Scenario 3	Valve request	1 POST call	10	60	93	99
Scenario 4	Consumer App home page	6 GET calls	60	120	581	77
Scenario 5	Login request	1 POST call	60	120	578	78
Scenario 6	Valve request	1 POST call	60	120	571	79
Scenario 7	Consumer App home page	6 GET calls	100	120	848.8	69
Scenario 8	Login request	1 POST call	100	120	798	69
Scenario 9	Valve request	1 POST call	100	120	917	62
Scenario 10	Consumer App home page	6 GET calls	200	200	1,752	57
Scenario 11	Login request	1 POST call	200	200	1,869	53
Scenario 12	Valve request	1 POST call	200	200	1,749	58

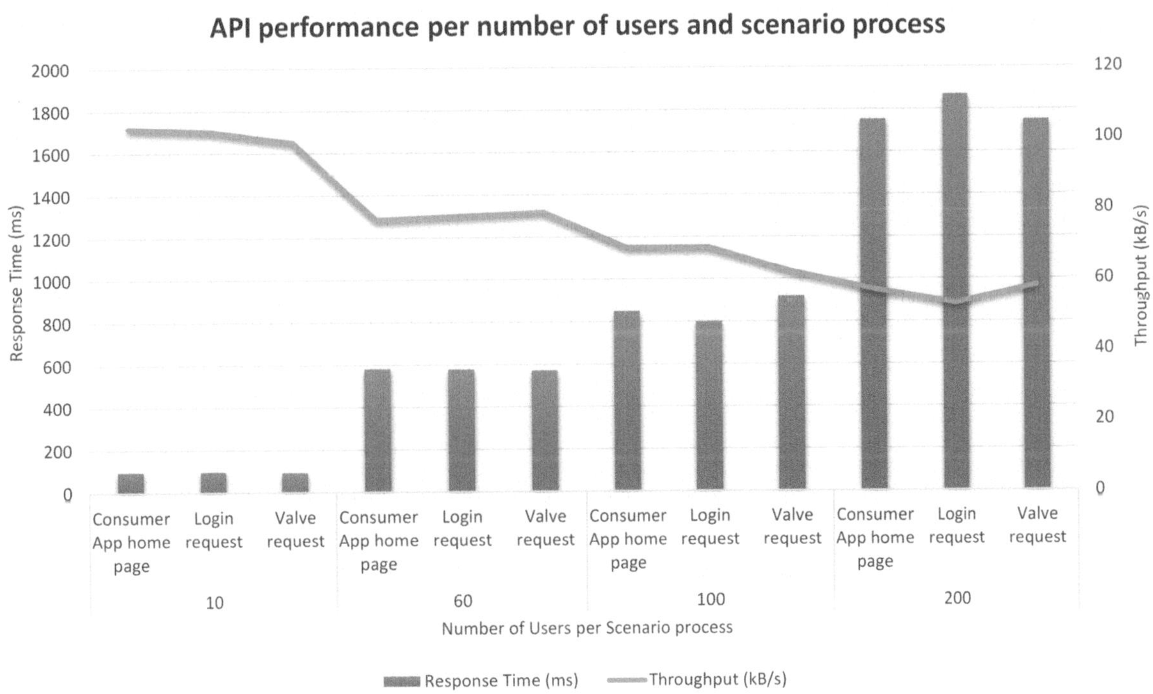

Figure 12 | Response time (ms) and throughput (kB/s) estimations using the Apache Jmeter performance tool, for each of the simulated test scenarios and the number of users.

propagation issues introduced by the dense urban environment. The time periods of the other two stages are considerably lower and the respective deviations denote the merest fluctuation. This stability implies that the system can handle all traffic generated by the end devices without delay (near real time).

Moreover, *Runtime Performance*, one of Google Chrome's Developer Tools, was employed to effectively evaluate the response times for the two web applications (the water utility company's and the consumer's). More specifically, the above tool describes and analyzes the way that a website appears and loads. In order to measure the response time of the consumer application, the browsing conditions on a mobile device should be simulated. Therefore, 4xCPU slow down for throttling and the Fast 3G for the network were selected.

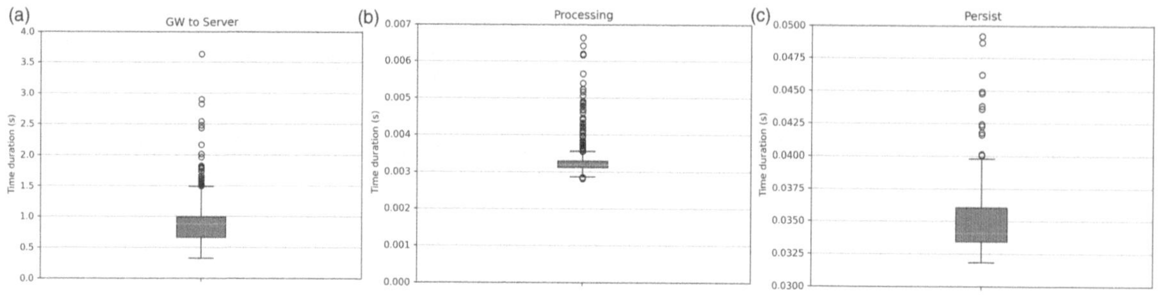

Figure 13 | Box plots for (a) 'Gateway to server' stage time; (b) 'Processing' stage time; (c) 'Persist' stage time.

Runtime performance analysis is categorized into four main groups of functionalities: *Scripting*, *Rendering*, *Painting* and *Loading*. Scripting is the time for JavaScript analysis and evaluation. Rendering and Painting are related to HTML compilation and CSS editing on objects that can be displayed on the user's screen. Finally, Loading refers to the interaction of the website with the network as well as the loading of HTML, CSS and Javascript (https://github.com/GoogleChrome/devtools-docs/blob/master/docs/timeline.md).

In Figure 14(a), *Response Time* for the water utility company's web application main components, measured over the group of functionalities. As expected, scripting is a significantly time-consuming operation in all the main components and it is followed by Rendering. Similarly, the *Response Time* of the main components of the consumer application is presented in Figure 14(b). The consumer's application includes the *Consumer Homepage* and *Consumer Dashboard*, consisting of the billing information page, water consumption and water valve actuation option.

Experiments and results of water consumption forecasting

Multi-step water consumption forecasting is performed using ARIMA and LSTM models. The goal is to compare the performance of the models over their ability to make accurate short-term predictions of total water consumption. The available set of data is related to the water consumption, corresponding to the period between 2020-02-01 T00:00:00 and 2020-05-18 T23:00:00 local time. The collected data were provided in flow rates measured in an hourly frequency from 75 AMRs (total: 2,587 hourly observations). Using the z-score approach, 30 outliers were detected and their values were substituted by propagating the last known observation each time. The average consumption is around 263.75 l/h and the maximum consumption ranged to 1,209 l/h. The distribution of the time series water consumption data grouped by days and hours exhibits high variability over the specific period (Figure 15).

The stationary ADF test exhibits that the time series is stationary as its value (−3.87) is less than the critical values 25, 50, and 75% (96, 209, and 371.5 respectively). The *p*-value is 0.002 less than 0.05, which indicates to reject the null hypothesis (H0), the data does not have a unit root and the time series is stationary. This is rational, as the limited size of the dataset did not imply any trend or seasonality.

In order to fit forecasting models for water consumption, the selected dataset needs to be transformed. The data should be re-scaled to values between 0 and 1 as described above. Applying the LSTM model, it is required that the data be

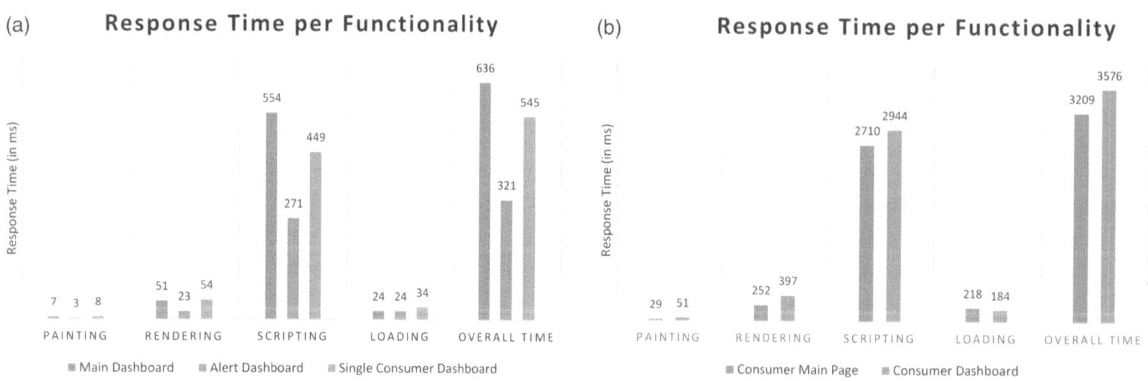

Figure 14 | Response times per main group of functionalities over the (a) Water Utility Dashboard; (b) consumer application components.

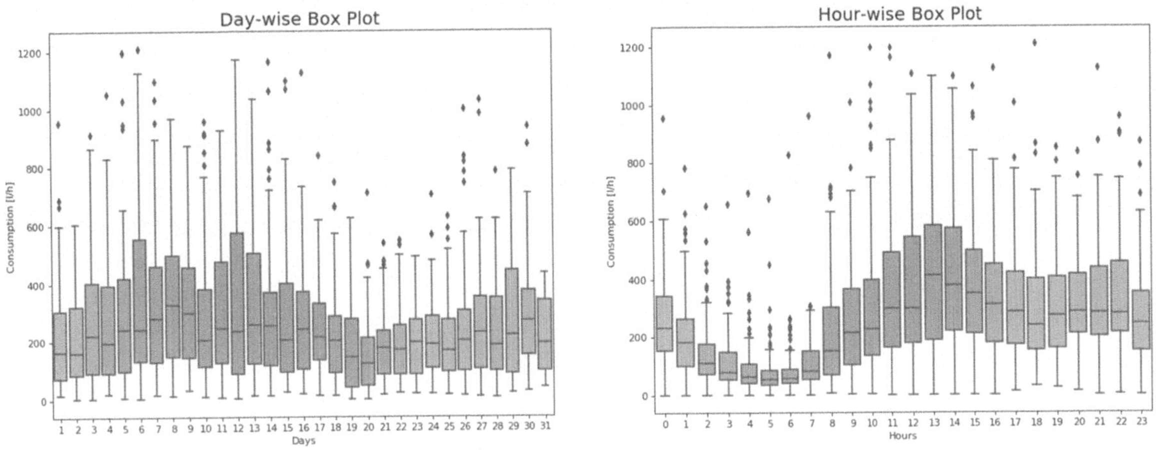

Figure 15 | Boxplots, which indicate the distribution of water consumption per day and hour in the examined time period.

within the scale of the activation function of the model. Then, the time series dataset was divided into input and output subsets enabling supervised learning to be undertaken. The past (*T-n*) time steps were used as input in order to predict the water consumption in the time step T + 1. Furthermore, the time series data was divided into 83.30% (2,155 hourly observations) for training and the remaining 16.70% (432 hourly observations) for validation/testing the models' performance in terms of the RMSE, MAPE and R^2 evaluation measures.

A series of experiments were realised to fine tune the training parameters of the LSTM network, namely the epochs, the number of neurons and the activation function. The RMSE is employed to evaluate the performance of the LSTM network for each set of parameters over the testing set. The best set of parameters, minimizing the RMSE, is {Epochs = 50, #neurons = 4, activation function = 'sigmoid'}. The performance of the LSTM network in terms of the RMSE over the training and testing set after 10 iterations fluctuated from the training set from 129.13 to 135.59 with mean ± standard deviation value equal to 130.96 ± 1.761 and for the testing set from 161.83 to 177.08 with mean ± standard deviation value equal to 166.55 ± 4.668. Similarly, the R^2 fluctuates from 53% to 58% in the training set (mean ± standard deviation = 56.40% ±1.35) and in the testing set from 57% to 64% (mean ± standard deviation = 62.20% ±2.15). Finally, the minimum MAPE value is 45.19% for training and 38.52% for testing and the maximum values are 65.17% and 50.21% respectively. The average MAPE is 54.59% ±7.186 for training and 43.23% ±4.401 for testing.

In the following figure (Figure 16), the performance of the LSTM network in the training and testing sets is exhibited in terms of its ability to learn the water consumption pattern hidden in the time series data (training) as well as to forecast the hourly water consumption in a forecasting horizon of the 18 days.

In order to compare the performance of the LSTM model with an ARIMA approach, we need to select an optimal time series model determining the appropriate set of

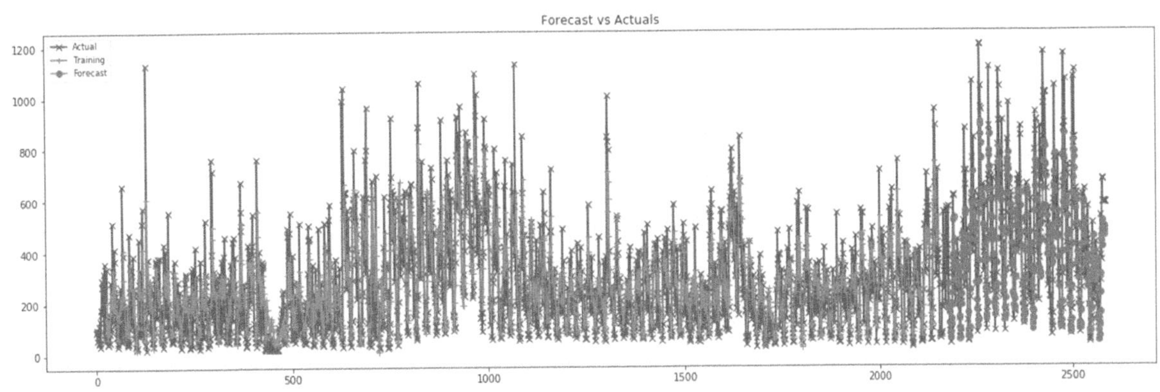

Figure 16 | Water consumption observed in the period from 2020-02-01 T00:00:00 to 2020-05-18 T23:00:00. Forecasts given by LSTM ({Epochs = 50, #neurons = 4, activation function = 'sigmoid'}) in the period from 2020-05-01 T00:00:00 to 2020-05-18 T23:00:00.

parameters *(p, d, q)*. Using the *auto_arima* function from the python *pmdarima* package, the ARIMA(6,0,0) model attains to minimise the Akaike information criterion (AIC). However, the performance of this model over the testing set is poor, as the RMSE was around 321.17 and MAPE climbed to 82.28%. Hence, the LSTM model seems to outperform the baseline models such as ARIMA in this set of time series data (Figure 17).

CONCLUSIONS

In this work, the main aspects of the on-going research project, called SMART-WATER, is presented. Specifically, SMART-WATER is a holistic water management solution that aims to offer innovative services to consumers and water utility companies. Pilot tests of the SMART-WATER system took place in the wider area of Thessaloniki, Greece. Real time telemetry and remote control of water supply are feasible utilizing state-of -the-art end devices on a fixed bidirectional network. High precision, wireless and M.I.D. certified water meters are used in conjunction with wireless water valves. Real time telemetry and remote-control services are viable through a fixed wireless and bidirectional network, which can manage messages from multiple protocols and standards (wM-Bus, LoRaWAN, NB-IoT, GSM). One of the key elements of the proposed system are the custom-made multi-protocol gateways (MPG), which unlock protocol and vendor binding issues. They can simultaneously interface with a wide set of end devices (water meters and water valves) of different vendors and protocols, providing bidirectional connectivity. Given the limited indoor penetration of the wM-Bus protocol, the gateways need to be located close to the end devices but at the same time remain connected to the central infrastructure via GSM or NB-IoT. Field tests also denoted that LoRaWAN technology is significantly superior to legacy wM-Bus technology regarding connectivity range.

At the other side of the architecture, a universal server can decode and decrypt wireless messages from subscribed wM-Bus and LoRa end devices regardless of the internet protocol used. The performance of the server and its API was tested in terms of processing time, response time and throughput. Simulation campaigns from real metering data demonstrated excellent data processing time upon gateway reception to aggregation (less than a second on average). Response time and throughput for different scenarios render the infrastructure able to serve numerous users simultaneously. As expected, a larger number of users introduces increased response time and moderate throughput values.

The short-term water demand forecasting approach adopts a seven-step schema, from the acquisition of the time-series data to the evaluation of the forecasting performance. The approach has been already tested on real data retrieved from the end devices installed in the field. Although the preliminary results for the operation of the

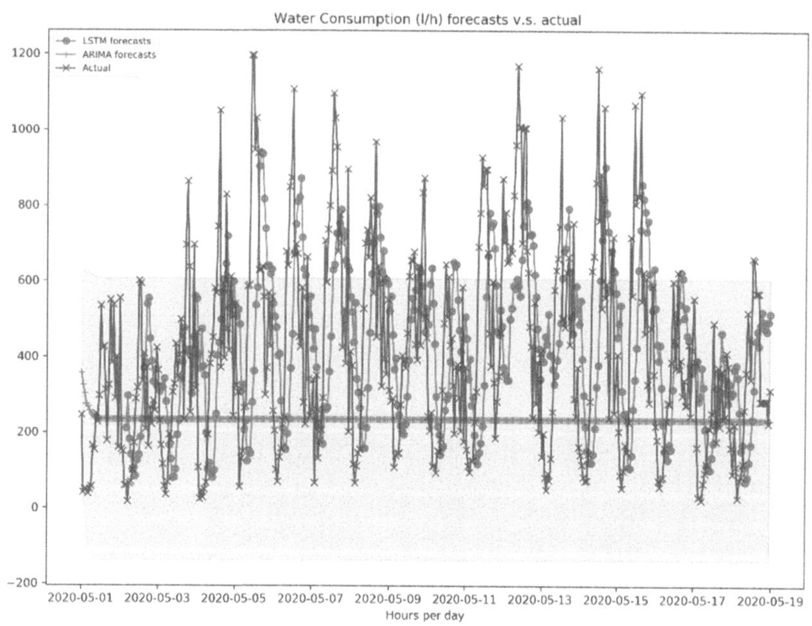

Figure 17 | Performance comparison between LSTM and ARIMA (6,0,0) models.

system are encouraging, however, further work should be done in order to evaluate its performance in the entire pipeline. Especially, in the advanced analytical perspective, as all the end devices will be deployed and start to collect data seamlessly, more robust, statistically sound and safe conclusions will emerge in terms of the short-term water demand/consumption estimations. Furthermore, clustering techniques and forecasting water consumption models will be applied, aiming at the determination of consumption profiles of the consumers.

ACKNOWLEDGEMENTS

This research has been co-financed by the European Union and Greek national funds through the Operational Program Competitiveness, Entrepreneurship and Innovation, under the call RESEARCH – CREATE – INNOVATE (project code: T1EDK- 04337).

DATA AVAILABILITY STATEMENT

Data cannot be made publicly available; readers should contact the corresponding author for details.

REFERENCES

3GPP Org. 2016 Standardization of NB-IOT completed. pp. 1. www.3gpp.org/news-events/3gpp-news/1785-nb_iot_complete (accessed 19 February 2016).

Alvisi, S., Casellato, F., Franchini, M., Govoni, M., Luciani, C., Poltronieri, F., Riberto, G., Stefanelli, C. & Tortonesi, M. 2019 Wireless middleware solutions for smart water metering. *Sensors* **19** (8), 1853. doi:10.3390/s19081853.

Amiruddin, A., Agung, A., Ratna, P., Harwahyu, R. & Sari, R. F. 2018 Secure multi-protocol gateway for Internet of Things. In *2018 Wireless Telecommunications Symposium (WTS)*, pp. 1–8. doi:10.1109/WTS.2018.8363934.

Anele, A., Hamam, Y., Abu-Mahfouz, A. & Todini, E. 2017 Overview, comparative assessment and recommendations of forecasting models for short-term water demand prediction. *Water* **9** (11), 887. doi: 10.3390/w9110887.

Antunes, A., Andrade-Campos, A., Sardinha-Lourenço, A. & Oliveira, M. 2018 Short-term water demand forecasting using machine learning techniques. *Journal of Hydroinformatics* **20** (6), 1343–1366. doi: 10.2166/hydro.2018.163.

Benítez, R., Ortiz-Caraballo, C., Preciado, J. C., Conejero, J. M., Sánchez Figueroa, F. & Rubio-Largo, A. 2019 A short-term data based water consumption prediction approach. *Energies* **12** (12), 2359. doi:10.3390/en12122359.

Boyle, T., Giurco, D., Mukheibir, P., Liu, A., Moy, C., White, S. & Stewart, R. 2013 Intelligent metering for urban water: a review. *Water* **5** (3), 1052–1081. doi:10.3390/w5031052.

Candelieri, A. & Archetti, F. 2014 Identifying typical urban water demand patterns for a reliable short-term forecasting – the icewater project approach. *Procedia Engineering* **89**, 1004–1012. https://doi.org/10.1016/j.proeng.2014.11.218.

Carratù, M., Ferro, M., Paciello, V., Pietrosanto, A. & Sommella, P. 2017 Performance analysis of wM-bus networks for smart metering. *IEEE Sensors Journal* **17** (23), 7849–7856. doi: 10.1109/JSEN.2017.2738102.

ChirpStack. 2017 www.chirpstack.io/ (accessed 5/1/2020).

Chowdury, M. S. U., Emran, T. B., Ghosh, S., Pathak, A., Alam, M. M., Absar, N., Andersson, K. & Hossain, M. S. 2019 IoT-based real-time river water quality monitoring system. *Procedia Computer Science* **155**, 161–168. doi:10.1016/j.procs.2019.08.025.

Cole, G., O'Halloran, K. & Stewart, R. A. 2012 Time of use tariffs: implications for water efficiency. *Water Science and Technology* **12** (1), 90–100. doi: 10.2166/ws.2011.123.

Diehl Stiftung & Co. KG. 2020 https://www.diehl.com/metering/en/portfolio/solutions/ (accessed 5/5/2020).

Donkor, E. A., Mazzuchi, T. A., Soyer, R. & Roberson, J. A. 2014 Urban water demand forecasting: review of methods and models. *Journal of Water Resources Planning and Management* **140** (2), 146–159.

Fantozzi, M., Popescu, I., Farnham, T., Archetti, F., Mogre, P., Tsouchnika, E., Chiesa, C., Tsertou, A., Castro Gama, M. & Bimpas, M. 2014 ICT for efficient water resources management: the ICeWater energy management and control approach. *Procedia Engineering* **70**, 633–640. doi: 10.1016/j.proeng.2014.02.069.

Gislason, D. 2008 *Zigbee Wireless Networking*. Elsevier Inc., Burlington, MA, USA.

Gomez, C., Oller, J. & Paradells, J. 2012 Overview and evaluation of bluetooth low energy: an emerging low-power wireless technology. *Sensors* **12** (9), 11734–11753. doi: 10.3390/s120911734.

Guancheng, G. & Shuming, L. 2018 Short-term water demand forecast based on deep neural network. In *1st International WDSA/CCWI 2018 Joint Conference*, July 23–25, 2018, *Canada*.

Gunasagarana, R., Kamarudina, L. M., Zakariaa, A., Kanagaraja, E., Alimona, M. S. A. M., Shakaffa, A. Y. M., Ehkanb, P., Visvanathana, R. & Razalia, M. H. M. 2015 Internet of Things: Sensor to Sensor Communication. In *2015 IEEE SENSORS*. doi:10.1109/ICSENS.2015.7370448.

Hochreiter, S. & Schmidhuber, J. 1997 Long short-term memory. *Neural Computation* **9** (8), 1735–1780.

Hua, Y., Zhao, Z., Li, R., Chen, X., Liu, Z. & Zhang, H. 2019 Deep learning with long short-term memory for time series prediction. In *IEEE Communications Magazine* **57** (6), 114–119. doi: 10.1109/MCOM.2019.1800155.

Hyndman, R. J. & Athanasopoulos, G. 2018 Forecasting: principles and practice. In: *OTexts*, 2nd edn (R. J. Hyndman ed.). Melbourne, Australia. OTexts.com/fpp2 (accessed on 02/02/2020).

Kaur, J. & Singh, M. 2013 Multiprotocol gateway for wireless communication in embedded systems. *International Journal of Computer Applications (0975–8887)* **72** (18), 27–31.

Kofinas, D., Mellios, N., Papageorgiou, E. & Laspidou, C. 2014 Urban water demand forecasting for the Island of Skiathos. *Procedia Engineering* **89**, 1023. doi: 10.1016/j.proeng.2014.11.220.

Kossieris, P., Kozanis, S., Hashmi, A., Katsiri, E., Vamvakeridou-Lyroudia, L. S., Farmani, R., Makropoulos, C. & Savic, D. 2014a A web-based platform for water efficient households. *Procedia Engineering* **89**, 1128–1135. doi: 10.1016/j.proeng.2014.11.234.

Kossieris, P., Panayiotakis, A., Tzouka, K., Gerakopoulou, P., Rozos, E. & Makropoulos, C. 2014b An eLearning approach for improving household water efficiency. *Procedia Engineering* **89**, 1113–1119. doi: 10.1016/j.proeng.2014.11.232.

Kouroupetroglou, C., van Slooten, J., Clifford, E., Coakley, D., Curry, E., Smit, S. & Perfido, D. 2015 WATERNOMICS: Serving diverge user needs under a single water information platform. In *e-proceedings of the 36th IAHR World Congress, The Hague, The Netherlands*.

Kusuma, S. S. & Anil, G. N. 2018 An IoT based water supply monitoring and controlling system. *International Research Journal of Engineering and Technology (IRJET)* **5** (2), 857–860.

Makropoulos, C., Kossieris, P., Kozanis, S., Katsiri, E. & Vamvakeridou-Lyroudia, L. 2014 From Smart Meters To Smart Decisions: Web-Based Support For The Water Efficient Household. In *11th International Conference on Hydroinformatics, New York City, USA*.

Masum, S., Liu, Y. & Chiverton, J. 2018 Multi-step Time Series Forecasting of Electric Load Using Machine Learning Models. In: *Artificial Intelligence and Soft Computing* (L. Rutkowski, P. Scherer, M. Korytkowski, W. Pedrycz, R. Tadeusiewicz & J. Zurada, eds), ICAISC 2018. Lecture Notes in Computer Science, vol 10841. Springer, Cham, Switzerland.

Mourtzios, C., Kourtesis, D., Papadimitriou, N., Antzoulatos, G., Kouloglou, I. O., Vrochidis, S. & Kompatsiaris, I. 2019 Work-In-Progress: SMART-WATER, a Novel Telemetry and Remote Control System Infrastructure for the Management of Water Consumption in Thessaloniki. In: *Proceedings of the 13th International Conference on Interactive Mobile Communication Technologies and Learning*, p. 1192, eBook.

Moy de Vitry, M., Schneider, M. Y., Wani, O., Manny, L., Leitão, J. P. & Eggimann, S. 2019 Smart urban water systems: what could possibly go wrong? *Environmental Research Letters* **14**, 081001.

Mudumbe, M. J. & Abu-Mahfouz, A. M. 2015 Smart water meter system for user-centric consumption measurement. In: *IEEE 13th International Conference on Industrial Informatics (INDIN), Cambridge*, pp. 993–998.

Novak, J., Melenhorst, M., Micheel, I., Pasini, C., Fraternali, P. & Rizzoli, A. E. 2018 Integrating behavioural change and gamified incentive modelling for stimulating water saving. *Environmental Modelling & Software* **102**, 120–137. doi: 10.1016/j.envsoft.2017.11.038.

Oyebode, O. & Ighravwe, D. E. 2019 Urban water demand forecasting: a comparative evaluation of conventional and soft computing techniques. *Resources* **8** (3), 156. doi: 10.3390/resources8030156.

Perfido, D., Messervey, T., Kouroupetroglou, C. & Costa, A. 2015 WATERNOMICS (ICT for water resource management) methodology and water information platform. *Procedia Environmental Science, Engineering and Management* **2** (4), 285–293.

Public Utilities Board Singapore (PUB). 2016 Managing the water distribution network with a smart water grid. *Smart Water* **1**, 1–13. doi: 10.1186/s40713-016-0004-4.

Ribeiro, R., Loureiro, D., Barateiro, J., Smith, J., Rebelo, M., Kossieris, P., Gerakopoulou, P., Makropoulos, C., Vieira, P. & Mansfield, L. 2015 Framework for technical evaluation of decision support systems based on water smart metering: the iWIDGET case. *Procedia Engineering* **119**, 1348–1355. doi: 10.1016/j.proeng.2015.08.976.

Rizzoli, A. E., Castelletti, A., Fraternali, P. & Novak, J. 2018 Demo abstract: smartH2O, demonstrating the impact of gamification technologies for saving water. *Comput Sci Res Dev* **33**, 275–276. doi: 10.1007/s00450-017-0380-5.

Sanchez, R. I. & Cano, M. D. 2016 State of the Art in LP-WAN solutions for industrial IoT services. *Sensors* **16** (5), 708. doi:10.3390/s16050708.

Saraswathi, V., Rohit, A., Sakthivel, S. & Sandheep, T. J. 2018 Water leakage system using IoT. *IJIREM* **5** (2), 67–69.

Semtech Corporation. 2015 AN1200.22 LoRa™ Modulation Basics Application Note Revision 2 (accessed 10/03/2020).

Sensus, Xylem brand. 2020 sensus.com/communication-networks/sensus-technologies/sensus-rf/ (accessed 15/04/2020).

SigFox. 2016 https://www.sigfox.com/ (accessed 15/11/2019).

Tsavdaridou, A. D., Kolokytha, E. G. & Mentes, A. K. 2012 Urban water consumption and new metering technologies: Application in the city of Thessaloniki. In: *Proceedings of the XI Protection and Restoration of the Environment (PRE) Conference*, 203–212.

Vijai, P. & Sivakumar, P. 2018 Performance comparison of techniques for water demand forecasting. *Procedia Computer Science* **143**, 258–266. doi: 10.1016/j.procs.2018.10.394.

Walker, D., Creaco, E., Vamvakeridou-Lyroudia, L., Farmani, R., Kapelan, Z. & Savic, D. 2015 Forecasting domestic water consumption from smart meter readings using statistical methods and artificial neural networks. *Procedia Engineering* **119**, 1419–1428. doi: 10.1016/j.proeng.2015.08.1002.

Wescott, B. 2013 Every Computer Performance Book: *How to Avoid and Solve Performance Problems on The Computers You Work With*, 1st edition. *Create Space Independent Publishing Platform*, SC, USA.

Xenochristou, M., Kapelan, Z, Hutton, C. & Hofman, J. 2018 Smart water demand forecasting: Learning from the data. In *13th International Conference on Hydroinformatics (HIC 2018)*. EasyChair Publications, Vol. 3, pp. 2351–2358. doi: 10.29007/wkp4.

First received 31 May 2020; accepted in revised form 29 July 2020. Available online 18 August 2020

© IWA Publishing 2020

Fault detection and diagnosis in water resource recovery facilities using incremental PCA

Pezhman Kazemi, Jaume Giralt, Christophe Bengoa, Armin Masoumian and Jean-Philippe Steyer IWA

ABSTRACT

Because of the static nature of conventional principal component analysis (PCA), natural process variations may be interpreted as faults when it is applied to processes with time-varying behavior. In this paper, therefore, we propose a complete adaptive process monitoring framework based on incremental principal component analysis (IPCA). This framework updates the eigenspace by incrementing new data to the PCA at a low computational cost. Moreover, the contribution of variables is recursively provided using complete decomposition contribution (CDC). To impute missing values, the empirical best linear unbiased prediction (EBLUP) method is incorporated into this framework. The effectiveness of this framework is evaluated using benchmark simulation model No. 2 (BSM2). Our simulation results show the ability of the proposed approach to distinguish between time-varying behavior and faulty events while correctly isolating the sensor faults even when these faults are relatively small.

Key words | BSM2, EBLUP, fault detection, fault isolation, incremental PCA, time-varying processes

Pezhman Kazemi (corresponding author)
Jaume Giralt
Christophe Bengoa
Universitat Rovira i Virgili,
Departament d'Enginyeria Química, Avda. Paisos Catalans, 26, 43007, Tarragona,
Spain
E-mail: *pezhman.kazemi@urv.cat*

Armin Masoumian
Universitat Rovira i Virgili,
Departament d'Enginyeria Informàtica i Matemàtiques, Avda. Paisos Catalans, 26, 43007, Tarragona,
Spain

Jean-Philippe Steyer IWA
INRAE, Univ Montpellier,
LBE, 102 avenue des Etangs, 11100, Narbonne,
France

HIGHLIGHTS

- A novel framework is proposed for adaptive fault detection and diagnosis based on IPCA.
- The validity and applicability of the proposed framework are examined using benchmark simulation model No.2 (BSM2).
- The realtime imputation of intricate patterns of missing values is performed using EBLUP.
- The proposed framework can detect small-magnitude drift faults while tracking time-varying behavior.

GRAPHICAL ABSTRACT

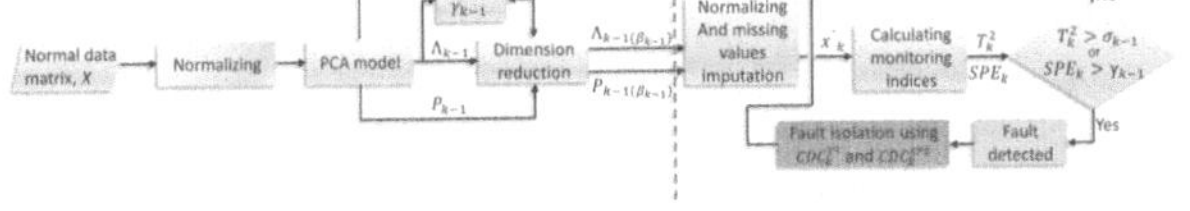

doi: 10.2166/wst.2020.368

ABBREVIATIONS

X	Data matrix
T	Score matrix
P	Loading matrix
E	Residual matrix
S	Covariance matrix
n	Data matrix row number
m	Data matrix column number
V_k	Eigenvector matrix for kth sample
Λ_k	Eigenvalues matrix for kth sample
λ_j	jth eigenvalues
β_k	Number of principal components to be retained for kth sample
T_k^2	Hotelling's T-squared for kth sample
SPE_k	Square prediction error for kth sample
x_k	kth sample
x'_k	Standardized kth sample
m_k	Mean of kth sample
f	Forgetting factor
δ_k	Standard deviation for kth sample
a_k	Projection of kth sample using the current eigenvector V_{k-1}
h_k	Residual vector for kth sample
$\hat{h}_k$	Normalized residual vector for kth sample
$\|\|h_k\|\|_2$	Euclidean norm of h_k matrix
R_k	Rotation matrix for kth sample
D_k	A matrix that decomposed to obtain R_k
$CPV(\beta_k)$	Cumulative percent variance with β_k retained principal components
χ^2_{α,β_k}	Chi-distribution with β_k degrees of freedom and a confidence limit α
σ_k	T^2 threshold for kth sample
γ_k	SPE threshold for kth sample
η_α	Normal deviate corresponding to the $(1-\alpha)$ percentile
$CDC_k^{T^2}$	Complete decomposition contribution for kth sample using Hotelling's T-squared
CDC_k^{SPE}	Complete decomposition contribution for kth sample using SPE
x_k^o	Vector of kth sample with observed values
x_k^m	Vector of kth sample with missing values
m_{k-1}^o	Mean of observed values of k – 1th sample
m_{k-1}^m	Mean of missing values of k – 1th sample
V_{k-1}^o	Eigenvector matrix of observed values of k – 1th sample
V_{k-1}^m	Eigenvector matrix of missing values of k – 1th sample
$\hat{x}_k^m$	Imputed missing values of kth sample
$\mathbb{E}$	Conditional expectation
S_{in}	BSM2 inorganic nitrogen concentration in anaerobic digester
v_s	BSM2 secondary clarifier exponential settling velocity function
μ_A	BSM2 growth rate for the autotrophs
PCA	Principal component analysis
IPCA	Incremental principal component analysis
EBLUP	Empirical best linear unbiased prediction
BSM2	Benchmark simulation model No.2
WRRF	Water resource recovery facilities
WWTP	Wastewater treatment plant
MSPCs	Multivariate statistical process controls
RPCA	Recursive principal component analysis
EWMA	Exponential weighted moving average
FOP	First-order perturbation
PCs	Principal components
TSS	Total suspended solid
PI	Proportional integral
AD	Anaerobic digestion
COD	Carbon-oxygen demand
FAR	False alarm rate
MDR	Missed detection rate

INTRODUCTION

Because of the complexity of modern industrial processes, the demand has increased to implement fault detection and a diagnosis framework for those processes. Water resource recovery facilities (WRRF), formerly called wastewater treatment plants (WWTP), are no exception (Sánchez-Fernández *et al.* 2018). Abnormal or faulty events are those that occur when the process indicates a deviation from its normal behavior. In WRRF, abnormal events include changes in influent quality (e.g. rainfall, industrial discharge), outbreaks of microorganisms (e.g. filamentous bacteria, algae) that impact treatment quality, irregularities or damage to treatment units (e.g. membranes, clarifiers), mechanical failures (e.g. pumps, air blowers) and sensor failure (e.g. drift, bias, electrical interference) (Newhart *et al.* 2019). All these faults can undermine process performance, so they must be detected and diagnosed as soon as they occur. Fault detection techniques are generally divided into three main groups: model-based methods, knowledge-based methods, and data-driven methods. To implement techniques in the first two groups, an in-depth knowledge of the system's behavior is required. However, obtaining in-depth knowledge of reactions or pathways phenomena (especially for WRRFs, which are very complicated processes) is both time-consuming and challenging.

Data-driven methods, on the other hand, rely only on historical and online data and do not require in-depth prior knowledge of the process (Kazemi *et al.* 2020). Thanks to the advanced process control framework, WRRFs collect large amounts of data using measurements from numerous online sensors (Wang & Shi 2010). Data-driven techniques have therefore received significant attention in process monitoring over the last few years. Multivariate statistical process controls (MSPCs) are a subset of data-driven methods used to analyze the performance of the processes and identify the parameters that govern them. One of the most widely used techniques for MSPC is principal component analysis (PCA) (Newhart *et al.* 2019). Various types of PCA methods, such as conventional PCA (Garcia-Alvarez *et al.* 2009; Sanchez-Fernández *et al.* 2015; Xiao *et al.* 2017), kernel PCA (Lee *et al.* 2004; Jun *et al.* 2006; Xiao *et al.* 2017) and dynamic PCA (Lee *et al.* 2006; Mina & Verde 2006) have been successfully used to monitor processes and detect faults in time-invariant processes. However, because of the complexity of the biological reactions and non-stationary plant influent, WRRFs have time-varying characteristics. Using conventional PCA to monitor such processes may therefore lead to excessive rates of false alarms and missing detections. An adaptive process monitoring framework is therefore needed. Although adaptive PCAs are more suitable for non-stationary processes, due to high computational costs, not every adaptive approach is applicable for realtime fault detection. Several adaptive approaches include recursive PCA (RPCA), moving window PCA and exponential weighted moving average (EWMA) have recently been used for fault detection (Li *et al.* 2000; Rosen & Lennox 2001; Choi *et al.* 2006; Elshenawy *et al.* 2010; Shang *et al.* 2015). Haimi *et al.* implemented a moving window PCA technique to detect and isolate the faults in WRRF (Haimi *et al.* 2016). These authors took into account the variable length of the historical data in the model's construction to prevent sub-optimal monitoring performance. Li *et al.* applied an RPCA monitoring framework that recursively updates the correlation matrix. These authors used two approaches, namely rank-one modification and Lanczos tridiagonalization, to compute the eigenvalues of the updated correlation matrix (Li *et al.* 2000). Recently, Elshenawy *et al.* proposed RPCA based on first-order perturbation analysis (FOP), which is a rank-one update of the eigenvalues and their corresponding eigenvectors from a sample covariance matrix (Elshenawy *et al.* 2010; Elshenawy & Mahmoud 2018). These authors stated that the computational cost of their approach is lower than that of previously used methods such as RPCA using Lanczos tridiagonalization and MWPCA.

In this paper we propose a new low-computational-cost adaptive fault detection framework for WRRF that uses incremental PCA (IPCA). This approach is based on the IPCA proposed by Hall *et al.*, which was motivated by developments in the field of computer vision (Hall *et al.* 1998; Brand 2002; Arora *et al.* 2012). The main benefit of IPCA over other adaptive PCA approaches is that it does not require all eigenvalues to be computed since only the largest ones are needed. This considerably accelerates computation time. Cardot and Degras (Cardot & Degras 2018) compared the computation time of adaptive PCA methods such as perturbation and stochastic approximation and concluded that IPCA outperforms other methods, particularly when the data have many features (sensor measurements). These authors also studied the accuracy of adaptive methods in determining the eigenspace. Their results revealed that IPCA is more accurate than methods such as RPCA, which is based on FOP theory. The IPCA algorithms offer an excellent compromise between statistical accuracy and computational speed. Another major drawback of PCA approaches are missing data during realtime monitoring. In practice, due to sudden mechanical breakdown, maintenance, hardware sensor failure or malfunction of the data acquisition system, etc., some sensor signals may become unavailable (Zhang & Dong 2014). To meet the requirements of realtime fault detection and diagnosis, therefore, the problem of missing data needs to be considered. To handle missing data, the empirical best linear unbiased prediction (EBLUP) approach is incorporated into the proposed IPCA method (Cardot & Degras 2018).

Once a fault is detected, it is essential to find out its primary source. The most common fault isolation method are contribution charts (Nomikos & MacGregor 1995). With this method, the variables that contribute to the T^2

(Hotelling's T-squared) and SPE (square prediction error) statistics are calculated and the variable with the highest contribution is considered the primary cause of the fault. Because of the recursive nature of IPCA, the contribution plots of T^2 and SPE are updated in accordance with the adaptive eigendecomposition. To test the proposed framework, several types of process parameters and sensor faults were simulated using Benchmark Simulation Model No. 2 (BSM2) (Jeppsson *et al.* 2006). This paper is organized as follows: first we provide a brief review of the theoretical background behind the conventional PCA method; then we discuss the IPCA algorithm and its realtime application framework for fault detection and diagnosis; and finally we analyze the performance of the monitoring framework on the complete and incomplete (i.e. with missing values) data sets using BSM2.

CONVENTIONAL PCA

Principle component analysis (PCA) is a well-known method designed to convert a data set with possibly correlated variables into a set of values of linearly uncorrelated variables called principal components (PCs) (Xiao *et al.* 2017). Let a data matrix $X \in \Re^{n \times m}$ be composed of n samples and m sensors. The X matrix is scaled to zero mean and unit variance (Z-score normalization) in order to avoid the scaling problem. The PCA model can be defined as:

$$X = TP^T + E \tag{1}$$

where $T \in \Re^{n \times m}$, $P \in \Re^{m \times m}$, and $E \in \Re^{n \times m}$ are score (PCs), loading, and residual matrices, respectively. The scores are generated by projecting X onto to the loading matrix. To solve the PCA model the covariance matrix, $S \in \Re^{m \times m}$, of X should be decomposed as follows:

$$S = \frac{1}{n-1} X^T X = P \Lambda P \tag{2}$$

The columns of P are eigenvectors of S, and $\Lambda \in \Re^{m \times m}$ is a diagonal matrix containing the eigenvalues $\Lambda = diag\,(\lambda_1, \lambda_2 \cdots \lambda_m)$ arranged in descending order. Usually, the number of (β) principal components, which sufficiently explain the variability of the process, are selected. Therefore, $P \in \Re^{m \times \beta}$ and its corresponding eigenvalues are $\Lambda \in \Re^{\beta \times \beta}$. For fault detection, two statistics (mainly T^2 and SPE) are normally used. Variations in the mean and covariance in PCs are measured using T^2 while the variation in the residual subspace is measured using SPE (Elshenawy & Awad 2012). These are given by:

$$T_k^2 = x_k^T P_{k(\beta_k)} \Lambda_{k(\beta_k)}^{-1} x_k P_{k(\beta_k)}^T \quad k = 1 \cdots n \tag{3}$$

$$SPE_k = x_k (1 - P_{k(\beta_k)} P_{k(\beta_k)}^T)^2 x_k^T \tag{4}$$

Here, as mentioned earlier, β_k is the number of PCs to retain. To perform fault detection, suitable thresholds for these two statistical indices must be obtained. This enables small faults to be detected with a minimum rate of false alarms. Process operation is normal if these statistical indices remain below these thresholds. The procedure for estimating the thresholds of T^2 and SPE will be discussed later.

INCREMENTAL PCA

Detecting and monitoring faults using conventional PCA is suitable for time-invariant processes. If it were used for time-variant processes, it would be difficult to monitor the typical time-varying characteristics caused by external disturbances and changes in operating conditions. To handle time-varying issues, an adaptive fault detection framework must be developed (Hu *et al.* 2012). The key idea behind IPCA is to update the eigenvector by incrementing the new data to the PCA. To update process monitoring, both the mean $m_k \in \Re^{1 \times m}$ and the standard deviation $\delta_k \in \Re^{1 \times m}$ need to be re-estimated whenever a new sample $x_k \in \Re^{1 \times m}$ data vector becomes available (Hall *et al.* 1998; Artač *et al.* 2002; Cardot & Degras 2018). These are given by:

$$m_k = (1-f)m_{k-1} + f x_k \tag{5}$$

$$\delta_k = \sqrt{(1-f)\delta_{k-1}^2 + f(x_k - m_k)^2} \tag{6}$$

where $0 \le f < 1$ is a forgetting factor. Let us assume that eigenvector P_{k-1} has already been derived using the data from X. Also, Λ_{k-1} is a diagonal matrix of the corresponding eigenvalues λ, and m_k and δ_k are the mean and variance vector of a new sample x_k, respectively. Each sample must be standardized according to:

$$x'_k = \frac{x_k - m_k}{\delta_k} \tag{7}$$

where x'_k is the standardized sample. Next, the projection of the new sample a_k is computed using the current eigenvector P_{k-1}:

$$a_k = P_{k-1}^T x'_k \tag{8}$$

Finally, residual vector h_k is estimated using the feature vector a_k. The residual vector is orthogonal to the eigenvector.

$$h_k = P_{k-1} a_k - x'_k \tag{9}$$

To update the eigenvector, h_k needs to be normalized according to the condition outlined in the equation below:

$$\hat{h}_k = \begin{cases} \frac{h_k}{\|h_k\|_2}, & if\ \|h_k\|_2 \neq 0 \\ 0, & otherwise \end{cases}, \quad \|h_k\|_2 = \sqrt{h_k^T . h_k} \tag{10}$$

where $\|h_k\|_2$ is the Euclidean norm of matrix h_k.

The new eigenvector P_k can be estimated by adding $\hat{h}_k$ to the current eigenvector P_{k-1} and rotating the result using rotation matrix R:

$$P_k = [P_{k-1},\ \hat{h}_k] R_k \tag{11}$$

R is calculated by solving the eigenvector decomposition problem as shown below:

$$D_k R_k = R_k \Lambda_k \tag{12}$$

where matrix D_k is composed as follow:

$$D_k = (1-f)\begin{bmatrix} \Lambda_{k-1} & \underline{0} \\ \underline{0}^T & 0 \end{bmatrix} + f(1-f)\begin{bmatrix} a_k a_k^T & \gamma a_k \\ \gamma a_k^T & \gamma^2 \end{bmatrix} \tag{13}$$

where $\underline{0}$ is a vector of zero and $\gamma = \hat{h}_k x'_k$. By updating the eigenvector P_k and the eigenvalue matrix Λ_k for each new sample, the process monitoring statistics must also be updated according to Equations (3) and (4).

Estimating the number of PCs

To fully implement IPCA, the number of PCs must be updated after a new sample becomes available. Numerous methods are available for calculating the number of PCs (Li *et al.* 2000). In this study we used cumulative percent variance (CPV), which can be obtained from:

$$CPV(\beta_k) = \frac{\sum_{j=1}^{\beta_k} \lambda_j}{\sum_{j=1}^{m} \lambda_j} 100\% \tag{14}$$

where β_k is a number of selected PCs. The number of PCs is chosen when CPV reaches a predetermined limit, e.g. 95%.

Process monitoring statistics thresholds

Because of the non-stationary behavior of water resource recovery facilities, the thresholds for the detection statistics (T^2 and SPE) change over time. For realtime monitoring, therefore, it is necessary to adapt these limits (Li *et al.* 2000). The T^2 threshold σ_k is approximated using the chi-distribution χ^2 with β_k degrees of freedom and a confidence limit α:

$$\sigma_k = \chi^2_{\alpha, \beta_k} \tag{15}$$

and the threshold γ_k for the SPE statistic is given by:

$$\gamma_k = \theta_1 \left[\frac{\eta_\alpha \sqrt{2\theta_2 h_0^2}}{\theta_1} + 1 + \frac{\theta_2 h_0 (h_0 - 1)}{\theta_1^2} \right]^{\frac{1}{h_0}} \tag{16}$$

where η_α is the normal deviate corresponding to the (1-α) percentile, and

$$h_0 = 1 - \frac{2\theta_1 \theta_3}{3\theta_2^2} \tag{17}$$

$$\theta_i = \sum_{j=\beta_k+1}^{m} \lambda_j^i; \qquad i = 1, 2, 3 \tag{18}$$

The fault is detected once the T^2 and SPE statistics exceed their respective adaptive thresholds, σ_k and γ_k.

Fault isolation

Once a fault has been detected, the variables that contribute to the deviation must be identified in order to determine the location of primary causes. The idea behind the contribution chart is that variables with the largest contributions to the fault detection statistics are probably the faulty variables (Alcala & Qin 2009). Many methods are available for calculating the contributions of variables. In this study we used complete decomposition contribution (CDC), which is widely used in the industry. The CDC for the T^2 and SPE statistics is defined as:

$$CDC_k^{T^2} = (x_k P_{k(\beta_k)} \Lambda_{k(\beta_k)}^{-1/2} P_{k(\beta_k)}^T)^2 \quad (19)$$

$$CDC_k^{SPE} = (x_k - (x_k P_{k(\beta_k)} P_{k(\beta_k)}^T))^2 \quad (20)$$

where $CDC_k^{T^2}$ and CDC_k^{SPE} are the vectors whose elements are the variables that contribute to the T^2 and SPE statistics for each column (sensor) of a sample, respectively.

Missing data imputation

Methods such as mean, regression, hot-deck, maximum likelihood, multiple imputations, etc., are suggested for imputing missing values in the realtime application of PCA (Cardot & Degras 2018). In this paper we used the empirical best linear unbiased prediction (EBLUP) method described by Brand (2002) to impute missing values in the new observation vector x_k. With this method, the missing observations in vector x_k are approximated using the mean value m_{k-1} and the eigenvector decomposition (P_{k-1} and Λ_{k-1}) of the current sample. The x_k can be split into two sub-vectors: x_k^o (observed values) and x_k^m (missing values), respectively. Similarly, m_{k-1} and P_{k-1} are split into m_{k-1}^o, m_{k-1}^m and p_{k-1}^o, p_{k-1}^m respectively. Let $\Lambda_{k-1}^{-1/2}$ be the diagonal matrix containing the square roots of the diagonal eigenvalues arranged in descending order (Brand 2002; Cardot & Degras 2018). By applying the conditional expectation for multivariate normal distribution, the EBLUP can be obtained as follows:

$$\begin{aligned}\hat{x}_k^m &= \mathbb{E}(x_k^m | x_k^o, m_{k-1}, P_{k-1}, \Lambda_{k-1}) \\ &= m_{k-1}^m + (p_{k-1}^m \Lambda_{k-1}^{-1/2})(p_{k-1}^o \Lambda_{k-1}^{-1/2})(x_k^o - m_{k-1}^m) \quad (21)\end{aligned}$$

where $\hat{x}_k^m$ is the imputed missing value.

ADAPTIVE FAULT DETECTION AND ISOLATION

In this section we discuss the complete structure of the adaptive fault detection and isolation scheme. The proposed adaptive fault detection and isolation framework can be divided into two stages: (i) offline training and (ii) online monitoring. The following steps summarize the overall procedure.

Offline training

(1) Collect the training data matrix $X \in \Re^{n \times m}$ and normalize it to zero mean m_{k-1} and unit variance δ_{k-1}.
(2) Use Equation (2) to compute S and its corresponding eigenpairs P_{k-1}, Λ_{k-1}.
(3) Calculate the number of retaining PCs (β_{k-1}) by applying Equation (14).
(4) Calculate the thresholds of the fault detection statistics σ_{k-1} and γ_{k-1} using Equations (15) and (16).

Online monitoring

(1) Collect a new data vector x_k and normalize it to zero mean and unit variance using Equations (5)–(7). If there are any missing values, compute them using Equation (21).
(2) Calculate the monitoring statistics T_k^2 and SPE_k using Equations (3) and (4).
(3) If the values of the monitoring statistics are below their corresponding thresholds, the process status is normal. The updating procedure continues as follows:
    (i) Update m_k and δ_k using Equations (5) and (6).
    (ii) Calculate the updated eigenpairs P_k and Λ_k using Equations (8)–(13).
    (iii) Calculate the number of retained PCs, β_k, using Equation (14).
    (iv) Update the monitoring statistics thresholds σ_k and γ_k using Equations (15) and (16).
    (v) Return to step one.
(4) If the values of the monitoring statistics exceed their corresponding thresholds, the process status is faulty and the updating procedure should be stopped. The fault isolation procedure (CDC) needs to be run to identify the process variables responsible for the detected fault.
    (i) The contribution statistics for the faulty points need to be calculated according to Equations (19) and (20).
    (ii) The process variables with the largest contributions are responsible for the fault.

The complete diagram for the proposed adaptive IPCA monitoring procedure is shown in Figure 1.

SIMULATION RESULTS

Description of the water resource recovery facility

In this section, the proposed adaptive IPCA framework is applied to Benchmark Simulation No. 2 (BSM2) (Jeppsson

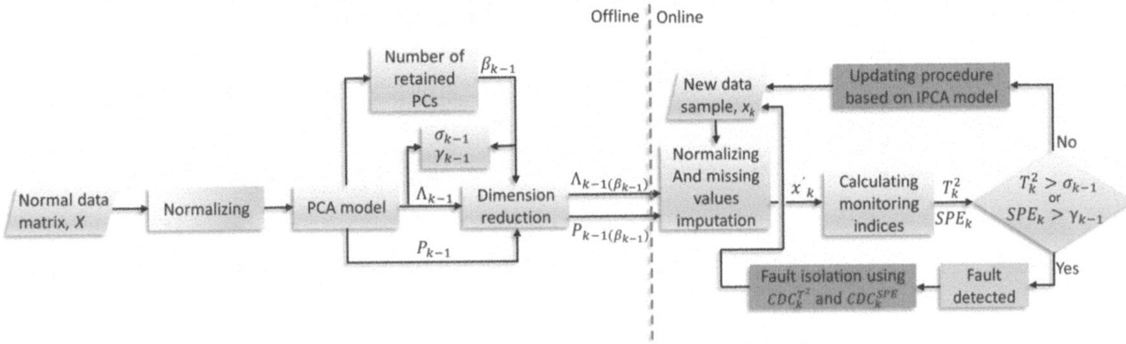

Figure 1 | Adaptive IPCA monitoring diagram.

et al. 2006; Nopens *et al.* 2010). BSM2 is a well-known and powerful dynamic mathematical simulation model able to simulate the physicochemical and biological phenomena in WRRF. BSM2 comprises numerous unit operations, including primary clarifier, activated sludge biological reactor, secondary clarifier, thickener, anaerobic digester, dewatering unit, and storage tank. Five reactors are included in the activated sludge process: two anoxic reactors (used for nitrification) with a total volume of 3,000 m^3 and three aerobic reactors (used for predenitrification) with a total volume of 9,000 m^3. The plant capacity is 20,648 $m^3\,d^{-1}$ with an average influent dry weather flow rate of 592 $mg\,L^{-1}$ of average biodegradable COD. Activated Sludge Model No. 1 (ASM1) and the Anaerobic Digestion Model (ADM1) were used to describe the biological phenomena that take place in the activated sludge and AD reactor, respectively. The influent characteristics consist of a 609-day dynamic influent data file (with sampling frequency equal to a data point every 15 min) that includes rainfall and seasonal temperature variations over the year (Jeppsson *et al.* 2006; Nopens *et al.* 2010).

Fault detection and isolation

BSM2 simulation was modified to obtain sensor data from various parts of the process by simulating different types of fault. By performing this modification, 320 data measurements (16 state variables × 20 measurement points) can be obtained. However, measuring all these variables is not common practice in a real WRRF. In this paper, therefore, we considered only realistic and commonly available sensor measurements. Twenty-eight process measurements obtained from various parts of BSM2 simulation (Figure 2) were recorded every 15 min as the inputs for the proposed monitoring framework. All these variables were corrupted with white noise to simulate real sensor measurements.

To build the offline adaptive IPCA process monitoring framework, 21 days of data measurements (days 435–456) were recorded under normal operating conditions. We chose this period because there were no rain or storm events on those days. The measurements from days 457 to 530 were used to test the performance of the IPCA during time-varying characteristics and abnormal behavior. On those days, there were three storm events, which we will discuss in the next section. The main sources of time-variant behavior in this simulation are drift and abrupt changes in the total suspended solid (TSS) concentration of the reject flow of the dewatering process. These changes in the TSS concentration of the dewatering unit are due to seasonal and diurnal patterns. The number of retained PCs, β_k, was estimated recursively using the CPV method in such a way that the retained PCs explained almost 99% of total variance. Several varieties of faults in the sensors, process variables and process parameters were simulated according to Table 1. Each fault began on day 470 and continued until day 530. The thresholds confidence limit for all charts was considered equal to 99%.

Normal operation of the WRRF

First, we simulated the normal operation of the process (i.e. with no fault added) to show that conventional PCA is unable to deal with the time-variant characteristics of WRRF. The monitoring statistics obtained by conventional PCA and IPCA are shown in Figure 3.

Before we discuss the monitoring results, note that the three severe peaks in this figure for both PCA and IPCA represent storm events (around days 490, 505 and 523) during normal operation of the WRRF. These storm events, which exist in all the fault detections studied in this paper, are classified as faults. As Figure 3 shows, the statistics for conventional PCA exceeded their thresholds for several normal

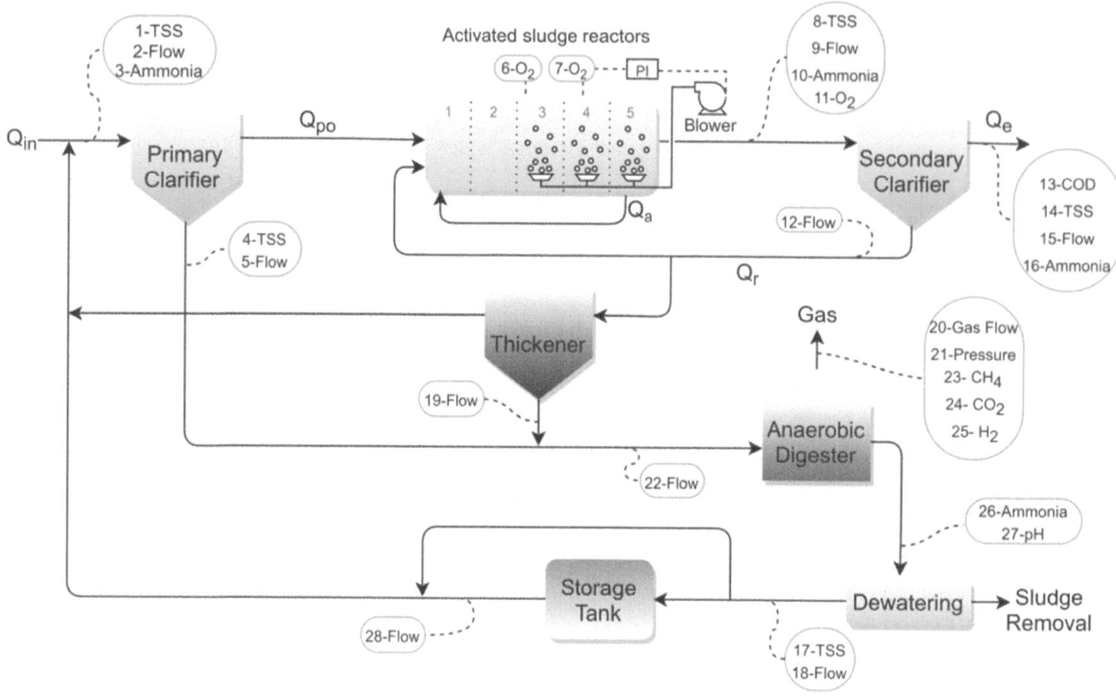

Figure 2 | Schematic diagram of BSM2 and locations of the measured variables.

Table 1 | Description of the simulated faults

Fault#	Description		Fault type
1	Sensor fault	Sensor no. 7	Drift fault ($0.1\ g{\cdot}m^{-3}{\cdot}d^{-1}$)
2	Change in inorganic nitrogen of anaerobic digester	S_{in}	Added $0.2\ kmol{\cdot}m^{-3}$ to its normal value
3	Secondary clarifier parameter	v_s	Decreased by 50%
4	Step change in bioreactor parameters	μ_A	Decreased from 0.5 to $0.1\ d^{-1}$

operational points in addition to these storm events. In other words, these results indicate the excessive rate of false alarms, which illustrates a major limitation of using conventional PCA for time-varying processes such as WRRF. What conventional PCA captures in this case is only the dynamic and statistical characteristics of the process supplied by the training data set. However, in time-varying processes, these dynamic or statistical characteristics change over time and must therefore be updated rather than considered constant. Unlike PCA (which produces an excessive rate of false alarms), IPCA produces a very low rate of false alarms. Apart from storm events, which are correctly labeled as faults, data points mislabeled by the T^2 and SPE statistics of IPCA are very few (shown in red). Also, since the correlation for T^2 did not change, the threshold for T^2 remained constant. On the other hand, due to adaptation, the threshold for SPE varied more although its range remained quite low, i.e. between 0.115 and 0.121 (see Figure 3).

Drift fault in the dissolved oxygen sensor

The first fault, shown in Table 1, is simulated by applying a drift in the oxygen sensor of the fourth reactor at day 470. This sensor is used in combination with a proportional-integral (PI) controller to manipulate the concentration of dissolved oxygen in the third, fourth, and fifth reactors. The size of the drift was $0.1\ g{\cdot}m^{-3}{\cdot}d^{-1}$. This was intentionally small in order to evaluate the adaptive behavior of IPCA during such a fault. The monitoring statistics and results of diagnosis for IPCA are shown in Figure 4. As we can see, both monitoring statistics detected the fault, though there were some delays. These delays in detection were due to the fact that, as the drift starts, the PI controller tries to compensate for them and brings the dissolved oxygen concentration back to the setpoint by decreasing the blower speed. The fault therefore cannot be detected directly from the faulty sensor at the initial stage of its

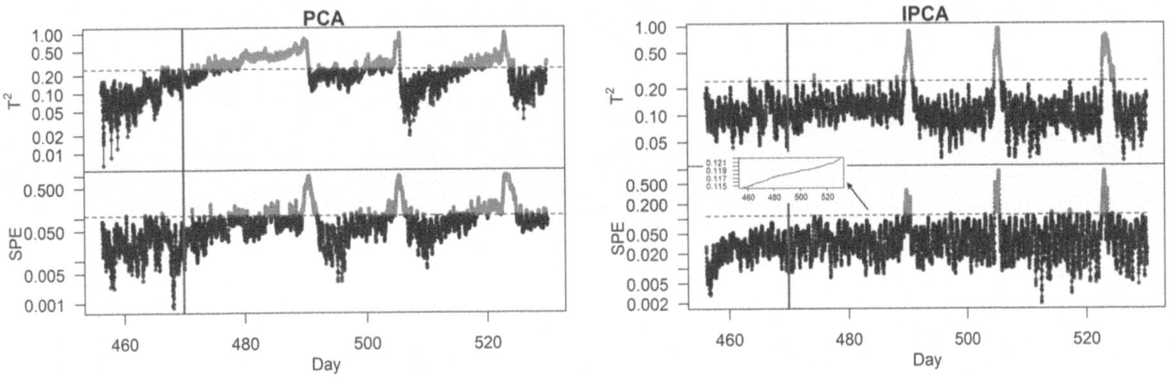

Figure 3 | Fault detection results of PCA and IPCA for normal operation (dashed blue line: 99% confidence limit; solid vertical blue line: onset of the fault). The small plot in the IPCA chart shows the adaptive threshold for SPE. Please refer to the online version of this paper to see this figure in color: http://dx.doi.org/10.2166/wst.2020.368.

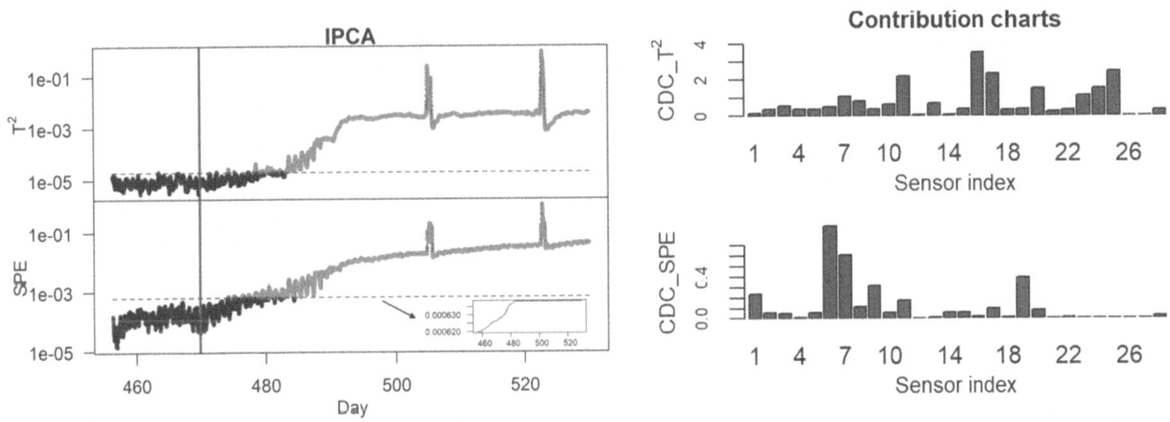

Figure 4 | Fault detection and diagnosis results of IPCA for drift fault in the dissolved oxygen sensor (dashed blue line: 99% confidence limit; solid vertical blue line: onset of the fault). The small plot inside the IPCA chart shows the adaptive threshold for SPE. Please refer to the online version of this paper to see this figure in color: http://dx.doi.org/10.2166/wst.2020.368.

occurrence. The impact of the PI controller action is more significant on dissolved oxygen in the third and fifth reactors at the initial stage of the fault occurrence.

The SPE statistic therefore begins to violate the threshold around day 475. This is due to the change in the dissolved oxygen concentrations in the third and fifth reactors. However, it is difficult to allocate the drift because it is very small. The T^2 statistic begins to violate the threshold significantly around day 483, which indicates that the PI controller reduces the blower speed to its minimum. At this point, most of the variables begin to deviate and the fault becomes more visible and easier to detect. The contribution chart for T^2 shows that the ammonia, H_2, TSS, and O_2 sensors (numbers 16, 25, 17, and 11, respectively; see Figure 2 for the type and location of sensors) contribute the most. Of these sensors, sensor 11, which represents the dissolved oxygen in the fifth reactor, is isolated correctly. However, the other sensors, such as the concentration of ammonia in the effluent stream (number 16), may provide indirect clues as to the root of the fault. As the concentration of dissolved oxygen decreases due to the action of the PI controller, the nitrification rate decreases, so the concentration of ammonia increases. The most contributing sensors for the SPE statistic are numbers 6 and 7, which represent the concentration of dissolved oxygen in the third and fourth reactors, respectively. Sensor 6 contributes more than faulty sensor number 7, which shows that the PI controller has more impact on the concentration of dissolved oxygen in the third reactor at the initial stage of the fault.

Step change in inorganic nitrogen in the anaerobic digester

The nitrogen level in the anaerobic digestion (AD) process is critical because of its inhibitory impact on microbial activity and needs to be monitored carefully. Fault number 2 (Table 1) is simulated by inducing a step change at day 470 equal to $+0.2\ kmol{\cdot}m^{-3}$ in the AD inorganic nitrogen

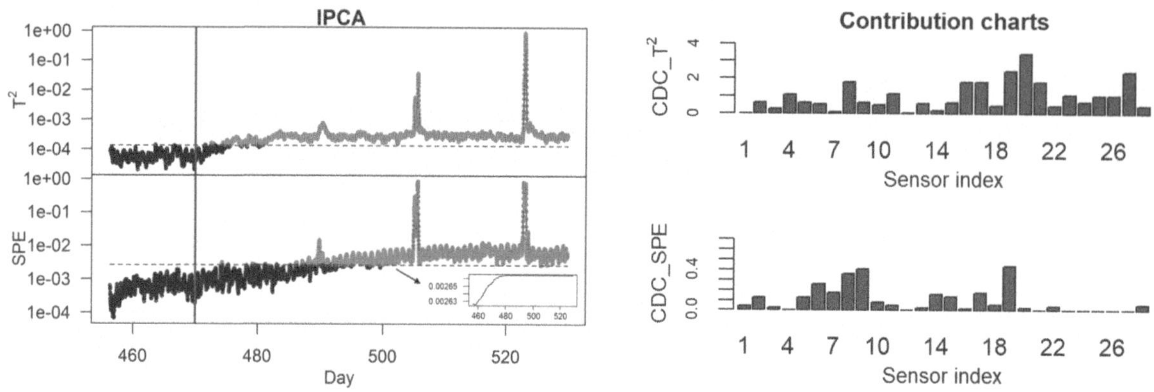

Figure 5 | Fault detection and diagnosis results of IPCA for a step change in AD inorganic nitrogen (dashed blue dashed: 99% confidence limit; solid vertical blue line: onset of the fault). The small plot inside the IPCA chart shows the adaptive threshold for SPE. Please refer to the online version of this paper to see this figure in color: http://dx.doi.org/10.2166/wst.2020.368.

level. The monitoring statistics contribution charts are shown in Figure 5.

As we can see, both monitoring statistics detect the fault, though with some delays. The delay is shorter in T^2 than in SPE. These delays are due to the small size of the faults and their lesser impact at the initial stage of their occurrence. Figure 5 shows that sensor 20 has the largest contribution to the T^2 statistic. Sensor 20 is the gas flow of AD, which is correctly isolated. Due to the inhibition phenomena caused by the increase in inorganic nitrogen, the activity of microbial communities decreases, which accounts for the reduction in gas flow. The contribution chart for SPE could not isolate the fault correctly.

Step change in the settling velocity of the secondary clarifier

To simulate fault number 3 (Table 1), the double exponential settling velocity function (v_s) of the second clarifier was reduced by 50%. This fault can be classified as a change in the process parameters. Figure 6 shows the detection performance and contribution charts for the T^2 and SPE statistics.

Figure 6 shows that both statistics can detect the fault instantly after its occurrence. However, SPE has a stronger detection than T^2. Clearly, both statistics provide accurate isolation. Sensors 13 (carbon-oxygen demand (COD)) and 14 (TSS) contribute the most to T^2 and SPE, respectively. Since there is no sensor to measure settling velocity, COD and TSS are considered the most relevant sensors (these variables are directly related to the settling velocity). Note that the smaller magnitude of this fault could still be detected, though the detection rate may be lower due to its lesser impact on the process.

Step change in the bioreactor parameters

One of the most crucial parameters in WRRF is the specific growth rate for the autotrophs (μ_A), which determines the speed of ammonia conversion into nitrite (nitrification

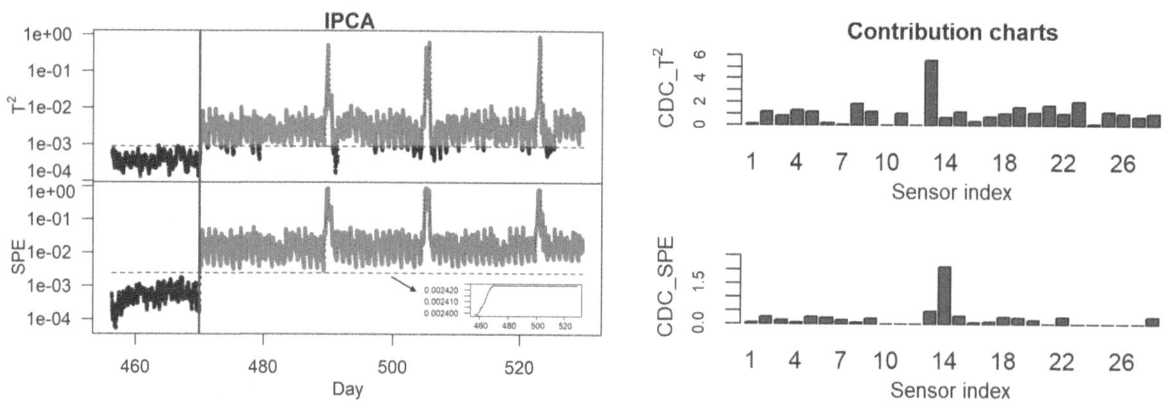

Figure 6 | Fault detection and diagnosis results of IPCA for a step-change in the settling velocity of the secondary clarifier (dashed blue line: 99% confidence limit; solid vertical blue line: onset of the fault). The small plot inside the IPCA chart shows the adaptive threshold for SPE. Please refer to the online version of this paper to see this figure in color: http://dx.doi.org/10.2166/wst.2020.368.

rate). If the inhibition (due to toxicity or changes in pH, etc.) occurs in the activated sludge reactors, the nitrification rate is altered due to the lower microbial activity. To simulate fault number 4 (Table 1), the value of μ_A was changed from 0.5 to 0.1 d^{-1}. The monitoring statistics and contribution charts are shown in Figure 7.

This figure shows that the T^2 statistic begins to violate the threshold earlier than the SPE statistic. Although T^2 begins violating the threshold in the initial stage of the fault occurrence, it took almost 30 days to be completely above the threshold. As the nature of this fault is very similar to the drift fault, the detection delay for both methods is very high due to the lesser impact of the faults at the initial stage of their occurrence. The variable contribution chart shows that sensor 10 (the concentration of ammonia in reactor 5) has clearly contributed the most to the deviation in the T^2 statistic. As the nitrification rate decreases, the concentration of ammonia increases. The contribution chart of the SPE statistic provides no information about the root cause of the fault.

Storm events

As we mentioned earlier, there are three storm events in the simulated data set. These events are visible in Figure 3 around days 490, 505, and 523. Figure 3 also shows that both the T^2 and SPE statistics were able to detect these events accurately. Moreover, as the storm events end, the statistical indices immediately go back to their normal values, which indicates the robustness of the method during such a fault. Figure 8 shows the variables responsible for the deviation in the T^2 and SPE statistics during the storm events. The sensors' contributions to these three storm events are quite similar. As well as the increasing influent flowrate, many other variables are influenced by storm events; therefore, it is difficult to isolate the fault just by looking at the contribution charts. The trends in all correlated variables therefore need to be examined by experienced process operators. The contribution chart shows that sensors 13 and 19 (effluent COD and thickener underflow flow rate) contribute the most to the deviation in T^2. During the

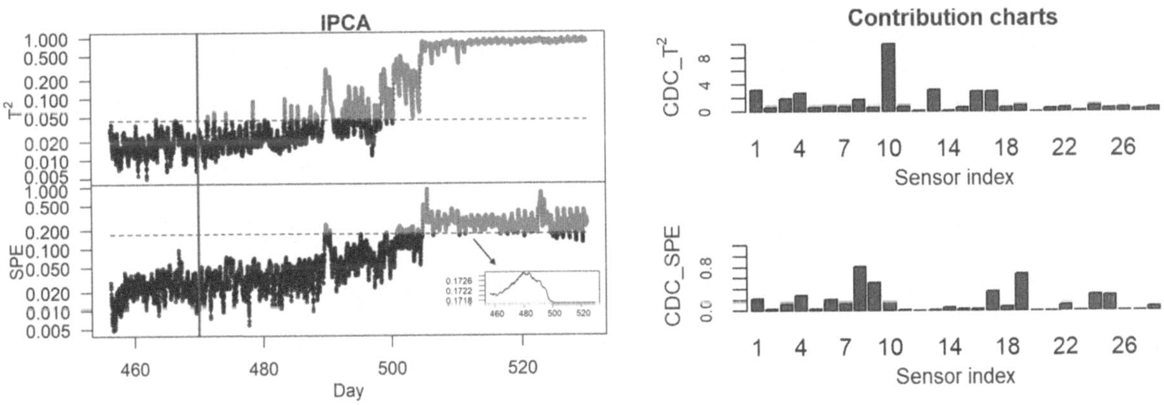

Figure 7 | Fault detection and diagnosis results of IPCA for a step change in the specific growth rate of the autotrophs (dashed blue line: 99% confidence limit; solid vertical blue line: onset of the fault). The small plot inside the IPCA chart shows the adaptive threshold for SPE. Please refer to the online version of this paper to see this figure in color: http://dx.doi.org/10.2166/wst.2020.368.

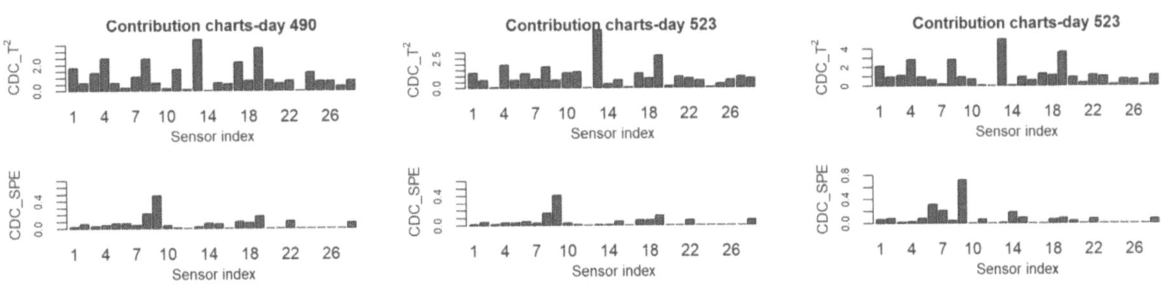

Figure 8 | Sensors' contribution to T^2 and SPE statistics at different days of operation under storm condition.

storm events, the COD also decreases due to the dilution effect caused by the increase in plant influent flow.

The thickener underflow flow rate also increases as a result of the increase in plant influent flow. Sensor 9 (inlet flow rate of the second clarifier) contributes the most to SPE deviation, which is directly related to the plant influent flow rate. Therefore, as the plant influent flow rate increases due to the storm, the thickener underflow flowrate also increases.

FAULT DETECTION WITH MISSING VALUES

Missing values in the input data vector of the monitoring framework is a complex and challenging issue. However, in practice, missing values due to sensor failure or maintenance activities are common. A robust monitoring framework must therefore be designed to handle these missing values. In this study, missing values are imputed using EBLUP. The advantage of this method is that these values can be introduced as they appear during the fault detection procedure, which is useful for realtime applications. To assess the accuracy of our proposed method in the presence of missing sensor signals, several measurements were deleted randomly from the data set. Figure 9 shows the pattern of missing values from day 457 to day 488 during normal operation of the WRRF. This period was chosen to avoid the storm events during the normal running of the plant. Sampled data are shown in a continuous grey/black color scheme (the darker the color, the higher the value of the sensor), while missing values are shown in red. As we can see, the pattern of the missing values is extremely intricate since many sensor signals can become unavailable at the same time, which is challenging for any fault detection and monitoring framework.

Table 2 shows the false alarm rate (FAR%) and missed detection rate (MDR%) of the monitoring framework for the complete and incomplete data sets during normal and faulty operation of the WRRF. These rates can be estimated as follows:

$$FAR\% = \frac{\textit{Number of normal samples incorrectly classified as faulty}}{\textit{Total number of normal samples}} \tag{22}$$

$$MDR\% = \frac{\textit{Number of faulty samples incorrectly classified as normal}}{\textit{Total number of faulty samples}} \tag{23}$$

We can see that the FAR% for the T^2 statistic is not affected by the missing values. On the other hand, the FAR% for the SPE statistic increases significantly in the presence of missing values. The most significant changes to the FAR% occurred with the largest level of missingness (10%). This may be because the imputation of missing values has a more significant impact on the residual subspace compared to the mean and covariance in PCs. Since the variation in the residual subspace is measured by SPE, it is more sensitive to the amount of imputed missing values.

CONCLUSION

We have proposed a novel fault detection and isolating framework based on incremental PCA (IPCA) and complete decomposition contribution (CDC). The advantage of IPCA over other adaptive PCA approaches is that it does not require all eigenvalues to be computed. This characteristic speeds up computation time. The proposed adaptive IPCA monitoring framework consists of (i) adaptive estimation of the mean and variance of data, (ii) adaptive estimation of selected PCs and threshold, (iii) estimation of eigenvalue decomposition by incrementing the new data to the PCA, and (iv) adaptive estimation of the contribution

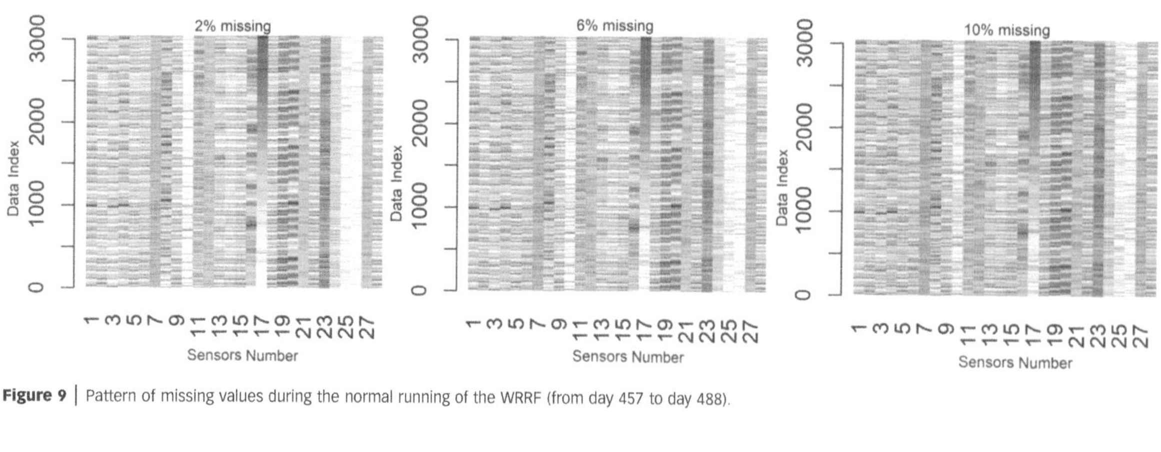

Figure 9 | Pattern of missing values during the normal running of the WRRF (from day 457 to day 488).

Table 2 | Fault detection performance for complete and uncompleted data set

		0%[a]		2%[a]		6%[a]		10%[a]	
	Fault#	**T^2**	**SPE**	**T^2**	**SPE**	**T^2**	**SPE**	**T^2**	**SPE**
FAR %	normal[b]	0.45	0.26	0.59	3.87	0.49	14.05	0.52	26.49
MDR%		–	–	–	–	–	–	–	–
FAR%	1	0.07	0	0.22	3.20	0.15	15.70	0.15	30.10
MDR%		18.53	14.70	18.4	12.99	18.5	10.33	18.58	8.80
FAR%	2	0.07	0	0.22	3.20	0.15	15.70	0.15	30.10
MDR%		8.56	34.26	8.57	29.95	8.44	24.07	8.68	19.86
FAR%	3	0.07	0	0.22	3.20	0.15	15.7	0.15	30.10
MDR%		7.91	0.06	8.45	0.08	9.72	0.08	11.41	0.06
FAR%	4	0.07	0	0.22	3.20	0.15	15.70	0.15	30.10
MDR%		37.86	55.99	37.84	55.40	38.10	54.16	38.34	53.90

[a]Percent of missing values; [b]from day 457 to day 488 during normal operation; MDR%, missed detection rate; FAR%, false alarm rate.

charts for T^2 and SPE. This framework was applied to BSM2. Our results, demonstrated with simulated faulty scenarios, show that IPCA is able to adapt time-varying process behavior while detecting and isolating faults. Although there are some delays in detecting the faults, the performance is acceptable since the faults under study were small at the initial stage of their occurrence. The contribution charts showed that this method was able to correctly isolate the faults. Sometimes, however, due to a lack of sensor measurements, it was not possible to isolate the sensor directly and only the impact of the fault on the other measurements could be detected. Our proposed framework can also handle in realtime highly intricate patterns of missing values (e.g. more than one sensor signal failure at the same time) by applying the EBLUP method. Our study of fault detection performance with missing values shows that SPE is more sensitive to the imputation of missing values than T^2. This is due to the significant impact of the imputed values on the residual subspace.

ACKNOWLEDGEMENTS

We gratefully thank the Spanish Ministry of Economy and Competitiveness for providing financial support through grants CTM2015-67970-P and for funding the doctoral scholarship (BES-2012-059675). Some authors are recognized by the Comissioner for Universities and Research of the DIUE (Department of Innovation, Universities and Business) of the autonomous government of Catalonia (2017 SGR 396) and supported by the Universitat Rovira i Virgili (2017PFR-URV-B2-33) and Fundació Bancària 'la Caixa' (2017ARES-06). The authors are also grateful to Dr Ulf Jeppsson of Lund University, Sweden, for kindly providing the BSM2 Matlab code.

DATA AVAILABILITY STATEMENT

All relevant data are included in the paper or its Supplementary Information.

REFERENCES

Alcala, C. F. & Qin, S. J. 2009 Reconstruction-based contribution for process monitoring. *Automatica* **45**, 1593–1600. www.elsevier.com/locate/automatica (Accessed April 10, 2020).

Arora, R., Cotter, A., Livescu, K. & Srebro, N. 2012 Stochastic optimization for PCA and PLS. 2012 50th Annual Allerton Conference on Communication, Control, and Computing. *Allerton* **2012**, 861–868.

Artač, M., Jogan, M. & Leonardis, A. 2002 Incremental PCA for online visual learning and recognition. In *Proceedings – International Conference on Pattern Recognition*, pp. 781–784.

Brand, M. 2002 Incremental singular value decomposition of uncertain data with missing values. Lecture Notes in Computer Science (including subseries Lecture Notes in Artificial Intelligence and Lecture Notes in Bioinformatics), 2350, 707–720.

Cardot, H. & Degras, D. 2018 Online principal component analysis in high dimension: which algorithm to choose? *International Statistical Review* **86** (1), 29–50.

Choi, S. W., Martin, E. B., Morris, A. J. & Lee, I. B. 2006 Adaptive multivariate statistical process control for monitoring time-varying processes. *Industrial and Engineering Chemistry Research* **45** (9), 3108–3118.

Elshenawy, L. M. & Awad, H. A. 2012 Recursive fault detection and isolation approaches of time-varying processes. *Industrial and Engineering Chemistry Research* **51** (29), 9812–9824.

Elshenawy, L. M. & Mahmoud, T. A. 2018 Fault diagnosis of time-varying processes using modified reconstruction-based contributions. *Journal of Process Control* **70**, 12–23. https://doi.org/10.1016/j.jprocont.2018.07.017.

Elshenawy, L. M., Yin, S., Naik, A. S. & Ding, S. X. 2010 Efficient recursive principal component analysis algorithms for process monitoring. *Industrial and Engineering Chemistry Research* **49** (1), 252–259.

Garcia-Alvarez, D., Fuente, M. J., Vega, P. & Sainz, G. 2009 Fault detection and diagnosis using multivariate statistical techniques in a wastewater treatment plant. *IFAC Proceedings Volumes* **42** (11), 952–957.

Haimi, H., Mulas, M., Corona, F., Marsili-Libelli, S., Lindell, P., Heinonen, M. & Vahala, R. 2016 Adaptive data-derived anomaly detection in the activated sludge process of a large-scale wastewater treatment plant. *Engineering Applications of Artificial Intelligence* **52**, 65–80. http://dx.doi.org/10.1016/j.engappai.2016.02.003.

Hall, P. M., Marshall, D. & Martin, R. R. 1998 Incremental eigenanalysis for classification. In: Proceedings of the British Machine Conference 1998. BMVA Press, September 1998, pp. 29.1-29.10. Available from: http://www.bmva.org/bmvc/1998/papers/d186/h186.htm (accessed 5 May 2020).

Hu, Z., Chen, Z., Hua, C., Gui, W., Yang, C. & Ding, S. X. 2012 A simplified recursive dynamic PCA based monitoring scheme for imperial smelting process. *International Journal of Innovative Computing, Information and Control* **8** (4), 2551–2561.

Jeppsson, U., Rosen, C., Alex, J., Copp, J., Gernaey, K. V., Pons, M.-N. & Vanrolleghem, P. A. 2006 Towards a benchmark simulation model for plant-wide control strategy performance evaluation of WWTPs. *Water Science and Technology* **53** (1), 287–295. Available from: https://iwaponline.com/wst/article/53/1/287/12474/Towards-a-benchmark-simulation-model-for-plantwide (accessed 2 September 2018).

Jun, B. H., Park, J. H., Lee, S. I. & Chun, M. G. 2006 Kernel PCA based faults diagnosis for wastewater treatment system. In *Lecture Notes in Computer Science (including Subseries Lecture Notes in Artificial Intelligence and Lecture Notes in Bioinformatics)*. Springer Verlag, Berlin, Germany, pp. 426–431.

Kazemi, P., Steyer, J. P., Bengoa, C., Font, J. & Giralt, J. 2020 Robust data-driven soft sensors for online monitoring of volatile fatty acids in anaerobic digestion processes. *Processes* **8** (1), 67.

Lee, J. M., Yoo, C. K., Choi, S. W., Vanrolleghem, P. A. & Lee, I. B. 2004 Nonlinear process monitoring using kernel principal component analysis. *Chemical Engineering Science* **59** (1), 223–234.

Lee, C., Choi, S. W. & Lee, I. B. 2006 Sensor fault diagnosis in a wastewater treatment process. *Water Science and Technology* **53** (1), 251–257. Available from: https://iwaponline.com/wst/article/53/1/251/12470/Sensor-fault-diagnosis-in-a-wastewater-treatment (accessed 23 March 2020).

Li, W., Yue, H. H., Valle-Cervantes, S. & Qin, S. J. 2000 Recursive PCA for adaptive process monitoring. *Journal of Process Control* **10** (5), 471–486.

Mina, J. & Verde, C. 2006 Fault detection for large scale systems using dynamic principal components analysis with adaptation. In *IFAC Proceedings Volumes (IFAC-PapersOnline)*. pp. 220–225.

Newhart, K. B., Holloway, R. W., Hering, A. S. & Cath, T. Y. 2019 Data-driven performance analyses of wastewater treatment plants: a review. *Water Research* **157**, 498–513. https://doi.org/10.1016/j.watres.2019.03.030.

Nomikos, P. & MacGregor, J. F. 1995 Multivariate SPC charts for monitoring batch processes. *Technometrics* **37** (1), 41–59.

Nopens, I., Benedetti, L., Jeppsson, U., Pons, M.-N., Alex, J., Copp, J. B., Gernaey, K. V., Rosen, C., Steyer, J.-P. & Vanrolleghem, P. A. 2010 Benchmark simulation model No. 2: finalisation of plant layout and default control strategy. *Water Science and Technology* **62** (9), 1967–1974. Available from: http://www.ncbi.nlm.nih.gov/pubmed/21045320 (accessed 22 August 2018).

Rosen, C. & Lennox, J. A. 2001 Multivariate and multiscale monitoring of wastewater treatment operation. *Water Research* **35** (14), 3402–3410.

Sanchez-Fernández, A., Fuente, M. J. & Sainz-Palmero, G. I. 2015 Fault detection in wastewater treatment plants using distributed PCA methods. In: *IEEE International Conference on Emerging Technologies and Factory Automation*, 2015 October 1. ETFA, Luxembourg.

Sánchez-Fernández, A., Baldán, F. J., Sainz-Palmero, G. I., Benítez, J. M. & Fuente, M. J. 2018 Fault detection based on time series modeling and multivariate statistical process control. *Chemometrics and Intelligent Laboratory Systems* **182** (July), 57–69.

Shang, L., Liu, J., Turksoy, K., Min Shao, Q. & Cinar, A. 2015 Stable recursive canonical variate state space modeling for time-varying processes. *Control Engineering Practice* **36**, 113–119.

Wang, L. & Shi, H. 2010 Multivariate statistical process monitoring using an improved independent component analysis. *Chemical Engineering Research and Design* **88** (4), 403–414.

Xiao, H., Huang, D., Pan, Y., Liu, Y. & Song, K. 2017 Fault diagnosis and prognosis of wastewater processes with incomplete data by the auto-associative neural networks and ARMA model. *Chemometrics and Intelligent Laboratory Systems* **161**, 96–107. Available from: https://www.sciencedirect.com/science/article/abs/pii/S0169743916301812 (accessed 6 March 2019).

Zhang, Z. & Dong, F. 2014 Fault detection and diagnosis for missing data systems with a three time-slice dynamic Bayesian network approach. *Chemometrics and Intelligent Laboratory Systems* **138**, 30–40.

First received 11 June 2020; accepted in revised form 22 July 2020. Available online 5 August 2020

 © IWA Publishing 2020

Use of real-time sensors for compliance monitoring of nitrate in finished drinking water

Christopher S. Jones, Tianyi Li, Alex Sukalski, Darrin A. Thompson and David M. Cwiertny

ABSTRACT

Across the Midwestern United States, Public Water Systems (PWSs) struggle with high levels of nitrate in source waters from intense agricultural activity. Leveraging a sensor network deployed across Iowa surface waters, we evaluated the potential of the Hach Nitratax SC Plus, which uses UV-light absorption to quantify dissolved nitrate-nitrite (NO_x-N) down to 0.1 mg-N L^{-1}, for real-time monitoring of NO_x-N in drinking water. For six different PWSs over multiple years, we compare NO_x-N levels in source waters (surface and groundwater under surface influence) to those measured via traditional methods (e.g., ion chromatography (IC)) for US EPA compliance monitoring. At one large PWS, we also evaluated sensor performance when applied to near-finished drinking water (filter effluent). We find good agreement between traditional analytical methods and *in situ* sensors. For example, for 771 filter effluent samples from 2006–2011, IC analysis averaged NO_x-N of 5.8 mg L^{-1} while corresponding sensor measurements averaged 5.7 mg L^{-1} with a mean absolute error of 0.23 (5.6%). We identify several benefits of using real-time sensors in PWSs, including improved frequency to capture elevated NO_x-N levels and as decision-support tools for NO_x-N management.

Key words | compliance, drinking water, monitoring, nitrate, regulation, sensor

Christopher S. Jones
David M. Cwiertny (corresponding author)
IIHR Hydroscience and Engineering,
University of Iowa,
Iowa City, IA,
USA
E-mail: *david-cwiertny@uiowa.edu*

Tianyi Li
Alex Sukalski
Darrin A. Thompson
David M. Cwiertny
University of Iowa Center for Health Effects of Environmental Contamination,
Iowa City, IA,
USA

Tianyi Li
Department of Earth and Environmental Science,
University of Iowa,
Iowa City, IA,
USA

Alex Sukalski
David M. Cwiertny
Public Policy Center,
University of Iowa,
Iowa City, IA,
USA

Darrin A. Thompson
Department of Occupational and Environmental Health,
University of Iowa College of Public Health,
Iowa City, IA,
USA

David M. Cwiertny
Department of Civil and Environmental Engineering,
University of Iowa,
4105 Seamans Center for the Engineering Arts and Sciences,
Iowa City, IA,
USA

HIGHLIGHTS

- Monitoring of drinking water nitrate using *in situ* sensors at six community water systems is evaluated.
- Sensors placed in source and finished water agree well with more traditional nitrate analyses.
- Sensors provide accurate, high frequency monitoring data responsive to temporal dynamics in agricultural watersheds.
- Automated nitrate sensors can improve monitoring for better compliance with health-based regulations.
- Sensors can also be used as decision-support tools to guide nitrate management strategies.

INTRODUCTION

Nitrate (NO_3^-) is one of the original parameters identified by the 1974 Safe Drinking Water Act (SDWA) (Tiemann 2014) for regulation in US Public Water Systems (PWSs). A limit (maximum contaminant level (MCL)) of 10 mg L^{-1} (as nitrogen, e.g. NO_3-N) has been enforced by the states and the US Environmental Protection Agency (USEPA) since 1977. Likewise, the World Health Organization (WHO) set a drinking water guideline of 50 mg L^{-1} for NO_3^- (11 mg L^{-1} NO_3-N) in 1970 (WHO 1970). These limits were established to be protective against methemoglobinemia, a human infant (<6 months old) disease resulting from inadequate blood oxygen levels. Nitrate consumed by

doi: 10.2166/wst.2020.365

the infant in water or infant formula prepared with high-nitrate water is converted to nitrite (NO_2^-) in the infant's digestive tract and is absorbed into the bloodstream. Nitrite reacts with hemoglobin to form methemoglobin, reducing the blood's ability to transport oxygen (Kross *et al.* 1992). This condition has been known in the USA since at least the 1940s (Comly 1945).

More recent research has suggested adverse health effects are possible for adults that consume high-nitrate water. Nitrite resulting from the reduction of nitrate consumed in drinking water can react with amides and amines in the bladder and the digestive system to form N-nitroso compounds with potent carcinogenity (Walker 1990; Bruning-Fann & Kaneene 1993). Weyer *et al.* (2001) found positive associations between drinking water nitrate levels and bladder and ovarian cancer in Iowa (USA) women for the period 1955–1988. An association between drinking water nitrate and female bladder cancer was also reported by Jones *et al.* (2016).

Other cancers have also been linked to drinking water nitrate. Schullehner *et al.* (2018) found statistically significant increased colorectal cancer risk for residents of Denmark consuming water with nitrate at levels above 0.9 mg/L (NO_3-N), less than one-tenth the US and WHO standards. A positive association between drinking water nitrate concentrations and non-Hodgkins lymphoma was reported by Gulis *et al.* (2002) for Slovakia over 1986–1995. Ward *et al.* (2018) reported that the most likely links between drinking water nitrate and negative health outcomes (not including methemoglobinemia) are increased risk of colorectal cancer, thyroid disease, and neural tube defects.

In the Midwestern-USA state of Iowa, PWSs face an ongoing struggle with nitrate contamination of surface and groundwater resources, an outcome of intense row crop and animal agriculture conducted on 80% of the state's land area (Hanson *et al.* 2016; Jones *et al.* 2018a, 2018b; Jones & Schilling 2019; McGill *et al.* 2019; Secchi & Cwiertny 2019). Approximately 60 of the state's drinking water utilities provide water to customers that exceeds 5 mg L^{-1} NO_3-N, and nearly 30% of the state's 880 PWSs have been deemed vulnerable to nitrate contamination (Eller 2015). Private wells are also affected, with about 12% containing NO_3-N concentrations greater than 10 mg L^{-1} (CHEEC 2009).

Enforcement of SDWA regulations in Iowa is a function of the Iowa Department of Natural Resources (IDNR). Regulations for nitrate, like those for most other regulated parameters, are enforced with compliance monitoring dictated by each utility's PWS permit. A utility's required monitoring frequency is dependent upon historical data for the utility, current source water nitrate concentrations, source water vulnerability, available treatment processes and source water blending capabilities. Sampling protocols call for grab sample collection of the finished (treated) water at the treatment plant, followed by delivery to a state-certified testing laboratory where analysis is conducted using Method 353.2, automated colorimetry (O'Dell 1993), or Method 300.0, ion chromatography (Pfaff 1993). A few PWSs have internal certified laboratories where results can be generated in minutes or hours; all others submit samples to the state laboratory or private entities where results may not be known for 48 h (or longer) following sample collection. Required sampling frequency ranges from daily to annually, depending upon the utility's permit.

Nitrate (and nitrate-nitrite in sum, NO_x-N) can be measured continuously *in situ* using devices that work on the principles of UV light absorption, as NO_x-N absorbs at 210 nm. Protocols for the deployment, operation, quality control and data management for these devices have matured rapidly over the past 15 years (Kirchner *et al.* 2004; Pellerin *et al.* 2016; Jones *et al.* 2018a), and data quality generated has been found to be accurate and dependable. Real-time monitoring has long been used by PWSs to track both turbidity and disinfectant levels as a barrier for microbiological contamination of drinking water, and the data has been deemed adequate and necessary for compliance with regulations. Utilities using surface water are required to continuously monitor filter effluent turbidity *in situ* and report exceedances of enforced limits; this data also informs treatment plant operators about impending needs for filter backwash or other maintenance processes (Pontius 1999). Similarly, continuous monitoring of residual chlorine is required in the USA to maintain the integrity of treated drinking water leaving the treatment plant and in the water distribution system (USEPA 1989). Real-time monitoring of parameters such as pH, specific conductance, oxidation-reduction potential and temperature has long helped the water supply industry optimize treatment processes and drinking water quality, protecting the public health. Continuous monitoring also provides an uninterrupted record to autopsy treatment plant upsets, and this information can be used to inform water customers about lapses in water quality.

In this study, we postulate that undetected violations of the drinking water standard (10 mg L^{-1} for NO_3-N) may be occurring in Iowa because sampling protocols, lab methods, and PWS permits are unable to account for episodically-high concentrations resulting from the dynamics associated with Iowa's agricultural landscape, extreme weather and altered hydrology. To examine this question in detail, we compared high frequency (15 min) *in situ* nitrate data generated by sensors deployed in Iowa's streams as part of a statewide sensor network with data generated by nearby PWSs according to their operating permit protocols. Our research objectives also included examining how predictive the sensor data was of PWS permit data and assessing the suitability of real-time, continuous, *in situ* sensors to supplement or possibly replace grab sampling and lab monitoring for drinking water compliance for nitrate.

METHODS AND MATERIALS

Drinking water nitrate data

Our approach was to compare laboratory-generated NO_x-N data from six Iowa PWSs with data generated by real-time, continuous, *in situ* NO_x-N sensors deployed in nearby streams, or sensors deployed within the PWS itself. SDWA data for drinking water NO_x-N were obtained through a database maintained by the University of Iowa's Center for Health Effects of Environmental Contamination (CHEEC). CHEEC is a state-funded research center that maintains a database of all compliance data submitted by PWSs in Iowa under the SDWA. Queries to this database provided NO_x-N data used for SDWA compliance from PWSs near deployed in-stream sensors, allowing comparison of sensor data to NO_x-N levels in finished drinking water.

In identifying PWSs for analysis, a priority was placed on those systems that rely on surface water and groundwater under direct influence of surface water (GWUDI). The six utilities used in this studied are identified as PWS1–PWS6. They are described in detail below and summarized in Table 1, and their approximate locations are shown in Figure 1. These PWSs and all others in Iowa are granted water use permits from IDNR that allow them to withdraw stream and alluvial groundwater in amounts that maintain a set minimum flow downstream of the utilities' treatment facilities.

Table 1 | Iowa public water supplies (PWS) evaluated in this study

Utility	Population Served	Water source	Nitrate mitigation strategy
PWS1	63,265	GWUDI	Blending with emergency sources
PWS2	26,684	Surface	Reverse osmosis, blending
PWS3	133,499	GWUDI	Blending
PWS4	12,700	GWUDI	Blending
PWS5	523,098	Surface, GWUDI	Ion exchange, blending
PWS6	25,594	Surface	Blending with emergency sources

GWUDI = Groundwater under direct influence of surface water.

PWS1

This utility draws water from radial collector wells in the Iowa River alluvium. The utility has the capacity to use Iowa River water directly in an emergency and water from a former sand pit as low nitrate dilution water. Treatment processes include lime softening, granular activated carbon filtration, fluoridation, and disinfection with chlorine. No nitrate removal treatment is available.

PWS2

Relying solely on Iowa River water except during high nitrate episodes, this utility has just recently implemented nitrate removal capacity using reverse osmosis when river water nears or exceeds the nitrate MCL. The treatment plant also has the capacity to use water from a deep confined aquifer, and this has been used in the past as low-nitrate dilution water. Other treatment processes include lime softening, rapid sand filtration, fluoridation, and chlorination.

PWS3

This utility uses radial collector wells under the influence of the Cedar River. PWS3 also operates shallow vertical wells in the Cedar alluvium and has an emergency connection to the Cedar River. A total of 50 wells serve two treatment plants. Lime softening, rapid sand filtration, fluoridation and chlorination comprise the treatment train, but nitrate removal is not available. High nitrate episodes in the wells are managed by aggressive management of blending volumes, but the utility is vulnerable to MCL exceedances, although none have occurred to date.

PWS4

Shallow groundwater (15 m) from the Skunk River alluvium is treated using coagulation, sedimentation, lime softening,

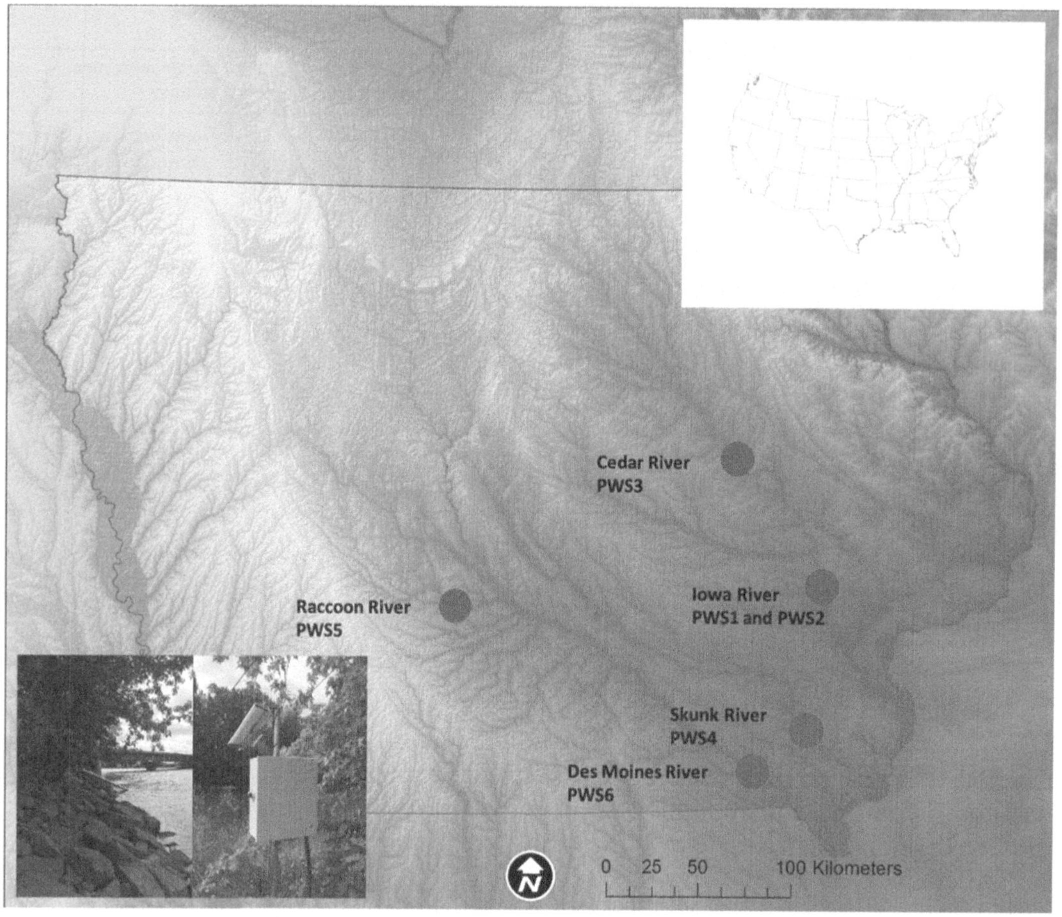

Figure 1 | Location of river sensors and PWSs in this study. In the upper right is the State of Iowa lying within the contiguous US; in the lower left is a typical sensor deployment in an Iowa river. Sensors are slid down the PVC housings into the moving current (left). The dry box on stream bank houses data logger, deep cycle batter, modem and other equipment (right).

filtration, chlorination and fluoridation. No nitrate removal capacity is available.

PWS5

This utility operates three treatment plants that process a diversity of water sources. Raccoon and Des Moines River water is used directly, along with GWUDI of both rivers. The largest treatment facility (pre-sedimentation, lime softening, rapid sand filtration) has ion exchange nitrate removal capacity. A second treatment plant (lime softening, rapid sand filtration) blends down high nitrate from radial collector wells with low nitrate lake water. The third facility (membrane treatment) has reverse osmosis treatment for both hardness and nitrate reduction. PWS5 monitors the filter effluent water from the first two plants using real time continuous nitrate sensors, with the data streamed onto the utility's SCADA system.

PWS6

Des Moines River water is used to provide the municipal supply. Treatment processes include coagulation/sedimentation, lime softening, sand filtration, chlorination and fluoridation.

Real time continuous nitrate measurement

Stream NO_x-N sensor data was aggregated from the University of Iowa's Water Quality Information System (IWQIS) (Jones *et al.* 2018a) and from US Geological Survey water quality stations (USGS 2019). At both the IWQIS and USGS sites, NO_x-N was quantified using the Hach Nitratax SC plus (Loveland, Colorado). Most of the sensors have a 5 mm optical cell path length that can accommodate NO_x-N in the range of 0.1–25 mg L^{-1}. Some of the devices have a 2 mm path length which can measure concentrations up to 50 mg L^{-1}. The Nitratax works on the principles of UV light absorption. As the NO_x-N concentration increases, absorption of UV at 210 nm also increases. A built-in photometer measures the primary beam, while a reference beam corrects for interference caused by dissolved organic material and turbidity, facilitating measurement of unfiltered river water. The sensor's optic windows are automatically cleaned by a wiper mechanism just prior to each measurement.

For the IWQIS sensors, the UV cell was immersed into the stream extending outward several centimeters from a polyvinyl chloride housing, and data were collected streamside with a custom-designed data logger at each site, logging continuously at 15-min intervals. Sensors are retrieved for periodic maintenance and to prevent ice damage during winter and thus some data gaps are inevitable. Each station was powered by a deep cycle 12 V marine battery, continuously charged during daytime by one 20 W solar panel. Each sensor was serviced and calibrated to a known concentration solution by the Hach Service Department according to the manufacturer's specifications before field deployment. A typical IWQIS deployment is shown in Figure 1.

Data generated by the USGS sensors were harvested by IWQIS from the USGS website (USGS 2019). Deployment of USGS sensors and data handling were somewhat different from IWQIS protocols (Pellerin *et al.* 2013). USGS sensors were designated as such within the IWQIS system. More details on the IWQIS and USGS monitoring systems, including data quality control protocols, can be found at Jones *et al.* 2018a.

The IWQIS design is based on the generalized water cyberinfrastructure developed at the University of Iowa (Demir & Krajewski 2013; Krajewski *et al.* 2017). It provides an integrated and scalable workflow environment to support end-to-end cyberinfrastructure for sharing scientific research and raw data (Duffy *et al.* 2012; Gil *et al.* 2016), in addition to scientific visualization and modeling results that support watershed management and research (Demir & Beck 2009; Demir *et al.* 2009, 2015).

Nitratax sensors were also deployed in the filter effluent water of PWS5. Plastic tubing transported water from the combined filter effluent pipe to the bottom of a PVC enclosure that housed the sensor. Water overflowed at the top of the enclosure and was sent to waste, with total water residence time in the housing of ~1 min. Grab samples were periodically collected from the overflow for laboratory analysis for the purpose of lab/sensor data comparison.

Assessing field deployed sensor performance

To assess the accuracy of the Nitratax sensors, additional water quality data were obtained through PWS1, the City of Iowa City Water Treatment Plant (CICWTP) and PWS5, the Des Moines Water Works (DMWW), for comparison to concentrations obtained by in-stream sensors. For PWS1, NO_x-N sensor data from the Iowa River (IWQIS) was compared to NO_x-N data generated by Iowa River grab samples collected by PWS1 staff and analyzed by spectrophotometry. Sensor data generated by continuous monitoring of PWS5 filter effluent water was compared to concentrations measured in grab samples collected by PWS5 staff and also analyzed by ion chromatography in the treatment plant laboratory. Both laboratories (PWS1 and PWS5) are certified by IDNR. This scheme provided the opportunity to assess how concentrations from field- and treatment plant-deployed sensors compared to more traditional forms of analysis for NO_x-N in source water samples.

RESULTS

Sensor versus lab data comparison of NO_x-N

We first intended to establish that the sensor data reliably and accurately reflected river- and drinking-water NO_x-N. Grab samples collected and analyzed by labs at PWS1 (using USEPA-approved Hach Company TNTplus 835/836 Nitrate Method 10206) and PWS5 (using ion chromatography and Method 300.0 (Pfaff 1993)) were compared with the instantaneous sensor measurement at the time of grab sample collection (DMWW-PWS5), or the daily average sensor measurement retrieved from the historical database (CICWTP-PWS1). In the case of PWS1, the grab samples were from the Iowa River just downstream from a Nitratax sensor site. These data are shown in Figure 2(a). Over a 7 year period (2012–2019), lab samples ($n = 346$) averaged 6.9 mg L^{-1} while daily average sensor measurements averaged 7.2 mg L^{-1}, with no temporal trends observed in the lab/sensor NO_x-N ratio. A linear regression line produced an R^2 value of 0.947, while the mean absolute error was 0.55 (16.5%). It should be noted that quality in the Iowa River is very dynamic with large daily and seasonal changes in turbidity, organic carbon, temperature and other physical and chemical parameters.

Figure 2(b) shows a similar comparison for the PWS5 Fleur Drive Treatment Plant. The measured water had undergone all treatment processes occurring in the plant except for NO_x-N removal, disinfection, and fluoridation. Thus it was very similar in character to the finished drinking water provided to the customers, and, unlike the Iowa River, of a very stable quality throughout the year with turbidity

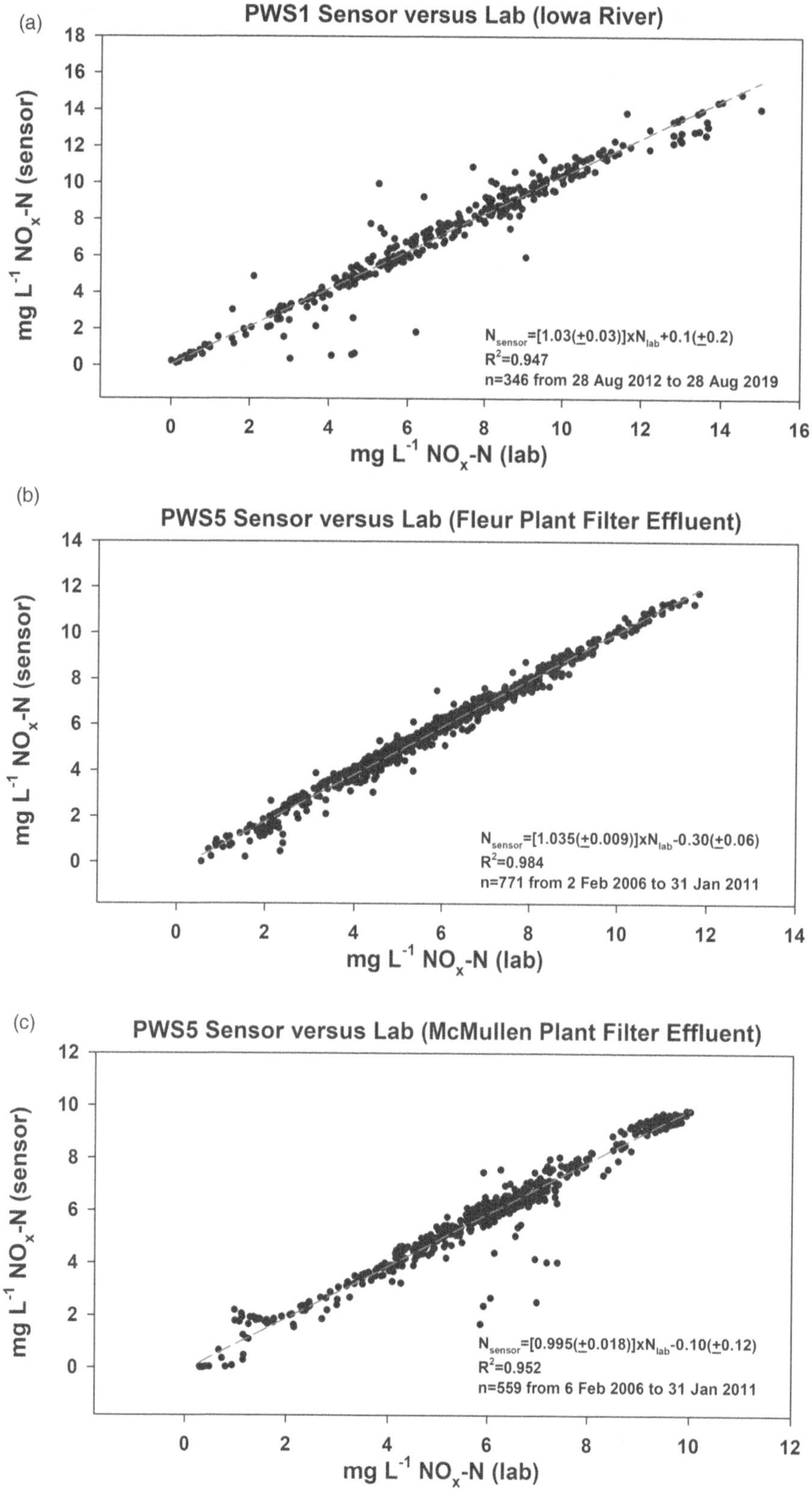

Figure 2 | Comparison of sensor-generated NO_x-N concentrations and data generated by more traditional analysis of water samples collected as grab samples by PWS staff. Sensor data represent instantaneous measurements at the time of collection. A 1-to-1 line (dashed) is shown for comparison. Also provided is the output equation from a best-fit linear regression analysis of the data, where uncertainties in fit coefficients correspond to 95% confidence intervals.

characteristically low following sand filtration. The data period evaluated here was 2006–2011, with the number of grab samples totaling 771. Lab NO_x-N averaged 5.8 mg L^{-1} while instantaneous sensor measurements at the time of grab sample collection averaged 5.7 mg L^{-1}, again with no temporal trends observed in the lab/sensor NO_x-N ratio. A linear regression line produced an R^2 value of 0.984, while the mean absolute error was 0.23 (5.6%).

A third lab-sensor comparison (Figure 2(c)) was conducted on data (2006–2011) generated from the PWS5 McMullen Treatment Plant (filter effluent water). Lab NO_x-N averaged 6.3 mg L^{-1} whereas instantaneous sensor measurements at the time of grab sample collection averaged 6.1 mg L^{-1}. A linear regression line produced an R^2 value of 0.952, while the mean absolute error was 0.29 (7.3%). A few samples produced differences in NO_x-N concentration of >2 mg L^{-1} (Figure 2(c)) with the sensor measuring higher than the lab; these all occurred early in the period of record (2006–2007).

Stream sensors versus SDWA data

Time series stream sensor NO_x-N data were tracked along with finished drinking water data for the nearby PWS with the presumption that the two would trend in unison because the PWS source waters were either the stream itself or GWUDI. These time series data are shown in Figure 3(a)–3(e). In all cases the sensor data shown is the daily average for measurements collected at 15-min intervals, whereas the PWS data is lab-generated from grab samples. The sensor data is much denser since the PWS permits do not require daily monitoring in most cases. Stream sensor data is summarized in Table 2. Maximum sensor NO_x-N ranged from 14.5 mg L^{-1} (Skunk R.) to 20.5 mg L^{-1} (Raccoon R.). The percentage of days exceeding the drinking water standard of 10 mg L^{-1} is also shown in Table 2, with lowest and highest values again represented by the Skunk (8.5%) and Raccoon (31.1%). Stream NO_x-N concentrations were characteristically high in the Apr-Jun period correlating with farm field work and when precipitation is abundant. Concentrations decline in late summer as crop water and nutrient uptake is robust, and increase again in late fall and winter before declining again in late winter during snowmelt. Concentrations are especially elevated in post drought conditions such as spring 2013 and lower during periods of extended dryness such as 2014.

Figure 3(a) illustrates the time series NO_x-N of PWS1 and PWS2 finished drinking water (lab) and the nearby Iowa River (sensor). PWS1 finished drinking water should in theory be the same as the shallow alluvial water used to supply the plant, since there is no treatment capacity for NO_x-N removal (dilution with sand pit water is possible in rare circumstances). Drinking water NO_x-N concentrations are lower but trend in unison with the river. PWS2 finished water NO_x-N may reflect dilution with deep well water, or some reverse osmosis removal of NO_x-N later in the record. Since the primary PWS2 source is the river itself, finished water data should in theory be the same as the sensor, assuming no blending or NO_x-N removal. Figure 3(b) illustrates time series sensor data for the Cedar River and finished water treated at the two PWS3 treatment plants, which both use shallow alluvial water under direct influence of this river. Skunk River sensor and PWS4 drinking water data for finished water are shown Figure 3(c). PWS4 uses alluvial water under the influence of the Skunk River. Figure 3(d) illustrates Raccoon River sensor data and SDWA data for PWS5-McMullen Plant, which uses GWUDI of that river. Finally, Figure 3(e) shows NO_x-N sensor data for the Des Moines River and reported finished drinking water NO_x-N concentrations for PWS6, which uses this river for its source supply.

DISCUSSION

Each of the utilities is vulnerable to high NO_x-N episodes in its source water, and while each has a strategy to keep finished water below the MCL, some are more vulnerable than others. PWS5, for example, has ion exchange NO_x-N removal capacity at its Fleur treatment plant and has not had a violation of the drinking water standard since this capacity was implemented in 1992, even though source water NO_x-N has been as high as 24 mg L^{-1} NO_x-N (Corrigan 2020). PWS2 recently installed reverse osmosis for the purposes of NO_x-N removal after struggling with the contaminant for many years and diluting high NO_x-N river water with water from a low quality deep well. The other four PWSs use a combination of well management and blending with emergency sources to stay in compliance with regulations. An effect of this can be increased treatment and energy costs and lower quality finished water for parameters such as hardness, taste and odor as the sources used for dilution are usually inferior from a treatment perspective. Further, consumer water demand can overwhelm the water yield needed from a blended source to dilute finished water NO_x-N below the MCL.

It has been previously demonstrated that continuous NO_x-N sensors can inform plant operations (Jones *et al.*

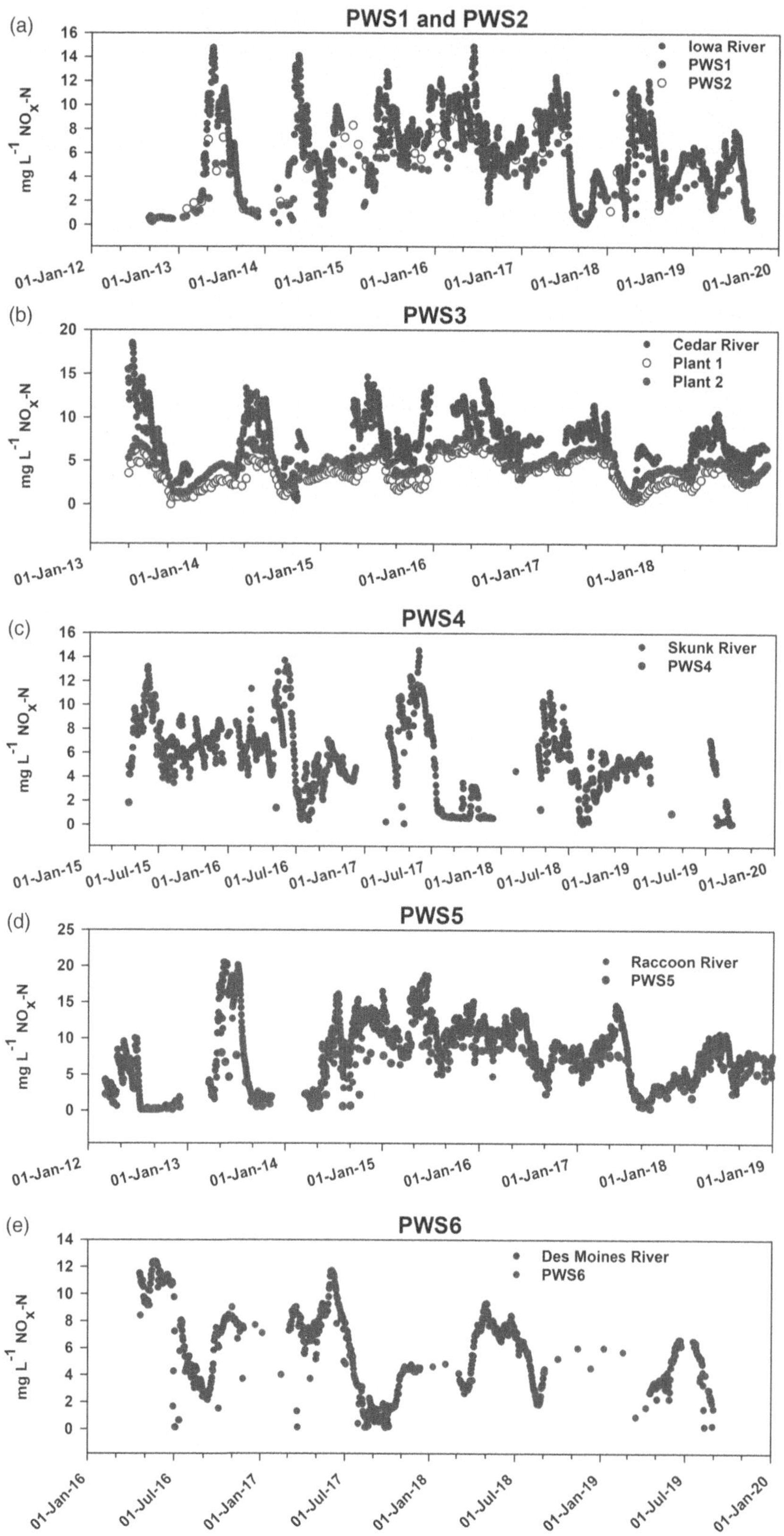

Figure 3 | Time series NO_x-N data for the finished water produced by PWSs in this study (blue), along with continuous data generated by sensors in nearby streams (brown). PWS5 finished water data is from the McMullen treatment plant.

Table 2 | Summary of NOx-N sensor data for the rivers evaluated in this study

River	n (days)	Average NO_x-N (mg L^{-1})	Maximum NO_x-N (mg L^{-1})	% of days >10 mg L^{-1}
Iowa	2,106	5.9	14.9	11.4
Raccoon	2,426	7.7	20.5	31.1
Cedar	1,607	7.3	18.5	17.1
Skunk	1,227	5.4	14.5	8.5

Data reflect daily averages, *n* = days.

2007) by providing utility staff with real-time data that can be used to manage diverse sources and implement removal treatment, and Figure 2 illustrates the accuracy of and reliability of the sensor data. Comparing the stream sensor data here with the SDWA data suggests continuous monitoring can also inform compliance and regulatory strategies. Few Iowa utilities are required to monitor for finished water NO_x-N more frequently than monthly, and only PWS5 here regularly reports compliance data on a daily basis, but only during known high source water NO_x-N episodes. The time series sensor data, shown within the backdrop of SDWA compliance monitoring of finished drinking water (Figure 3), shows that the frequency of required compliance monitoring may currently be insufficient to capture episodes where exceedances of the drinking water standard are possible. Using the data shown in Figure 3, we calculated the overall ratio of sensor NO_x-N to finished water NO_x-N (Table 3). From these ratios we estimated the sensor NO_x-N which conceivably could cause finished water NO_x-N at the treatment plant to exceed the MCL (10 mg L^{-1}), assuming no removal or blending (Table 3). Four of the treatment plants shown in Figure 3 had days of unreported SDWA data when their finished water could have exceeded 10 mg L^{-1}. Each of the PWSs here has strategies to cope with high NO_x-N episodes and so we do not assert that unreported violations are occurring at these utilities. Rather, we wish to point out that the frequency of required compliance monitoring may be insufficient to capture high NO_x-N episodes, and the resulting data may give the public an inaccurate picture of the threat the contaminant poses to safe drinking water in each of these communities.

Monitoring costs

It appears likely that widespread continuous, real-time *in situ* monitoring of both municipal source supply and treated water could help improve drinking water quality and reduce risks presented to the public health by high-NO_x-N water. Many of the Iowa PWSs coping with high NO_x-N water are small, rural communities that typically struggle to maintain adequate water and wastewater infrastructure (National Research Council 1997) and extra water monitoring costs could be financially burdensome. However, NO_x-N is regulated as an acute contaminant, and adequate and timely monitoring data is necessary for robust protection of public health. Fees for NO_x-N analysis at a certified lab are around \$20, but many other costs are also associated with this traditional monitoring: labor for sample collection, bottles, and shipping or transporting the samples, to name a few. Total costs for one sample could easily double the lab cost, and by the time results are received, finished water NO_x-N concentrations may have exceeded or retreated below the MCL. At \$40 total cost per sample, one year of daily monitoring would total \$14,600. This compares favorably with the costs associated to deploy a Nitratax sensor (\$20,000), which can generate data in real time for many years with minimal maintenance (Jones *et al.* 2007; Jones *et al.* 2018a).

It is worth emphasizing that in our study, even with the abundance of compliance data collected in the larger community water systems (i.e. PWS1, PWS3, PWS5), daily sensor data (Figure 3) are far more abundant over the time intervals considered. This provides higher temporal resolution for how water quality is changing at the treatment plants across seasons, which is particularly valuable during the high rates of NO_x-N increase (and decline) that coincide with weather and the start and stop of intense agricultural activity in Iowa. Thus, aside from their ability to provide insights into finished water quality when NO_x-N management systems are not being used, the real time data provided by the sensor should be valuable as a decision-support tool for such PWS regarding the vulnerability of their source water to nitrate contamination and

Table 3 | Ratio of NO_x-N in finished water to that measured by nearby river sensor, along with the sensor NO_x-N that would in theory produce finished water of 10 mg L^{-1}

Utility	PWS1	PWS2	PWS3 Plant 1	PWS3 Plant 2	PWS4	PWS5 McMullen	PWS6
Ratio of finished/sensor NO_x-N	0.66	0.97	0.51	0.62	0.26	0.68	0.93
Sensor NO_x-N that would generate 10 mg L^{-1} in finished water	15.1	10.3	19.7	16.1	38.5	14.8	10.8

when best to use nitrate control strategies to maintain SDWA compliance.

Finally, the time series data, like those illustrated in Figure 3, can help regulatory agencies design, implement and enforce compliance monitoring requirements for PWSs vulnerable to NO_x-N violations. Monthly compliance monitoring (e.g. PWS4, PWS6 and likely many others) or less may be inadequate to ensure the integrity of drinking water and protect the public health of the affected communities.

CONCLUSIONS

Our analysis shows good agreement between sensors and more traditional modes of NO_x-N analysis. Thus, when NO_x-N control strategies are not being employed, sensor measurements in source waters can be a good estimate for finished drinking water quality. We also anticipate that they would provide accurate data if deployed directly in a higher quality water source, such as finished drinking water prior to distribution.

Accordingly, we contend that ideal use of these sensors is via deployment at/near the intake (e.g. a PWS where there is no nitrate control) or directly in finished drinking water such as during clearwell storage. We would advocate for the use of these sensors for compliance monitoring as an alternative to more traditional laboratory measurements such as ion chromatography. For such applications, acquisition and use of sensors may be most practical and economically viable for large community water systems with sufficient resources to purchase, install and maintain this equipment. However, sensor deployment would reduce existing costs (e.g. personnel and analytical) associated with more traditional monitoring and reporting, and these savings may make their use more reasonable for smaller, more resource-constrained systems. Beyond compliance monitoring, these sensors would also be a useful decision support tool for PWSs with source water at risk from NO_x-N contamination. Real time quality data would be instructive on the timing of when best to activate NO_x-N management systems, as well as when source quality is acceptable and such systems are no longer necessary. This in turn would help to reduce treatment costs, as in some cases (e.g. ion exchange and reverse osmosis), NO_x-N treatment systems can be costly and ultimately result in higher water rates for consumers.

Broader use of these sensors in drinking water quality monitoring would also yield benefits for public health. Greater frequency of data collection would be more likely to identify violations when seasonal and temporal dynamics result in spikes in NO_x-N concentration that may otherwise go undetected during gaps in the compliance sample collection schedule. Further benefit to the public health community would be derived from the improved knowledge regarding exposure to NO_x-N below the current USEPA MCL, especially amid growing concern about chronic NO_x-N health effects from exposure below 10 mg L^{-1} (Temkin *et al.* 2019). Sensor data would help to more accurately determine the extent and duration of NO_x-N exposure at levels where adverse chronic effects are now being suggested, and this information would help regulators in assessing the benefits and costs of lowering the NO_x-N MCL from its current value of 10 mg L^{-1}.

A final benefit of these sensors is that the real-time output of water quality data could be used to better inform consumers. Data from a sensor is easily exported and displayed (Jones *et al.* 2018a), allowing results to be shown in real time at the utility homepage. In moving toward 'smarter water systems', drinking water providers need to ensure that data relevant to the consumer and their public health is readily available and easily shared. These sensors would help to accomplish that goal.

ACKNOWLEDGEMENTS

The authors would like to acknowledge Mr Jon Durst and Mr Craig Meacham at the City of Iowa City Water Treatment Plant, and Mr Ted Corrigan and Mr Jeff Mitchell at the Des Moines Water Works for their assistance with NO_x-N data. Support for CSJ was provided through the Iowa Nutrient Research Center.

DATA AVAILABILITY STATEMENT

All relevant data are included in the paper or its Supplementary Information.

REFERENCES

Bruning-Fann, C. S. & Kaneene, J. B. 1993 The effects of nitrate, nitrite and N-nitroso compounds on human health: a review. *Veterinary and Human Toxicology* **35**, 521–538.

CHEEC, Center for Health Effects of Environmental Contamination, University of Iowa 2009 *Iowa Statewide Rural Well Water Survey Phase 2 (SWRL2): Results and*

Analysis. Submitted to Iowa Department of Natural Resources Watershed Monitoring and Assessment Section. https://cheec.uiowa.edu/sites/cheec.uiowa.edu/files/SWRL2%20results.pdf.

Comly, H. H. 1945 Cyanosis in infants caused by nitrates in well water. *Journal of the American Medical Association* **129** (2), 112–116.

Corrigan, T. 2020 (4/1/20). Personal Communication (telephone call).

Demir, I. & Beck, M. B. 2009 GWIS: A prototype information system for Georgia watersheds. In *Georgia Water Resources Conference: Regional Water Management Opportunities, UGA*, Athens, GA, USA.

Demir, I. & Krajewski, W. F. 2013 Towards an integrated flood information system: centralized data access, analysis, and visualization. *Environmental Modelling & Software* **50**, 77–84.

Demir, I., Jiang, F., Walker, R. V., Parker, A. K. & Beck, M. B. 2009 Information systems and social legitimacy scientific visualization of water quality. In *2009 IEEE International Conference on Systems, Man and Cybernetics*. IEEE, pp. 1067–1072.

Demir, I., Conover, H., Krajewski, W. F., Seo, B. C., Goska, R., He, Y., McEniry, M. F., Graves, S. J. & Petersen, W. 2015 Data-enabled field experiment planning, management, and research using cyberinfrastructure. *Journal of Hydrometeorology* **16** (3), 1155–1170.

Duffy, C., Gil, Y., Deelman, E., Marru, S., Pierce, M., Demir, I. & Wiener, G. 2012 Designing a road map for geoscience workflows. *Eos, Transactions American Geophysical Union* **93** (24), 225–226.

Eller, D. 2015 60 Iowa cities see high nitrates. *Des Moines Register*, July 4, 2015. Available at: https://www.desmoinesregister.com/story/money/agriculture/2015/07/04/high-nitrates-iowa-cities/29720695/ (Accessed Feb. 1, 2020).

Gil, Y., David, C. H., Demir, I., Essawy, B. T., Fulweiler, R. W., Goodall, J. L., Karlstrom, L., Lee, H., Mills, H. J., Oh, J. H. & Pierce, S. A. 2016 Toward the geoscience paper of the future: best practices for documenting and sharing research from data to software to provenance. *Earth and Space Science* **3** (10), 388–415.

Gulis, G., Czompolyova, M. & Cerhan, J. R. 2002 An ecologic study of nitrate in municipal drinking water and cancer incidence in trnava district, Slovakia. *Environmental Research* **88** (3), 182–187.

Hanson, M. J., Keller, A., Boland, M. A. & Lazarus, W. F. 2016 The debate about farm nitrates and drinking water. *Choices* **31** (316-2016-7799), 1–7.

Jones, C. S. & Schilling, K. E. 2019 Iowa statewide stream nitrate loading: 2017–2018 update. *The Journal of the Iowa Academy of Science* **126**, 6–12.

Jones, C. S., Hill, D. & Brand, G. 2007 Use a multifaceted approach to manage high source-water nitrate. *Opflow* **33** (6), 20–22.

Jones, R. R., Weyer, P. J., DellaValle, C. T., Inoue-Choi, M., Anderson, K. E., Cantor, K. P., Krasner, S., Robien, K., Freeman, L. E. B., Silverman, D. T. & Ward, M. H. 2016 Nitrate from drinking water and diet and bladder cancer among postmenopausal women in Iowa. *Environmental Health Perspectives* **124** (11), 1751–1758.

Jones, C. S., Davis, C. A., Drake, C. W., Schilling, K. E., Debionne, S. H., Gilles, D. W., Demir, I. & Weber, L. J. 2018a Iowa statewide stream nitrate load calculated using in situ sensor network. *JAWRA Journal of the American Water Resources Association* **54**, 471–486.

Jones, C. S., Nielsen, J. K., Schilling, K. E. & Weber, L. J. 2018b Iowa stream nitrate and the Gulf of Mexico. *PloS one* **13** (4), e0195930. https://doi.org/10.1371/journal.pone.0195930.

Kirchner, J. W., Feng, X., Neal, C. & Robson, A. J. 2004 The fine structure of water-quality dynamics: the (high-frequency) wave of the future. *Hydrological Processes* **18**, 1353–1359.

Krajewski, W. F., Ceynar, D., Demir, I., Goska, R., Kruger, A., Langel, C., Mantilla, R., Niemeier, J., Quintero, F., Seo, B. C. & Small, S. J. 2017 Real-time flood forecasting and information system for the state of Iowa. *Bulletin of the American Meteorological Society* **98** (3), 539–554.

Kross, B. C., Ayebo, A. D. & Fuortes, L. J. 1992 Methemoglobinemia: nitrate toxicity in rural America. *American Family Physician* **46**, 183–188.

McGill, B. M., Burgin, A. J., Hively, W. D., Loecke, T., Dagnew, A. T. & Beeson, P. 2019 Farming for human health and a smaller dead zone: how agricultural conservation practices, climate change, and subsurface drainage intensification affect nitrate loads in Iowa. *AGUFM* **2019**, GC33D-03.

National Research Council 1997 *Safe Water From Every Tap: Improving Water Service to Small Communities*. The National Academies Press, Washington, DC. https://doi.org/10.17226/5291.

O'Dell, J. W. 1993 *Method 353.2: Determination of Nitrate-Nitrite Nitrogen by Automated Colorimetry*. Environmental Monitoring Systems Laboratory. US Environmental Protection Agency, Cincinnati, OH, USA.

Pellerin, B. A., Bergamaschi, B. A., Downing, B. D., Saraceno, J. F., Garrett, J. D. & Olsen, L. D. 2013 *Optical Techniques for the Determination of Nitrate in Environmental Waters: Guidelines for Instrument Selection, Operation, Deployment, Maintenance, Quality Assurance, and Data Reporting*. Techniques and Methods, 1-D5, United States Geological Survey, California Water Science Center, Sacramento, CA, USA.

Pellerin, B. A., Stauffer, B. A., Young, D. A., Sullivan, D. J., Bricker, S. B., Walbridge, M. R., Clyde Jr, G. A. & Shaw, D. M. 2016 Emerging tools for continuous nutrient monitoring networks: sensors advancing science and water resources protection. *JAWRA Journal of the American Water Resources Association* **52**, 993–1008.

Pfaff, J. D. 1993 *Method 300.0 Determination of Inorganic Anions by ion Chromatography*. US Environmental Protection Agency, Office of Research and Development, Environmental Monitoring Systems Laboratory, Cincinnati, OH, USA, pp. 28.

Pontius, F. W. 1999 Complying with the interim enhanced surface water treatment rule. *Journal-American Water Works Association* **91** (4), 28–188.

Schullehner, J., Hansen, B., Thygesen, M., Pedersen, C. B. & Sigsgaard, T. 2018 Nitrate in drinking water and colorectal cancer risk: a nationwide population-based cohort study. *International Journal of Cancer* **143**, 73–79.

Secchi, S. & Cwiertny, D. M. 2019 *Iowa's Grants to Counties Program: A Valuable but Underutilized Program for Protecting the Public Health of Private Well Users*. University of Iowa Institutional Repository. Available at: https://iro.uiowa.edu/view/delivery/01IOWA_INST/12716559630002771.

Temkin, A., Evans, S., Manidis, T., Campbell, C. & Naidenko, O. V. 2019 Exposure-based assessment and economic valuation of adverse birth outcomes and cancer risk due to nitrate in United States drinking water. *Environmental Research* **176**, 108442.

Tiemann, M. 2014 *Safe Drinking Water act (SDWA): A Summary of the act and its Major Requirements*. Congressional Research Service, Washington, DC, USA, pp. 7–5700.

USEPA, United States. Environmental Protection Agency. Office of Drinking Water. Science and Technology Branch 1989 *Guidance Manual for Compliance with the Filtration and Disinfection Requirements for Public Water Systems Using Surface Water Sources: Draft*. Science and Technology Branch, Criteria and Standards Division, Office of Drinking Water, US Environmental Protection Agency, Washington, DC, USA.

USGS 2019 *National Water Information System, Current Water Data for Iowa*. Available at: https://waterdata.usgs.gov/ia/nwis/rt.

Walker, R. 1990 Nitrates, nitrites and N-nitrosocompounds: a review of the occurrence in food and diet and the toxicological implications. *Food Additives & Contaminants* **7** (6), 717–768.

Ward, M. H., Jones, R. R., Brender, J. D., De Kok, T. M., Weyer, P. J., Nolan, B. T., Villanueva, C. M. & Van Breda, S. G. 2018 Drinking water nitrate and human health: an updated review. *International Journal of Environmental Research and Public Health* **15**, 1557.

Weyer, P. J., Cerhan, J. R., Kross, B. C., Hallberg, G. R., Kantamneni, J., Breuer, G., Jones, M. P., Zheng, W. & Lynch, C. F. 2001 Municipal drinking water nitrate level and cancer risk in older women: the Iowa Women's Health Study. *Epidemiology* **12** (3), 327–338.

WHO 1970 *European Standards for Drinking Water*, 2nd edn. World Health Organization, Geneva, Switzerland.

First received 18 May 2020; accepted in revised form 22 July 2020. Available online 6 August 2020

© IWA Publishing 2020

Data integration for infrastructure asset management in small to medium-sized water utilities

N. Carriço, B. Ferreira, R. Barreira, A. Antunes, C. Grueau, A. Mendes, D. Covas, L. Monteiro, J. Santos and I. S. Brito

ABSTRACT

Water utilities collect, store and manage vast data sets using many information systems (IS). For infrastructure asset management (IAM) planning those data need to be processed and transformed into information. However, information management efficiency often falls short of desired results. This happens particularly in municipalities where management is structured according to local government models. Along with the existing IS at the utilities' disposal, engineers and managers take their decisions based on information that is often incomplete, inaccurate or out-of-date. One of the main challenges faced by asset managers is integrating the several, often conflicting, sources of information available on the infrastructure, its condition and performance, and the various predictive analyses that can assist in prioritizing projects or interventions. This paper presents an overview of the IS used by Portuguese water utilities and discusses how data from different IS can be integrated in order to support IAM.

Key words | data integration, data interoperability, infrastructure asset management, water utilities

N. Carriço (corresponding author)
B. Ferreira
R. Barreira
Polytechnic Institute of Setubal, Barreiro School of Technology,
Rua Américo da Silva Marinho, 2839-001 Lavradio, Portugal
E-mail: *nelson.carrico@estbarreiro.ips.pt*

A. Antunes
C. Grueau
Polytechnic Institute of Setubal, Setubal School of Technology,
Campus do IPS Estefanilha, 2914-508 Setubal, Portugal

A. Mendes
Polytechnic Institute of Setubal, School of Business Administration,
Campus do IPS Estefanilha, 2914-508 Setubal, Portugal

D. Covas
L. Monteiro
Instituto Superior Técnico, Lisbon University,
Avenida Rovisco Pais 1, 1049-001 Lisbon, Portugal

J. Santos
I. S. Brito
Polytechnic Institute of Beja, School of Technology and Management,
Rua Pedro Soares, 7800-295 Beja, Portugal

HIGHLIGHTS

- How small and medium-sized water utilities can integrate their own solutions.
- How it is possible to develop data integration of diverse systems to facilitate data analytics.
- Due to inefficient information management, some water utilities are not set up for data integration solutions.
- A project in Portuguese as a case study.
- A possible solution for data integration for IAM.

INTRODUCTION

Water utilities (WU) manage a large number of assets (e.g., pipes, valves, pumps), many of which were installed several decades ago and are therefore reaching the end of their expected useful life. These deteriorating assets may lead to problems such as leakages, burst pipes and service disruptions, which increase the systems' maintenance requirements and operational costs. WU face the challenge of keeping their systems operational, efficient, and reliable to provide good-quality service. Therefore, a WU should adopt infrastructure asset management (IAM) processes not only to plan asset interventions (e.g., repair or replacement), but also to ensure its own sustainability (Alegre & Coelho 2011).

According to ISO 55000:2014, asset management is the coordinated activity carried by an organization to realize value from its assets involving a balancing of costs, risks, opportunities, and performance benefits. In the context of water services, the concept of IAM is adopted instead and defines the framework to pursue and achieve sustainable infrastructure, in which utilities manage their assets and the associated performance, risks and costs over the life-cycle (Alegre & Coelho 2011).

doi: 10.2166/wst.2020.377

In this context, many methodologies for urban water IAM can be applied. In Portugal, the National Water and Waste Services Regulator (ERSAR) has published two technical guides, one recommending a methodology for water supply infrastructure and the other for wastewater infrastructure. Both present the same methodology organized into three levels of management and planning (i.e., strategic, tactical and operational). They are structured as a loop comprising the following stages: (i) definition of objectives and targets; (ii) diagnosis; (iii) plan production, including the identification, comparison and selection of alternative solutions; (iv) plan implementation; and (v) monitoring and review (Alegre & Coelho 2011).

A key aspect for a WU in the application of any IAM methodology is having reliable data about their assets. These data can be roughly divided into two groups, namely: (i) cadastral and operational information (e.g., components' technical specifications, location, condition and information about failures in the various assets) and (ii) billing and account information (i.e., network-related maintenance and operational costs, energy and personnel costs, amortizations and interest, revenues, among others) (Alegre & Coelho 2011). Therefore, good data is the foundation for the application of any IAM methodology.

In Portugal, the provision of water services is divided in two: bulk activities, called 'high service', comprise water abstraction, treatment, and transmission; retail activities, called 'low service', comprise water storage and distribution. The responsibility for high service is shared between the 308 municipalities and the state through the public holding company group Águas de Potugal (AdP). The low service is exclusively the responsibility of the municipalities, which can manage the service either directly or indirectly through concessions. According to the 2018 Annual Report on Water and Waste Services, there are 15 high service utilities providing water to 73% of the population, and 306 low service, mostly managed directly by the municipalities (i.e., $202 \approx 66\%$) representing about 52% of the population (ERSAR 2019). Municipalities can create municipal services that operate independently, or municipal companies, or intermunicipal services with other municipalities. So, the diversity of WU is pronounced, due to the institutional and organizational framework and the geographical context, among many other aspects (Amaral *et al.* 2017).

Over the last few decades, many municipalities have made investmented significantly in implementing information systems (IS) to address the increasing complexity of their daily control, operation, management and planning tasks (Halfawy 2008), namely, enterprise resource planning system (ERP), geographic information system (GIS), customer relationship management (CRM) and supervisory control and data acquisition (SCADA), among others (Alegre & Coelho 2011).

These IS collect, store, and manipulate data which, after formatting, are transmitted, making the information available to specific recipients and users located, generally, in different departments or divisions.

Digitalization in the water industry has brought more sophisticated solutions, such as the internet of things (IoT), big data, artificial intelligence (AI), augmented and virtual realities, and digital twins, which help them to manage their activities (Laranjo & Martinho 2019). All these solutions face some barriers, such as lack of integration and interoperability of the different existing IS, people's resistance to changing their cultural mindset, and cybersecurity issues (Sarni *et al.* 2019). To solve the integration and interoperability problem some middle software such as NAVIA may be acquired, which focuses particularly on system operations and maintenance (Tavares *et al.* 2019).

Information management efficiency in municipalities often falls short of desired results since management is usually structured according to local government models. In these municipal WU, and due to lack of technical capacity and human resources, it is not usually possible to generate the critical information through daily routines (e.g., well planned and organized service orders that can ensure useful, good-quality information). According to Silva *et al.* (2019) 52% of low service WU have low infrastructural knowledge indices, which means that most of the WU do not know the assets they are managing. Additionally, available data may not benefit the various sectors of the utility (e.g., billing data are not linked to cadastral information) (Silva *et al.* 2012). As result, insufficient planning leads to a lack of integration, which affects information fluxes and its uses.

Despite that, IS have been further developed in order to promote efficient information management, with direct implications for the improvement of the utility's efficiency and effectiveness. These extraordinary developments in IS are motivated by the growing need to make WU more resilient and flexible, with greater transparency and rationality in the decision-making process, as well as enabling them to respond to challenges more quickly and efficiently. Nevertheless, and regardless of these developments, engineers and managers make their decisions based on information that is often incomplete, inaccurate, or out-of-date. One of the main challenges faced by engineers and managers in most WU (and particularly in those with more diversified data) is integrating the several, often conflicting, sources of information available on the infrastructure, its condition, its performance, and the various predictive analyses to assist in prioritizing asset interventions.

The DECision support platform for urban water InfrastructurEs (DECIdE) is a Portuguese R&D project aimed at the development of an information technology (IT) platform that integrates data from different IS for small and medium-sized municipal utilities. Additionally, this platform includes tools to support decision-making regarding IAM (e.g., performance system assessment, water and energy balances). This paper presents the project's global framework, the main findings of the WU characterization and the survey of its existing IS. Important issues are highlighted regarding data and information management required to successfully integrate data from these different systems, ultimately aiming at IAM.

METHODOLOGY

Information systems characterization

Five WU, four municipalities and one municipal company participated in the DECIdE project. The project's first task of was to characterize the five WU's IS and collect their needs, which were intended to be solved with the platform. The degree of maturity of the IS management was different between the five, essentially due to the management model adopted, and the WU's geographical context and financial capacity. Regardless of their size or maturity, utilities perform similar tasks both at an operational and accounting level. In the less mature utilities, some of the IS was still be based on paper records or extensive spreadsheets, while in utilities of great maturity more sophisticated IS were used.

In 2016, Reis *et al.* (2017) carried out a survey of 171 of the 309 Portuguese WU (representing 55% of the total), and showed that 80% had a GIS and 89% had a cadastral record of their assets. In addition, about 89% of the surveyed WU used customer management systems and 95% used ERP systems for the accounting and financial aspects, as well as to manage human resources. According to the same survey, some IS were used by less than 50% of the WU surveyed. SCADA systems were present in only 43% of the WU surveyed, with maintenance support systems in 24% and energy management systems in only 6.4%.

In a municipality, water services are just one of the several services to be managed, namely, parking, transport and its infrastructure, urbanisation, conservation and green spaces, municipal equipment management, social action, administration, consulting, and evaluation, among others. Considering this vast number of services to be managed, the human, financial and technological resources are usually scarce and unevenly distributed among services, with the aggravating factor that leaders are elected every four years, making it difficult to establish long-term priorities such as the acquisition and implementation of IS.

The diversity of IS management is therefore pronounced. Some utilities lack certain IS while others have duplicated IS in different departments (e.g., some municipalities have different GIS for water services and urban planning services and sometimes only one of the departments has GIS specialized personnel). The former is due to a lack of need, since either they serve few customers or they manage small systems, but in most cases it is due to the management model adopted. The latter is due to lack of inter-departmental communication and pre-established goals.

As stated above, the DECIdE project characterized five different low service utilities, from small to medium sized, in order to cover a range of utility maturities (see Table 1). In order to highlight the representativeness of the five utilities selected, it should be stated that in 2018 (ERSAR 2019), of the total of 306 low service WU existing in Portugal, approximately 184 are managed directly by municipalities, 19 are municipal services, 48 are managed by parish councils, 23 are municipal corporations, 2 are public-public partnerships, 1 is state delegated, 28 are municipal concessions and 1 is an intermunicipal concession. Of those WU, about 84% are small to medium sized (i.e., approximately 256 of the total). Of the five WU that participated in this project, WU 1–4 (2% of the total) accounted for services provided directly by the municipalities, and WU 5 (4% of the total) is a municipal corporation. Despite this small number of participants, they can be considered representative of the diversity of governance models in the water sector in Portugal.

Since most of the assets in a WU are buried, geospatial information (i.e., asset location and characteristics) plays a very important role in any IAM methodology. Thus, GIS is indispensable to manage spatial data in any WU. The survey carried out showed that GIS was used in all five WU but with different degrees of implementation. WU 1 and 5 had complete information about their assets on a GIS, motivated by several cadastral surveys that had been

Table 1 | Characteristics and aspects of the five WU

	WU 1	WU 2	WU 3	WU 4	WU 5
Type of WU	Municipality	Municipality	Municipality	Municipality	Municipal corporation
Population in last census (2011)	78,764	62,831	10,828	17,437	Varying from 3,000 to 14,500[a]
Area (km^2)	36.69	465.12	464.00	1,232.97	6.45
Territorial typology	Urban	Mostly rural	Predominantly rural	Predominantly rural	Urban and golf courses
Network extension (km)	309.5	591.5	159.2	189.7	74.7
Number of service connections	18,173	21,303	6,816	8,259	1,941
Water source	Groundwater abstraction	Import and groundwater abstraction	Groundwater abstraction	Groundwater abstraction	Import

[a]WU 5 manages the water supply system of a tourist area, where the population changes from around 3,000 inhabitants in winter to 14,500 inhabitants in the summer season.

carried in recent years. In both WU, the GIS was maintained and regularly updated by a dedicated technician. The use of a GIS by WU 2 was a recent practice and as such a technician dedicated to the management of geospatial information had not yet been assigned. As consequence, data was not routinely updated as it should have been and, as consequence, data was out of date. WU 3 and 4 used an open-source GIS whose data were collected during a project aimed at a systematic and standardized cadastral survey of the water supply and sanitation assets of 13 Portuguese municipalities. In both WU, the available information was quite complete but, due to the lack of a dedicated technician to manage and update the geographic information, data have become outdated.

Water services require high investment costs for the construction and maintenance of their assets, as well as significant and continuous operational costs (i.e., costs related to energy consumption, water and wastewater treatment or failure repairs, among others). These costs are covered by the tariffs charged to users. In order to guarantee the quality and continuity of the provided service, WU rely on ERP systems and CRM to manage their financial and billing-related data, respectively. The survey conducted revealed the generalized use of ERP and CRM systems to manage and store financial and billing data, with all WU having a dedicated department for this task. In WU 2, 4 and 5, the same IS was used to manage both financial and billing-related data, while the remaining utilities use separate IS.

Assets deteriorate with time, and anomalies or failures will happen. Analysis of the failure history may reveal recurrent problems due to unforeseeable events. Therefore, utilities should use service work orders (SWOs) to record and manage their interventions. The survey revealed that the use of an IS to manage the SWO-related data was restricted to WU 1 and 5. In both utilities, the SWO-related data was uploaded *in situ* via mobile devices to its GIS. As a result, the detailed database of their assets' anomalies and failures was maintained (e.g., location, cause, duration, date and time, material, among others). In the remaining WU, instead of GIS, a paper form was filled in *in situ*. This form was subsequently archived in digital and paper formats, in WU 2 and 3 respectively, which hinders straightforward access to the data as it always requires human intervention. In WU 4 the paper forms were transcribed directly onto spreadsheets designed for the purpose. Although a major part of the essential information was recorded, the full location of the SWO may not have been recorded since no association was made with the element that had been repaired in GIS. As such, only a partial location was recorded (e.g., name of the street). Additionally, this method is more vulnerable to data becoming outdated and to insertion flaws (e.g., data may be inserted twice).

SCADA systems are used by utilities to monitor their systems and to control processes through real-time acquisition of data from several sensors (e.g., flow, pressure, energy consumption, pH). The survey showed a generalized use of SCADA systems to manage the input water volumes or flows in their systems, with different degrees of implementation. WU 1 and 5 had their system completely monitored, as result of years of investment in SCADA systems. In the remaining WU a different paradigm was applied, as only their primary systems or district metered areas (DMAs) were monitored.

System hydraulic anomalies and failures are usually not directly and easily identified. Sometimes, utility managers or engineers realize there is a problem because of indirect signs (e.g., user complaints of lack of water pressure). The detailed diagnosis of defects by direct inspection are usually too costly, leading to the need for modelling and analysis tools. Simulation models are tools that allow, with an estimable margin of error, the analysis and prediction of the hydraulic behaviour of a system, based on the characteristics of its assets, its operation mode and the time-varying demands. Thus, hydraulic simulation models allow the rapid and efficient assessment of system hydraulic behaviour for given conditions and with a reasonable margin of error. The survey revealed that only WU 1 and 5 had hydraulic simulation models of their complete systems. WU 4 had a simulation model for their main system while the remaining WU did not have any kind of hydraulic simulation model.

Data integration and interoperability

In order to gain a better understanding of their assets and systems' performance through data science and analytics, WU need to somehow aggregate all data into a single system. Thus, data integration and interoperability are fundamental processes for a water utility.

The terms integration and interoperability, although used interchangeably, have different meanings. Integration refers to connecting different IS so that data from one system can be accessed by the others (Roberts 2017). As such, it involves a third-party software called middleware that translates data and makes it readable for the receiving system. Interoperability refers to the basic ability of an IS to connect and communicate with another IS promptly, even if they were developed by different manufacturers (O'Connor 2017). When systems are interoperable, they can not only share information, but also interpret incoming data and present it as it was received, preserving its original context (Roberts 2017). The complexity of data integration and interoperability is especially the case at the levels of data storage, structure, and the levels at which the data can be integrated and operated as a single entity (Kadadi *et al.* 2014).

The application of an IAM methodology may require the use of different analitical techniques for system diagnosis (e.g., water and energy balances, system performance assessment, simulation of hydraulic behaviour, and more complex AI techniques). These techniques use on data stored in multiple IS and often distributed across several decentralized databases. The required integration and interoperability of the different IS faces many barriers, such as the spread and variety of data sources and formats, resulting from diverse software systems and departments, as well as the outdatedness of major technological trends (i.e., use of old technologies based on spreadsheets). Allied to the lack of machine communication protocols, common data formats or the common meaning of exchanged content, the application of information and communication technologies in WU is often restricted, restricting the ability to share data and knowledge and hence interoperate between the people and software involved (Howell *et al.* 2016).

Several software solutions have been developed to solve the data integration and interoperability and the analytics problems in WU. The development of this type of solution may start within the utility itself, if the required human and technology resources are available to develop a system suited to their needs. If no human resources are available at the WU then the typical option is to acquire commercial software.

In Portugal, a limited number of utilities (i.e., the biggest) have the resources to develop their own solutions. This capacity is not, however, achievable for the majority of WU. At least two WU have developed their data integration and analytics through GIS-driven solutions, in which it is possible to easily access data related to infrastructure, fleets, water demand, billing, among others (Silva *et al.* 2017; Sousa *et al.* 2017). Another Portuguese utility developed a platform aimed at integrating data gathered from different sources (i.e., sensors, GIS, equipment automation signals, among others), as well as the results from hydraulic simulation models, aiming ultimately to produce predictive models of network hydraulic behaviour and decision support tools (Vieira *et al.* 2018). Some of the largest water companies operating in Portugal have designed and developed their own platform, connecting the different IS to support data management and information analysis, aimed at better IAM (Feliciano *et al.* 2015). Nonetheless, the use of these platforms is restricted to the utility itself.

Different commercial software (e.g., Baseform, WaterSmart) claim to provide comprehensive, actionable analytics to assist WU-specific efficiency issues and key business outcomes. Although integration of real-time sensor data is possible with these solutions, the integration of raw data from the scattered data sources may not be possible or requires processing treatment to import and integrate these data, which may not be justified.

RESULTS AND DISCUSSION

DECIdE platform

IAM is considered by the Portuguese regulator (ERSAR) to be a key issue to increase the efficiency of WU. This fact has been transposed into Portuguese legislation, which requires WU providing a service to populations above 30,000 inhabitants to have IAM plans. Additionally, ERSAR evaluates the quality of the water service provided by all WU annually. A performance assessment system is used, requiring WU to send a huge amount of data in spreadsheets. As mentioned above, most of Portuguese utilities are municipalities that lack resources.

Any WU needs to collect vast data sets from their assets to manage them more efficiently, gain insights into their problems and plan their interventions. These data are transformed into useful information only when they are put together into a single technology (e.g., platform) on which deep analyses can be carried out. For example, pipe failure could be caused not only by age, but also by the pipe material, surrounding soil, water pressure, and street traffic.

In this context, DECIdE was undertaken, between 2018 and 2020, to develop a platform to integrate data from the different IS of the five participating WU. This data integration was to help WU to report data to the regulator more quickly and efficiently. Since WU also need to reduce water losses and energy costs, the platform includes tools to carry out water and energy balances (Alegre *et al.* 2000; Mamade *et al.* 2017). These balances use a broad set of data (e.g., input volumes and energy consumption, billed and unbilled consumptions, elevations). If sectorial balances need to be calculated, the complexity of data collection and management carried out by the utility manager can be overwhelming. As such, WU need a system to assess and compare the performance of their water supply systems. The basic question is therefore: how can we provide a unified approach to integrate and compare data organized and stored in a plethora of IS?

The selection of the integration method is of utmost importance as several features or analysis techniques depend on the data upload and update regularity. Taking into account the available technology and state-of-the-art techniques, the integration module could have been developed as a continuous connection between utility's IS databases and the platform databases (i.e., remote database access), or through a raw data file submission and setup module (i.e., manual raw data file upload and semi-automatic mapping between database attributes). Although the first option could lead to a better platform usability, as data is constantly and automatically uploaded and updated, it requires initial setup and continuous assistance by a utility's IT technician. It was concluded that remote database access could not be adopted in the short term due to the unavailability of IT technicians in four WU. So a compromise solution based on a raw data file submission and setup module was considered, since it is comprehensive and would allow the platform to be used by all WU.

Therefore, the platform can parse uploaded files and update the database with the parsed data. The challenge was to implement a file parsing mechanism, through a common interface and a semi-automatic process that minimized user intervention. File parsing starts by trying to determine file content. Identification is based on file extension and content analysis. Previous knowledge obtained by analysis of partner WU sample data files is included in the detection algorithm. After detecting the file type and content, the platform can identify the related data model elements, database tables and formats to target. An initial mapping is created between data model elements and file content. The user can correct and/or validate the detected mapping. The validated mapping is registered in the system and used the next time a similar file from that entity is uploaded. After validation, the import process starts.

The data model is set up to maintain a historical record of infrastructure and other time-sensitive data. When updating such data elements, the previous state is timestamped and kept in the database, allowing historical analysis of data evolution. As such, WU can assess the status of the several water distribution network components (e.g., valve status) for a given past date.

Data integration into the DECIdE platform may allow utilities to carry out advanced AI analysis for solving several problems, namely, pattern recognition and classification (Henriques *et al.* 2015), water demand prediction modelling, burst pipe detection and location (Covas & Ramos 2006) and anomalous event detection (Marchionni *et al.* 2016). In this context, another R&D project aimed at the development of algorithms and models that allow the extraction of knowledge from the data, supporting the WU in decision-making and thus improving the management of their systems. This project has been undertaken to continue the development of the platform.

Some points about water utilities data management

The DECIdE project provided some interesting findings about water utilities data management. Firstly, the collected data are stored in different IS in several departments (e.g., water department, financial department, urban planning department). These specific data depend on specialized human resources for the required treatment and validation process. As an example, one of the surveyed WU needs two engineers full-time for almost a month to collect and process the required data for the spreadsheet for an ERSAR report. As many municipalities lack human resources, that activity takes their attention away from other important activities (e.g., geospatial information updates). Additionally, some data has a high degree of uncertainty associated with them. As an example, SWO-related data should be collected by an operational technician at the field. However, most of the times they has no time to collect all the information. Additionally, this technician is often not aware of the importance of collecting those data.

One surprising finding is about IS that they are contracted for a particular objective, such as the billing system. These systems use cloud-based databases that are not controlled by the utility. When data access is required, a requisition must be made to the developer company, which leads to questions relating to the ownership of the data. Should the developer company or the utility own the data? Additionally, if a utility wants to apply advanced algorithms to those data for detailed analysis, using for instance AI techniques, they do not have direct or straightforward access. Another point is the time required to access and retrieve specific data. Due to obsolete network infrastructure, it can take about 30 min to obtain the billing data for about 45,000 clients. If a more complete analysis is to be carried, it can take a day or more to retrieve the required data, making the computer inoperable during this time.

Regarding SWOs, they should be linked to GIS and should be as detailed as possible about the cause that originate that order. Some utilities have restricted categorizations of their interventions in a certain IS, which can distort a broader assessment of the real problem.

Currently, long-term monitoring and systems modelling are required for a better understanding of the utilities' systems. However, these data obtained through the sensors installed in the network present some problems, namely, null values, outlier values, irregular acquisition frequency and non-calibrated sensors. These data needs to be treated (i.e., identified and corrected) preferably in an automatically, namely by removing the nulls and the outlier values, eliminating nonreliable data, and filling the gaps created. So, the data integration process should also consider statistical data and analysis methodologies for evaluating these huge data sets. The platform's future developments aim to integrate time-series, considering a treatment and validation process.

CONCLUSIONS

This paper presents an overview of the IS used by Portuguese utilities as well as a reflection on how data from different IS can be integrated. The diversity of IS in a WU is challenging for bringing data together into a single technology (e.g., platform) for analytics. With this in mind, the DECIdE platform was developed for data integration, in collaboration with a representative set of Portuguese WU. Data integration on the DECIdE platform is done through a two-step import process. The first step consists of the import of all the necessary files to integrate the GIS data and the data related to water meters. After the parsing of all these data, the user is informed about the possible missing information, which is needed to calculate the set of metrics. After that the user has the option of validating or completing the missing data. This two-step validation process will help WU with fewer data to continue to take advantage of the platform even though they cannot provide all the information.

Once the assets' physical data have been integrated (e.g., pipe, pumps, water meters), monthly water and energy measurements can be imported and assigned to the respective assets. Additionally, SWOs can be integrated, assigned to a specific DMA or system.

Due to inadequate resources (e.g., human, financial), data integration faces several limitations in most WU. Some WU lack skilled professionals but have financial resources, and choose to acquire expensive software that claims data integration. However, in reality commercial software has several limitations and may not match WU needs. Data integration may also require mapping components to a new schema that will make the existing ones uniform, with new attributes. Nonetheless, the new database may have scalability problems depending on the chosen technology.

The DECIdE platform was the first step in the development of a reference tool to support IAM in urban water services. Future work will focus on data science, with a special focus on the detection, categorization, and location of anomalous events.

ACKNOWLEDGEMENT

The authors want to acknowledge Fundação para a Ciência e a Tecnologia, Programa Operacional Regional Lisboa 2020 and Programa Operacional Regional Alentejo 2020 (grant number SAICT-POL/24135/2016 | LISBOA-01-0145-FEDER-024135 | ALT20-03-0145-FEDER-024135) for supporting the DECIdE project. The authors also thank the engineers from the five participating water utilities for their contribution.

DATA AVAILABILITY STATEMENT

All relevant data are available from an online repository or repositories. The link is https://github.com/DECIdE-RDProject/Platform-test-utility.

REFERENCES

Alegre, H. & Coelho, S. T. 2011 Infrastructure asset management of urban water systems. In: *Water Supply System Analysis: Selected Topics* (A. Ostfeld, ed.). IntechOpen, London, United Kingdom, pp. 49–74.

Alegre, H., Hirner, W., Baptista, J. M. & Parena, R. 2000 *Performance Indicators for Water Supply Services*, 1st edn. Manual of Best Practice Series. IWA Publishing, London. p. 160.

Amaral, R., Alegre, H. & Matos, J. S. 2017 Highlights of key international water infrastructure asset management initiatives, and trends, challenges and developments in Portugal. *Water Policy* **19**, 128–146.

Covas, D. & Ramos, H. 2006 Case studies of leak detection and location in water pipe systems by inverse transient analysis. *Journal of Water Resources Planning and Management* **136** (2), 248–257.

ERSAR 2019 *Annual Report of Water and Waste Services in Portugal (Volume 1): Sector's Characterization (in Portuguese)*. Entidade Reguladora dos Serviços de Águas e Resíduos, Lisboa, Portugal.

Feliciano, J., Almeida, R., Santos, A., Ramalho, P., Ganhão, A., Covas, D. & Alegre, H. 2015 Finding new trends through data analytics in infrastructure asset management – rehabilitation investments versus operational management. In *LESAM – Strategic Asset Management of Water and Wastewater Infrastructures*, 17–19 November 2015, *Yokohama, Japan*.

Halfawy, M. 2008 Integration of municipal infrastructure asset management processes: challenges and solution. *Journal of Computing in Civil Engineering* **22** (3), 216–229.

Henriques, R., Antunes, C. & Madeira, S. 2015 A structured view on pattern mining-based biclustering. *Pattern Recognition* **48** (12), 3941–3958.

Howell, S. K., Rezgui, Y., Beach, T., Zhao, W., Terlet, J. & Li, H. 2016 Smart water system interoperability: integrating data and analytics for demand optimized management through semantics. *ICCCBE* **2016**, 1–9.

Kadadi, A., Agrawal, R., Nyamful, C. & Atiq, R. 2014 Challenges of data integration and interoperability in big data. In *2014 IEEE International Conference on Big Data, Washington, USA*.

Laranjo, N. M. & Martinho, S. 2019 A digital transformation route – the case of INOVA municipal company of Cantanhede (In Portuguese). In *Proceedings of ENEG2019, Ílhavo, Portugal*.

Mamade, A., Loureiro, D., Alegre, H. & Covas, D. 2017 A comprehensive and well tested energy balance for water supply systems. *Urban Water Journal* **14** (8), 853–861.

Marchionni, V., Cabral, M., Amado, C. & Covas, D. 2016 Estimating water supply infrastructure cost using regression techniques. *Journal of Water Resources Planning and Management* **142** (4), 04016003-1–04016003-10.

O'Connor, S. 2017 *What Is Interoperability, and Why Is it Important?* Available from: https://www.adsc.com/blog/what-is-interoperability-and-why-is-it-important (accessed 21 April 2020).

Reis, P. C., Ferreira, E. & Tenório, C. 2017 Specialized APDA information systems commission – presentation of the survey results of characterization of the sector in terms of information systems (in Portuguese). In *ENEG 2017, Évora, Portugal*.

Roberts, B. 2017 *Integration vs Interoperability: What's the Difference?* Available from: https://blog.sisfirst.com/integration-v-interoperability-what-is-the-difference (accessed 21 April 2020).

Sarni, W., White, C., Webb, R., Cross, K. & Glotzbach, R. 2019 *Digital Water: Industry Leaders Chart the Transformation Journey*. International Water Association and Xylem Inc.

Silva, M. S., Coelho, S. T. & Moreira, M. 2012 Efficient information management in urban water services (in Portuguese). In *Proceedings of the National Meeting of Basic Sanitation*, 10–12th October 2012, *Évora, Portugal*.

Silva, M., Vilela, J. & Morais, M. 2017 Integration between information systems. Contribution to the provision of quality service (In Portuguese). In *Proceedings of ENEG 2017, Évora, Portugal*.

Silva, A. T., Germano, R. & Monteiro, A. J. 2019 Development of sustainable solutions for collection and management of infrastructure management (In Portuguese). In *Proceedings of ENEG2019, Ílhavo, Portugal*.

Sousa, C., Marques, A. & Fernandes, J. 2017 The contribution of GIS information for an integrated and efficient management systems (In Portuguese). In *Proceedings of ENEG 2017, Évora, Portugal*.

Tavares, J., Costa, C., Barrocas, G., Nogueira, S., César, S. & Castro, F. 2019 The integral digitalization of urban cycle water management – a pragmatic approach of water industry 4.0 (In Portuguese). In *Proceedings of ENEG2019, Ílhavo, Portugal*.

Vieira, P., Costa, F., Cunha, R., Marques, R. & Fernandes, M. 2018 Technological platform for the urban water cycle. From concept to implementation (In Portuguese). In *Proceedings of 18thENASB, Porto, Portugal*.

First received 30 April 2020; accepted in revised form 28 July 2020. Available online 11 August 2020

© IWA Publishing 2020

Multi-objective optimization of energy and greenhouse gas emissions in water pumping and treatment

Iliana Cardenes, Afreen Siddiqi, Mohammad Mortazavi Naeini and Jim W. Hall

ABSTRACT

A large part of operating costs in urban water supply networks is usually due to energy use, mostly in the form of electricity consumption. There is growing pressure to reduce energy use to help save operational costs and reduce carbon emissions. However, in practice, reducing these costs has proved to be challenging because of the complexity of the systems. Indeed, many water utilities have concluded that they cannot practically achieve further energy savings in the operation of their water supply systems. This study shows how a hybrid linear and multi-objective optimization approach can be used to identify key energy consumption elements in a water supply system, and then evaluate the amount of investment needed to achieve significant operational gains at those points in the supply network. In application to the water supply system for the city of London, the method has shown that up to 18% savings in daily energy consumption are achievable. The optimal results are sensitive to discount rate and the financial value placed on greenhouse gas emissions. Valuation of greenhouse gas emissions is necessary to incentivise high levels of energy efficiency. The methodology can be used to inform planning and investment decisions, with specific focus on reducing energy consumption, for existing urban water supply systems.

Key words | energy efficiency, energy intensity, energy operations, optimization, urban infrastructure, water-energy nexus

Iliana Cardenes (corresponding author)
Mohammad Mortazavi Naeini
Jim W. Hall
Environmental Change Institute, Oxford University Centre for the Environment,
University of Oxford,
S Parks Road,
Oxford,
UK
E-mail: *ilianacardenes@gmail.com*

Afreen Siddiqi
Institute for Data, Systems, and Society,
Massachusetts Institute of Technology,
Cambridge, MA,
USA
and
Belfer Center for Science and International Affairs,
Harvard Kennedy School,
Cambridge, MA,
USA

HIGHLIGHTS

- A hybrid linear and multi-objective optimization approach is used to evaluate investment costs within an existing water supply network to identify operational savings, finding that, for the city of London, up to 18%% savings in daily energy consumption can be made through this method.
- There are persisting constraints, including security of supply and water quality, in operating and upgrading legacy water supply infrastructure for new objectives of reduction of energy and emissions.

INTRODUCTION

Urban water supply networks have high energy costs (usually electricity) associated with their operation, and high associated greenhouse gas (GHG) emissions (Brandt *et al.* 2011). Around the world, the total energy consumption and energy intensity (energy used per unit volume of water) in urban water supply systems has been steadily increasing over the last decade (Water UK 2006, 2008, 2009, 2012; Rothausen & Conway 2011; Kenway 2013; Siddiqi & de Weck 2013; Spang & Loge 2015). As growing urban populations and climate change place greater strains on existing systems to provide more, better quality water without increasing associated costs or harmful environmental impacts, there are increasing

doi: 10.2166/wst.2020.507

requirements for existing supply systems to upgrade and adapt to new conditions. However, there are typically few cost-effective options available for improving large urban supply networks operating with legacy systems.

The question of energy efficiency in water use has attracted growing attention, though for the time being the focus has been on abating water (and energy) demand by installing high efficiency appliances in homes. New tools have been developed for creating cost abatement curves which have informed demand-side management for addressing the water-energy nexus challenge (Chini *et al.* 2016). From the supply side, however, there has been limited work on developing practical approaches to determine optimal strategies for improving existing large-scale urban water systems. This study aims to address this gap and develops a multi-objective optimisation approach for informing investment decisions for reducing energy consumption in existing water supply systems. The methodology is demonstrated for London's water supply system within a context of future uncertain energy prices and GHG emission costs.

This work offers two distinct contributions: First, it demonstrates a computationally tractable modelling and optimization approach (through system aggregation and bi-level optimization) for analysing large complex urban supply systems such that optimal options for operational energy reduction can be identified by evaluating trade-offs between capital expenditure (CAPEX), operating expenditure (OPEX) and GHG emissions. While optimization techniques have been extensively used for studying water networks' operation, the emphasis has largely been on improving pumping operations for cost savings from multi-pattern electric tariffs (Schaake & Lai 1969; Ulanicki & Kennedy 1994; Broad *et al.* 2010; Perelman *et al.* 2013; Puleo *et al.* 2014; Ghaddar *et al.* 2015; Sarbu 2016). Operational optimization of the whole supply system (that includes pumping and water treatment) has not received as much attention (Ulanicki & Kennedy 1994). Here, the focus of optimization is expanded to include key elements in water supply networks. This enables a more holistic view of the energy consumption hotspots and allows for identifying a wider variety of options for increasing energy efficiency.

Second, the method is applied on a very large real-world network of the city of London. This is an important distinguishing feature of the work as most published studies use theoretical networks in optimization analysis and therefore have had limited adoption in practice (Marques *et al.* 2012; Maier *et al.* 2014, 2015). In addition, there are often practical limitations to the implementation of solutions that are derived from optimisation studies. By using a real-world network, we are able to provide insights into the practical limitations that may constrain realization of theoretical energy savings.

The approach developed and demonstrated here can be used by water providers and public-sector decision makers to identify optimal strategies for operational cost savings (by reducing energy costs), and to investigate the relationship of GHG emission costs that can have important implications on feasibility of capital investments. It is also of relevance to researchers by pushing the limits of optimisation of water supply system operation, whilst validating the feasibility of the results in the context of a practical system.

BACKGROUND

Existing water supply networks have high operating energy costs. Previous studies have found that direct consumption accounts for the largest proportion of electricity use, upwards of 87% (Parsons & Marcet 2012). It is therefore in the interest of utilities to consider options for reducing energy consumption. Furthermore, with emerging regulations and growing awareness of the need to curb emissions, there are two additional incentives converging: there may be carbon emission costs that the utilities may directly or indirectly bear, and to be better stewards of the environment, energy consumption should be reduced. While there is extensive research taking place on developing new technologies (Cook *et al.* 2012; Kartalidis *et al.* 2015; Napoli & Rioux 2015), such as desalination or more efficient pumping, this paper considers the case of action in large-scale water-supply legacy systems where radical changes are expensive, and there is a need to identify opportunities for improvement with strategic cost effective interventions and investments.

Water utilities are responsible for planning and investing to achieve reliable supplies of water. Uncertainties abound in water resources management with pressures from climate change, urbanization, population growth and increasing demand (Roach *et al.* 2016). Water planners and agencies are thus under pressure to make water resources planning decisions that can meet demand while ensuring affordability, and which will be reasonably robust and adaptable to future uncertainties. When investments are needed to enhance security of supply, there is a trade-off between affordability for the customers, who ultimately pay for investment via their water bills, and security of supply.

Traditionally, water infrastructure costs are dealt with in whole life terms; that is, costs are estimated as discounted capital expenditure (CAPEX), as well discounted operational expenditure (OPEX), though the operational costs have tended to be dealt with rather crudely. Given the very different energy implications of alternative infrastructure options, there is a strong case for a more sophisticated treatment of energy costs in decision-making processes. Requirements to consider greenhouse gas emissions associated with energy use in the water sector are a further motivating consideration of energy efficiency.

There is also growing recognition of the need to address uncertainty and its impact on water infrastructure planning (Ray & Brown 2015), which has led to increasing uptake of various methods for decision-analysis under uncertainty (Lempert 2003; Ben-Haim 2006; Brown 2011; Groves *et al.* 2013; Matrosov *et al.* 2013). Optimization studies have focused on deterministic approaches to optimizing both the design and operation of water networks (Perelman *et al.* 2013), thus providing limited applicability to real-world network where there is usually uncertainty in the data and parameters. Recent studies have highlighted that uncertainty in model parameters has rarely been incorporated into water network optimization studies (D'Ambrosio *et al.* 2015), and so remains a research challenge for the field (Mala-Jetmarova *et al.* 2017). Developing a further understanding on how energy consumption looks like for water infrastructure systems, and its associated uncertainties, will provide more insight to the water sector, as it seeks to both to supply more affordable water to customers and to do so in a more sustainable way.

MATERIALS AND METHODS

The data used for this study, including the source and details of datasets can be found in the Supporting Information (section 1). The method consists of creating three main analytical components: First, the supply system is modelled as a network, for which a linear optimization problem is formulated. The decision variables represent flows over each edge (link or pipeline), and capacity constraints and conservation constraints are applied. The objective of the optimization problem is to find daily flow rates that minimise total energy costs in the network, whilst meeting demand. While the linear program (LP) formulation guarantees an optimal solution (Broad *et al.* 2010), it requires some simplification for tractability. The linear optimization model determines the optimal routing (in volumetric flow rates) for distributing the water through the supply network at minimum cost.

The second component consists of another (second) level of optimization based on a multi-objective evolutionary algorithm (MOEA), in which optimal investment strategies are identified. To do this, 'bottlenecks' are identified using the upper Lagrange multipliers from the optimal solution in the LP. A MOEA is run that aims to meet water demand whilst minimizing total expenditure costs for a set of investments that are identified based on the Lagrange multipliers obtained from the LP.

The third component is of uncertainty analysis in which changes in population, per capita and industrial water demand, water available for supply, discount rates, energy prices and carbon prices are stochastically modelled and provided as inputs. Several network investment strategies are then evaluated to determine if any of these configuration changes improve the optimal solution enough to warrant investment. The final target is to identify a strategy with the lowest energy consumption that best meets water demand under uncertainty for a given budget.

Mathematical formulation of linear model

The problem addressed in this study can be stated as follows:

Let $G = (T, P)$ be a directed network with n nodes and m arcs, where T is the set of nodes that represent a source (such as a river) or a storage element (such as a reservoir or water treatment works (WTW)) and P is the set of arcs (representing a set of network pipes). Each arc represents a pipe $(i, j) \in P$, *with* a cost c_{ij} that denotes the energy cost in kilowatt hours (kWh) of a unit of water flowing in the pipe. Researchers have previously approximated the energy consumed by pumps as a linear function of the pumping station water flow, and comparisons with more complex hydraulic models have demonstrated that they are capable of solving real-life pumping problems in a comparable and more computationally efficient manner (Giacomello *et al.* 2013; Puleo *et al.* 2014). Each P are also associated with an amount of x_{ij} water flow in meters cubed (m^3) per day. Each element of T and P are associated with a lower bound l_{ij} and an upper bound u_{ij} of the flow; b_i represents the available amount of supply or demand of water.

$$Minimize\ J = \sum_{(i,j)\in P} c_{ij}x_{ij}, \tag{1}$$

Subject to:

$$\sum_{j:(i,j)\in P} x_{ij} - \sum_{j:(i,k)\in P} x_{jk} = b_i \qquad \textit{for all } i \in T, \tag{2}$$

Flow capacities in pipelines:

$$l_{ij} \leq x_{ij} \leq u_{ij}, \qquad \textit{for all } (i,\ j) \in P. \tag{3}$$

The first equation is for the objective function J, where the objective is to minimise the total daily operational energy costs in the network. The second equation is a constraint that assures that the summation of supply and demand around each demand node is zero. The third equation is a constraint imposed to assure that there are no negative water volumes and to establish the capacities of each water source, reservoir and treatment works. Following other studies (Puleo *et al.* 2014), this LP formulation assumes that the network hydraulics are embedded in the empirical energy consumption relationship and are represented in the cost coefficients c_{ij} for each pipeline.

Implementation of multi-objective evolutionary algorithm

The MOEA is implemented to evaluate a set of investment options. In water supply systems (as in many other infrastructure systems) there are typically a finite and discrete set of options (such as upgrades, retrofits and expansion) that can be considered for investment decisions. Here, using the city of London as a case, a set of 12 options was considered (outlined in Table 1). As the MOEA searches for the optimal solution, it uses the LP to obtain optimal flow volumes in the distribution network for each case. Figure 1 illustrates the computational setup.

The optimization is carried out using a multi-objective evolutionary algorithm that evaluates trade-offs between the objectives of capital expenditure and operational expenditure and builds a 'Pareto approximate set', which is the best-known approximation to the Pareto optimal set (Kasprzyk *et al.* 2013). An ϵ-dominance multi-objective evolutionary algorithm (ϵMOEA) was used, which has been applied to water resources problems in the past (Mortazavi *et al.* 2012; Mortazavi-Naeini *et al.* 2014, 2015; Borgomeo *et al.* 2016). ϵMOEA uses the ϵ-dominance concept to divide the objective space into hyper boxes of size ϵ and allows only one non-dominated solution to reside in each

Table 1 | Short-term investment strategies derived from optimal LP model

	Type of investment	Location of investment in network	Capacity (m^3/day)	Investment details	CAPEX (£000 s)	OPEX (kWh/m^3)
1	Capacity upgrade	Chingford WTW	27,300	Expand up to 32,760 m^3/d	278[a]	0.68
2	Energy efficiency improvement	Walton WTW	85,900	Use new technology to increase efficiency. Reliability factor associated to the technology included	160[b]	2.57 – improved up to 1.7
3	Capacity upgrade	Hornsey Reservoir and WTW	30,000	Increase the capacity of reservoir and WTW up to 33,000 m^3/day	303[a]	0.15
4	Capacity upgrade	Ashford WTW	682,700	Expand up to 750,970 m^3/day	4,723[a]	0.12
5	Energy efficiency improvement	Groundwater pumps	286,600	Groundwater pump replacement results in a 2–10% gain in the energy intensity of groundwater pumping[c], there are 31 stations.	Thames Water process model	1.02 – improved by up to 10%
6–12	Energy efficiency improvement	Main distribution pumps in system	NA	Improve pumping energy use by 2%: variable speed drives (VSDs) increasing the efficiency of the pumps in the water distribution system implemented, improving the energy intensity from 0.117 kWh/m^3 to 0.116 kWh/m^3	Thames Water Process model	0.167–0.105 improved by up to 10%

[a]Estimated from plant capacity construction cost curve for existing plants; originally established by Hinomoto (1977).
[b]Estimated from the capital cost of installing an efficient UV technology in a medium sized water treatment (WT) plant (National Research Council 1999).
[c]From Nogueira Vilanova & Perrella Balestieri (2014).

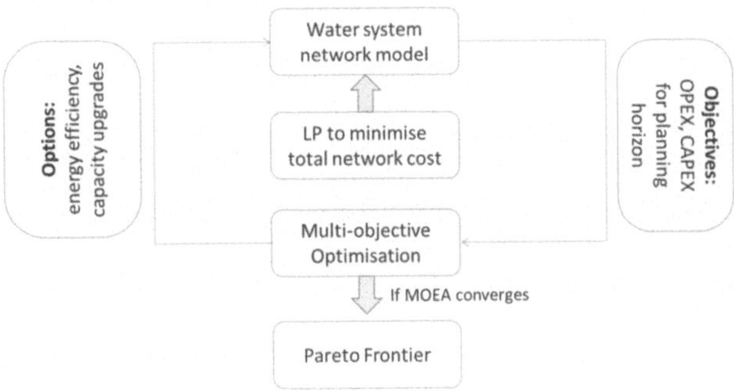

Figure 1 | Relationship of two-stage LP-MOEA optimization.

box (Laumanns *et al.* 2002). Inclusion of this concept in an evolutionary algorithm produces a method capable of maintaining a diverse and well-distributed set of solutions with a small algorithmic computational cost (Deb *et al.* 2003).

Pareto efficient solution

The final pareto efficient solution of each pareto approximation curve was calculated by dividing the change in operational energy costs over the change in capital investment cost, where the maximum E is the pareto efficient solution for a given multi-objective trade-off:

$$E = MAX\left\{\frac{-\Delta OPEX}{\Delta CAPEX}\right\} \tag{4}$$

This is the point on the pareto curve at which the capital expenditure provides the largest operational saving per unit of capital investment. It is important to note that OPEX usually includes more than energy costs such as staff costs, maintenance costs, insurance, vehicles and others which are not included in the model. In this study, OPEX is used to represent operational energy cost, as the largest operational cost for utilities, usually upwards of 87%. Furthermore, within OPEX, energy costs are variable (depending on the price of electricity), while other costs (such as salaries) are largely fixed.

Application: London water supply system

To demonstrate the proposed method, the model was applied to the London urban water supply system, located in the Thames River Basin in the South East of England, and operated by Thames Water Utilities. The region is under serious water stress and climate change is projected to result in a decrease of water availability (Environment Agency 2008; Burt *et al.* 2016). The area is also known as the London Water Resource Zone (WRZ). A WRZ is the standard geographical unit for water resources planning and is defined by the UK Environment Agency (2009) as: 'The largest possible zone in which all resources, including external transfers, can be shared, and hence the zone in which all customers experience the same risk of supply failure from a resource shortfall.' It is thus an appropriate regional scale at which to make management decisions for the daily operations of the water network. The WRZ is mostly supplied by surface water abstractions from the river Thames via pump storage reservoirs, and by some groundwater abstractions, delivering clean water to approximately 7 million people every day (Thames Water 2015). Water abstractions are subject to limits to maintain environmental flows. A previous study (Borgomeo *et al.* 2014) identified population growth and abstraction allowance reductions as two of the major uncertain factors posing the greatest pressures on the supply system.

The London WRZ has the challenge of continuing to supply water to growing populations while ensuring affordability. Because it is in a heavily urbanised and densely populated area, the opportunities for operational improvement lie mostly in making the operation of the system more efficient, and in investments at key network points that can reduce the energy consumption of the system.

Figure 2 shows the London WRZ. The ovals over the reservoirs shows how they have been grouped in the model. The yellow diamonds show the water treatment works and how they are connected to the reservoirs and the ring main, all of which are represented in the model. A simplified representation of how the model is connected can be found in the Supporting Information (section 2). The optimisation model uses an aggregated (simplified) representation of the actual network in order to reduce the number of nodes and reduce the computational burden for optimization (following a similar approach from other studies) (Mala-Jetmarova *et al.* 2017). The treatment works included in the model represent all surface water treatment works in London. The groundwater treatment work (GWTWs) is a conglomerate of all the groundwater treatment works, in which the capacity of all the groundwater treatment works has been summed and an average cost of abstraction used. The groundwater sources have a minimum daily supply requirement of at least 40% of their capacity. This was incorporated after discussions with Thames Water modelling experts to represent the system more accurately, as some London demand must be met through groundwater sources. The x## numbers represent the

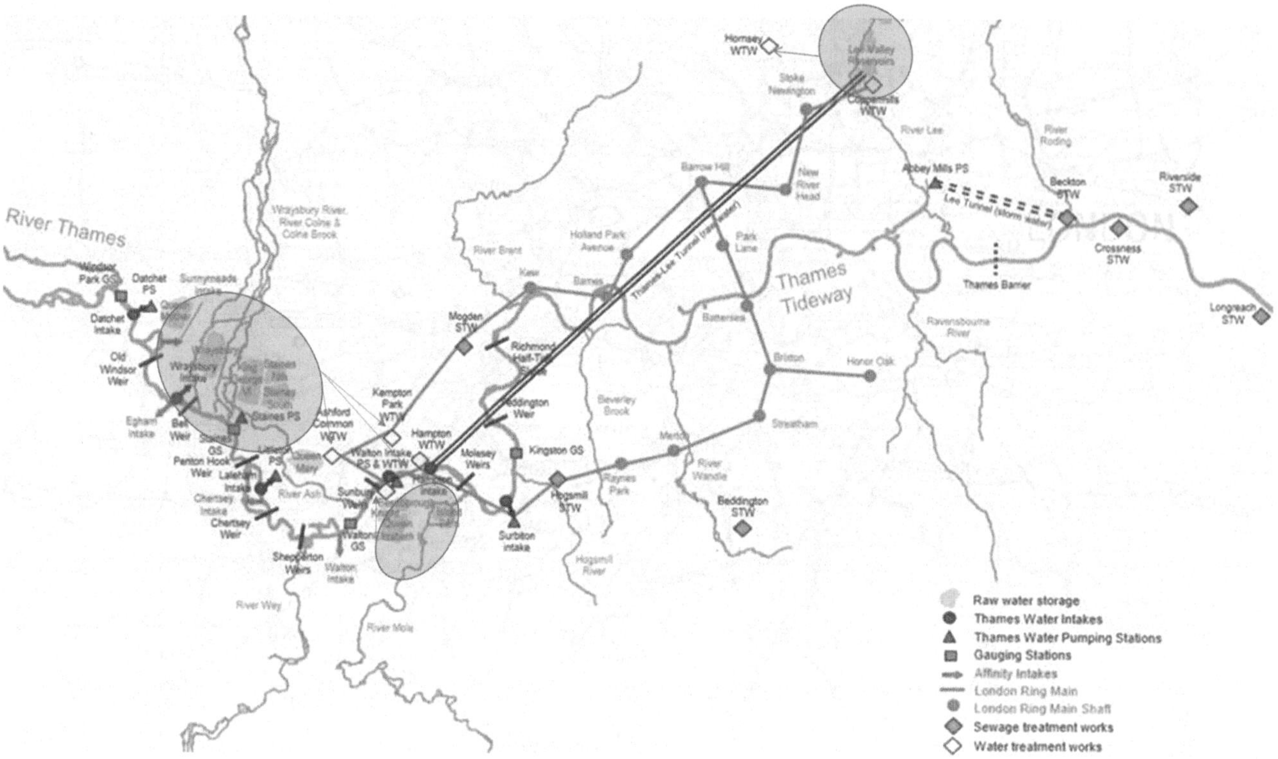

Figure 2 | The London Water Resource Zone and simulation model (Thames Water 2014).

assigned names for the optimization for each node and arc of the network. This configuration was used to illustrate the ability of the modelling technique to provide useful investment solutions.

Each reservoir has abstraction pumping costs (River to Res connections in network diagram). All treatment works also have transmission pumping costs that are equal for all stations (WTW to demand centre connections). It is assumed that the water pressure in the system is constant and there are no flow variations due to pressure differences. Operational water pressure for the Thames Water (TW) network was not available, so the effect of head loss is not included in the model. It is assumed that the pressure variations are captured in the empirical cost numbers used in the model. Each treatment work has a unique cost per unit of water treated, which is representative of the average cost to treat each unit of water in these treatment works in London. Reservoirs are assumed to have daily capacities matching daily demand unless otherwise stated. Reservoir 4 has a lower capacity than the others (30,000 m^3/d), while the others have virtually unlimited capacity. This was due to lack of data for daily capacities of reservoirs. Discussions with Thames Water experts highlighted that this was not a problem for modelling of daily supply because service reservoirs have capacities far in excess of daily demands. Each treatment work has a capacity which matches the real capacity in these treatment works in London.

Uncertainty modelling

A number of different input parameters are varied in Monte Carlo simulations (20,000 runs) to ascertain how changes in different factors would influence the operational costs. The parameters, represented by a set of distributions, were:

1. Water demand: change in total water demand, which is a combination of change in per capita demand and change in population
2. Electricity prices (£/kWh)
3. Carbon prices (£/tCO_2e)
4. Discount rates: the rate at which the value of investment is discounted over time.

According to the Thames Water Resources Management Plan (WRMP), the average demand for the London WRZ is approximately 2,035 megalitres per day (ML/d). The mean total demand in the optimization was thus 2,035 ML/d (which is broken down into a multiplication of the population with the per capita demand plus the industrial and commercial demand for London). Per capita demand was generated from a normal distribution, with a mean of 160 L/p/d, which was the average Thames Region per capita demand, and a standard deviation of 1 L/p/d to capture small variations over the 5-year period under investigation. The 5-year period was deemed as a useful time-frame as it matched the water sector's price review time periods. This is the period over which the regulator, Ofwat, assesses each company's ability to finance its activities (National Audit Office 2015; Ofwat 2017). Gaussian distribution probability distributions are commonly used for Monte Carlo simulations of water demand where relatively accurate estimations of demands for a region can be made (Babayan *et al.* 2005). A right skew Pearson distribution (skew: −0.75, kurtosis: 3.4) was applied to population, as population was expected to increase in the basin. This was done so that the distribution of resulting simulations included a larger number of higher than average demand figures than lower than average, as demand was expected to increase, and thus the uncertainty was rightly skewed. Other population forecast studies have traditionally used rightly skewed distributions to represent the probability that the population will increase (Alho 1997). Demand was calculated as follows:

$$D_T = (P * PC_D) + I_D + C_D \tag{5}$$

where D_T represents total demand, P represents the population, and PC_D, I_D and C_D represent, per capita demand, industrial demand and commercial demand respectively.

The water available for use (WAFU) was derived from the maximum abstraction allowed for the London WRZ published in the TW WRMP of 2,385 ML/d. The availability in the LP was generated from a normal distribution with $\mu =$ 2,385 ML/d and $sd = 10$ ML/d. The LP was optimized to meet London's daily demand, and to not exceed the maximum abstraction available for the region. The LP was run for each of the 20,000 simulations.

Electricity prices were derived from forecasted figures for the next 5-year horizon from the UK Government (DECC 2016). The electricity price was thus generated from a normal distribution with $\mu = 0.125$ £/kWh and $sd = 0.01$ £/kWh. It is important to note that electricity prices are not just uncertain, but also variable on short time-scales, and that is why an uncertainty analysis around electricity price changes was carried out.

Greenhouse gas emissions

The operational greenhouse gases were calculated by multiplying the daily kWh consumed by a unit rate (emissions factor) to generate a tCO_2e from a kWh. The emissions factor was varied in the Monte Carlo simulations to assess the impact of a decarbonising grid, with a mean of 0.212 $kgCO_2e$/kwh and a standard deviation of 0.05, to incorporate higher end emissions factors closer to today's rates (around 0.4) and projected changes of down to 0.1. Capital investment embodied greenhouse gases were calculated from Thames Water's 2017 driven rates for embodied carbon of proposed capital investment options in tCO_2e/£, provided by the utility. Carbon prices were also used to calculate the monetary cost of greenhouse gas emissions. These were derived from the UK government's carbon price values used for UK public policy appraisal (DECC 2015). They published three scenarios: low, central and high, with approximate values of 4, 21 and 46 £/tCO_2e. The starting multi-objective optimization was carried out with the central value.

Investment strategies

For the ϵMOEA optimization, several investment strategies were tested to evaluate potential operational energy savings. A pareto approximation cost curve was derived that gives an indication of how much capital investment would be needed to warrant specific operational gains. These results can be used to evaluate the costs of different technologies or capacity expansions versus the potential savings that such technologies may provide, for a given budget.

Opportunities for improving the performance of the system were identified with the outputs from the LP, using the Lagrange multipliers. The optimal LP solution was used as a basis to identify strategic options for CAPEX investment solutions. The Lagrange multipliers show how much the optimal (minimum) cost value would change if a constraint of capacity is relaxed. They show the elements of the network where if the constrains are relaxed, the solution can be further improved. This is a rigorous systematic approach to identifying potential for cost reduction, which has traditionally been done through expert knowledge. The investment options were first identified based on values of the Lagrange multipliers, and were then discussed with the utility experts to determine the feasibility of the options, which were then refined and are listed in Table 1.

The capital costs for proposed investments that included pumping were derived from Thames Water's in-house process model for pumping stations. The final capital costs in the model depend on the ML/d required to be pumped, for which the total associated cost was provided by Thames Water from output from their inhouse model. The costs depend on the type of pump, as abstraction pumps (e.g. river to reservoir) have a different cost curve to pressurized distribution pumps.

These investment options are by no means exhaustive, and some may not be feasible due to other external constraints, such as land limitations; but they illustrate the capacity of this model set-up to provide an indication of where costs can be saved within a water distribution and treatment network, and provide a cost trade-off for each investment option. The cost of the investments was calculated for a five-year return on investment period, following current water industry regulation, and industry discount rate of 3.5%.

RESULTS

Linear programme

The LP results show that optimizing the system based on energy cost provides an approximate 18% daily saving in energy consumption of the water distribution and treatment system, from an average daily total energy consumption of 1.15 million kWh to 939,000 kwh per day. It also provides an average optimal annual cost in pounds of approximately £42.8 million per year to run the network. Thames Water currently spends approximately £50 million per year to operate the water network in London, thus using optimising techniques for the network could provide significant annual savings. TW reports to Ofwat show that their annual average operating expenditure for water delivery for their whole region is £327 million a year (Ofwat 2010). This includes other areas outside of London, and contains other costs such as staff costs, maintenance costs, insurance, vehicles and other miscellaneous costs.

Discussions with experts within the case study utility to validate and discuss the model output place this result within the context of the daily operations of a water utility. Such savings are currently not achieved in practice due to competing needs that must be fulfilled. The first is security of supply. Even though some sources of water within the network are very costly, for example, groundwater supply, there are sections of the network that require water from these sources to meet demand, and thus they need to be used daily at a high cost. Some of these are represented in the model, such as groundwater source requirements, but not all. The second priority is ensuring high water quality. Due to the variable nature of water quality, for example, the emergence of algal blooms seasonally, some more efficient parts of the network sometimes need to be turned off to avoid contamination, thus requiring more costly areas of the network to be run. Such modifications occur on a regular basis, requiring expensive resources to be turned on often. The third is the avoidance of leakage. If an efficient section of the network suffers from increased leakage, for example, from a burst pipe, supply is drawn away from that area and more costly water supply options may be used that would not be used otherwise, to avoid water losses. The reduction in energy consumption is thus currently the fourth priority for the case study utility, and most likely for the whole sector. These priorities added together mean that, in practice, it is difficult for utilities to make large energy savings based on existing operations.

However, discussions with experts from the utility did bring out that there are opportunities for more coordination between priorities, to improve the understanding of the connections between operating goals and energy costs, and benefit from potential savings. Studies such as this one can support such coordination and contribute to cohesion between objectives by showcasing methodological approaches that could be used in conjunction with existing processes for operating networks.

Table 2 presents a comparison between the volumes of water treated in the existing system (and associated empirical operational energy costs provided by Thames Water), with the optimized results for the water treated (in m^3) in each treatment work and associated energy costs (in kWh) without any additional investments. The figures in this table are lower than the overall system results above because they focus only on the water treatment works, and the overall result also includes pumping figures.

As can be seen, even though the actual treatment works provide less water, on average, than the modelled results (around 35,000 m^3/day, most likely due to daily variations in water supplied), the total energy consumption is significantly lower for the modelled results (476,170 kWh/day) than the current system (552,798 kWh/day), but the results are highly comparable.

Multi-objective evolutionary algorithm results

Figure 3 shows the εMOEA trade-off between capital investment costs and operational energy costs. Each point in the figure represents a unique combination of investments.

Table 2 | Comparison of LP optimization results for the main WTWs with operational figures for validation

	Treatment works	Current average water provision (m^3/d)	Current average daily energy use (kWh/d)	Model average water provision (m^3/d)	Model average daily energy use (kWh/d)
GWTWs	GW Sources	200,800	154,104	114,000	116,394
WTW 1	Chingford WTW	7,698	7,885	0	0
WTW 2	Coppermills WTW	458,372	105,056	523,606	120,429
WTW 3	Hamptom WTW	520,377	106,134	664,300	146,146
WTW 4	Ashford WTW	617,381	75,381	682,700	81,924
WTW 5	Kempton WTW	138,688	44,224	21,176	6,776
WTW 6	Walton WTW	26,780	55,609	0	0
WTW 7	Hornsey WTW	30,000	4,404	30,000	4,500
Total	–	2,000,096	552,798	2,035,782	476,170

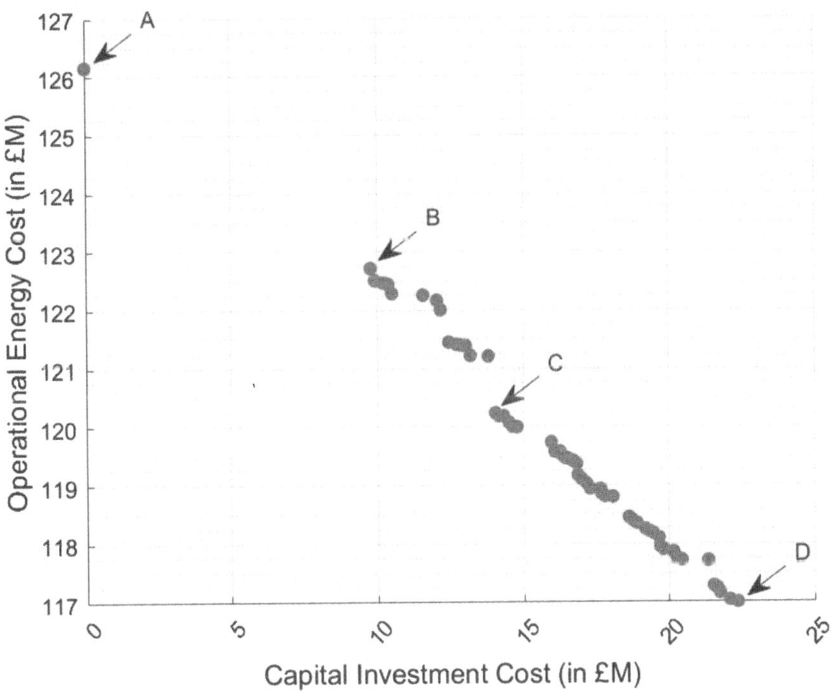

Figure 3 | Multi-objective trade-off between CAPEX and OPEX costs. Solution A represents no capital investment; Solutions B and D are the lowest and highest combination of capital investment respectively; and solution C is the pareto efficient solution. There is a gap between solution A and B because the capital investment at solution B is the minimum amount needed to reach an optimal combination of capital investment and operational energy costs.

The solution A on the x axis is the LP optimal result. The LP resulted in an optimal average 5-year discounted operating cost of approximately £126.2 million pounds. This is the single optimal solution for minimizing operational energy consumption in the network without additional investment, and for a set of specific parameters, which are later treated as uncertain – see Supporting Information (section 3) for list of parameter values. The solution C is the Pareto efficient solution in the Pareto-approximation curve, as defined in section 2.3.1. It is the point at which the relative operational energy savings are the best in relation to the capital investment. The ϵMOEA results show the trade-off between the capital costs of different combinations of investment strategies and the operational 5-year discounted costs for each of these optimal combinations. The ϵMOEA, even though computationally time-consuming, allows for a visualization of a Pareto-approximation frontier and trade-off between multiple objectives, showing how operational savings can be made with more investment. This permits a more in-depth evaluation of options to the water utility based on available investment budgets. Furthermore, it allows the water utility to calculate the recoupment period for each investment combination.

The Pareto curve in Figure 3 appears linear, but is in fact many small Pareto curves made up of a combination of increasing capital investments which happen to assimilate a linear increment when seen together. Figure 4 shows a close-up of the location of solutions B, C and D to visualize the small Pareto curves at those locations; the Pareto efficient solutions can be seen in a different color.

It may appear that operational cost savings are not large enough when compared to capital investment. However, the figures shown only show the savings over a five-year period, but the life cycle of investments are in the range of 30–40 years, or even more. Thus, operational savings of about ~£5 million in a 5-year period for a ~£14 million investment, as seen in Figure 6. Solutions A and C can provide a large saving over the life-time period of the asset. In this case, the evaluation is 5-years because of the budget cycles in the UK, as outlined above, but water utilities also plan in longer-term frameworks.

Solutions B and D are the lowest and highest combination of capital investment, respectively, and are chosen with C (the Pareto efficient solution) to explain in detail in

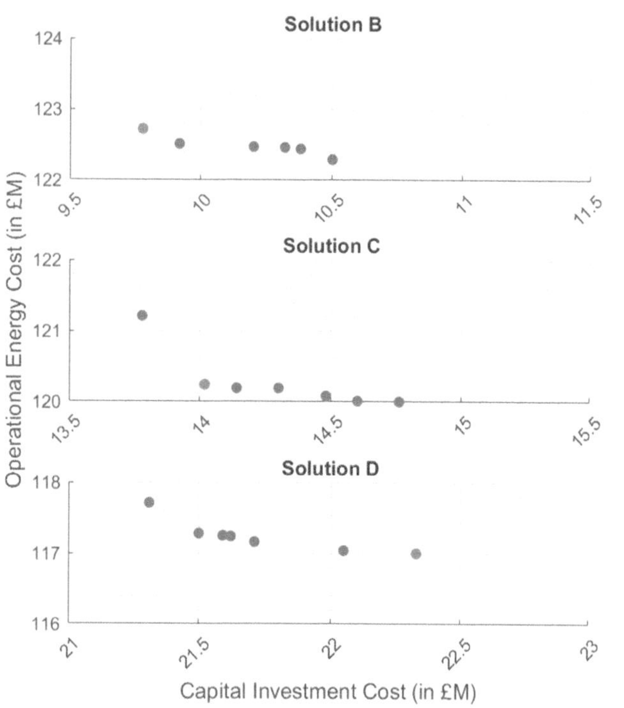

Figure 4 | Slices of Figure 3 showing the Pareto curves of specific solutions.

Table 3. Solution A has no capital investments. There is a gap between solutions A and B because the capital investment at solution B is the minimum amount needed to reach an optimal combination of capital investment and operational energy costs.

All the solutions in the Pareto-approximation curve include using groundwater pumping in the investment solution. This is because the model has a constraint where a minimum amount of water needs to be extracted from groundwater systems, as occurs in the actual system, which is a more expensive source than some surface water treatment works. All the solutions on the Pareto-approximation curve also involve investment in capacity expansion at Ashford WTW. Ashford is already a large

Table 3 | Optimal solutions showing Pareto efficient solutions (E) in the results section

Solution	Capital investment (million £)	5-year discounted operational energy (million £)	Investment solution details
A	0	126.20	NA
B	9.78	122.72	– Ashford expansion – GW pumping – 5/12 distr. pumps
C(E)	14.02	120.24	– Ashford expansion – GW pumping – 6/12 distr. pumps
D	22.33	117	– Ashford expansion – Walton efficiency – GW pumping – 12/12 distr. pumps
Change in discount factor			
E(E)	5.97	99.06	– 8/12 distr. pumps
F(E)	9.41	151.01	– GW pumping – 7/12 distr. pumps
Inclusion of GHG emissions			
G(E)	14.46	124.85	– Ashford expansion – 8/12 distr. pumps
Change in carbon price			
H(E)	11.45	121.67	– 9/12 distr. pumps
I(E)	10.5	131.27	– 8/12 distr. pumps

All solutions are presented in this table for ease of comparison. Solutions A–D are described above this table, and solutions E–I are described in subsequent text.

and efficient WTW, so re-routing more water through this location improves the energy consumption of the system significantly.

Water utility decision-makers can use this trade-off and options to evaluate whether they would be cost-effective, for example within 5-year or 10-year investments cycle. The above-optimal solution C would save £5.9 million within a 5-year investment cycle. Undoubtedly, this trade-off evaluation does not consider additional costs such as staff costs, land costs, insurance costs, capital maintenance costs and other externalities. But with more complete data-sets available in-house to the utilities, such tools could provide very useful information to improve performance of water networks at key nodes.

Discount rate

To test the sensitivity of the solution to changes in the discount rate. Two additional Pareto-approximation curves were generated with different discount rates (2% and 5%) in addition to the industry standard of 3.5%. Figure 5 shows the original trade-off curve at 3.5% and two additional Pareto-approximation curves, with their corresponding Pareto efficient solutions E and F. As can be seen, a change in the discount rate affects the costs in the trade-off curves and the optimal solution. Higher discount rates reduce the operational cost more and lower discount factors mean higher operational costs. The capital investments are unchanged as they are Net Present Value figures.

Solution E does not include the groundwater efficiency improvement (as in previous solutions), including Pareto efficient solutions C and F, which is why the capital investment is comparatively smaller (Table 3). This is probably because the investment cost of groundwater efficiency improvements is higher than the discounted operational cost of the pumps. It is interesting to point out that the 'recouping' percentage – how much of the capital investment is made back over the 5-year period – is quite different for the lower discount factor over the two other solutions. For optimal solution C it is 41%, E is 42% and F has a saving of 62% of the investment. This is most likely to do with the lower discount factor allowing for a larger saving over the 5-year period.

Greenhouse gas emissions

The analysis is continued with the central discount factor (3.5%) to evaluate the difference in the optimal solution and multi-objective optimization results when the capital

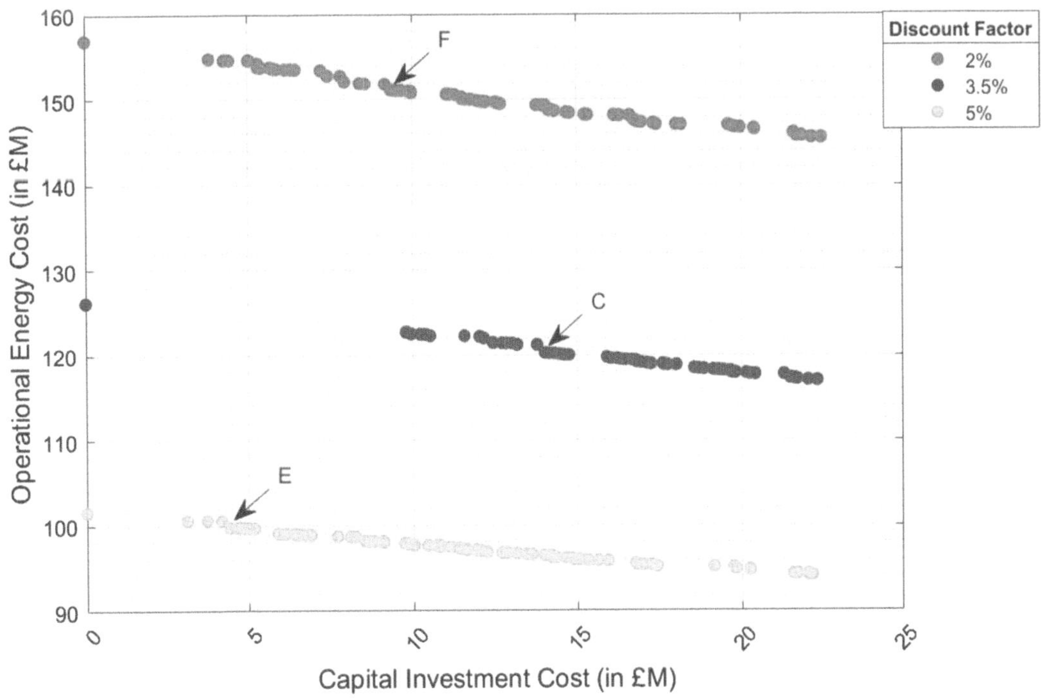

Figure 5 | Multi-objective trade-off between CAPEX and OPEX costs with different discount rates. Solutions E and F represent higher and lower discount rates respectively. Solution C remains as the original Pareto efficient solution in Figure 3.

and operational costs of greenhouse gas emissions are considered in the analysis.

Figure 6 shows the original 3.5% trade-off curve between capital investments and operational energy costs and a trade-off curve that includes capital investments costs plus the embodied capital investment costs from GHG emissions on the y axis, and operational energy costs plus operational GHG emission costs on the x axis. The two curves show the difference between just considering direct capital and operational energy costs and adding the costs of GHG emitted to the calculations. It is important to note that there are other costs involved in the building, maintaining and operating of water supply options, but this analysis aimed to investigate how the addition of GHG emissions may alter optimal solutions, particularly when considering carbon price changes. Thus, the graph includes direct capital and operational energy costs plus the greenhouse gas emission costs involved in the proposed CAPEX and the GHG emissions in the operation of the network.

Solution G is the Pareto efficient solution when GHG emissions are considered in the trade-off (Table 3). The investment in solution G is like solution C in that it also includes an expansion at Ashford WTW, but it does not include an improvement in the energy efficiency of groundwater treatment. This is probably due to the high embodied capital costs of improving so many pumps. It instead includes an upgrade in eight out of the 12 large water distribution pumps in the system, instead of the six in the original solution, denoting how an inclusion of greenhouse gas emissions changes the best investment choice. The embedded greenhouse gas costs in the capital investment are small compared to those in the energy costs but could become significant if the carbon intensity of electricity is reduced to close to zero.

Carbon price

The Pareto-approximation curve including greenhouse gas emissions costs was subjected to a sensitivity analysis to evaluate the influence of a change in carbon prices. Figure 7 shows three curves, each representing a multi-objective optimization trade-off between capital costs and embodied capital GHG emissions on the *y* axis and operational energy and GHG emissions costs on the *X* axis. The central curve is the same as the GHG emissions curve in the previous figure and has a carbon priced of 21 £/tCO2e. The other two curves represent the trade-offs with a carbon price of 4 £/tCO2e and 46 £/tCO2e and the details

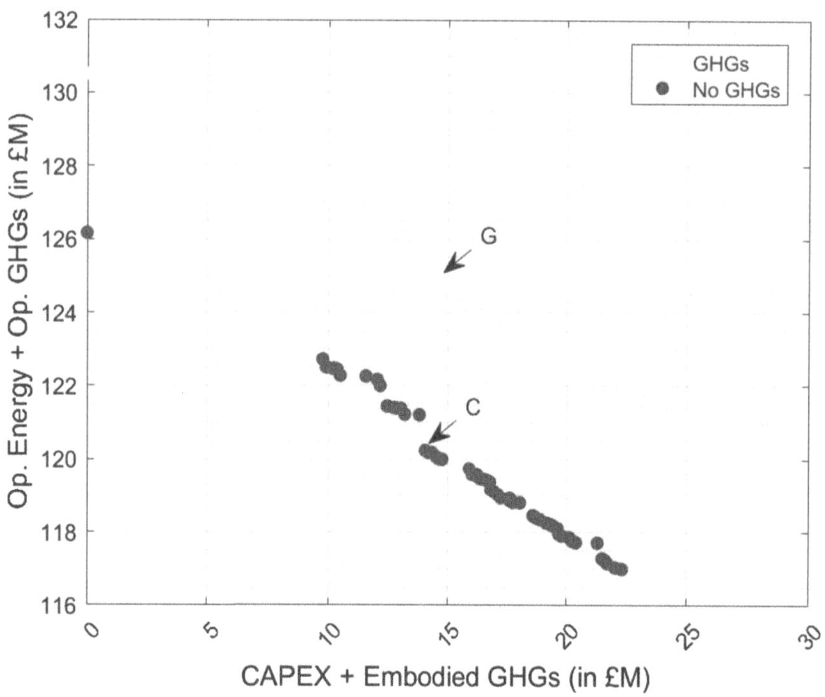

Figure 6 | Multi-objective trade-off between CAPEX and OPEX, and CAPEX and OPEX with GHGs. Solution C represents the original pareto efficient solution in Figure 3 and point G represents the pareto efficient solution when GHG emissions are considered.

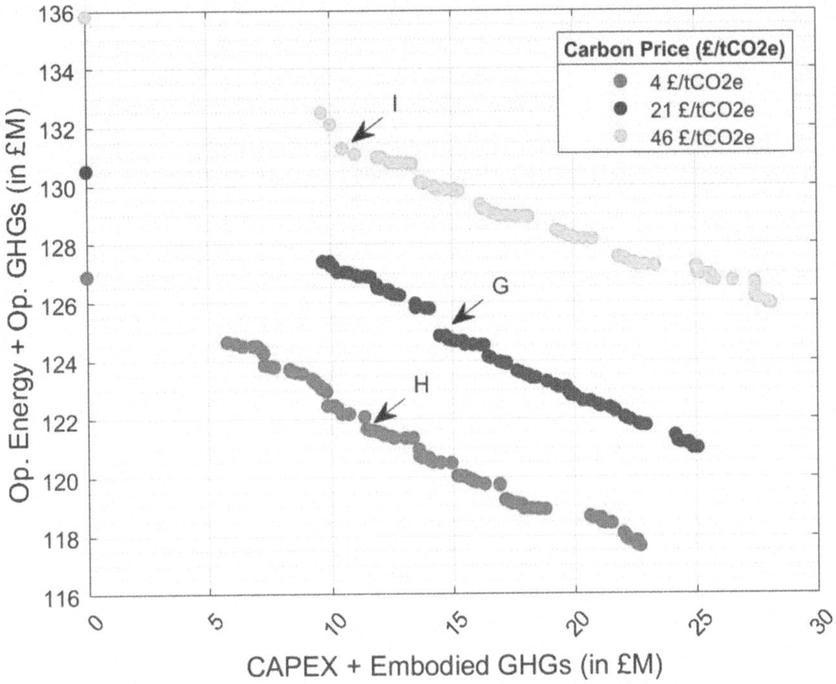

Figure 7 | Comparison of CAPEX and embodied capital GHGs with operational energy and GHG costs. Solution G is the original pareto efficient solution in Figure 6 which includes the cost of GHGs; Solutions H and I represent higher and lower carbon prices respectively.

of the Pareto efficient solutions H and I can be seen in Table 3.

As can be seen, carbon prices change the optimal investment combinations, highlighting the importance of setting a carbon price so that decision-makers can use the information to make the most efficient choices for their businesses.

Sensitivity of optimal solution

There are two additional sources of exogenous uncertainty in the model that may influence the optimal solution. As mentioned previously, electricity prices are not just uncertain, but also variable on short timescales, so an uncertainty analysis to the variability of energy prices can provide a more accurate representation of how costs in the system may vary in the short term. Figure 8 shows how optimal solution G (which includes the costs of GHG) would vary with changes in the additional two sources of uncertainty: change in water demand (made up of population change and per capita demand) on the x axis and change in energy price on the y axis. Each dot on the figure represents one scenario out of the 20,000 runs. The figure is

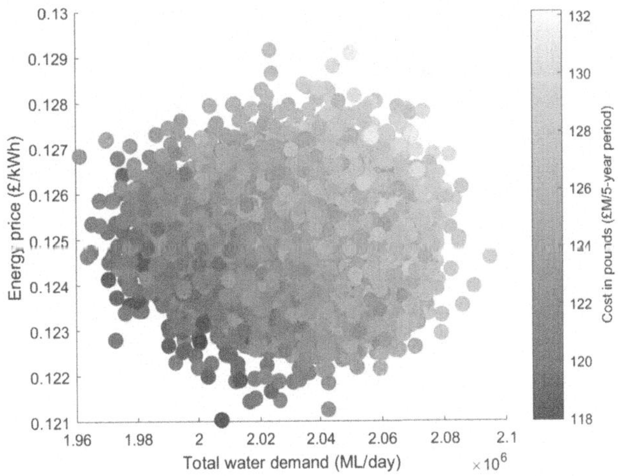

Figure 8 | Cost in pounds change with total water demand change and energy price change.

coloured by the total 5-year discounted cost in pounds to see how the sources of uncertainty affect the change in cost.

The figure shows how increases in demand raise the average operational energy costs in pounds, as does an increase in energy price. The results suggest that both small percentage changes in water demand and energy price could increase operational energy costs by between 2% and 5%. Additional simulations could be done in future work where larger uncertainty is looked at, in order to further analyse and refine the results.

CONCLUSIONS

Major cities around the world with large existing water supply systems are struggling to identify options to reduce operational energy use. In this study we have introduced a new a hybrid linear and multi-objective optimization approach. We have shown how this objective optimization approach can help to evaluate trade-offs between operational energy costs and capital investments within a water supply network. Visualizing the generated sets of Pareto-approximate optimal solutions helped to evaluate how different uncertainties (in discount rates and carbon prices) could affect projected costs and recoupment of investments within a network.

For the city of London, the optimization study shows that daily energy consumption can be reduced by up to 18% by optimally routing water distribution and treatment through the network. Other studies attempting the optimization of water networks through operational changes have achieved results in a range of 5–25% in energy cost savings (Cherchi *et al.* 2015). As well as being at the higher end of this range of efficiency savings, we have been able to incorporate a wider set of system requirements, particularly provision of high-quality water.

The case study highlights how small changes in the regulatory and planning framework, such as discount rates or carbon prices can have a significant effect on the cost of investment within water supply networks and are thus important elements to consider when evaluating the costs of possible solutions. Sensitivity analysis has demonstrated how a change in discount rates could change the daily operating costs by ~19% (from £126 million to £151 million), that, in turn, leads to a different set of optimal investment options. GHG prices and changes in discount rates have an impact on the cost of water system operations, and there are regions at which the optimal solution for the utility changes. For example, at a 5% discount rate, investment in groundwater pump efficiency improvements as part of the efficient Pareto set of investments disappears as compared with the 3.5% discount rate set, most likely because the investment cost of groundwater efficiency improvements is higher than the discounted operational cost of the pumps. However, the 'recouping' percentage – how much of the capital investment is made back over the 5-year period – remains the same, at approximately 41%, due to the higher discounting.

Under cases where additional external costs such as GHG prices are not considered, there may not be enough of a justification to invest in higher energy saving approaches because the solutions are not cost-effective. When wider costs are included, a more realistic recoupment of investment can be evaluated and the case for higher investment made. If carbon prices are imposed in the future, this can impact investment decisions, and thus, uncertainty within this policy area can lead to inefficiencies in the economy.

The study also demonstrated how changes in water demand, including per capita use and population can have an important effect on the operational energy costs of water supply systems. Conservation measures and behavioural change are key for reducing water demand and associated energy implications. The results of this study quantified this impact for the city of London, showing that the fraction of reduction in water demand is almost one to one with the reduction in operational energy costs: 3% water demand reduction results on average in a 3% reduction in operational energy costs, including GHG emissions costs.

These results identify and highlight the persistent constraints and challenges in operating and upgrading legacy water supply infrastructure for new objectives of reduction of energy and emissions. Nonetheless, the optimisation methods we have proposed help to elucidate these constraints and thus facilitate discussions between utilities, regulators and governments about how they can be addressed.

ACKNOWLEDGEMENTS

Iliana Cardenes is funded by the La Caixa Foundation and Thames Water Utilities Ltd We would like to thank Terence Winfield and Dr Keith Colquhoun from Thames Water for their support throughout the development of the study. The views expressed in this study are those of the authors and do not necessarily reflect the views or policies of Thames Water Utilities Ltd. The authors would like to acknowledge the use of the University of Oxford Advanced Research Computing (ARC) facility in carrying out this work. http://dx.doi.org/10.5281/zenodo.22558.

SUPPORTING INFORMATION

The Supporting information includes further details on:

1. Data sources
2. The simplified model
3. Starting linear programme parameters
4. Key equations

DATA AVAILABILITY STATEMENT

Data cannot be made publicly available; readers should contact the corresponding author for details.

REFERENCES

Alho, J. M. 1997 Scenarios, uncertainty and conditional forecasts of the world population. *J. R. Stat. Soc. Ser. A. Stat. Soc.* **160** (1), 71–85.

Babayan, A., Kapelan, Z., Savic, D. & Walters, G. 2005 Least-cost design of water distribution networks. *J. Water Resour. Plan. Manag.* **131** (October), 375–382.

Ben-Haim, Y. 2006 *Info-Gap Decision Theory*. Academic Press, London, UK.

Borgomeo, E., Hall, J. W., Fung, F., Watts, G., Colquhoun, K. & Lambert, C. 2014 Risk-based water resources planning: incorporating probabilistic nonstationary climate uncertainties. *Water Resour. Res.* **1**, 4840–4847.

Borgomeo, E., Mortazavi-Naeini, M., Hall, J. W., O'Sullivan, M. J. & Watson, T. 2016 Trading-off tolerable risk with climate change adaptation costs in water supply systems. *Water Resour. Res.* **52** (2), 622–643.

Brandt, M. J., Middleton, R. A. & Wang, S. 2012 *Energy Efficiency in the Water Industry: A Compendium of Best Practices and Case Studies*, Volume 11. IWA Publishing, London, UK. https://doi.org/10.2166/9781780401348.

Broad, D. R., Maier, H. R. & Dandy, G. C. 2010 Optimal operation of complex water distribution systems using metamodels. *J. Water Resour. Plan. Manag.* **136** (4), 433–443.

Brown, C. 2011 *Decision-Scaling for Robust Planning and Policy Under Climate Uncertainty*. World Resour. Report, Washington, DC, USA.

Burt, T., Boardman, J., Foster, I. & Howden, N. 2016 More rain, less soil: long-term changes in rainfall intensity with climate change. *Earth Surf. Process. Landforms* **41** (4), 563–566.

Cherchi, C., Badruzzaman, M., Gordon, M., Bunn, S. & Jacangelo, J. G. 2015 Investigation of cost and energy optimization of drinking water distribution systems. *Environ. Sci. Technol.* **49** (22), 13724–13732.

Chini, C. M., Schreiber, K. L., Barker, Z. A. & Stillwell, A. S. 2016 *Quantifying Energy and Water Savings in the U.S.* Residential Sector.

Cook, S., Hall, M. & Gregory, A. 2012 *Energy use in the Provision and Consumption of Urban Water in Australia: an Update Prepared for the Water Services Association of Australia*. Sydney, Australia.

D'Ambrosio, C., Lodi, A., Wiese, S. & Bragalli, C. 2015 Mathematical programming techniques in water network optimization. *Eur. J. Oper. Res.* **243** (3), 774–788.

Deb, K., Mohan, M. & Mishra, S. 2003 Towards a quick computation of well-Spread pareto-optimal solutions. *Evolutionary Multi-Criterion Optimization* **2632**, 68, 222–236.

DECC 2015 *Updated Short-Term Traded Carbon Values Used for UK Public Policy Appraisal*. Department of Energy & Climate Change, UK Government, London, UK.

DECC 2016 *Energy and Emissions Projections*. https://www.gov.uk/government/collections/energy-and-emissions-projections (accessed August 1, 2016).

Environment Agency 2008 Water resources in england and wales – current state and future pressures. *Bristol Environ. Agency*. December 2008.

Environment Agency. 2009 *Delivering Water Neutrality : Measures and Funding Strategies*. Environment Agency, Bristol, UK.

Ghaddar, B., Naoum-Sawaya, J., Kishimoto, A., Taheri, N. & Eck, B. 2015 A lagrangian decomposition approach for the pump scheduling problem in water networks. *Eur. J. Oper. Res.* **241** (2), 490–501.

Giacomello, C., Kapelan, Z. & Nicolini, M. 2013 Fast hybrid optimization method for effective pump scheduling. *J. Water Resour. Plan. Manag.* **139** (2), 175–183.

Groves, D. G., Bloom, E., Johnson, D. R., Yates, D. & Mehta, V. K. 2013 *Addressing Climate Change in Local Water Agency Plans: Demonstrating A Simplified Robust Decision Making Approach in the California Sierra Foothills*. Rand Corporation, Santa Monica, CA, USA. https://www.rand.org/pubs/research_reports/RR491.html.

Hinomoto, H. 1977 Capital cost functions of a surface water treatment plant and its components. *Water Resour. Res.* **13** (6), 873–876.

Kartalidis, A., Tzen, E., Kampragkou, E. & Assimacopoulos, D. 2015 Responding to water challenges in Greece through desalination: energy considerations. *Int. J. Water Resour. Dev.* **627** (December), 1–14.

Kasprzyk, J. R., Nataraj, S., Reed, P. M. & Lempert, R. J. 2013 Many objective robust decision making for complex environmental systems undergoing change. *Environ. Model. Softw.* **42**, 55–71.

Kenway, S. 2013 The water-Energy nexus and urban metabolism – connections in cities. *Urban Water Secur. Res. Alliance*. No. 100. Urban Water Security Research Alliance, Brisbane, Australia.

Laumanns, M., Thiele, L., Deb, K. & Zitzler, E. 2002 Combining convergence and diversity in evolutionary multiobjective optimization. *Evol. Comput.* **10** (3), 263–282.

Lempert, R. J. 2003 *Shaping the Next One Hundred Years: New Methods for Quantitative, Long-Term Policy Analysis*. Rand Corporation, Santa Monica, CA, USA.

Maier, H. R., Kapelan, Z., Kasprzyk, J., Kollat, J., Matott, L. S., Cunha, M. C., Dandy, G. C., Gibbs, M. S., Keedwell, E., Marchi, A., Ostfeld, A., Savic, D., Solomatine, D. P., Vrugt, J. A., Zecchin, A. C., Minsker, B. S., Barbour, E. J., Kuczera, G., Pasha, F., Castelletti, A., Giuliani, M. & Reed, P. M. 2014 Evolutionary algorithms and other metaheuristics in water resources: current status, research challenges and future directions. *Environ. Model. Softw.* **62**, 271–299.

Maier, H., Kapelan, Z., Kasprzyk, J. & Matott, L. 2015 Thematic issue on evolutionary algorithms in water resources. *Environ. Model. Softw.* **69**, 222–225.

Mala-Jetmarova, H., Sultanova, N. & Savic, D. 2017 Lost in optimisation of water distribution systems? A literature review of system operation. *Environ. Model. Softw.* **93**, 209–254.

Marques, J., Cunha, M. C., Sousa, J. & Savić, D. 2012 Robust optimization methodologies for water supply systems design. *Drink. Water Eng. Sci.* **5** (1), 31–37.

Matrosov, E. S., Woods, A. M. & Harou, J. J. 2013 Robust decision making and info-Gap decision theory for water resource system planning. *J. Hydrol.* **494**, 43–58.

Mortazavi, M., Kuczera, G. & Cui, L. 2012 Multiobjective optimization of urban water resources: moving toward more practical solutions. *Water Resour. Res.* **48** (3), 1–15.

Mortazavi-Naeini, M., Kuczera, G. & Cui, L. 2014 Application of multiobjective optimization to scheduling capacity expansion of urban water resource systems. *Water Resour. Res.* **50** (6), 4624–4642.

Mortazavi-Naeini, M., Kuczera, G., Kiem, A. S., Cui, L., Henley, B., Berghout, B. & Turner, E. 2015 Robust optimization to secure urban bulk water supply against extreme drought and uncertain climate change. *Environ. Model. Softw.* **69**, 437–451.

Napoli, C. & Rioux, B. 2015 Evaluating the economic viability of solar-powered desalination: Saudi Arabia as a case study. *Int. J. Water Resour. Dev.* **627** (January), 1–16.

National Audit Office. 2015 The economic regulation of the water sector, October, HC 487.

National Research Council 1999 *Identifying Future Drinking Water Contaminants*. The National Academies Press, Washington DC, USA.

Nogueira Vilanova, M. R. & Perrella Balestieri, J. A. 2014 Energy and hydraulic efficiency in conventional water supply systems. *Renew. Sustain. Energy Rev.* **30**, 701–714.

Ofwat 2010 *Future Water and Sewerage Charges 2010–15: Final Determinations*. Ofwat (The Water Services Regulation Authority), Birmingham, UK.

Ofwat 2017 *Price Reviews – Ofwat*. http://www.ofwat.gov.uk/regulated-companies/price-review/ (accessed August 16, 2017).

Parsons, D. & Marcet, E. C. 2012 *Carbon Sensitive Urban Water Futures*. EU Tansitions to the Urban Water Services of Tomorrow (TRUST).

Perelman, L., Housh, M. & Ostfeld, A. 2013 Robust optimization for water distribution systems least cost design. *Water Resour. Res.* **49** (10), 6795–6809.

Puleo, V., Morley, M., Freni, G. & Savić, D. 2014 Multi-stage linear programming optimization for pump scheduling. *Procedia Eng.* **70**, 1378–1385.

Ray, P. A. & Brown, C. 2015 *Confronting Climate Uncertainty in Water Resources Planning and Project Design The Decision Tree Framework*. World Bank Group, Washington DC, USA.

Roach, T., Kapelan, Z., Ledbetter, R. & Ledbetter, M. 2016 Comparison of robust optimization and info-Gap methods for water resource management under deep uncertainty. *J. Water Resour. Plan. Manag.* **142**, 4016028.

Rothausen, S. & Conway, D. 2011 Greenhouse-gas emissions from energy use in the water sector. *Nat. Clim. Chang.* **1** (4), 210–219.

Sarbu, I. 2016 A study of energy optimisation of urban water distribution systems using potential elements. *Water (Switzerland)* **8** (12), 1–19.

Schaake, J. & Lai, F. H. 1969 *Linear Programming and Dynamic Programming Application to Water Distribution Network Design*. M.I.T. Hydrodynamics Laboratory, Cambridge, MA, USA.

Siddiqi, A. & de Weck, O. 2013 Quantifying end-use energy intensity of the urban water cycle. *J. Infrastruct. Syst.* No. December, **19** (4), 474–485.

Spang, E. S. & Loge, F. J. 2015 A high-Resolution approach to mapping energy flows through water infrastructure systems. *J. Ind. Ecol.* **19** (4), 656–665.

Thames Water 2014 *Thames Water Water Resources Management Plan 2015–2040*. Reading, UK.

Thames Water 2015 *Executive Summary Thames Water Water Resources Management Plan 2015–2040*, pp. 1–43.

Ulanicki, B. & Kennedy, P. R. 1994 An optimization technique for water network operations and design. *Proc. 1994 33rd IEEE Conf. Decis. Control* **4** (December), 4114–4115.

Water UK 2006 *Towards Sustainability 2005–2006: UK Water Industry Sustainability Indicators*. Water UK, London, UK.

Water UK 2008 *Sustainability Indicators 2007/08*. Water UK, London, UK.

Water UK 2009 *Sustainability Indicators 2008/09*. Water UK, London, UK.

Water UK 2012 *Sustainability Indicators 2010/2011*. Water UK, London, UK.

First received 9 June 2020; accepted in revised form 5 October 2020. Available online 23 October 2020

© 2020 The Authors

Development of a biogas distribution model for a wastewater treatment plant: a mixed integer linear programming approach

Harry Laing IWA, Chris O'Malley, Anthony Browne, Tony Rutherford, Tony Baines and Mark J. Willis

ABSTRACT

In this paper, we propose a realistic model for gas distribution of an advanced municipal wastewater treatment works and through minimisation of the total cost of gas distribution we perform retrospective optimisation (RO) using historical plant data. This site is the first in the UK with a mixed operational strategy for biomethane produced on site: to burn in combined heat and power (CHP) engines to create electricity, burn in steam boilers for onsite steam use or inject the biomethane into the National Grid. In addition, natural gas can be imported to make up shortfalls in biomethane if required. Implemented using a novel mixed integer linear programming (MILP) approach, to ensure a fast and robust solution, our results indicate the plant operated optimally within accepted tolerance 98% of the time. However, improving plant robustness (such as reducing unexpected breakdown incidents) could yield a significant increase in gas revenue of 7.8%.

Key words | advanced anaerobic digestion, biomethane, gas distribution, MILP, optimisation, wastewater

Harry Laing IWA (corresponding author)
Chris O'Malley
Mark J. Willis
School of Engineering, Merz Court,
Newcastle University,
Newcastle upon Tyne, NE1 7RU,
UK
E-mail: *h.j.laing@ncl.ac.uk*

Anthony Browne
Tony Rutherford
Tony Baines
Northumbrian Water Limited,
Northumbria House, Abbey Road, Pity Me,
Durham, DH1 5FJ,
UK

HIGHLIGHTS

- Wastewater treatment sites produce biomethane.
- Optimisation of on-site biomethane distribution using MILP.
- Retrospective optimisation of historic plant operations.
- Optimal daily economic operational strategy for operators.

GRAPHICAL ABSTRACT

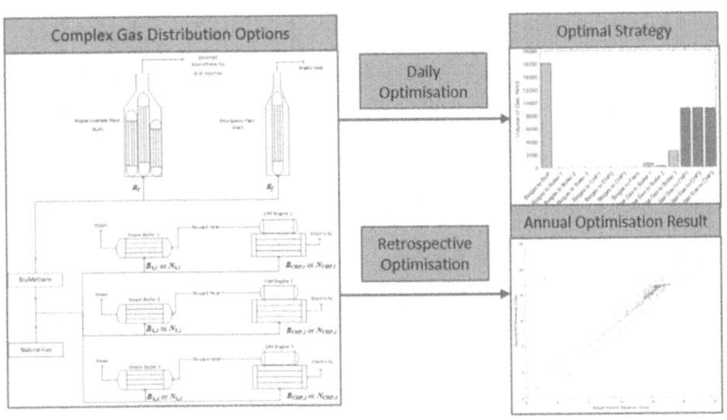

This is an Open Access article distributed under the terms of the Creative Commons Attribution Licence (CC BY 4.0), which permits copying, adaptation and redistribution, provided the original work is properly cited (http://creativecommons.org/licenses/by/4.0/).

doi: 10.2166/wst.2020.363

INTRODUCTION

In 2015, Northumbrian Water Limited (NWL) was the only wastewater company in the UK to use all sludge after wastewater treatment to produce renewable electricity (Northumbrian Water 2015). Recently it has added a gas-to-grid injection plant at its sewage treatment works (STW) in Howdon, Tyneside, whereby biomethane/biogas produced from anaerobic digesters is purified prior to injection into the National Grid (Martins & Wilson 2016). The site typically digests up to 40,000 tonnes of sewage sludge (dry solids) annually producing renewable biogas. The STW can be considered to have two sections: effluent water treatment, and sludge treatment. Sludge treatment occurs at the advanced anaerobic digestion (AAD) plant. The relatively new processing techniques (CAMBI thermal hydrolysis) involved in the AAD plant results in significant operational challenges to maximise the economic performance.

Wastewater treatment plants (WWTPs) are continuously being driven towards increased efficiency of plant operation due to regulatory pressures (O'Brien *et al.* 2011; Mulas *et al.* 2015), and it is widely thought that there is considerable room for improving the efficiency of WWTPs (Longo *et al.* 2016). If companies are able to schedule their operations successfully around supply and demand of energy, there are opportunities for annual OPEX savings of 2–5% (Merkert *et al.* 2015), and the use of energy recovery techniques and biogas can reduce a WWTP carbon footprint by 10% (Singh *et al.* 2016). In addition, WWTPs have an abundance of low-grade thermal energy and organic substances in sludge (Zhang *et al.* 2017) which, if harnessed, can reduce carbon emissions and energy cost/consumption.

Typically, biogas produced on a wastewater anaerobic digestion (AD) plant in the UK is used to generate electricity only. The AAD plant at Howdon has the ability to use the biomethane in different ways. The AAD plant is rare in that it has three possible uses for biogas produced on site, shown in Figure 1: upgrade for injection into the national gas grid; burning it in combined heat and power (CHP) engines; or burning it in steam boilers for the thermal hydrolysis plant. There are three CHP engines, three steam boilers and one gas upgrade/injection plant on site (Figure 2). If required, the plant may flare excess biogas under emergency circumstances, though operators minimise this as much as possible. To ensure overall sludge processing remains unimpeded the plant may draw natural gas from the national gas grid to be used in the CHP engines or steam boilers. Nationally, many similar sites are looking to upgrade to produce biomethane for injection into the National Gas grid to take advantage of the government's renewable heat incentive (Hale 2017).

CHP electricity generation is an effective way to reduce energy consumption and carbon emissions (Mago & Chamra 2009; Bischi *et al.* 2014); these systems typically take a fuel source and efficiently convert it into usable heat and energy (Andersen & Lund 2007) and in the case of the Howdon WWTP the CHP units are 'gas engines' provided by MWM (Martins & Wilson 2016) that burn the non-purified biogas from the digesters to generate heat and electricity.

According to M. Katebi *et al.*, 'The barrier to the successful implementation of the control system is not the control software or hardware, but rather the problem of designing control systems that are integrated with the plant operation and management and have a high degree of local autonomy, flexibility and reliability' (Katebi *et al.* 1998). Typically, in the water industry, advanced process control is uncommon and standard control schemes use on-off control via PLC (programmable logic controller) and SCADA (supervisory control and data acquisition) systems to control localised processes to within specified limits, without consideration of upstream plant behaviour (O'Brien *et al.* 2011). Widespread use of SCADA-type system technology permits

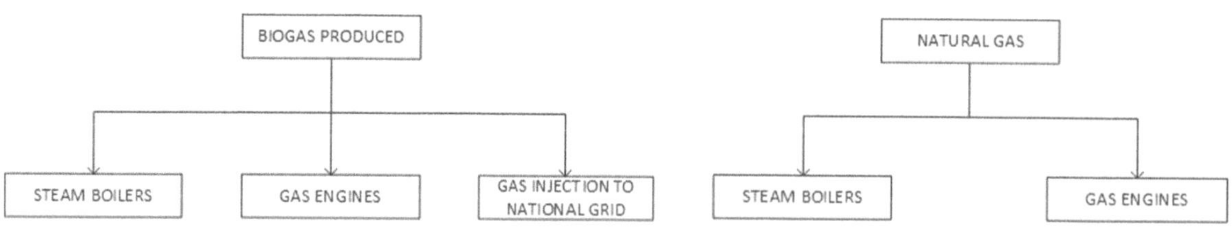

Figure 1 | Possible gas distribution across AAD plant.

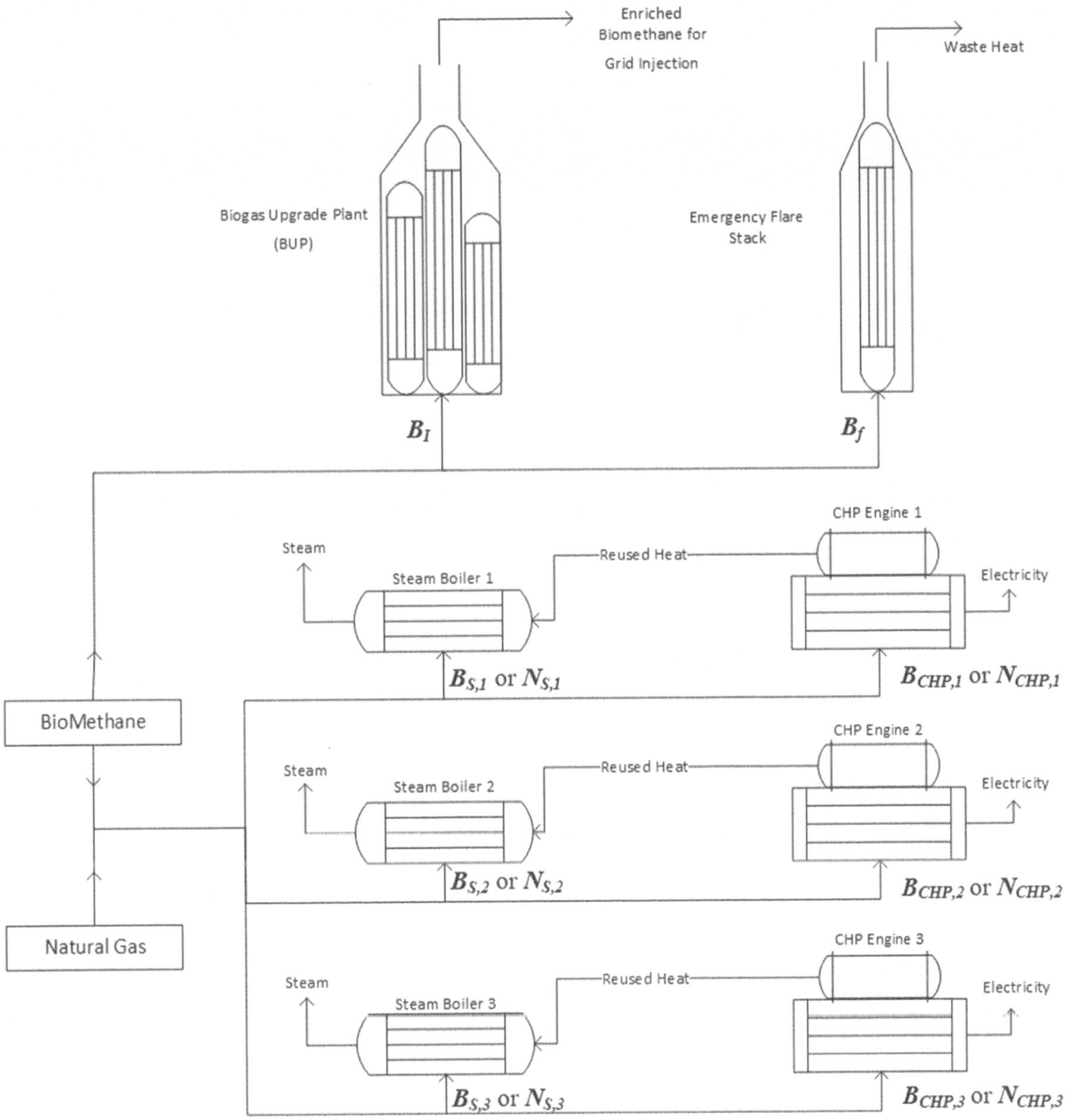

Figure 2 | Process diagram for gas distribution on site. Biomethane source is from anaerobic digestion on site, natural gas is from the National Grid. Pipelines shown are for illustration purposes and do not reflect physical constraints or locations on site.

the exploitation of more advanced supervisory concepts and system control (Ocampo-Martinez 2010).

SCADA based systems can store vast amounts of historical data regarding plant operations. Such large amounts of data can be used for retrospective analysis and learning to make improvements to future operations. An example of the application of retrospective learning is in the aviation industry (Dambier & Hinkelbein 2006), where analysis and learning after incidents creates improved safety procedures. Typically, retrospective learning techniques are not used in regard of operational aspects of a process (Cummings *et al.* 2017). However, in work conducted by T. Cummings *et al.* in 2017, retrospective optimisation (RO) and learning was used to develop improved scheduling procedures for multiple nitrogen liquefier units and the development of electricity spot pricing for a pricing predictor (Cummings *et al.* 2017). In the case of Howdon WWTP, site systems store several years of operational data which is not currently being used for any RO or analysis. To the authors' knowledge, there has been no development of models of WWTPs based on RO in literature, though it should be noted that previous work has been done developing predictors or forecasts of energy pricing using grey prediction models (Gu *et al.* 2015).

To perform RO for this site, we develop a model of the process and use mixed integer linear programming (MILP) to determine optimal historic performance that can be compared to actual historic operations. MILP takes a series of linear relationships made up of equalities, inequalities, integer defined parameters and upper and lower bound constraints to minimise a linear objective function. Some examples of various MILP applications include: scheduling of a polymerisation reactor (Kelley *et al.* 2018), multiproduct milk processing (Touil *et al.* 2016), multi-week scheduling for an ice cream processing facility (Wari & Zhu 2016) and scheduling of a multiple cryogenic air separation unit and compressor plant (Adamson *et al.* 2017a, 2017b), where the later was performed within a Microsoft Excel spreadsheet. This highlights the effectiveness and ease of application of being able to model a process in a MILP form, if the problem can be posed in such a way to take advantage of MILP software.

There is a clear gap in literature of designing control schemes, optimisation techniques and models for gas distribution of AAD sections of WWTP, as literature-discussed here focuses on developing control strategies or models of the effluent treatment side of a WWTP. It should be noted that whilst there are many studies on improving the yield of biomethane produced from anaerobic digesters, such as using calcium or enzymatic pre-treatment of sewage (Agabo-García *et al.* 2019; Salama *et al.* 2020), this paper focuses on the optimal way to use the biomethane produced, not improving yield or developing a model of the anaerobic digesters.

Currently, there is no model on site to advise the optimal distribution of biogas produced or how much natural gas must be purchased to ensure optimal cost and sludge processing performance, nor is there evidence of such a model in use at a WWTP internationally. Here, for the first time, we report the development and application of a realistic model and its solution using MILP, which is used to advise the optimal daily operational strategy that will minimise of gas distribution costs. Using the model, we perform RO using historic data to determine how optimal previous operational strategies were on a cost minimisation basis.

The remainder of this paper is structured as follows. In the methods section we outline the site constraints, formulation of the MILP model and develop the objective function for optimisation. In the results we show example optimisation results from the optimiser, show the results of the RO performed, consider the importance of maintenance and improved breakdown robustness, and consider current model limitations. Finally, our conclusions are drawn and presented alongside proposals for further work and improvements to this model.

METHODS

Unit processing limits

To model the site and perform RO, processing limitations of gas flow for each unit must be known. The daily processing limits (Table 1) were determined through retrospective analysis of historical plant operational data and through discussions with operational managers on site. Flow limits are different for biogas and natural gas volumes. The minimum gas flow for an engine to operate is 50% the maximum flow. Currently, should there not be enough biogas produced to satisfy operation of a CHP engine, operators require that an engine run on natural gas and have the gas flow rate to that engine set to the maximum.

Nomenclature and constraints of biogas distribution model:

Symbol	Parameter	Units
B_I	Biogas flow to grid injection (biogas upgrade plant, *BUP*)	Nm^3
$B_{CHP,i}$	Biogas flow to CHP engine 'i'	Nm^3
$B_{S,i}$	Biogas flow to steam boilers	Nm^3
B_f	Biogas flow to flare stack	Nm^3
B_{Total}	Total biogas produced from digesters	Nm^3
$N_{CHP,i}$	Natural gas flow to CHP engines	Nm^3
$N_{S,i}$	Natural gas flow to steam boilers	Nm^3
$S_{min/max}$	Gas flow constraint for boilers to ensure some gas is used for steam production	Nm^3
C_B	Biogas cost to burn on site	£/Nm^3
C_N	Natural gas cost to burn on site	£/Nm^3
C_I	Biogas injection cost/revenue	£/Nm^3
C_f	Flare 'costs' for biogas on site	£/Nm^3
T_c	Total operational cost	£
T_B	Total cost of biogas usage	£
T_N	Total cost of natural gas usage	£
T_P	Total cost of penalty terms	£
z_i	Binary variable used by optimiser to determine fuel used in associated CHP engine 'i' or steam boiler 'i'	–

[Nm^3 = normalised cubic meter] (subscript 'i' represents parameter for multiple units, e.g. engine 1, 2 or 3).

Table 1 | Operational constraints for units on site

Limiting parameters	Operational constraints
$B_{I,\,max}$	Max flow: 40,000 Nm^3/day
$B_{I,\,min}$	Min flow: 0 Nm^3/day
$B_{CHP,\,max}$	Max flow: 16,000 Nm^3/day
$B_{CHP,\,min}$	Min flow: 50% of max flow
$B_{S,\,max}$	Max flow: 4,000 Nm^3/day
$B_{S,\,min}$	Min: 200 Nm^3/day
$B_{f,\,max}$	No max
$B_{f,\,min}$	Min: 0 Nm^3/day
$N_{CHP,\,max}$	Max flow: 9,000 Nm^3/day
$N_{CHP,\,min}$	Min flow: 9,000 Nm^3/day
$N_{S,\,max}$	Max flow: 2,500 Nm^3/day
$N_{S,\,min}$	Min flow: 200 Nm^3/day
S_{max}	Upper limit: 5,700 Nm^3/day
S_{min}	Lower limit: 3,300 Nm^3/day

The gas holders on site that capture and intermediately store the biogas produced are typically able to store up to an hour's production of biogas, should the biogas flow downstream be interrupted. Therefore, for the purpose of this model the daily volume of biogas produced must all be utilised in the available units on site.

The cost parameters for each gas flow were taken from the OPEX reporting features available on the onsite SCADA system. Costs tend to be in units of p/kWh or equivalent, so the cost parameters in the model are combined with standardised conversion factors to enable a simple cost in £/Nm^3 (GBP per normalised cubic metre) of gas; gas flows on site are reported in Nm^3. For the purposes of confidentiality, the cost function parameters cannot be reported here.

MILP EQUATIONS/PARAMETERS

Objective function

The objective function (Equation (1)) is defined as the minimisation of the total costs of gas distribution use on site, minimising total Biogas, Natural Gas and any penalty terms:

$$T_c = T_B + T_N + T_P \tag{1}$$

The total cost for Biogas and Natural Gas use on site, T_B and T_N respectively, are defined by summating the gas flows to each CHP Engine or Steam Boiler and the Biogas flow to the Biogas Upgrade Plant (BUP). Here, B represents a Biogas flow (Nm^3, variable), N represents a Natural Gas flow (Nm^3, variable) and C represents the cost of using the associated gas (£/Nm^3, constant):

$$T_B = \sum_{i=1}^{3}\{B_{CHP,i} \cdot C_B\} + \sum_{j=1}^{3}\{B_{S,j} \cdot C_B\} + B_I \cdot C_I \tag{2}$$

$$T_N = \sum_{i=1}^{3}\{N_{CHP,i} \cdot C_N\} + \sum_{j=1}^{3}\{N_{S,j} \cdot C_N\} \tag{3}$$

For a realistic optimisation, both costs and revenues of gas streams on site are considered; injecting biomethane into the National Grid through the BUP creates a revenue as the gas is sold (C_I), burning biogas in the boilers and CHP engines has no associated cost (C_B) whilst purchasing natural gas for use anywhere has a cost associated (C_N).

Whilst there is no cost associated with burning biogas on the flare under occasional emergency use, overuse is discouraged by regulatory bodies and is considered a waste. To prevent the optimiser from sending gas to the flare stack a penalty term, T_P, on flaring biogas was applied to the objective function. By setting the cost function for flaring, C_f, high the optimiser would allow flaring only as a last resort. The site requires minimising flaring in order to satisfy its environmental and regulatory commitments.

$$T_P = B_f \cdot C_f \tag{4}$$

Therefore, the total cost objective function is:

$$T_c = \sum_{i=1}^{3}\{C_b\,B_{CHP,i} + C_n N_{CHP,i}\} + \sum_{i=1}^{3}\{C_b\,B_{S,i} + C_n N_{S,i}\} + C_I B_I + C_f B_f \tag{5}$$

Should the optimiser advise that flaring gas should be done, it will affect the actual minimised cost; after optimisation, when calculating the total daily operational cost on site, if required the penalty term is subtracted from the minimised cost to reflect actual operational costs, and for more accurate RO.

Mass balance constraint – model input

Whilst individual process units have their own sets of constraints for gas usage, the overriding site constraint is given by the overall mass balance of biogas distributed across the site: the volume of biogas produced from the anaerobic digesters must equal that of biogas distributed across site. There is no constraint of natural gas of this form, as natural gas is readily available if required. The total volume of biogas produced, B_{Total}, is the only variable that is input into the model.

$$B_{Total} = \sum_{i=1}^{3} \{B_{CHP,i} + B_{S,i}\} + B_I + B_f \quad (6)$$

CHP engines constraints

The CHP engines can only utilise one fuel type at a time: biogas or natural gas. Therefore, the binary variable $z_i \in \{0, 1\}$ is introduced to ensure only one of each gas type is used by each unit. The total gas flows to any engine, $B_{CHP,i}$ and $N_{CHP,i}$, is between the maximum and minimum flows:

$$B_{CHP,min} \cdot z_i \leq B_{CHP,i} \leq B_{CHP,max} \cdot z_i \quad (7)$$

$$N_{CHP,min} \cdot (1 - z_i) \leq N_{CHP,i} \leq N_{CHP,max} \cdot (1 - z_i) \quad (8)$$

Under these conditions, when z_i takes the value of 1 then only biogas may flow to engine 'i', whereas a value of 0 denotes a natural gas flow. z_i is a variable determined by the model to optimise gas distribution. The binary variable z_i provides the gas selection functionality by allowing z_i to alter the upper and lower constraints of the inequalities.

Steam boilers constraints

The steam boilers can also only have one fuel source at a time. Similarly to the CHP inequalities in the 'CHP engines constraints' section above, the steam boiler parameters make use of a binary variable, $z_j \in \{0, 1\}$ subject to:

$$B_{S,min} \cdot z_j \leq B_{S,j} \leq B_{S,max} \cdot z_j \quad (9)$$

$$N_{S,min} \cdot (1 - z_j) \leq N_{S,j} \leq N_{S,max} \cdot (1 - z_j) \quad (10)$$

There are additional constraints for the steam boilers that differ from the CHP engines. Unlike the CHP engines, the steam boilers must always be producing enough steam to satisfy site process requirements and therefore do not all have high gas flow or low gas flow at the same time. After retrospective analysis of historic data, for any given day the three boilers operate on a total daily flow where one operates near maximum, one near minimum and one in between. This is in order to supply enough steam for use on site. As such, the additional constraint on the boilers is:

$$S_{min} \leq \sum_{j=1}^{3} \{B_{S,j} + N_{S,j}\} \leq S_{max} \quad (11)$$

Retrospective analysis of the different fuel flows to the boilers also revealed no distinguishable difference in processed volume of biogas or natural gas flows, hence each gas type has equal weighting in this constraint.

Gas to grid injection (biogas upgrade plant) constraint

The BUP takes the raw biogas and enriches it such that the resulting biomethane can be injected into the National Grid as a renewable energy source. As there is only one fuel source, the constraints of sending biogas to the BUP are:

$$B_{I,min} \leq B_I \leq B_{I,max} \quad (12)$$

In an ideal setting there would be no limit to the volume of biomethane that can be injected into the National Grid. However, the total volume that can be injected is subject to local demand and gas network pressures; if too much biomethane is injected too quickly the pressure in the grid could rise too high for continued injection, whereby Northern Gas Networks (the local gas network operator) would shut off grid injection from the site; this is known as going into a 'reject' state. Operators have discovered that the maximum daily volume of biogas that can be processed through the BUP is currently around 40,000 Nm^3, which was validated through retrospective analysis of BUP processing volumes, grid injection volumes and 'reject' state instances.

Flare stack constraint

Due to the volumes of biomethane produced on site and the safety considerations from site design, there is considered to be no upper limit for the total volume of biogas the flare stack can take:

$$0 \leq B_f \quad (13)$$

Solver

The optimiser model was developed using MATLAB using function *intlinprog*, which is a MILP algorithm in the

Optimisation Toolbox package. The optimiser model was implemented using this function and contains 24 constraints, 14 possible gas flows with 6 binary variables. The optimal gas flows and binary variable values are obtained through minimisation of the cost function to give optimal gas distribution on an economic basis, whilst maintaining site operability.

The total historic daily biogas volume produced on site is the only historic value passed to the optimiser. The model then calculates the volumes of biogas and natural gas that are required to be distributed to each unit to satisfy daily operation at the minimum cost, subject to the constraints outlined.

DISPLAYING OPTIMISATION RESULTS

The function *intlinprog* will always aim to minimise the objective function. As such, all revenue parameters are negative and cost parameters are positive inside the optimiser; a negative optimal value provided by the optimiser indicates a potential revenue, whilst a positive value indicates a cost. After optimisation, the resulting 'optimal value' is multiplied by −1 so that a positive value represents revenue and negative a cost, as one would expect. When displaying results of optimisations visually, total daily revenue is used for a direct comparison to site operations.

PERFORMING RETROSPECTIVE OPTIMISATION

To perform RO, historic plant operational data was used for the 12-month period. For a given date, the historic total daily production of biogas on site is passed to the optimiser which then provides operators with an optimised minimum cost and the optimal daily strategy for operating the plant for that date. The optimal strategy provided can then be compared to actual gas distribution on site, using historic data. For each day in the 12-month data set, the optimised minimum cost for RO was compared to historical site operation (and the subsequent cost of operating using that strategy) to compare how the site has performed in the past.

During RO, the model constraints are adjusted to reflect planned site maintenance; for example, if a CHP engine was offline on a specific date for annual safety inspections, then the upper and lower gas flow constraints for biogas and natural gas for that engine would be set to 0 Nm^3. This allows the model to perform optimisation with the correctly available units on site.

RESULTS AND DISCUSSION

Optimiser visual results

For a given daily volume of biogas produced on site, the optimiser provides the operator with a visual operation strategy for the optimised minimum cost; Figure 3 (left graph) shows the results to maximise cost reductions for a daily biogas production of 40,000 Nm^3 (a typical production level for the site). The rightmost graph in Figure 3 displays a version, provided by the optimiser, showing the percentage daily utilisation of each unit.

For a daily biogas production of 40,000 Nm^3, the site should be operated according to the strategy in Figure 3 for optimal cost efficiency: to inject all biomethane into

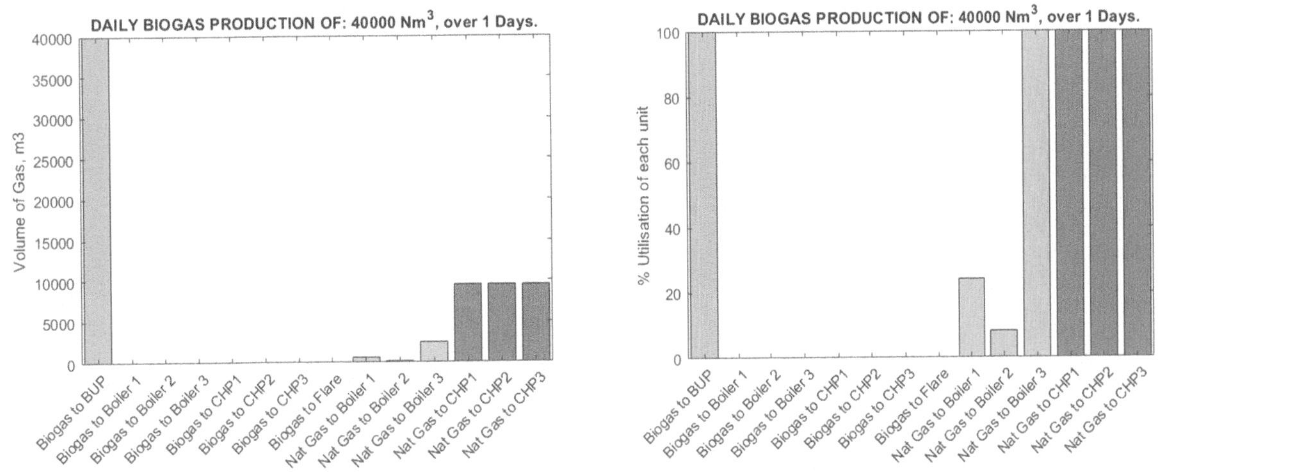

Figure 3 | Example optimised gas distribution for daily total biogas production of 40,000 Nm^3. Left: Optimised daily gas flow volumes, Right: Normalised optimised daily utilisation of each unit.

the National Grid, use natural gas in the CHP engines at 100% load to generate electricity on site and use natural gas in the steam boilers to create steam as required. For a given daily volume of biogas, the optimiser provides a fast and reliable result in a matter of seconds.

Retrospective optimisation (RO)

Performing RO of the plant over multiple days is difficult to achieve if the operators must sift through multiple graphs for each day that is analysed. For an easier visual indication of how the optimiser suggests the plant should be operated with regards to fuel types for each unit, a Gantt chart showing the gas type selected for is presented – Figure 4 shows an example date range from the full RO and compares this to historical operation.

The colour of each block represents a particular fuel type for each unit, typically biomethane or natural gas. However, in reality there may be times where both fuels were recorded historically, as shown in yellow. When switching an engine from biomethane to natural gas, historical operations will show both fuel types were used but for clarity each fuel would have been combusted separately, not as a blend. For example, Figure 4 suggests both fuel types were used historically in the CHP engines on some days, but for CHP engines this actually show days where a gas type switchover took place.

The historical fuel type shown in the boilers is inferred from historical gas flow data, and on occasion the flow meters indicate that two fuel types were used when this may not have been the case, hence the multiple 'yellow' days for boilers. Specifically, in the case of Boiler 1 it may be an indication of an issue with the gas flow meters on site. It may also be an indication that fuel switchovers in the boilers happen more often.

In this figure, it may be seen that the optimiser advised that, between days 6 and 11, the plant should have swapped some CHP engines and boilers to run on biomethane, which the plant mostly managed to achieve with the exception of the boilers.

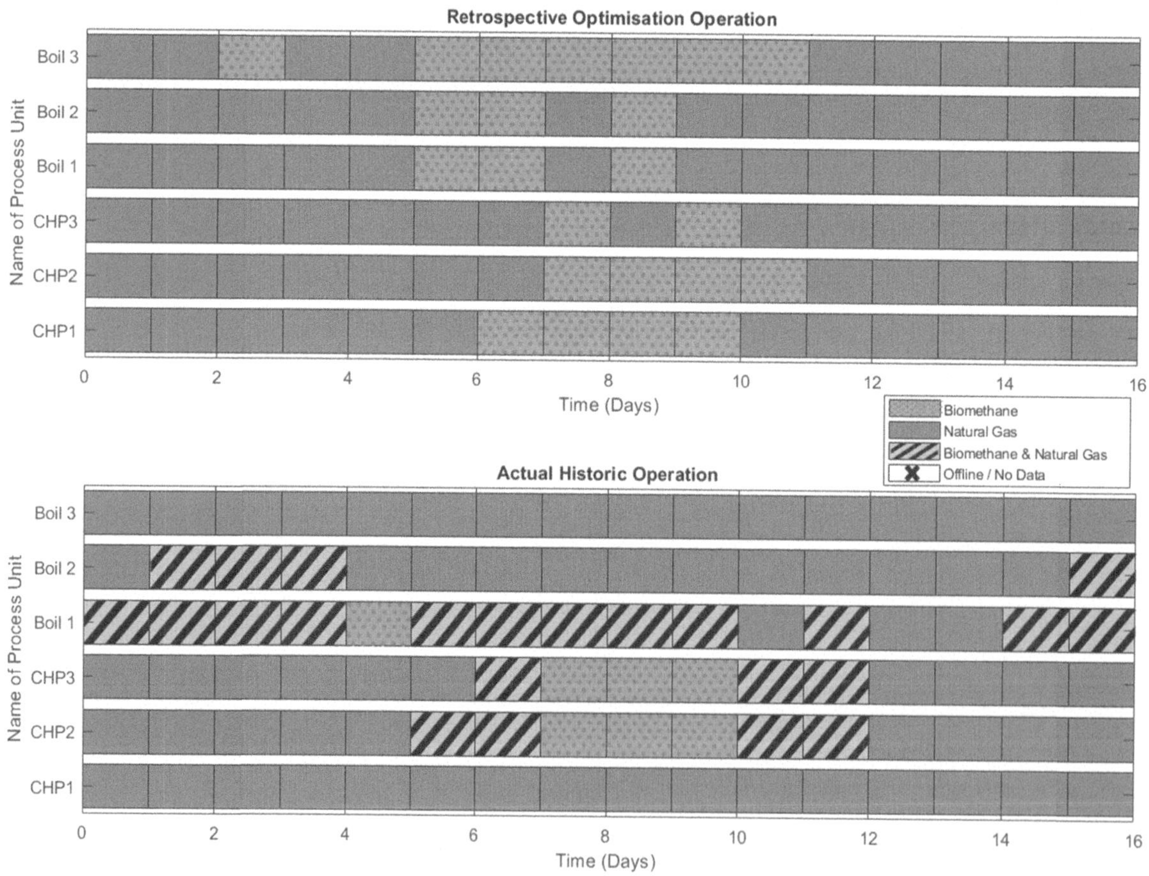

Figure 4 | Gantt Chart for an example period showing the type of gas used on site for each unit that can use either biomethane or natural gas. Results from retrospective optimisation are presented alongside historical operation.

Each daily optimisation has an associated minimised operational cost, which is used to evaluate RO of plant operations. The minimum cost found by the optimiser is compared to the cost associated with actual historical distribution on site by plotting one against the other. The daily optimal operational cost was compared to the actual historical cost over the 12-month period November 2017 to October 2018, the historical plant data available, and is shown in Figure 5.

In order that RO is representative of site operations, planned site maintenance is included in the RO. Using site maintenance logs, if a unit was taken offline for a full day for scheduled repairs or safety inspections then the optimiser will not have that unit available for the given day of RO.

It should be noted that, whilst the parameters of the model are static, for each 3-month moving period, the average maximum gas flows to each unit (biogas and natural gas) are recorded and compared to the gas flow parameters of the model. If the model parameters are more than 10% away from the 3-month average, a warning is displayed during RO such that the user can determine and select an improved parameter. As of yet, the warning parameters have not been triggered, indicating that the initial parameters are valid across the whole RO horizon. It is also why there is a 10% band on the RO graphs.

A band of 10% of the maximum revenue achieved was chosen as the tolerance for optimality as the model parameters do not deviate more than 10% from the historic use of the units on site. Of the 366 days analysed, based on a tolerance of ±10% of the highest revenue, RO indicates the plant operated to within tolerance 98% of the time (332 days). This provides some validation to operators that their historic operational strategy was optimal. However, the 2% non-optimal operational days have a potential lost revenue of almost £350,000 (£349,291); if the site was optimally operated 100% of the time, there is the potential for an increase in revenue of 7.8%. The non-optimal operational days are likely due to unplanned breakdowns and maintenance on site, which can be investigated by comparison with site maintenance logs.

Importance of maintenance logs

It is paramount that planned shutdowns are logged and programmed into the optimiser to allow for the best comparison of RO and historic operation. However, unplanned shutdowns cannot be pre-configured in the model. RO will highlight non-optimal days and, typically, will highlight the severity (if any) of unplanned shutdowns

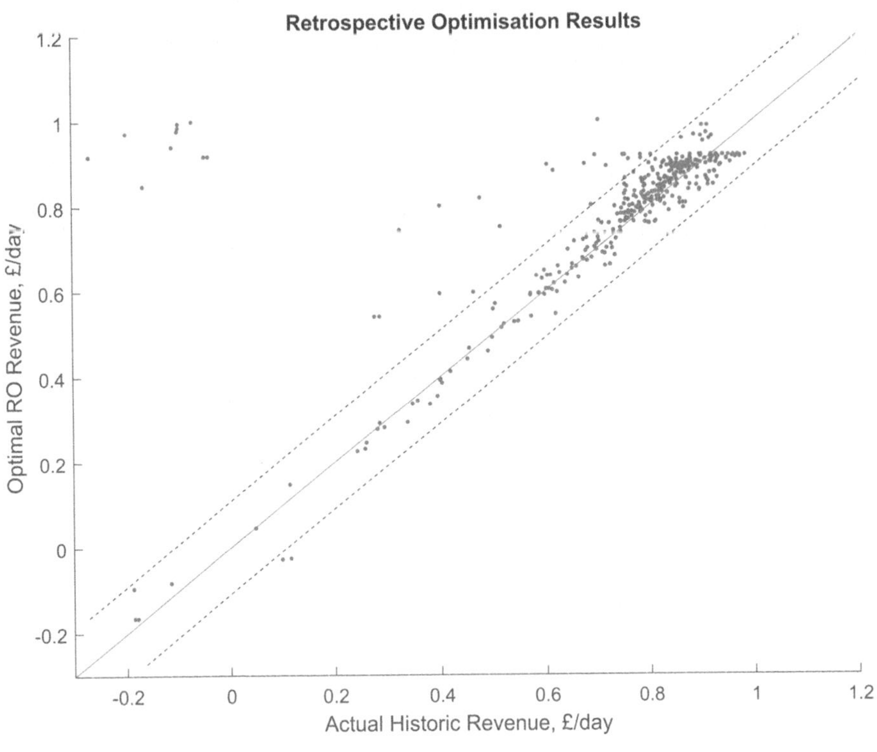

Figure 5 | Actual historic gas revenue vs optimiser gas revenue, normalised. Solid line represents shows y = x, and the dashed lines represent the tolerance band (±10% of the max revenue). **Note*: Wherever noted with '*', the figures presented use the actual revenue in £/day for plotting but for confidentiality the axis have been normalised based on the highest revenue and do not represent the actual financial figure.

across the site and will produce a monetary value attributed to each shutdown. However, should an operator wish to change the number of available units (e.g. CHP engines) on site for scenario modelling, they are able to do so.

Figures 6 and 7 demonstrate the importance of including planned shutdowns in the model; by not including them, the model would assume all units were available and thus it would appear the plant operated sub-optimally yet this was not the case historically. It should be noted here that the model has no preference of which engine should be in operation, hence on the example date shown in Figure 7 the optimiser selects a different engine used for biogas – the actual engine used is not important to the model, and is an operator decision to maintain the health of the engines.

Investigating unplanned outages

Using the site maintenance log, planned and unplanned downtime of units on site can be shown on the annual RO graph in Figure 5. The boiler, CHP engine and the BUP unplanned outages are plotted on Figure 8. Based on the distribution of outages on Figure 8, boiler downtime has no major impact on the historic operability of the site. The CHP engines would appear to have more of an impact on site operations as more of the outliers are associated with unplanned outages.

However, almost every non-optimal outlier on Figure 5 is associated with an unplanned outage (full or part day) with the BUP shown on Figure 8. Gas injection into the grid is where the plant makes the majority of gas revenue,

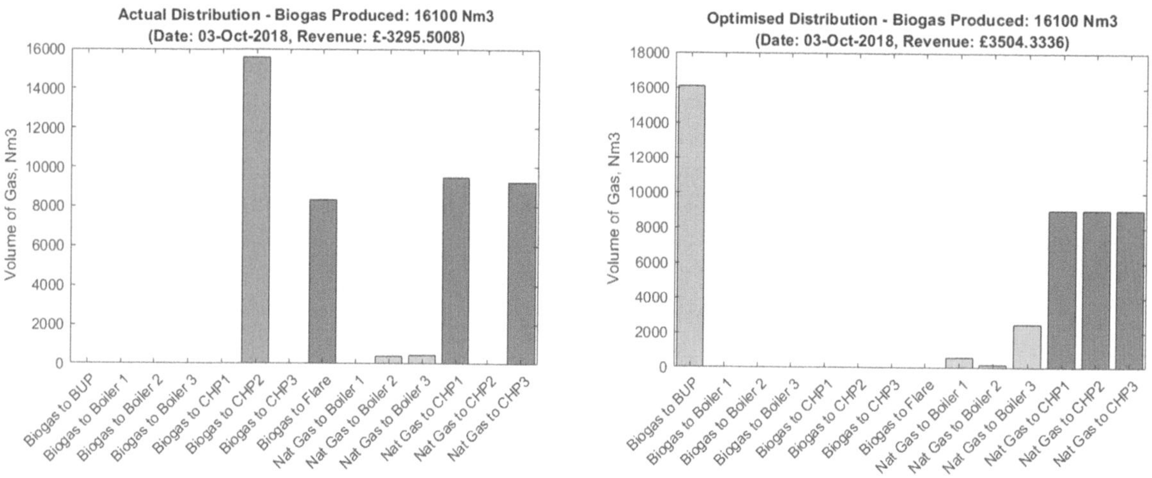

Figure 6 | Actual (left) and optimised (right) gas distribution on site, for example date. Optimised distribution assumes all units available, but RA showed the BUP unit was actually offline for planned maintenance.

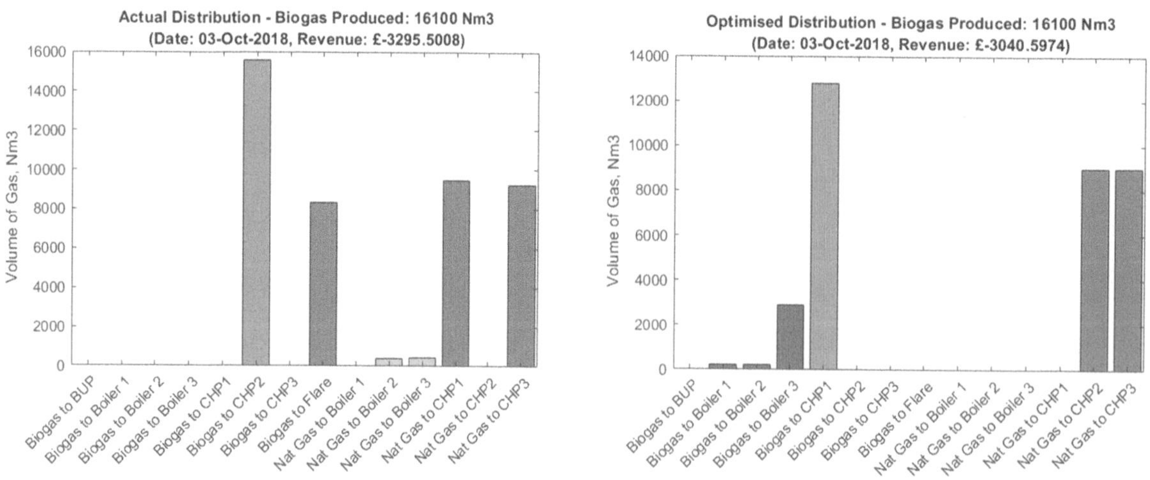

Figure 7 | Actual (left) and new optimised (right) gas distribution for same example date as Figure 7, but optimiser model now accounts for the planned offline BUP unit.

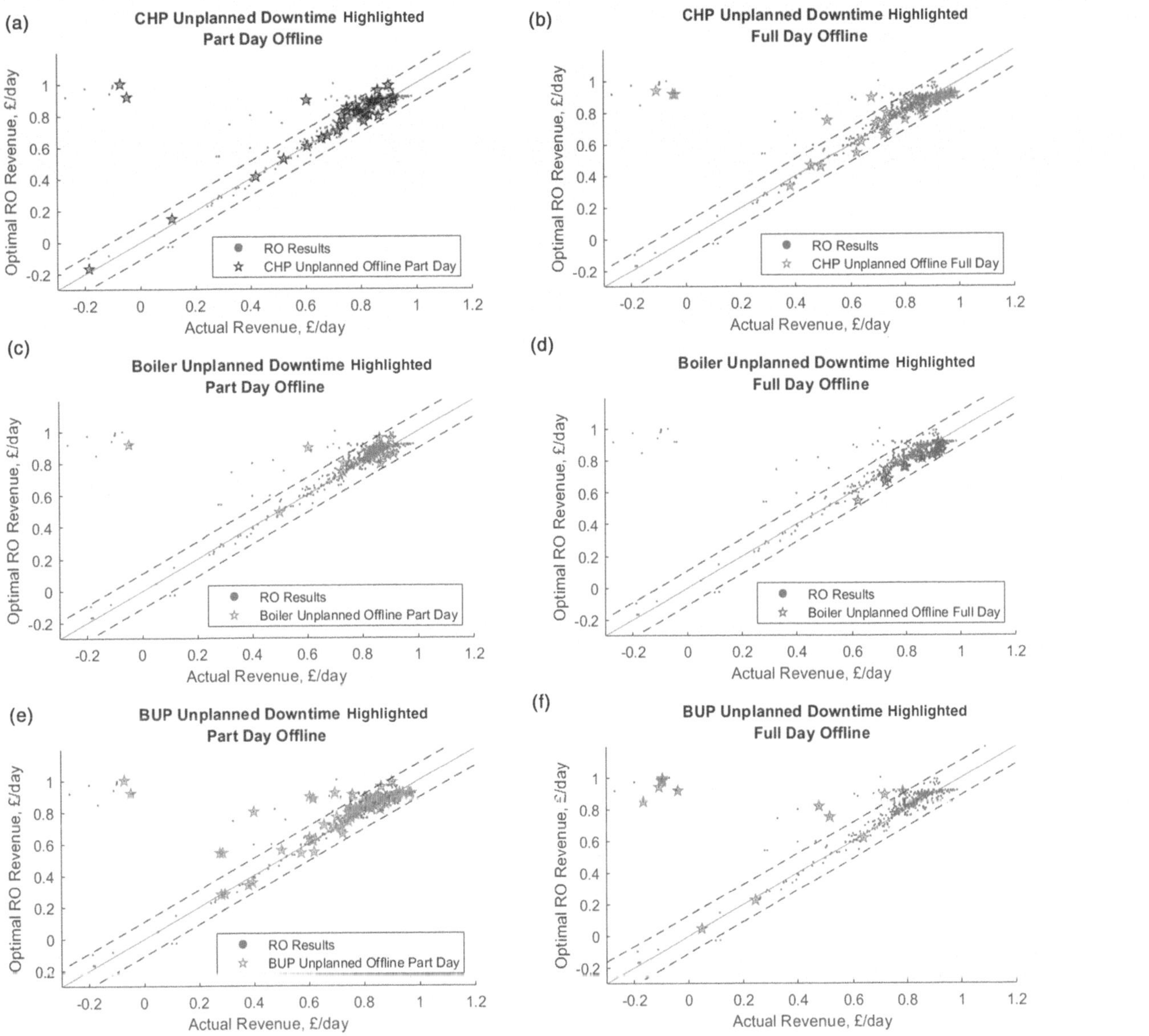

Figure 8 | Annual RO but days with unplanned downtime of units are highlighted. 'Part Day' refers to a unit being available for less than the 24-h optimisation period, and typically ranges from 1 to 6 h. 'Full Day' refers to a unit being offline for use for the entire 24-h period. Solid line represents shows y = x, and the dashed lines represent the tolerance band (±10% of the max revenue). (See the footnote for '*' in Figure 5 caption).

and the fluctuations of daily gas injection volumes account for most of the variation in the plant operation; whilst a maximum biogas processing volume of 40,000 Nm^3 is typical, this limit is influenced by external factors and on occasion the site may be able to inject more biomethane (and subsequently process more biogas) than the model accounts. It also stands to reason that unplanned outages with this unit will likely cause the plant to operate sub-optimally.

Crucially, full day planned maintenance outages do not cause the site to operate sub-optimally – the plant operates within optimal tolerance on each of the Full Day planned maintenance plots on Figure 9. Currently, the model can only handle planned maintenance where a unit is due to be offline for the full day – planned maintenance taking less than a full day cannot accounted for in the current model. The potential shortfall in revenue of ~£350 k is mostly due to due to unplanned breakdowns and planned maintenance that does not take a full day.

Achievable increase in revenue

In an ideal world, process plants would always run optimally without any unexpected breakdowns or unforeseen circumstances preventing optimal operation. In reality, this

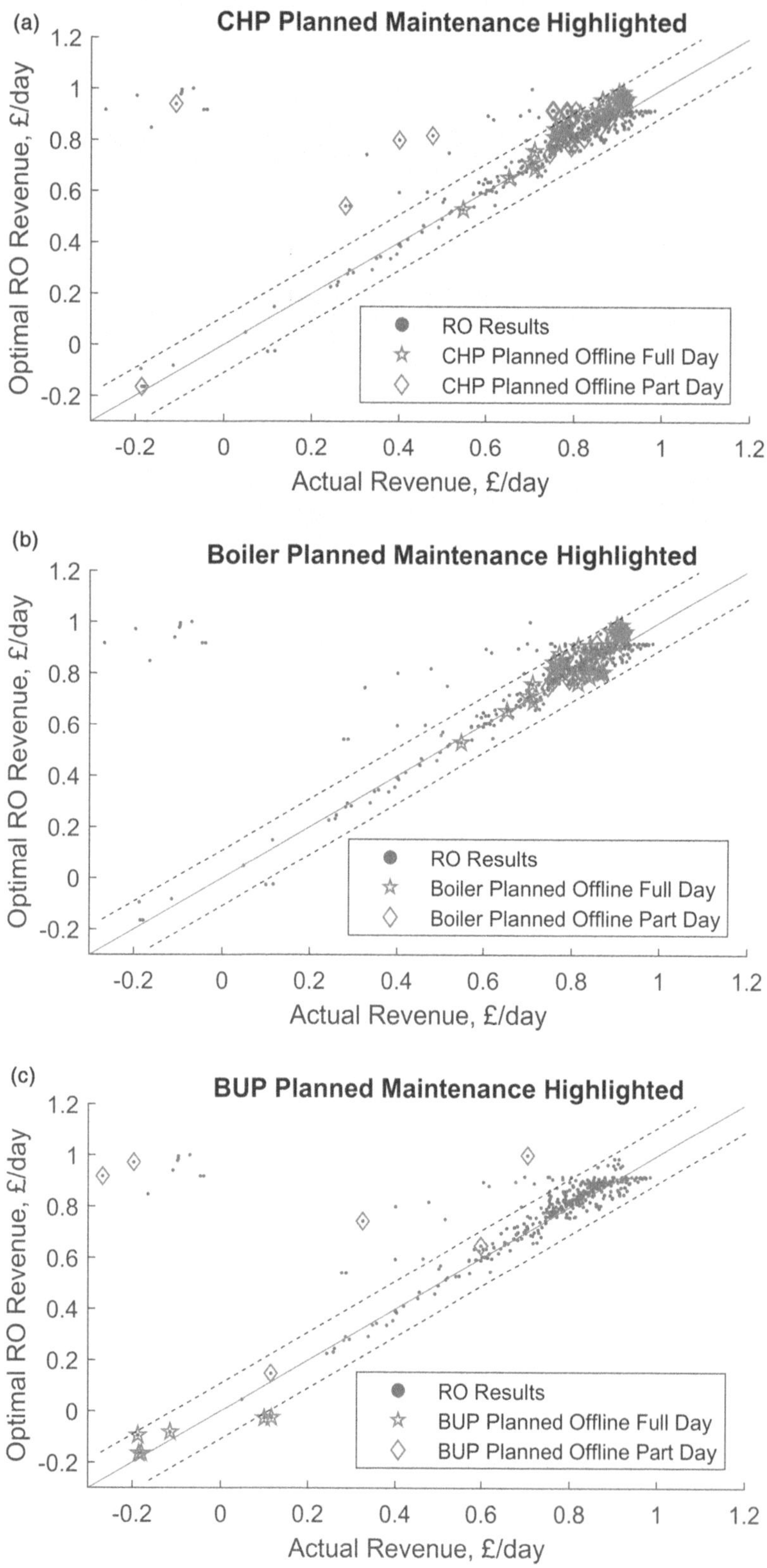

Figure 9 | Annual RO but days with planned maintenance of units are highlighted. 'Part Day' refers to a unit being available for less than the 24-h optimisation period, and typically ranges from 1 to 6 h. 'Full Day' refers to a unit being offline for use for the entire 24-h period.Solid line represents shows y = x, and the dashed lines represent the tolerance band (±10% of the max revenue). (See the footnote for '*' in Figure 5 caption).

is not the case and should be noted when considering the potential 7.8% increase in revenue available. If the plant operated to within optimal tolerance 98% of the time, then to what extent could this be practically increased further? Whilst undesirable, it is expected that process plants will ultimately have unexpected downtimes, especially with new technologies and installations. Acceptable levels of lost revenue due to unexpected maintenance is a decision for site managers; the RO presented here is supporting investment decision making – allowing operational managers to quantify the benefit of maintenance expenditure

After investigating the possible cause of outliers from RO (Figure 5), it is clear the main focus for improvement should be the BUP. Each of these outliers corresponds to a day where the BUP failed (full day or part day, Figure 8) or whether the BUP was planned to be offline for maintenance for part of the day (Figure 9). Failures in the BUP could be mechanical (such as a component breaking) or 'gas-spec' failures; if the gas composition does not meet the correct standard for injection it will enter a 'failed' state and reject gas until the specification target is met.

It is worth noting that the BUP part of the AAD plant is a recent addition to the site, with installation completing late 2015. It is a new operation for operators and for the local gas network to manage correctly; the data used for RO is still within the BUP's early years on site, so a higher number of unexpected downtime would be anticipated due to the complexity and age of the technology. It is believed by operators that the BUP has operated more robustly in recent years, but RO has not been performed over a later date yet so this cannot be confirmed in this paper.

The highlighted data on Figure 8 is also a good indicator of the number of unexpected instances on site, their origin and how impactful breakdowns of individual units are with regards to economic operations. As the number of planned part-day maintenance instances is low, RO of the plant using this model is a good indication of the potential improvement for plant robustness and could help provide justification for future investments.

Model limitations and future work

Currently, the Biogas Distribution Model is a retrospective optimiser only, as a future prediction of biogas production is currently unavailable. Work will be carried out modelling the digesters to predict this biogas production, to provide operators with a full sludge processing optimisation model to allow for improved operational forecasting.

In addition, the model does not consider electrical costs associated with the CHP engines, electrical import tariffs or site load with regards to electricity and heat and assumes that CHP engines must be in operation provided they are not offline for maintenance. Electricity generated through the CHP engines is used on site to reduce electrical consumption from the National Grid and heat produced from them is reused to reduce gas requirements for the steam boilers on site (the steam is used as part of the thermal hydrolysis process). Clearly, shutting down a CHP engine will increase the electrical demand on the National Grid, with site electrical costs increasing due to reduced CHP electrical output, as well as increasing the heat load on the steam boilers.

To incorporate electricity generation, it will require splitting the model into smaller optimisation steps. The model currently uses daily values for processing limits, however electrical import/export costs on site are 'fixed-variable' (vary half hourly but are fixed and known) meaning that each half hour the CHP electricity generated on site will offset a different import cost, thus will require its own optimisation step. This would present unique and interesting challenges, as each optimisation step would require linking together, to prevent excessive ramping (turning on and off) of CHP engines, or fuel type switching every half hour – fuel types for CHP engines cannot be switched over throughout the day. Such improvements to the model could be developed further to include rotational scheduling to improve the longevity of equipment on site.

CONCLUSIONS

This paper proposes a MILP model for gas distribution and optimisation of an AAD plant with multiple options available for biogas use on site. The optimisation model takes a single input of daily biogas volume produced, in m^3, by the anaerobic digesters to provide operators with an optimal daily operational strategy; the strategy is provided to the operators in visual form for each day, though weekly strategies are also possible if required. The model also includes a penalty term on the flare stack for preventing unnecessary use of the emergency flare.

RO techniques were performed with historical operational data to determine the economic effectiveness of previous operational strategies. The model proposed uses MILP optimisation techniques to perform RO across the historical data horizon (one year) in a matter of seconds, with individual daily optimisations taking far less computational time.

Comparing the summated daily gas revenues used for RO of the plant indicates a potential increase in revenue of 7.8% (~£0.35 m) for the period optimised if the plant was run optimally all of the time. Upon discussion with operators and analysis of the maintenance and breakdown logs, the 'lost revenue' and non-optimal performing days (points on Figure 5 not within tolerance band) are mostly due to unit malfunctions, indicated on Figure 8, and are mostly due to the loss in revenue from lower biomethane injection due to BUP unplanned outages (Figure 8, graphs 'e' and 'f').

The potential increase in annual revenue from gas distribution could be achieved through improved operational strategies or (more importantly) improved plant robustness, such as reducing the number of incidents of unplanned outages, with a specific focus on the biomethane injection plant. The potential increase in revenue could be used to justify future plant improvements.

Finally, improving the proposed model to include electrical costs and heat loads, as discussed previously, would transform the model into a full energy management model for the AAD plant: a possible 'digital twin', or a decision support system (DSS), that could be used for RO and to validate future operational strategies.

ACKNOWLEDGEMENTS

This work has been collaboratively funded by Newcastle University and Northumbrian Water Ltd (under grant EPN5095281) and also the Engineering and Physical Science Research council (EPSRC) as part of a grant of Centre for Doctoral Training in Engineering for the Water Sector (STREAM, EP/L015412/1). The authors wish to express thanks to the staff at Northumbrian Water (with special thanks to Andrew Moore, Luke Dennis and Ken Black), who have been supportive through this ongoing work and without whom this project would not be possible.

DATA AVAILABILITY STATEMENT

Data cannot be made publicly available; readers should contact the corresponding author for details.

REFERENCES

Adamson, R., Hobbs, M., Silcock, A. & Willis, M. J. 2017a Integrated real-time production scheduling of a multiple cryogenic air separation unit and compressor plant. *Computers and Chemical Engineering* **104**, 25–37. http://dx.doi.org/10.1016/j.compchemeng.2017.04.001.

Adamson, R., Hobbs, M., Silcock, A. & Willis, M. J. 2017b Steady-state optimisation of a multiple cryogenic air separation unit and compressor plant. *Applied Energy* **189**, 221–232. Available from: https://linkinghub.elsevier.com/retrieve/pii/S0306261916318244.

Agabo-García, C., Pérez, M., Rodríguez-Morgado, B., Parrado, J. & Solera, R. 2019 Biomethane production improvement by enzymatic pre-treatments and enhancers of sewage sludge anaerobic digestion. *Fuel* **255**, 115713. Available from: https://linkinghub.elsevier.com/retrieve/pii/S0016236119310658.

Andersen, A. N. & Lund, H. 2007 New CHP partnerships offering balancing of fluctuating renewable electricity productions. *Journal of Cleaner Production* **15** (3), 288–293.

Bischi, A., Taccari, L., Martelli, E., Amaldi, E., Manzolini, G., Silva, P., Campanari, S. & Macchi, E. 2014 A detailed MILP optimization model for combined cooling, heat and power system operation planning. *Energy* **74** (C), 12–26. http://dx.doi.org/10.1016/j.energy.2014.02.042.

Cummings, T., Adamson, R., Sugden, A. & Willis, M. J. 2017 Retrospective and predictive optimal scheduling of nitrogen liquefier units and the effect of renewable generation. *Applied Energy* **208**, 158–170. Available from: http://linkinghub.elsevier.com/retrieve/pii/S0306261917314678.

Dambier, M. & Hinkelbein, J. 2006 Analysis of 2004 German general aviation aircraft accidents according to the HFACS model. *Air Medical Journal* **25** (6), 265–269. Available from: https://linkinghub.elsevier.com/retrieve/pii/S1067991X06001477.

Gu, C., Xie, D., Sun, J., Wang, X. & Ai, Q. 2015 Optimal operation of combined heat and power system based on forecasted energy prices in real-time markets. *Energies* **8** (12), 14330–14345.

Hale, M. 2017 Maximizing biomethane use in sewage plants. *Filtration & Separation* **54** (2), 26–27. http://dx.doi.org/10.1016/S0015-1882(17)30082-4.

Katebi, M. R., Johnson, M. A., Wilkie, J. & McCluskey, G. 1998 Control and management of wastewater treatment plants. *IFAC Proceedings Volumes* **31** (20), 847–852. Available from: http://linkinghub.elsevier.com/retrieve/pii/S1474667017419033.

Kelley, M. T., Pattison, R. C., Baldick, R. & Baldea, M. 2018 An efficient MILP framework for integrating nonlinear process dynamics and control in optimal production scheduling calculations. *Computers and Chemical Engineering* **110**, 35–52. https://doi.org/10.1016/j.compchemeng.2017.11.021.

Longo, S., d'Antoni, B. M., Bongards, M., Chaparro, A., Cronrath, A., Fatone, F., Lema, J. M., Mauricio-Iglesias, M., Soares, A. & Hospido, A. 2016 Monitoring and diagnosis of energy consumption in wastewater treatment plants. A state of the art and proposals for improvement. *Applied Energy* **179**, 1251–1268. http://dx.doi.org/10.1016/j.apenergy.2016.07.043.

Mago, P. J. & Chamra, L. M. 2009 Analysis and optimization of CCHP systems based on energy, economical, and environmental considerations. *Energy and Buildings* **41** (10), 1099–1106.

Martins, S. v. & Wilson, M. 2016 *Howdon STW – Biogas Upgrading Project*. Water Projects Online, 2016 **Editi**(Northumbrian Water). Available from: http://www.waterprojectsonline.com/case_studies/2016/NWL_Howdon_Biomethane_2016.pdf.

Merkert, L., Harjunkoski, I., Isaksson, A., Säynevirta, S., Saarela, A. & Sand, G. 2015 Scheduling and energy – industrial challenges and opportunities. *Computers and Chemical Engineering* **72**, 183–198. http://dx.doi.org/10.1016/j.compchemeng.2014.05.024.

Mulas, M., Tronci, S., Corona, F., Haimi, H., Lindell, P., Heinonen, M., Vahala, R. & Baratti, R. 2015 Predictive control of an activated sludge process: an application to the Viikinmäki wastewater treatment plant. *Journal of Process Control* **35**, 89–100. http://dx.doi.org/10.1016/j.jprocont.2015.08.005.

Northumbrian Water 2015 *Powering Ahead with gas and Electricity 'Power From Poo'*. Available from: https://www.nwl.co.uk/media-centre/611_5059.aspx (accessed 11 June 2018).

O'Brien, M., Mack, J., Lennox, B., Lovett, D. & Wall, A. 2011 Model predictive control of an activated sludge process: a case study. *Control Engineering Practice* **19** (1), 54–61. http://dx.doi.org/10.1016/j.conengprac.2010.09.001.

Ocampo-Martinez, C. 2010 *Model Predictive Control of Wastewater Systems*. Springer London, London. Available from: http://link.springer.com/10.1007/978-1-84996-353-4.

Salama, E.-S., Jeon, B.-H., Kurade, M. B., Patil, S. M., Usman, M., Li, X. & Lim, H. 2020 Enhanced anaerobic co-digestion of fat, oil, and grease by calcium addition: boost of biomethane production and microbial community shift. *Bioresource Technology* **296**, 122353. Available from: https://linkinghub.elsevier.com/retrieve/pii/S0960852419315834.

Singh, P., Kansal, A. & Carliell-Marquet, C. 2016 Energy and carbon footprints of sewage treatment methods. *Journal of Environmental Management* **165**, 22–30. http://dx.doi.org/10.1016/j.jenvman.2015.09.017.

Touil, A., Echchatbi, A. & Charkaoui, A. 2016 An MILP model for scheduling multistage, multiproducts milk processing. *IFAC-PapersOnLine* **49** (12), 869–874. http://dx.doi.org/10.1016/j.ifacol.2016.07.884.

Wari, E. & Zhu, W. 2016 Multi-week MILP scheduling for an ice cream processing facility. *Computers and Chemical Engineering* **94**, 141–156. http://dx.doi.org/10.1016/j.compchemeng.2016.07.025.

Zhang, Q., Sun, D., Wang, M. & Yin, C. 2017 Analysis of typical energy saving technology in the sewage treatment plant. *Energy Procedia* **142**, 1230–1237. https://doi.org/10.1016/j.egypro.2017.12.511.

First received 30 April 2020; accepted in revised form 22 July 2020. Available online 4 August 2020

© 2020 The Authors

Predicting failures in electronic water taps in rural sub-Saharan African communities: an LSTM-based approach

N. M. Offiong, Y. Wu and F. A. Memon

ABSTRACT

There is a growing need to sustain solar-powered water taps in most parts of the sub-Saharan Africa. The frequent failure of the water taps gives rise to intermittent water supply and poor service delivery by the water service providers. The challenge is to foresee and predict the failure of these water systems before they occur. This study develops a scalable machine-learning model for failure prediction in electronic water taps to ensure timely maintenance of the taps. Specifically, we develop a model based on long short-term memory (LSTM) to efficiently make failure predictions with noisy heterogeneous time-series data from rural water taps. Results from the experiment prove that the proposed model can effectively classify activities and patterns in various time-series datasets. With the proposed model, the failures of the solar-powered taps due to abnormal events can be successfully predicted well in advance, with an accuracy of 78.54%. Based on the data analyses, common causes of failures are presented.

Key words | anomaly detection, deep learning, failure prediction, LSTM, time-series data

N. M. Offiong (corresponding author)
F. A. Memon
The University of Exeter, Centre for Water Systems,
University of Exeter,
Exeter, EX4 4QF,
UK
E-mail: *no270@exeter.ac.uk*

Y. Wu
Department of Computer Science,
EMPS, University of Exeter,
Exeter, EX4 4QF,
UK

HIGHLIGHTS

- Detecting failures in solar-powered standalone taps using machine learning.
- Adaptive framework for data feature extraction.
- Failure prediction in rural water supply taps.
- Coding errors to check their frequency of occurrence.
- Sustaining rural water supply taps.

INTRODUCTION

Rural villages in most of sub-Saharan Africa depend on hand pumps and solar-powered water taps for clean domestic water supply (Foster & Cota 2014). These solar-powered taps, which are not adequate for the growing rural population, sometimes break down and typically cause a shortage in the water supply. The failure of these taps affects both water users and water service providers, and is a considerable concern for efficient ways to manage them for continuous service delivery. Hence, this research focuses on the need to develop a useful model for failure prediction in solar-powered water taps deployed in the sub-Saharan region.

This research employs a data-driven approach for failure prediction; this involves the use of the data generated from the case study to predict possible failures in solar-powered water taps. Some studies have used statistical analysis to investigate time series (i.e., data points with temporal ordering) based on the data acquired from an operational system (Cabrera 2007; Mazumder *et al.* 2019). Recently, the application of machine learning (ML) paradigms to manage

This is an Open Access article distributed under the terms of the Creative Commons Attribution Licence (CC BY 4.0), which permits copying, adaptation and redistribution, provided the original work is properly cited (http://creativecommons.org/licenses/by/4.0/).

doi: 10.2166/wst.2020.542

water infrastructures has produced significant improvements in failure prediction by modelling temporal datasets acquired from the systems of study (Mounce *et al.* 2015). The advancements in ML paradigms have made it possible (Mounce *et al.* 2015), and the advancement in computing architecture has also made it possible for researchers to carry out failure prediction on time-series data (Jordan & Mitchell 2015).

Our failure prediction approach uses long short-term memory (LSTM), an ML-based approach, to analyse the time-series data obtained from stand-alone solar-powered electronic water taps. Besides the LSTM model, there are traditional methods for time series forecasting, which include the autoregressive (AR), autoregressive integrated moving average (ARIMA), K nearest neighbour (KNN) and the support vector Machines (SVM) models. These traditional methods can be categories into two groups, namely model-based and distance-based.

The model-based time-series classification approaches include the AR model, which is one of the simplest methods that have been used for time-series analyses by a lot of analysts (Kini & Sekhar 2013). However, the AR model has a disadvantage, which is that it requires the time-series to satisfy stationary assumptions. In practice, this requirement is always breached. Also, the ARIMA harnesses previously observed data (AR) and the moving average of past data points to model output indirectly by integrating differenced output values. Stationary time series are derived from this process, and are needed because of their ability to extrapolate information generated from a one-time step to the next (Sina & Thomas 2019). ARIMA also has some drawbacks, including the fact that it is limited when dealing with complex datasets (Chen & Wang 2019).

Time-series models based on distance include KNN and SVM. The two models can be directly applied to time-series classification. However, these models' optimization problem falls into non-convex optimization and can quickly encounter local minima. Optimizing these models can cause overfitting and may prove to be a difficult problem to solve (Fu *et al.* 2016). Another challenge with SVM is that it shows less efficiency for a complex and massive dataset, which is a characteristic of the dataset of this research.

However, the choice of LSTM hinges on recent advancements in computational power and its ability to learn from massive datasets with longer temporal sequences. Therefore, LSTM offers a better classification accuracy over these traditional methods when dealing with large amounts of time-series data (Nakisa *et al.* 2018). LSTMs also have a special memory built into the LSTM architecture, which allows it to store information for a more extended period (Shewalkar *et al.* 2019). The time-series data are generated through daily usage of the water taps; the data are collected and sent remotely to a base station where the water installations monitoring takes place. The data collected from the electronic water supply points (EWSP) are various time-series datasets that possess latent information about the system's behaviour and this information needs a systematic interpretation with the use of the LSTM approach (Laucelli & Giustolisi 2015).

Overall, the main target of this study is to identify technical failures in the EWSP by analysing the collected data and using the information obtained to predict EWSP failures. The failure prediction is made to support timely detection and system maintenance through the development of a data-driven warning system. We harnessed case studies from smart water setups in sub-Saharan Africa to validate the effectiveness of the developed system. Most of the recently conducted research focuses on the use of hydraulic and water-quality data to monitor failures (Wu & Rahman 2017; Mamun *et al.* 2020). In this study, we label anomalies in water withdrawal data and use the labels to determine which anomaly is likely to cause failures in a water dispensing system through the use of the LSTM model. This was achieved through preprocessing of historical water usage data, observing the possible causes of failure and using the LSTM model to predict their occurrence. Some of the potential failures considered in this paper include (i) failure due to lack of voltage in the system (indicating a failure in the solar energy supply system), (ii) malfunction of the electronic tap (due to mechanical failure), (iii) low flow error (due to shortage of water) and (iv) master tag error (due to chipset miscommunication).

The contribution of this study centres on the following:

- Anomaly prediction from solar-powered taps through the application of the LSTM paradigm. The proposed model is a classification framework based on LSTM and capable of adaptive data feature learning.
- The detection of different failure events by labelling anomalies in a stationary real-world dataset; in turn, the labelled anomalies are used to determine the type of failures that may occur and their frequency of occurrence.

Sustaining water withdrawal systems by predicting the water system's failures in rural areas of sub-Saharan Africa is one of the crucial issues that many researchers are diligently investigating (Behailu *et al.* 2017; Yuanyuan *et al.* 2017; Foster *et al.* 2018). In this section, we review already published theories and methods for failure prediction and sustainability; and give a description of time series to guide the reader.

Time series, as described by Malhotra *et al.* (2015), is a set $X = \{x^{(1)}, x^{(2)}, \ldots, x^{(n)}\}$, where each of the points $x^{(1)} \epsilon R^m$ in the time series is an m-dimensional matrix $\{x_1^{(t)}, x_m^{(t)}, \ldots, x_m^{(t)}\}$ whose elements correspond to the input variables. Time series are datasets produced over time. The goal of forecasting procedures is to extract a meaningful amount of information from the time-series data during the training process (i.e., the time where a model learns trends in the data) and use the obtained information to predict the behaviour of the next time point with the aid of ML (Ram 2017). ML gives insight into the pattern created by the time series. In most cases, these are patterns that are not clear to humans because they are latent and too complicated for a standard statistical method to understand, hence the need to use the LSTM approach. (Muharemi *et al.* 2019).

Wu *et al.* (2015) used ML to manage smart water supply networks by simulating and optimizing the control of water distribution systems. The study showed that ML is capable of extracting salient data features from a large dataset through a layered architecture (Najafabadi *et al.* 2015; Chollet 2018). The study saw the development of a framework for ML to carry out predictive analysis and anomaly detection in time-series data with a specific focus on pressure, flow and water consumption.

In another study, three variants of ML techniques, including the LSTM method, were used to study a water system based on the time-series data generated by their case study (Wei *et al.* 2020). The study used the root mean square error (RMSE) and the coefficient of determination (R^2) to evaluate the model. The study showed that the LSTM model could provide a more precise and robust prediction than most of the other ML models by characterizing the lag in time between the external inputs and responses from the investigated studies.

Similarly, the research carried out by Xu *et al.* (2020) applied LSTM to predict failure related to pressure and some other abnormal operational conditions in a water distribution system. In that study, they considered three model inputs and these include the pressures at measuring points, the water supply pressure and the entry point flow of the water supply system. Liu *et al.* (2019) analysed and predicted water quality with the use of a LSTM network. In that study, they designed a model for drinking-water quality and used the model to make future predictions on the quality of water based on past datasets.

The related works reviewed from other researches concentrated more on flow, pressure and water quality. In our research, we are interested in labelling some of the anomalies associated with the electronic tap (e-tap) and make predictions based on the labels.

MATERIALS AND METHODS

Dataset

The datasets used in this study are historical data (January to December 2018) of real usage of the EWSP installed by a commercial water services provider in the Gambia. The data contains a record with 1,047,114 rows representing time-series samples from 27 different EWSPs and 22 columns describing the features of the dataset. The data were collected over one year and transmitted to a remote server where they are kept for processing and planning purposes. The data contains both relevant features and other features that are not relevant to this study. From the dataset, we extracted the relevant feature for the analysis and they are shown in Table 1.

The three most important data columns that helped to answer the overall aims of the project are *ErrorCode*, *Voltage* and *FlowRate*. We split the data into two parts (80% for training (part of which we used for validation) and 20% for testing. Following this technique, we trained the developed model (explained later) on all the available data by selecting the time series on a step-by-step basis. During the implementation, we designed the training set to expand after each iteration while the test set remains fixed at a one-time step. The training and validation of the model occurred inside the loop of each iteration. With each iteration, a new model is created. This procedure repeats for

Table 1 | Extracted relevant data feature for the time-series analysis

Column header	Column description
AssetID	Unique identification for the EWSPs (tap number)
ErrorCode	Different codes to indicate issues with water tap usage (e.g. F1, F2, F3, F4)
FlowCount	Derived value to indicate value for flow rate
FlowTime	Time at which flow occurred
Litres	Volume of water withdrawn
Voltage	The battery voltage capacity detected at each instance the water tap accessed
DateTime	Time of event (water withdrawal date and time)
FlowRate	Volume of water collected per unit time

all the sequences and the average taken to give the overall effectiveness of the model. The splitting of the data was done as follows:

1. Training set (for the network to learn from)
2. Testing set (to evaluate the model)
3. Validation set (for early stopping and optimization of the parameters that the proposed model cannot learn by itself)

The original dataset is a noisy dataset with some missing information and mismatched datatypes, and preprocessing was done on the data to normalize the dataset and prepare it for the analysis. The data normalization method used in this study was the min-max method, which transforms the minimum value of the data to 0 and the maximum value to 1. The values between the minimum and maximum values are transformed to decimal values between 0 and 1. The formula for min-max normalization is shown in Equation (1) below:

$$x_{(0,1)} = \frac{x - \min}{\max - \min} \tag{1}$$

where $x_{(0,1)}$ and x are the normalized and original datasets, respectively. Min and max are the minimum and maximum values of the whole dataset, respectively.

The min-max normalization algorithm is widely used (Patro & Sahu 2015). The drawback of the method includes inefficiency in handling outliers in datasets. To solve this problem, we expanded the min-max range to accommodate any outliers in the dataset. Before the proposed LSTM-based model was built, the dataset was transformed to match a supervised learning format needed to solve the time-series modelling problem, where the time steps were structurally transformed to input and output values.

An initial analysis was done on the data with a conventional data manipulation to show how the two columns, *Voltage* and *FlowRate*, relate to one another, and more importantly, how both columns affect the behaviour of the e-taps. The preliminary analysis in Figure 1 shows the failures and normal functioning of the investigated taps.

Figure 1 reveals some failures and normal functioning of taps, but our interest is in failure trends. It can be seen from the preliminary analysis that at a point where the voltage of the system is 0, there was a considerable amount of flow from the EWSPs. It can also be seen that at a point where the voltage was at a peak, there was a failure and the tap did not dispense water. To uncover other latent behaviour of the system, we applied our model to investigate the daily water withdrawal data, which covers the period of January to December 2018 (one year). There are a few other interesting observations that can be made from the plot in Figure 1. A few striking observations are:

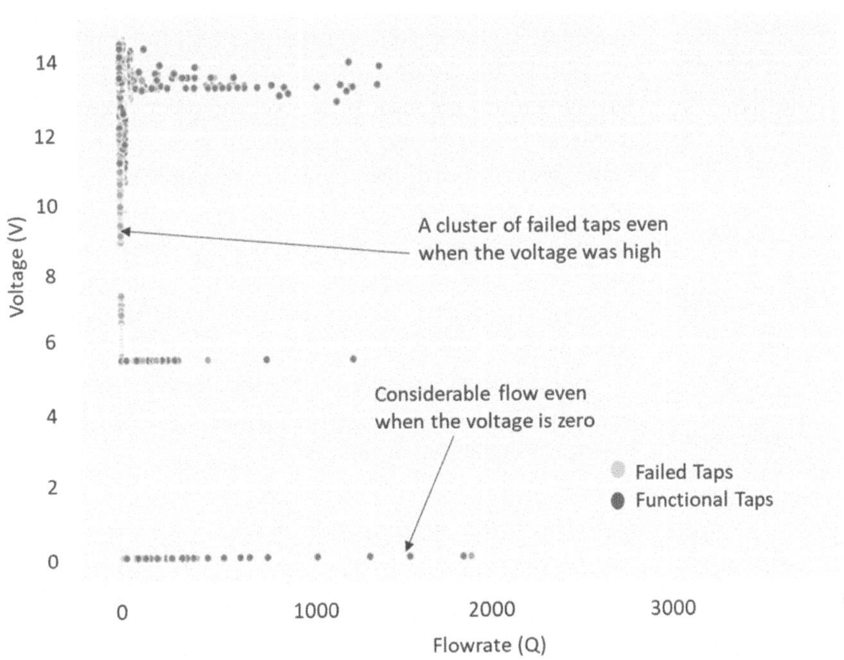

Figure 1 | Initial analysis of the dataset with a conventional method to show functionality and non-functionality of the EWSPs in the case study.

- Most of the water tap failures occur in the extreme right side of the plot, where *FlowRate* is the lowest. This may indicate a flow-related problem.
- There are two instances of failure where the *FlowRate* was recorded at the value of 1,900, but the voltage remained 0. This could be indicating a battery failure.
- There is one failure that occurred when the voltage was around 12.8 volts, but the *FlowRate* was around 15 litres, and it was recorded as a latent failure as well.
- There is another inspection that was performed to check if an error may be incorrectly indicating failures in the e-tap. The inspection shows that when the voltage was 0.00, there was still a reasonable quantity of flow (of up to about 13 litres). This may indicate a chipset error.

Based on the observations above, we coded some of the errors and used them as labels to investigate how they affect the sustainability of the electronic water withdrawal points. Table 2 below shows some of the sample errors and their associated codes (F1–F6).

Model development

As mentioned in the introduction section, we propose a novel predictive model based on the LSTM technique (Chollet 2018), which is capable of learning good representations of the available input time-series data. The representations can get closer to the expected output information for decision making.

After the preliminary investigation, we developed the proposed model (Figure 2) with Keras (one of the packages in the Python programming language that is widely used and supports LSTM modelling). Each of the LSTM memory cells has a 3D input (Wei *et al.* 2020). The choice of Keras in this study was because it gives room for consistency and provides useful feedback whenever there is a user error (Chollet 2018). The Keras application programming interface (API) is flexible and straightforward to use for a wide range of ML tasks. The LSTM layers were stacked so that the setup of the previous layer can be changed to give a 3D output. This was achieved by making the argument of the 'return sequence' on the last layer to True. The proposed model used for this study is a stacked LSTM network with eight hidden layers for 3D input in time processing. The hidden layers in the study are connected to a single output layer to produce a 2D output. The loss function used in the model design is the RMSE.

Table 2 | Some of the errors investigated in the research

Failure definition	Failure code
No flow error	F1
Proxy error	F2
Voltage error	F3
Faulty valve error	F4
Master tag error	F5
Host valve error	F6

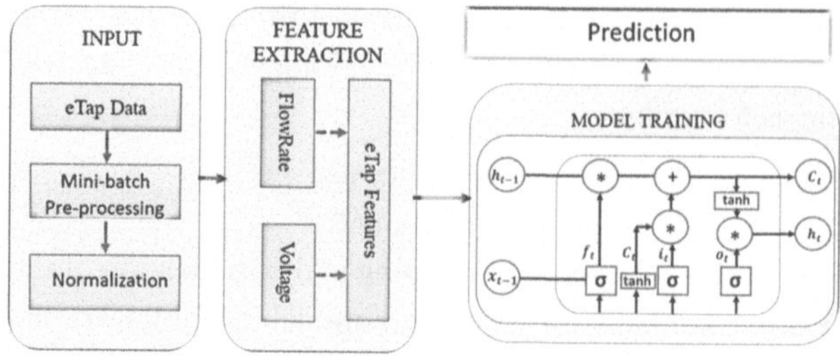

Figure 2 | The proposed model based on the LSTM network.

The central part of the proposed LSTM-based model is the cell state (which is the horizontal line running through the top in the 'MODEL TRAINING' box in Figure 2) down to the entire chain, with some linear interactions. The cells of the proposed model can remember values over a long period. The training model contains three gates that manage the flow of information in and out of the cell. The three gates in the model are: the input gate ($\boldsymbol{i_t}$), the forget ($\boldsymbol{f_i}$) gate and the output ($\boldsymbol{o_t}$) gate.

The input gate makes a decision on which new input to update to the current state of the cell; the forget gate determines the number of the previous states H_{t-1} that are allowed to go through the cell or what information to discard from memory. The output gate, on the other hand, makes the decision on the output based on the current state of the cell. The memory is then updated to hold the most current information based on the combination (or aggregation) of the old memory through the forget gate and the new memory state through the input gate.

These gates are introduced into the sigmoid function (σ) in Figure 2 to solve the vanishing gradient problem, and at time t it is computed as follows:

$$\boldsymbol{i_t} = \boldsymbol{\sigma}(\boldsymbol{W_i} * (\boldsymbol{W_h}, \boldsymbol{WH}_{t-1}) + \boldsymbol{b_i}) \tag{2}$$

$$\boldsymbol{f_t} = \boldsymbol{\sigma}(\boldsymbol{W_f} * (\boldsymbol{W_h x_t}, \boldsymbol{WH}_{t-1}) + \boldsymbol{b_f}) \tag{3}$$

$$\boldsymbol{o_t} = \boldsymbol{\sigma}(\boldsymbol{W_o} * (\boldsymbol{W_h x_t}, \boldsymbol{WH}_{t-1}) + \boldsymbol{b_o}) \tag{4}$$

$$Ç = \boldsymbol{Softmax}\ (\boldsymbol{W_c} * (\boldsymbol{W_h x_t}, \boldsymbol{WH_{t-1}}) + \boldsymbol{b_c}) \quad (5)$$

$$\boldsymbol{C_t} = \boldsymbol{f_t} * \boldsymbol{C_{t-1}} + \boldsymbol{i_t} * \boldsymbol{C_t} \quad (6)$$

$$\boldsymbol{H_t} = \boldsymbol{o_t} * \text{softmax}\ (\boldsymbol{C_t}) \quad (7)$$

where

$\boldsymbol{i_t}$ and Ç – input gate values and the state of the memory cell at time $\boldsymbol{t}$, respectively.
$\boldsymbol{f_t}$ – the value of the forget gate;
$\boldsymbol{o_t}$ – the output gate of the model;
$\boldsymbol{C_t}$ –the current state of the model;
$\boldsymbol{W_i}$, $\boldsymbol{W_f}$, $\boldsymbol{W_o}$ – randomly generated weight vectors for all three gates;
$\boldsymbol{b_i}$, $\boldsymbol{b_f}$, $\boldsymbol{b_o}$ – bias vectors for all three gates; and
$\boldsymbol{H_t}$ – the value of the memory cell at time $\boldsymbol{t}$.

The sigmoid function (σ) serves to control the output between the values of 0 and 1 based on the current input of x and the prior output WH_{t-1}. At the computational phase of Equations (2)–(7), the weight and biases get trained by the model. This training is achieved by minimizing the loss between the outputs and the actual training samples of the proposed model.

Model implementation and validation

The proposed LSTM-based model is designed with a single input layer and eight LSTM-type hidden layers. A few other LSTM structures were examined as well, but the structure used for this study proved to perform better for the dataset. The number of neurons in the proposed model was initially set to 200 neurons, which was run against four hidden layers. However, the resulting predictions showed a terrible convergence of the model. Based on some of the limitations found in the dataset and the inconsistencies found in the time steps and date, the better option was to focus on every *AssetID* (tap) independently. Two different time steps that need to be programmed into the proposed model were considered. First is the *n_steps_in* parameter, which indicates the number of past time steps taken by the model to predict the corresponding output values (in time steps). The number of output time steps produced is dependent on the second parameter, which is *n_steps_out*. The final model includes the last 30 time steps from the dataset for different assets (taps) to predict future time steps.

The model fitness was validated using the technique called the walk-forward validation technique because of its ability to preserve temporal data (Falessi *et al.* 2018). In this case, the dataset gets split into two units (the training set and the testing set) that can be ordered. Then a decision is made on the minimum number of observations needed for the model training. This is done to obtain a configuration for the test setup. The training of the model then begins at the start of the time series with a design that enables it to make decisions for the next time step. Subsequently, the prediction is evaluated against the next time step in the time series. The predicted values were then measured using RMSE scores.

The hyper-parameters were tuned with the use of Talos, a library built for automated hyper-parameter tuning; it can be installed to work with Keras and it helped to reduce procedural redundancy.

The proposed model was built such that hyper-parameters will not be manually tuned. The accuracy of the model was another critical study criterion that was taken into consideration. For the model to be accepted as a useful prediction model, the accuracy has to be high enough. The proposed model was built on training sets to determine its accuracy; then, the testing sets were used as moderator samples to test the *trained* model. The prediction results were then compared with the actual values (ground truth) by calculating the accuracy based on Equation (8):

$$\text{accuracy} = \frac{\text{correctly predicted classes}}{\text{total testing class}} \times 100\% \quad (8)$$

While the Python programming language takes the weight of writing complex ML algorithms, we make a decision on the hyper-parameters to use in the model tuning, and this greatly influences the model's accuracy. Figure 3(a) and 3(b) show the results of model accuracy and loss from the experiment.

Figure 3(a) shows the proposed model's accuracy. The model's accuracy was measured using the RMSE, which gives the differences between the actual and the predicted values. Equation (9) below gives the formula used to calculate the RMSE:

$$\text{RMSE} = \sqrt{\frac{1}{N}\sum_{i=1}^{N}(x_i - \hat{x}_i)^2} \quad (9)$$

where

N – number of total observations
x_i – actual value
$\hat{x}_i$ – predicted value

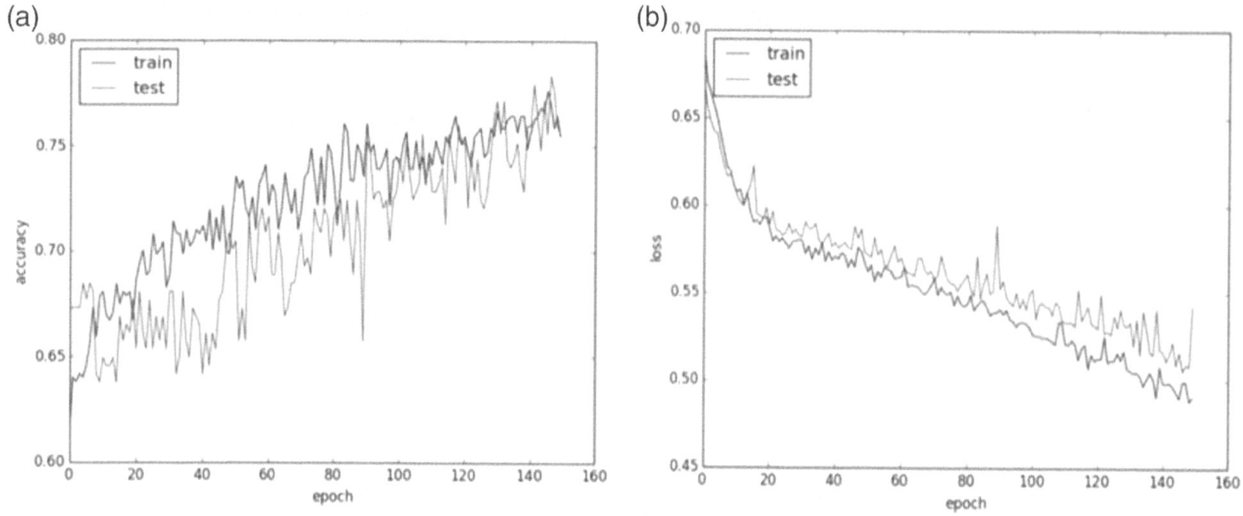

Figure 3 | Models showing accuracy and loss of the model. (a) Model accuracy, (b) Model loss.

The benefit of using RMSE is that it scales the scores and in the same unit as the values of forecast (i.e., hourly in our study). As seen from the diagram (Figure 3(a)), the accuracy of the model is increasing, with a current accuracy of 78%. The testing set shows this increment for categorical predictions only. So when different failures are categorized, the proposed model can easily predict the failure that is likely to occur in the future in the water taps. Figure 3(b) shows the training of the model's loss values for continuous predictions only. To make the model fit easily, we used fewer hyper-parameters. The epoch in the figure is the number of passes (times) that the proposed model's training vectors are used to update the weight.

Training the model is computationally very expensive. Our proposed model was built on a cloud-GPU platform, which hosts the Quadro P5000 graphics card. The time to complete a 100-epoch training (per tap) is, on average, 6 hours. Another GTX 1050 GPU was used for model training in parallel. It took roughly 8 hours to execute the task for one tap; this is because the program implementation needs a massive memory space to work effectively. With a cloud server, the training time of the model will reduce significantly.

RESULTS AND DISCUSSION

We proposed a predictive model for time-series prediction based on LSTM that is capable of predicting failures in rural water withdrawal taps. The model was built using the most fundamental principles and techniques, taking into consideration the many missing and incorrect values in the dataset (since the data is a very noisy dataset). The methodology presented in this paper is data-driven, and it is capable of self-learning to detect anomalies in the data. Concerning the above, the system can dynamically redesign itself to work for different time-series datasets.

The experimental model involves the use of a real dataset acquired from real solar-powered water taps. The data were preprocessed to remove impurities and further used as input for the prediction. The core of the LSTM model is the cell state and the three gates (the input, forget and the output gates as described in the 'Model development' section). The cell is the memory that holds information about the EWSPs. As the data are processed, information gets added or removed via the gates of the LSTM architecture. The gates, on the other hand, are different neural networks that are capable of making a decision about which information to allow into the memory or which one to forget during model training. Each of the gates has the activation function that scrunches values between 0 and 1 to help the cell state decide which data to allow or forget based on their importance. The model can categorize and predict failures for a real dataset and the model design can be further adjusted to suit the demand of future time series. A *complete* dataset would allow for higher quality predictions, accurately reflecting the future outcome for specific water tap assets that were analysed earlier. Two critical parameters that may be important to pay extra attention to would be: *n_steps_in* and *n_steps_out* (these parameters enable the model to take the past time steps *in* as a learning input, to produce predictions). The number of output

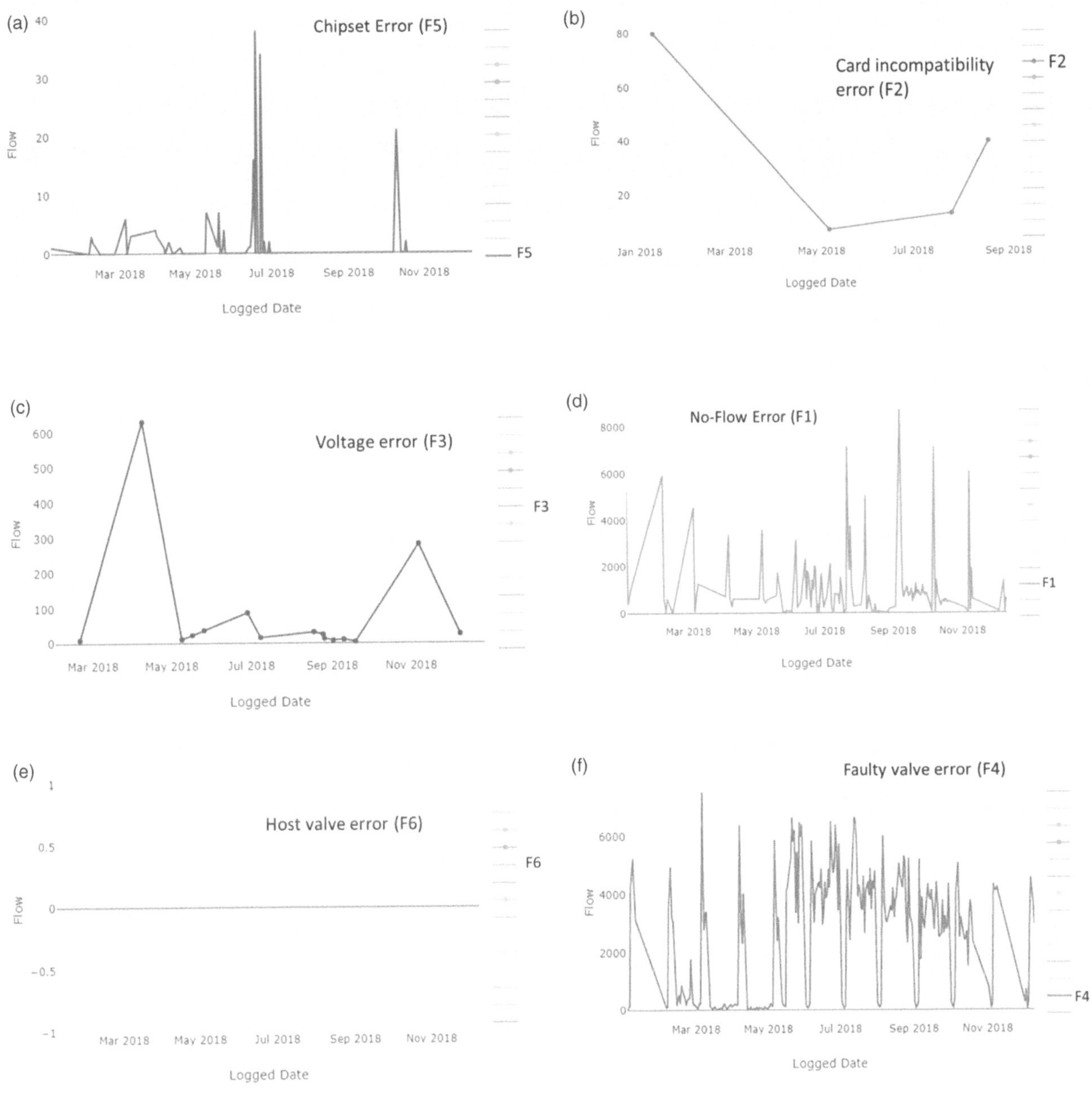

Figure 4 | Predicted errors/failures on some of the e-taps.

time-step predictions of the proposed model will depend on the value of *n_steps_out* (this parameter represents the number of time steps that are outputted as predictions).

The sample results for some of the failures are shown in Figure 4. Figure 4(a) shows a chipset error (F5), Figure 4(b) shows a card incompatibility error (F2), Figure 4(c) shows a voltage error (F3). Figure 4(d) represents error code F1, which is a flow-related error; Figure 4(e) shows a typical trend for failure type F6, which is a host valve error type. The figure shows that the error is predicted to remain at zero, which means that the error may not occur and cause the tap to fail. Figure 4(f), on the otherhand, shows a faulty valve error (F4). We observed that when there is a peak error, the tap stops functioning.

For all the results presented in this study, the two main empirical features used are the *Voltage* and *Flow-Rate* features, which were observed against the different failure trends in the *ErrorCode* column of the dataset. Figure 4(a) shows a monthly prediction of chipset error, which can cause the EWSPs to not dispense water. This kind of failure is caused when there is a technical mis-communication between a user token (pre-paid tag) and an EWSP or a temporary malfunction of the EWSP. The proposed model is able to determine when the fault will occur and show its severity. Figure 4(b) represents a card incompatibility failure, which can occur when a token has been restricted or, mainly, because it was lost by a user. This failure can also arise due to the mechanical or technical malfunctioning of the EWSP. This kind of failure occurs but may not last for an extended period of time. Figure 4(c) captures the failure that can be caused by the voltage. When the voltage is below 6.0 volts, the EWSP will not have the capacity to power the e-tap. This failure can occur mostly during the rainy season, where there is less sunlight to charge the battery component of the EWSP system. Figure 4(d) shows the frequency of the failure caused by flow. Figure 4(e) shows a host valve error, and it can be noted that there was no error with the host valve throughout the period of study. Figure 4(f) shows a faulty valve error.

CONCLUSION

This research paper presents a model based on the LSTM paradigm for fault prediction in EWSPs in rural sub-Saharan Africa. It is actually the first time the method has been used to investigate e-taps. The proposed model takes into consideration two crucial pieces of input information from the solar-powered electronic taps, which are the flow rate and the voltage of the e-tap. The dataset used for the research is a real historical dataset with some inconsistencies.

From the result of the experiments, it has been shown that the LSTM model can discover latent information from the time-series dataset and can make informed predictions based on the results of the predictions. The proposed model's performance based on LSTM proves to have superior accuracy and efficiency for massive datasets (typically generated from daily use of the EWSPs in our case study) compared to the traditional statistical methods mentioned in the introductory section. The user of the model decides the best number of outputs needed. The size of these parameters influences the computation cost of the training of the proposed model.

From the results and the accuracy of the model, we conclude that the proposed model can provide reasonable predictions for the investigated tap failures. In the future, we hope to compare LSTM with other ML tools so as to generate a hybrid model for the development of an early-warning system for smart water taps.

ACKNOWLEDGEMENTS

This research has been achieved through a Ph.D. scholarship provided by TETFund Scholarship Grant (Nigeria), which is thankfully acknowledged. The authors would also like to thanks Rob Hygate, Roger Godwin and other members of staff of eWaterPay for providing access to the e-tap data.

DATA AVAILABILITY STATEMENT

Data cannot be made publicly available; readers should contact the corresponding author for details.

REFERENCES

Behailu, B. M., Hukka, J. J. & Katko, T. S. 2017 Service failures of rural water supply systems in Ethiopia and their policy implications. *Public Works Management & Policy* **22** (2), 179–196. https://doi.org/10.1177/1087724X16656190.

Cabrera, E. 2007 Statistical modeling for failure prediction in water supply networks. In *International Conference on Dependable and Quality Management*, June.

Chen, Y. & Wang, K. 2019 Prediction of satellite time series data based on long short term memory-autoregressive integrated moving average model (LSTM-ARIMA). In *2019 IEEE 4th International Conference on Signal and Image Processing, ICSIP 2019*. pp. 308–312. https://doi.org/10.1109/SIPROCESS.2019.8868350.

Chollet, F. 2018 *Deep Learning with Python*. Manning Publications Co., New York.

Falessi, D., Narayana, L., Fong, J. & Turhan, B. 2018 *Preserving Order of Data When Validating Defect Prediction Models*. pp. 1–20.

Foster, R. & Cota, A. 2014 Solar water pumping advances and comparative economics. *Energy Procedia*. (January). https://doi.org/10.1016/j.egypro.2014.10.134.

Foster, T., Willetts, J., Lane, M., Thomson, P., Katuva, J. & Hope, R. 2018 Risk factors associated with rural water supply failure: a 30-year retrospective study of handpumps on the south coast of Kenya. *Science of the Total Environment* **626**, 156–164. https://doi.org/10.1016/j.scitotenv.2017.12.302.

Fu, R., Zhang, Z. & Li, L. 2016 Using LSTM and GRU neural network method for traffic flow prediction. 31st Youth Academic Annual Conference of Chinese Association of Automation (YAC), Wuhan, pp. 324–328, doi: 10.1109/YAC.2016.7804912.

Jordan, M. I. & Mitchell, T. M. 2015 Machine learning: trends, perspectives, and prospects. *Science* **349** (6245), pp. 255–256.

Kini, B. V. & Sekhar, C. C. 2013 Large margin mixture of AR models for time series classification. *Applied Soft Computing Journal* **13** (1), 361–371. https://doi.org/10.1016/j.asoc.2012.08.027.

Laucelli, D. & Giustolisi, O. 2015 *Detecting Anomalies in Water Distribution Networks Using EPR Modelling Paradigm, (October 2016)*. https://doi.org/10.2166/hydro.2015.113.

Liu, P., Wang, J., Sangaiah, A. K., Xie, Y. & Yin, X. 2019 Analysis and prediction of water quality using LSTM deep neural networks in IoT environment. *MDPI – Sustainability* 1–14. https://doi.org/10.3390/su11072058.

Malhotra, P., Vig, L., Shroff, G. & Agarwal, P. 2015 Long short term memory networks for anomaly detection in time series. In *Artificial Neural Network, Computational Intelligence and Machine Learning*, April. pp. 22–24.

Mamun, M., Kim, J. J., Alam, M. A. & An, K. G. 2020 Prediction of algal chlorophyll-a and water clarity in monsoon-region reservoir using machine learning approaches. *Water (Switzerland)* **12** (1). https://doi.org/10.3390/w12010030.

Mazumder, R. K., Salman, A. M., Li, Y. & Yu, X. 2019 Reliability analysis of water distribution systems using physical probabilistic pipe failure method. *Journal of Water Resources Planning and Management* **145** (2), 04018097. https://doi.org/10.1061/(asce)wr.1943-5452.0001034.

Mounce, S. R., Pedraza, C., Jackson, T., Linford, P. & Boxall, J. B. 2015 Cloud based machine learning approaches for leakage assessment and management in smart water networks. *Procedia Engineering* **119** (1), 43–52. https://doi.org/10.1016/j.proeng.2015.08.851.

Muharemi, F., Logofătu, D. & Leon, F. 2019 Machine learning approaches for anomaly detection of water quality on a real-world data set. *Journal of Information and Telecommunication* **3** (3), 294–307. https://doi.org/10.1080/24751839.2019.1565653.

Najafabadi, M. M., Villanustre, F., Khoshgoftaar, T. M., Seliya, N., Wald, R. & Muharemagic, E. 2015 *Deep Learning Applications and Challenges in Big Data Analytics*. pp. 1–21. https://doi.org/10.1186/s40537-014-0007-7.

Nakisa, B., Rastgoo, M. N., Rakotonirainy, A., Maire, F. & Chandran, V. 2018 Long short term memory hyperparameter optimization for a neural network based emotion recognition framework. *IEEE Access* **6**, 49325–49338. https://doi.org/10.1109/ACCESS.2018.2868361.

Patro, S. G. K. & Sahu, K. K. 2015 Normalization: a preprocessing stage. *International Advanced Research Journal in Science and Technology* **2** (3), 20–22. https://doi.org/10.17148/iarjset.2015.2305.

Ram, K. 2017 *Machine Learning Tools to Time Series Forecasting (December 2007)*. https://doi.org/10.1109/MICAI.2007.42.

Shewalkar, A., Nyavanandi, D. & Ludwig, S. A. 2019 Performance evaluation of deep neural networks applied to speech recognition: RNN, LSTM and GRU. *Journal of Artificial Intelligence and Soft Computing Research* **9** (4), 235–245. https://doi.org/10.2478/jaiscr-2019-0006.

Sina, D. & Thomas, B. 2019 Anomaly detection in univariate time series: an empirical comparison of machine learning algorithms. In *ICDM*. pp. 1–15.

Wei, X., Zhang, L., Yang, H. Q., Zhang, L. & Yao, Y. P. 2020 Machine learning for pore-water pressure time-series prediction: application of recurrent neural networks. *Geoscience Frontiers* (September 2019). https://doi.org/10.1016/j.gsf.2020.04.011.

Wu, Z. Y. & Rahman, A. 2017 Optimized deep learning framework for water distribution data-driven modeling. *Procedia Engineering* **186**, 261–268. https://doi.org/10.1016/j.proeng.2017.03.240.

Wu, Z. Y., El-Maghraby, M. & Pathak, S. 2015 Applications of deep learning for smart water networks. *Procedia Engineering* **119** (1), 479–485. https://doi.org/10.1016/j.proeng.2015.08.870.

Xu, Z., Ying, Z., Li, Y., He, B. & Chen, Y. 2020 Pressure prediction and abnormal working conditions detection of water supply network based on LSTM. *Water Supply* **20** (3), 963–974. https://doi.org/10.2166/ws.2020.013.

Yuanyuan, W., Ping, L., Wenze, S. & Xinchun, Y. 2017 A new framework on regional smart water. *Procedia Computer Science* **107** (Icict), 122–128. https://doi.org/10.1016/j.procs.2017.03.067.

First received 17 July 2020; accepted in revised form 26 October 2020. Available online 6 November 2020

© 2021 The Authors

Experiments of an IoT-based wireless sensor network for flood monitoring in Colima, Mexico

O. Mendoza-Cano, R. Aquino-Santos, J. López-de la Cruz, R. M. Edwards, A. Khouakhi, I. Pattison, V. Rangel-Licea, E. Castellanos-Berjan, M. A. Martinez-Preciado, P. Rincón-Avalos, P. Lepper, A. Gutiérrez-Gómez, J. M. Uribe-Ramos, J. Ibarreche and I. Perez

ABSTRACT

Urban flooding is one of the major issues in many parts of the world, and its management is often challenging. One of the challenges highlighted by the hydrology and related communities is the need for more open data and monitoring of floods in space and time. In this paper, we present the development phases and experiments of an Internet of Things (IoT)-based wireless sensor network for hydrometeorological data collection and flood monitoring for the urban area of Colima-Villa de Álvarez in Mexico. The network is designed to collect fluvial water level, soil moisture and weather parameters that are transferred to the server and to a web application in real-time using IoT Message Queuing Telemetry Transport protocol over 3G and Wi-Fi networks. The network is tested during three different events of tropical storms that occurred over the area of Colima during the 2019 tropical cyclones season. The results show the ability of the smart water network to collect real-time hydrometeorological information during extreme events associated with tropical storms. The technology used for data transmission and acquisition made it possible to collect information at critical times for the city. Additionally, the data collected provided essential information for implementing and calibrating hydrological models and hydraulic models to generate flood inundation maps and identify critical infrastructure.

Key words | digital water network, early warning systems, flooding, tropical storms

HIGHLIGHTS

- IoT-based wireless sensor network for hydrometeorological data collection and flood monitoring has been developed for Colima-Villa de Álvarez in Mexico.
- Real-time data is transferred over 3G cellular network and Wi-Fi.
- Tests of the network on three separate tropical storm events showed robustness of the network during extreme events.

O. Mendoza-Cano (corresponding author)
J. López-de la Cruz
M. A. Martinez-Preciado
P. Rincón-Avalos
J. M. Uribe-Ramos
Facultad de Ingeniería Civil,
Universidad de Colima,
Km. 9 Carretera Colima-Coquimatlán, Col. Jardines del Llano,
C.P. 28400 Coquimatlán, Colima, Mexico
E-mail: *oliver@ucol.mx*

R. Aquino-Santos
E. Castellanos-Berjan
Facultad de Telemática,
Universidad de Colima,
333. C.P. 28045 Colima, Mexico

R. M. Edwards
5G Research Centre,
Woflson School, Loughborough University,
Loughborough LE11 3TU, UK

A. Khouakhi
School of Water, Energy and Environment, Centre for Environmental and Agricultural Informatics,
Cranfield University, Cranfield, UK

I. Pattison
Heriot-Watt University, Institute of Infrastructure and Environment,
School of Energy, Geoscience, Infrastructure and Society, Edinburgh, UK

V. Rangel-Licea
A. Gutiérrez-Gómez
Facultad de Ingeniería,
Universidad Nacional Autónoma de México (UNAM),
México, México

P. Lepper
School of Mechanical, Electrical and Manufacturing Engineering,
Loughborough University,
Wolfson Building, Ashby Road,
Loughborough LE11 3TU, UK

J. Ibarreche
I. Perez
Tairda, S.A. de C.V.,
111-B Canario Street,
C.P. 28017 Colima, México

This is an Open Access article distributed under the terms of the Creative Commons Attribution Licence (CC BY 4.0), which permits copying, adaptation and redistribution, provided the original work is properly cited (http://creativecommons.org/licenses/by/4.0/).

doi: 10.2166/hydro.2021.126

INTRODUCTION

Recent developments of digital cities as the exploration of cyberspace to smart cities as the exploitation of the physical space resulted in proposing that the next stage is a networked society based on cyber-physical systems (Ishida 2017). This conceptualisation advocated in recent decade such as digital twins for a range of infrastructures is currently being implemented in water systems (Lee 2008). The coupling between the physical and cybernetic layers of water systems and related infrastructure can provide new opportunities to increase the efficiency of water infrastructure and can increase resilience in water planning and services (Makropoulos & Sávic 2019). For example, using multiple sensors, water systems and related infrastructure can be controlled in real-time and data can be continuously integrated into models. Such pairing of the virtual and physical components of the water systems will also be an important tool in applications such as urban flood management, which is one of the main challenges facing urban areas (Bachmann *et al.* 2016; Lund *et al.* 2018).

Water in cities is a critical resource, and its management is associated with many challenges; for example, monitoring freshwater quality and quantity is a local and global concern (Artiola 2004; Camara *et al.* 2020). Water can also present a threat to the population when flooding occurs (Olya & Alipour 2015; Plümper *et al.* 2017; Ryan 2018). Tackling flooding disasters in urban areas is a primary concern in many cities around the globe (Davidsson 2020). Therefore, smart digital infrastructure to monitor and forecast flood events is critical for improving the resilience of urban areas to hazards such as those related to flooding.

Climate change is set to increase the frequency and severity of floods in many parts of the world (Miller & Hutchins 2017). Extreme hydroclimatic events will continue to disproportionately affect vulnerable populations and low-income countries (Enenkel *et al.* 2020). In past years, the number of reported flood events has increased substantially (Tanoue *et al.* 2016), affecting people and causing significant economic loss and damage to affected environments over time, including Mexico (López-Marrero & Tschakert 2011; Kundzewicz *et al.* 2014; Khouakhi *et al.* 2019). On a global scale, records show that between 1900 and 2010, floods have a more significant and increasing number in comparison with other disasters both in terms of recorded events and people affected (EM-DAT 2020).

Floods in Mexico are a major challenge for managers as flood events are increasing in regularity and severity (De *et al.* 2012). Climate-related disaster management in Mexico has been focussing more on the actions of repairing, emergency assistance than on the prevention (Constantino *et al.* 2011). However, this form of intervention does not reduce the possibility of disasters with the corresponding negative effect on the social functioning of the geographical areas that usually face such events. In the light of current knowledge, it is important to develop a preventive approach to mitigating the risk of flooding in Mexico. One effective way to reduce the risk of floods lies in developing and implementing an early warning system. This early warning system is one of the main tools in the preventive approach; unfortunately, many developing countries have poor flood monitoring infrastructure (Matthews *et al.* 2011). The density of the river and climate monitoring network in Mexico is severely deficient now. Urban areas are particularly unmonitored currently, and only the cities of Tuxtla Gutierrez, Tijuana, Monterrey, Acapulco, Chalco, and Mexico City have a hydrometeorological alert system.

The state of Colima (west-central Mexico) is often affected by the cyclonic activity from the Pacific basin. Colima-Villa de Alvarez is the most urbanised area of the state. Its proximity to the coastline makes it exposed to tropical cyclones and their associated hazards. In the last decade, flooding is becoming more frequent in Colima and in the context of climate change; it is likely that this increasing trend will continue in the coming decades.

Different flood forecasting systems and hydrological models have been developed worldwide (e.g. Noymanee & Theeramunkong 2019; Uteuov *et al.* 2019; Cheval *et al.* 2020). Some of them use real-time data for their assessments and predictions (Alfieri *et al.* 2019; Echeverribar *et al.* 2019; Lindenschmidt *et al.* 2019; Zhang *et al.* 2019; Deng *et al.* 2020; Khalid & Ferreira 2020; Ritter *et al.* 2020). To develop a real-time flood forecasting system, several areas of research are needed. These include expertise in data sensing at the appropriate location and time, wireless transmission of flood data, sensor data fusion, model generation, prediction at the remote weather station, and flood resilience expertise. This project combined different groups and expertise to design a cost-effective flood monitoring and forecasting system (Mendoza-Cano *et al.* 2019). Resilience to floods in the study area has been evaluated by Mendoza-Cano *et al.* (2019) in 10 critical areas of Colima-Villa de Álvarez. Resilience is an indicator of a system's ability to withstand disruption within acceptable degradation parameters and their recovery time (Nguyen & James 2018). The study was performed on public knowledge of the existing security protocols for floods. It evaluated the public perception of critical services' availability, such as freshwater, electricity, food, drainage, communications and public transport during flash flood events. This research identified populated low resilience zones that can be considered a priority for resources and effort to mitigate floods and their impacts. These results demonstrated the need to implement strategies that strengthen the community capacity to cope with the effects of extreme hydrometeorological phenomena in Colima-Villa de Álvarez.

In this paper, we show how expertise in water engineering, embedded electronics, radio communications and web development have come together to develop and test an Internet of Things (IoT)-based digital water solution (1) allowing collection of high-resolution hydrometeorological data, (2) monitoring of urban fluvial water level, and (3) issuing potential flood level alerts for the public in Colima. Section 2 describes the study area, field survey, sensor network structure, types of nodes implemented and their specifications, the acquisition and transmission of data and the graphical user interface. Tests and Result section show some early results of the network monitoring of three main storm events which occurred during the tropical cyclone season of 2019.

METHODS

Study area

The Colima River catchment is located in the north of the Colima state (west-central Mexico). It is a sub-catchment of the Armeria's river basin (Figure 1). The catchment area is about 150 km^2, and the main Colima River originates from the slopes of the Colima volcano and has a dynamic river bank vegetation. The state of Colima is often affected by the Eastern Pacific cyclonic activity during the tropical cyclones season. Colima-Villa de Alvarez is the most urbanised area of the state, and its proximity to the Pacific coastline makes it exposed to tropical storms that frequently cause damage and disruptions. Some of the most destructive storms that made landfall near Colima and created significant damage to infrastructure, include *Jova* in 2011 and *Patricia* in 2015 (category 4 hurricane).

Figure 2 shows the monthly temperature and precipitation in the municipality of Colima for the period 1950–2017. A unimodal precipitation regime can be observed, with a significant wet period from June to October and a marked dry period from November to May. Most storms occur in September with about 17 storm events; some of them are of cyclonic origin. The percentage of days with a storm in the municipality of Colima is 18%.

Network design and sensor locations

Several field campaigns were conducted across the catchment area to select suitable sites for data collection and sensor deployment. First, rivers and their tributaries in the Colima river basin was mapped and analysed before the survey. The main goal was to assess the optimal sites needed for flood monitoring and prediction and assess the representativeness of the selected locations following the previous historical floods. Other objectives of the survey included (1) assessment of the security of the data loggers and base stations (fixed nodes), (2) measurement of the strength of the 3G network, and (3) assessment of the accessibility to sites for deploying and maintaining the sensors. The tributary interactions and critical locations for the data collections needed for the monitoring and modelling from the hydrological perspective were also considered. Sixteen locations have been defined hereafter represented as static nodes. These nodes are composed of eight nodes of water level and soil moisture and eight weather station nodes. Each node has a data logger, communication module, power module, local backup storage and corresponding sensors.

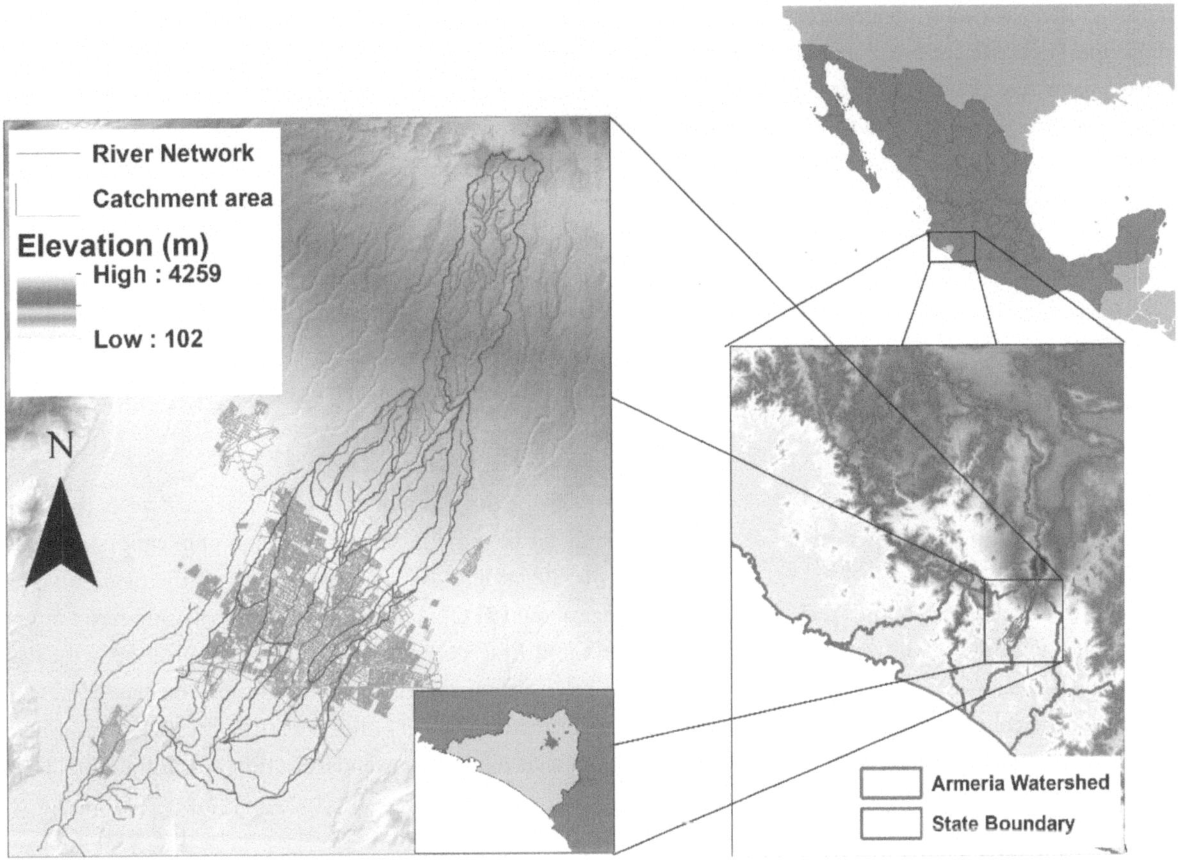

Figure 1 | Study area of Colima-Villa de Álvarez, Mexico. Left panel shows the city of Colima, the catchment area and its elevation and Colima river network.

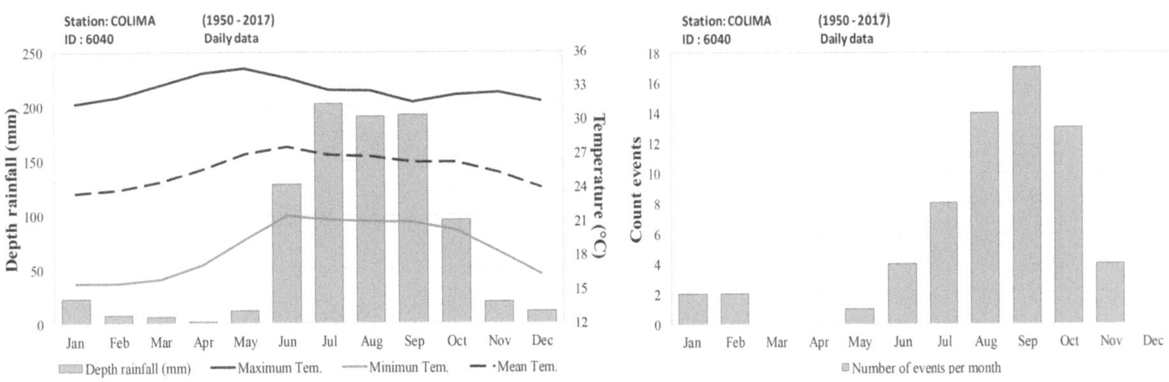

Figure 2 | Distribution of rainfall and temperature (left panel) and storm events (right panel) from 1950 to 2017.

We worked with different stakeholders during the site selection process, the most significant being the local water authority (CONAGUA-Colima) in addition to the public schools, universities and other public institutions. Several meetings with the above stakeholders were held to establish the most suitable and secure locations for the sensors. These collaborations provided valuable information on the suitability and security of the final selected locations. For example, public schools and universities across Colima city provided their support to deploy the weather stations on their building roofs where the weather station is exposed, safe and easy to maintain. For the water level nodes, non-contact ultrasonic sensors were placed under bridges with data loggers and solar panels placed on towers of 6 m high. Figure 3 shows the locations of the selected nodes along with the signal strength of the 3G/4G network in the study area.

Hardware and network development

The network is based on four building blocks (Figure 4):

- The weather station nodes use Ethernet technology for their communication with the server; the nodes are also equipped with local storage.
- Water level nodes (hereafter RiverCore) encompassing ultrasonic water level sensors (Toughsonic). RiverCore communicates with the server using 3G cellular communication technology, and it is also equipped with a local data storage.
- The Drifters which are floater designed to map water velocity for specific events, have a GPS and local storage; they use 900 MHz Long-Range RF Module (XBEE XSC PRO) technology to communicate with RiverDrones.
- RiverDrone is a drone used to locate the drifters along the rivers. RiverDrone uses 900 MHz Long-Range technology to communicate with the Drifters and 3G to connect to the central server.

Each of the components has its respective embedded system based on the energy and communication requirements. Regarding the software design, all the devices were programmed in C ++ for microcontrollers using serial interrupts and time control tasks. All communication with the central server uses the IoT publish-subscribe-based messaging protocol (MQTT) that includes messages in JSON format that allows the standardisation of information. This

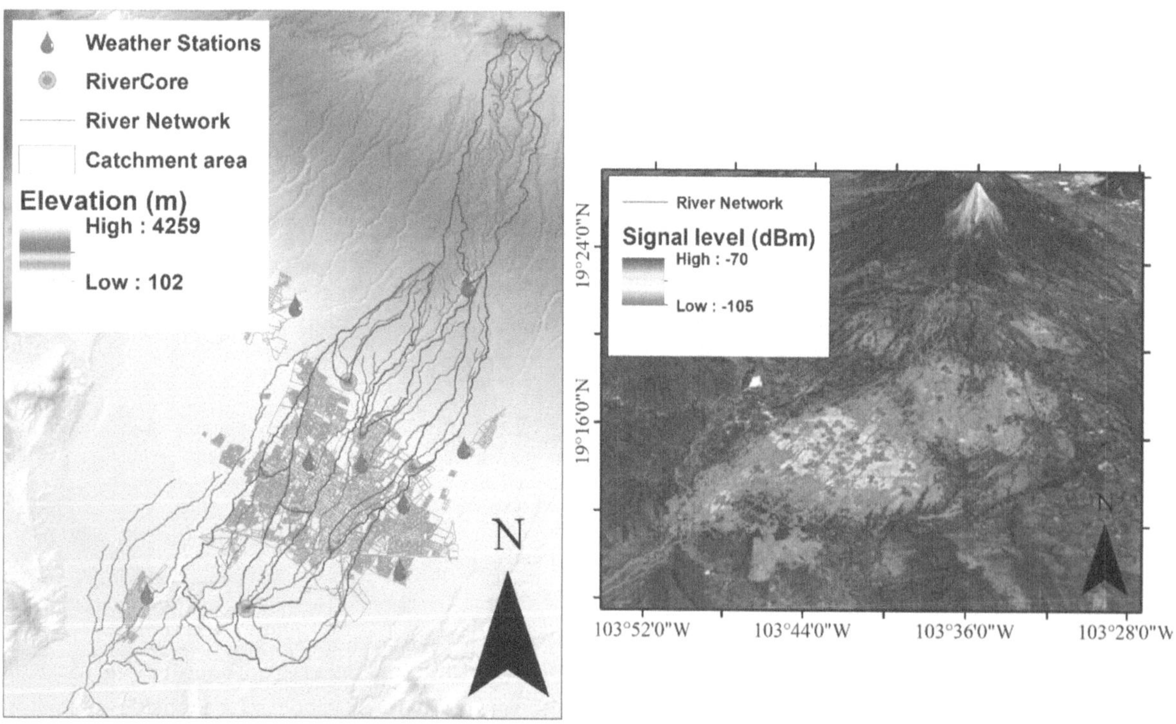

Figure 3 | Location of the fixed nodes in Colima-Villa de Álvarez, Mexico (left). The right panel map shows the 3G/4G signal strength in the study area.

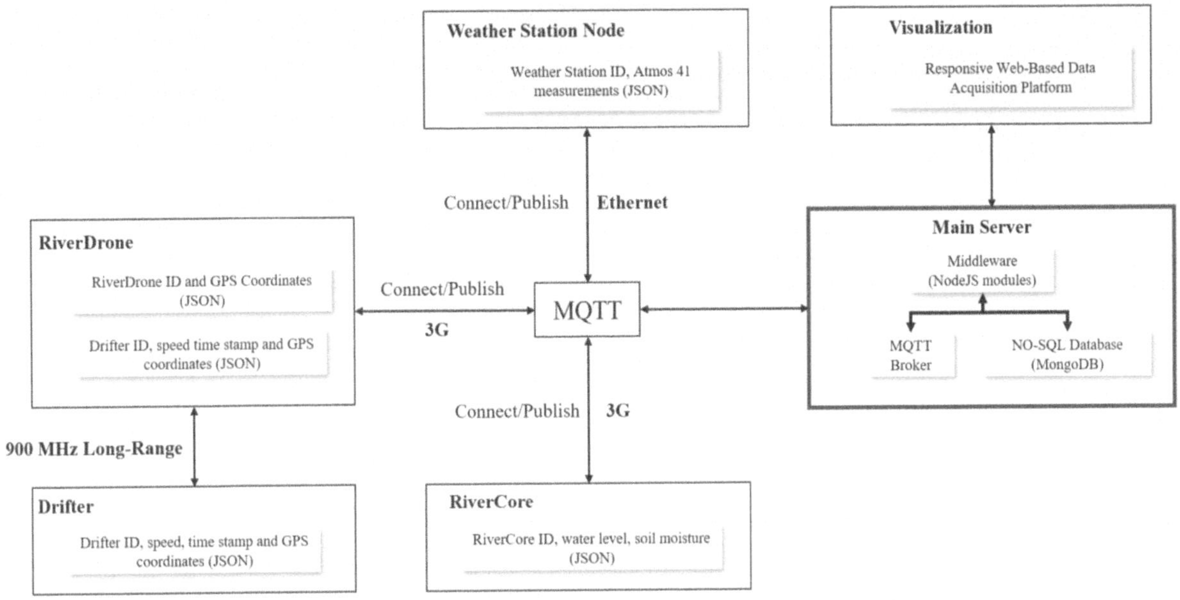

Figure 4 | General diagram of the implemented network.

allows the website to display the data in near-real-time and store the data in a database.

Static nodes

Weather station nodes. The weather station node retrieves information from the commercial multisensory Atmos 41, which includes 12 weather sensors (solar radiation, precipitation, vapour pressure, relative humidity, air temperature, humidity sensor temperature, barometric pressure, horizontal wind speed, wind gust, wind direction, tilt, lighting and lighting average distance). Atmos 41 is a three-wire interface following the SDI-12 protocol for communicating sensor measurements (Table 1). We designed a shield/daughterboard for Arduino MEGA with an Ethernet communication (ENC28J60), a local storage unit with a capacity of 32G allowing data backup during potential breaks of communication with the central server and a real-time clock to timestamp mark.

RiverCore nodes. The RiverCore node is composed of an ultrasonic water level (Toughsonic) and soil moisture (Teros10) used for water level monitoring, high water level warnings and for hydrological modelling a shield/daughterboard with own design using Arduino DUE with a 3G cellular communication (SIM5320A), RS-485 transceiver and a data logger; details are shown in Table 2.

Table 1 | Weather station node specifications

Devices	Manufacturer name	Description
Ethernet breakout	Microchip	ENC28J60
MicroSD card breakout board	Adafruit	microSDHC
Case	Steren	IP65
Weather station	METER Group	ATMOS41
PCB	Tres Ríos	Own design
Microcontroller	Microchip	ATMEGA2560
RTC breakout	Adafruit	DS1302

Dynamic nodes

The fixed static hydrometeorological nodes are supported by dynamic sensors (*Drifters and RiverDrones*) that can collect data during specific events to understand how the catchment behaves during floods in terms of tributary relative timing and interaction (Pattison *et al.* 2014). Drifters and RiverDrone are both designed to work together, where RiverDrone is used to locate the Drifters.

Table 2 | *RiverCore* specifications

Devices	Manufacturer name	Description
3G Radio	SIMCom	SIM5320A
MicroSD card breakout board	Adafruit	microSDHC
Case	ABB	IP65
Battery	LTH	12v 80Ah
Water level sensor	Senix	ToughSonic
Soil moisture sensor	METER Group	TEROS 10
Antenna	Taoglass	Multiband
PCB	Tres Ríos	Own design
Microcontroller	Microchip	AT91SAM3X8E

Drifters are equipped with a shield/daughterboard with own design using Arduino NANO with GPS (4 clicks), 900 MHz Long-Range RF Module (XBEE XSC PRO) to communicate with RiverDrone and a data logger. The objective of Drifters is to map the river flow velocity during specific river conditions and measure other parameters such as water temperature (Table 3).

RiverDrone has a shield/daughterboard with own design using Arduino MEGA with a 3G cellular communication (SIM5320A) to connect to the central server, and a 900 MHz Long-Range RF Module (XBEE XSC PRO) to communicate with the Drifters. More information and specifications are summarised in Table 4. The objective of RiverDrone is to locate drifters by sending a frame in broadcast so that each drifter responds to the request. The drifter's response message will contain GPS location, speed and

Table 3 | *Drifter* specifications

Devices	Manufacturer name	Description
GPS	MikroE	4 Click
MicroSD card breakout board	Adafruit	microSDHC
Case	Pelican	1010
Battery	Generic	lipo
900 MHz Long-Range RF Module	DIGI	Xbee XSC PRO
Antenna	Taoglass	Flexible
PCB	Tres Ríos	Own design
Microcontroller	Microchip	ATmega328

Table 4 | *RiverDrone* specifications

Devices	Manufacturer name	Description
3G Radio	SIMCom	SIM5320A
Case	Steren	IP65
Battery	Generic	lipo
900 MHz Long-Range RF Module	DIGI	Xbee XSC PRO
Antenna	Taoglass	Multiband
PCB	Tres Ríos	Own design
Microcontroller	Microchip	AT91SAM3X8E

signal strength of the 900 MHz radio. The signal strength is used to estimate the distance between drifter and RiverDrone. After this, the RiverDrone will search and communicate with the drifter to recover it later.

Web application development

Water level, soil moisture and weather data are transmitted through 3G/4G network and Wi-Fi (for the weather stations on public buildings) to the server and are uploaded and displayed on a web application (screenshot in Figure 5; http://tairda.siteldi.mx/ewin). Data are transmitted every 3 min, but the rate can be changed for a coarser temporal resolution from the server. From the web application dashboard, information of the last 2 h can also be viewed in charts. It is also possible to download historical data across different sensor nodes as csv or text files. The web application runs on a 64-bit Debian operating system. Both the front end and the back end were developed in NodeJS. The MongoDB database was used for its ease of use in IoT solutions, and the communication protocol used between the server and the nodes is MQTT.

TESTS AND RESULTS

A part of the web application platform is open to the public where real-time water level and weather parameters can be visualised. Another part of the application allows access and download of historical information using

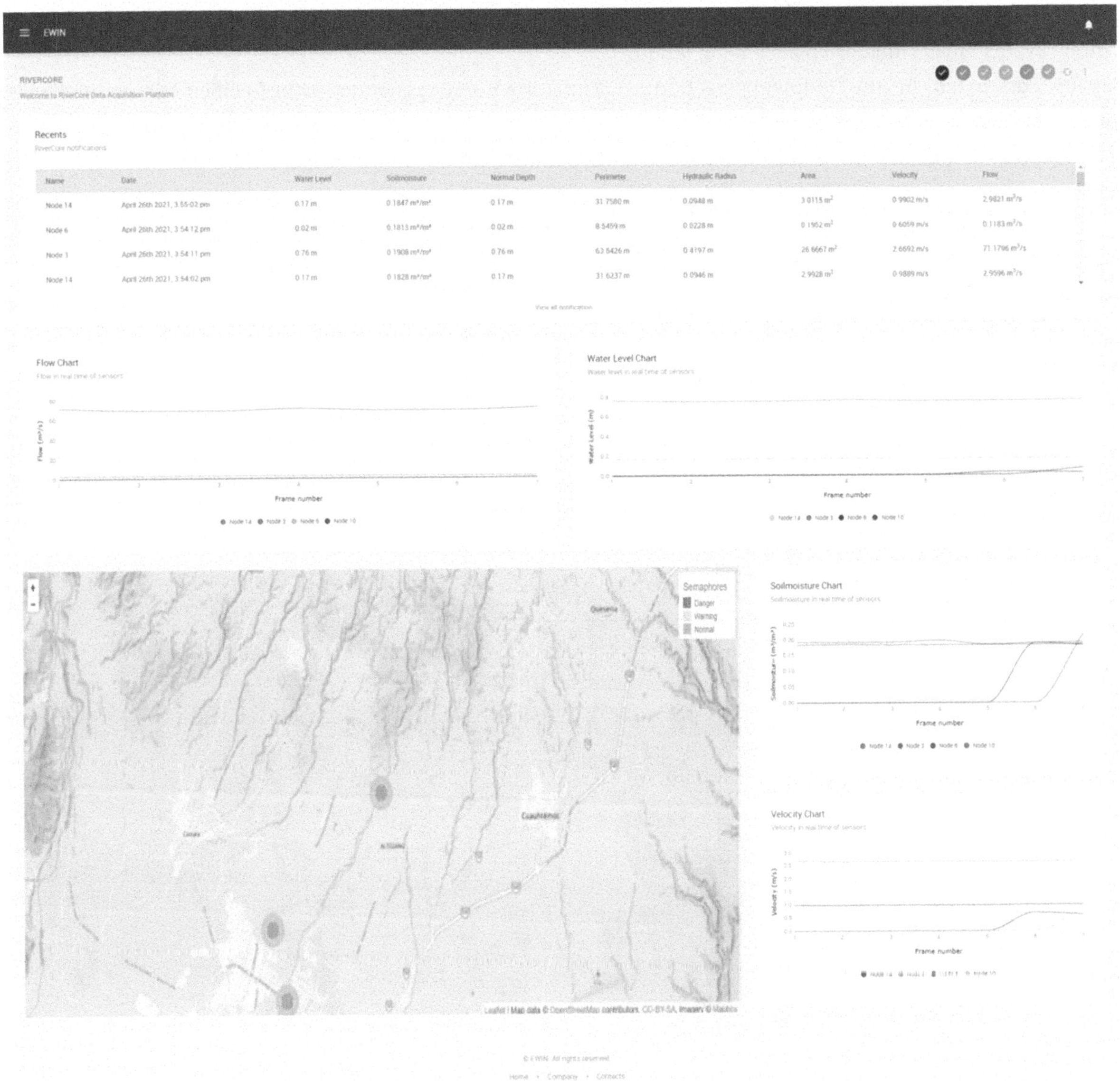

Figure 5 | Screen shot of the web application dashboard.

specific access key. At the moment, access key is generally given to specific collaborators such as experts, policy-makers and stakeholders mentioned above. Using historical data, analyses of the hydrometeorological events can be performed.

To demonstrate the network performance and its ability to collect and transfer the data in real-time during flood events, we used three different tropical storm events that occurred during the wet season of 2019: tropical cyclone Lorena, Narda and Priscilla (Table 5). These events generated significant rainfall and damage such as collapse of bridges, the destruction of roads and the flooding of critical infrastructure (e.g. schools, hospitals and

Table 5 | Tropical cyclone storms that affected Colima-Villa de Álvarez metropolitan area in 2019

Storm name	Date
Lorena	18–22 September 2019
Narda	27 September–1 October 2019
Priscila	19–20 October 2019

government offices). Our network monitoring system captured these three events' hydrometeorological data, and their analyses are discussed in the following section. During the event periods, the data were collected and sent to the server every 5 min but for practical reasons, the data were aggregated to 10 min in the following analyses.

Figure 6 shows the tracks and storm locations of the three tropical cyclones; only the tropical storm Narda entered the state of Colima as a hurricane. Though Lorena made landfall in the neighbouring state of Jalisco, its hydrometeorological effects were significant throughout the state of Colima. Priscilla was generated off the coast of Colima, and it quickly degraded and entered the state territory as a tropical depression.

Lorena storm

The storm Lorena generated in the Pacific basin and its effects on Colima-Villa de Álvarez metropolitan area occurred on 18 September 2019 at 4:00 pm Figure 7(a) shows the hyetograph recorded by the weather station nodes across the network. The maximum accumulated precipitation during the event was 157.82 mm recorded at node 6.

During the Lorena event, six weather station nodes out of eight were continuously in operation and did not lose the connection during the peak of the event. Figures 8(b)–8(d) show the spatial distribution of the precipitation across the area interpolated using kriging. The spatial interpolation showed that the network is well distributed spatially and allowed the capture of rainfall differences across the

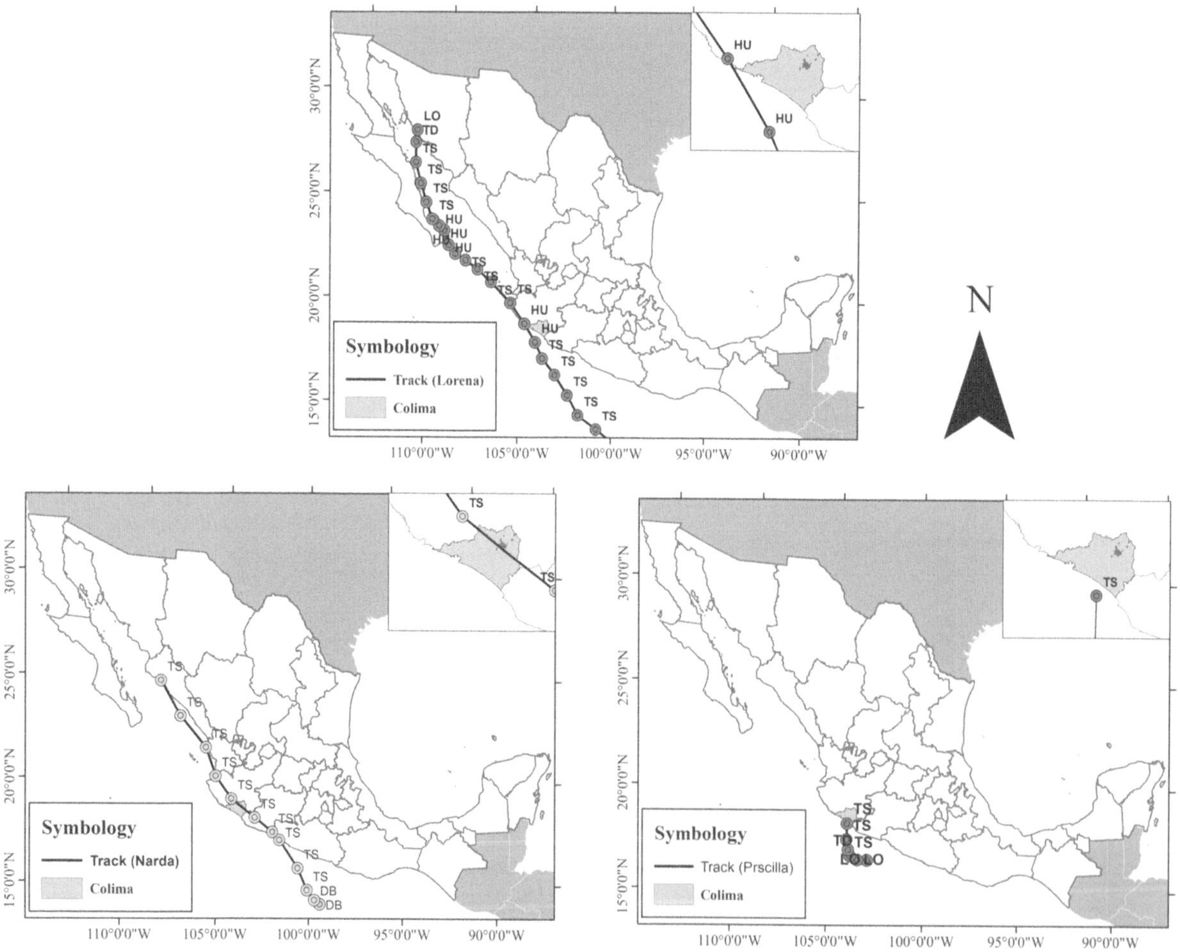

Figure 6 | Tropical cyclone tracks of Lorena (upper map) Narda (left) and Priscilla (right).

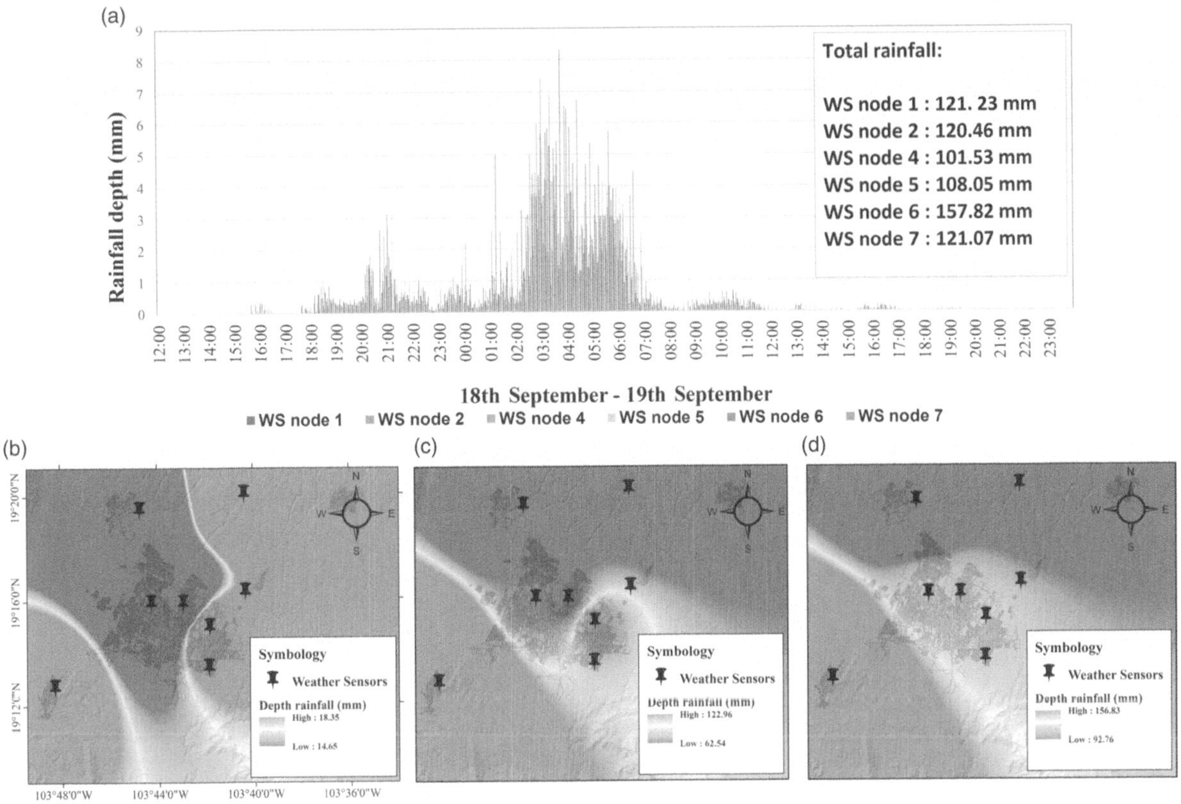

Figure 7 | Upper panel: (a) hyetograph of Lorena Tropical Cyclone. Bottom panel: (b) cumulative rainfall between 15:00 and 22.00 on 18 September, (c) cumulative rainfall between 15:00 on 18 September and 05:00 on 19 September and (d) total rainfall depth in mm.

metropolitan area. The precipitation's pattern coincides with the trajectory of Lorena, which was parallel to the coast of Colima. The maximum rainfall in the state of Colima was recorded at the Tecomán 343 mm station, which is located near the coast. This information was provided by the local water agency's network of stations, which have records of daily accumulated rainfall.

Narda storm

On September 29 at 10:00 am, the National Meteorological Service (SMN) reported the formation of the Potential Tropical Cyclone '16-E', south-southwest of Guerrero, which would travel parallel to the coasts of the Mexican Pacific. On the same day at 22:00, the SMN reported that from Potential Tropical Cyclone '16-E', tropical Storm 'Narda' had formed off the coast Guerrero and that it maintained its displacement parallel to the shores of the Mexican Pacific, causing heavy rains as it passes.

The effects of Narda started to be felt in Colima area on September 29 at 2:10 pm; Figure 8(a) shows the precipitation height hyetographs recorded by six of the eight meteorological stations. The effects of Narda in the city of Colima were felt from 14:10 from September 29 to September 30 at 10:00 am.

The spatial distribution of accumulated precipitation is shown in Figure 8 (bottom panels). Higher precipitation is recorded in the southern part of the city. A critical point detected was that 7 h after the storm started, flooding problems began to appear in the city. The storm caused significant damage to infrastructure such as flooding and damage to roads.

Priscilla storm

Tropical cyclone Priscilla was the most destructive event of the 2019 wet season. It triggered many emergencies for the population and the National Water Commission. The storm was formed on Sunday, 20 October, off the coast of western

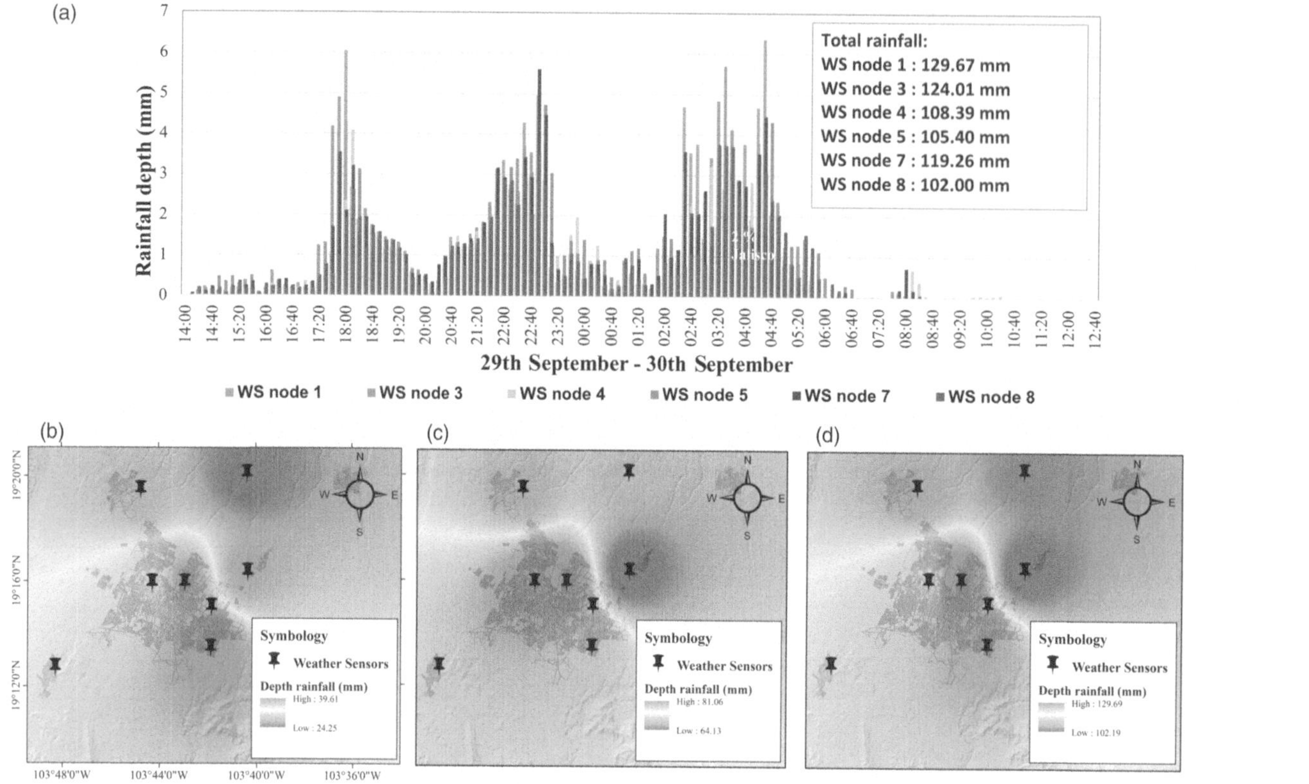

Figure 8 | Upper panel: (a) hyetograph of Narda Tropical Cyclone. Bottom panel: (b) cumulative rainfall in the first 7 h (29 September), (c) cumulative rainfall in the early 12 h (29–30 September) and (d) total rainfall (29–30 September).

Mexico. Due to its proximity to the coastal populations of the state of Colima, the event was closely monitored. However, Priscilla ceased to be a tropical storm as it weakened after making landfall. The government cancelled a tropical storm warning for the state of Colima.

The analysis of the precipitation records generated, as a result of this event, allowed to identify increases in the water level of urban rivers appeared since a day before (19 October) a mesoscale phenomenon that registers up to 40 mm the Colima-Villa Álvarez metropolitan area (Figure 9(a)). During the Priscilla pass, the maximum precipitation recorded across the network was 21 mm (20 October). Figure 9 (bottom panel) shows cumulative rainfall on 18 October and 19 October.

Hydrological modelling

The preliminary hydrological model for the Colima river system was tested as well. We simulate the rainfall–runoff process for extreme hydrometeorological events. The model was built in HEC-HMS in version 4.4.1. The rainfall–runoff process simulation was generated on an event scale for the period between 18 September 2019 and 20 September 2019 (Figure 10). The infiltration method used was the curve number, while the transformation model was the synthetic unit hydrograph developed by the Soil Conservation Service of the US Department of Agriculture. Based on the water level records at Node 14, located in the upper part of the catchment, the curve number (N) and the initial abstraction (Ia) were calibrated. These parameters are related to the loss by infiltration. Flow rates were obtained from the levels recorded with the sonic sensor and a rating curve (flow-discharge) generated based on a two-dimensional hydraulic model. We used Iber which is an open-source program and the software used to construct the hydraulic model in each node.

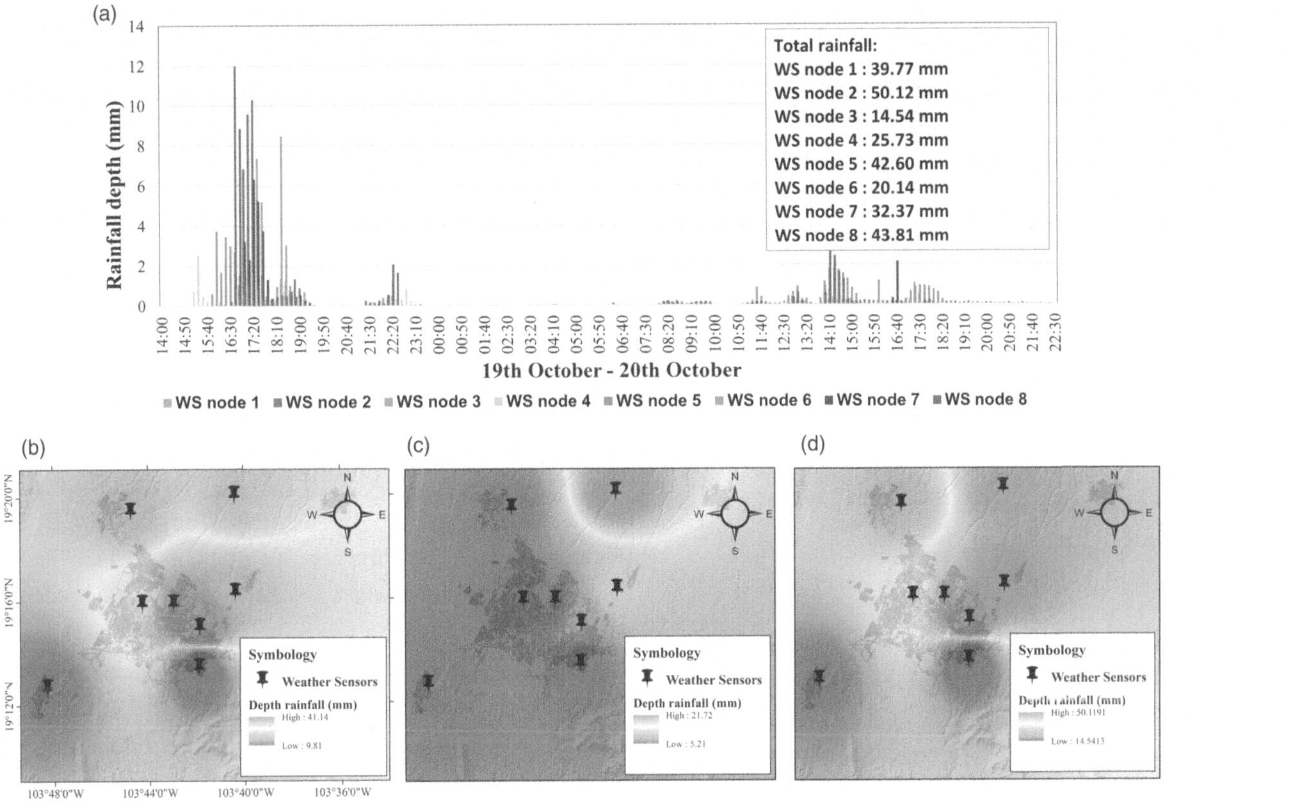

Figure 9 | Upper panel: (a) hyetograph of Priscilla Tropical Cyclone. Bottom panel: (b) cumulative rainfall on 18 October, (c) cumulative rainfall on 19 October and (d) total rainfall (18–19 October).

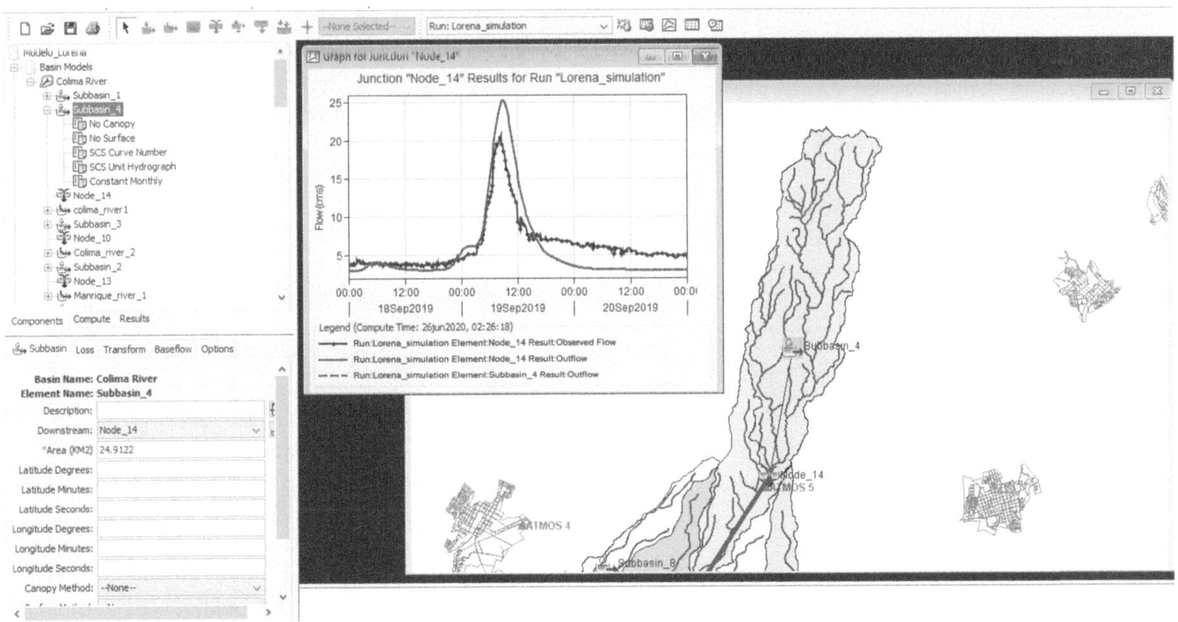

Figure 10 | Simulation of the rainfall–runoff process of the Colima river system (Node 14).

Drifters and *RiverDrone* experiments

The Drifter experiments were performed on 28 February 2020, in Coquimatlán, Colima, Mexico, between approximately 9 am and 12 pm (Figure 11). The site was selected for its suitability, easy accessibility and safety of the team. The following steps were performed to conduct the experiment:

- define the suitable river stretch for releasing and collecting the drifters,
- release the three drifters at the same time,
- follow and captured the drifter at the exit point, and extract the microSD card, and
- check the web application for data upload.

The Drifter path and its associated data are obtained and visualised on the web application's map as illustrated in Figure 12.

Table 6 summarises the data collected and related to the water velocity and the distance of Drifter journey.

RiverDrone experiment was performed on 25 January 2020, in Manzanillo, Colima, Mexico. The objective of the experiment was to examine the ability of the RiverDrone to search and find the drifters. The experiment was conducted in a first place in an open sea, and Drifter-Sniffer was tested to obtain real-time coordinates of the drifter location for their respective collection:

- The drifter released in the sea about 2 km from the Drifter-Sniffer.
- The location of the Drifters displayed on the web application is when the Drifter-Sniffer detects a drifter using the 3G network.
- Drifter are manually collected and data are uploaded to the server.

Figure 13 shows the drifter path along with its locations and velocities.

Dissemination of results

To disseminate the results and emphasise the importance of such systems for water monitoring in an urban environment, three workshops were conducted throughout the project life and a final workshop was organised in January 2020 (Figure 14). In this workshop, participants include local authorities, decision-makers, policymakers, stakeholders, academics and students to discuss the results and ways of supporting the project and improving the system towards an operational system for flood monitoring in Colima. The official presenting video of the project can be found here: https://www.youtube.com/watch?v=7MHIVOkaf-c.

CONCLUSIONS

This paper presents development phases and experiments of an IoT-based network of sensors for fluvial water level monitoring and food warning system for the urban area of Colima-Villa de Álvarez in Mexico. The network is composed of (1) eight fixed nodes (RiverCore) collecting fluvial water level across the main rivers in Colima city, (2) eight

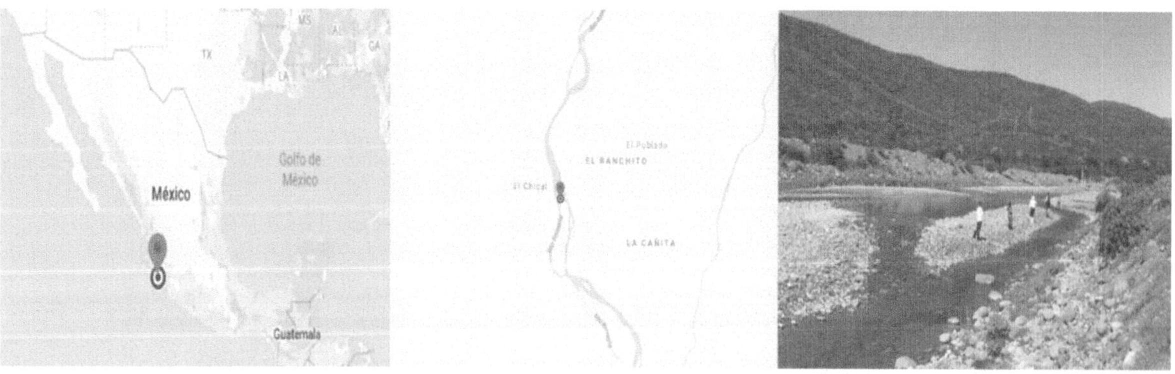

Figure 11 | Location of the drifter experiment.

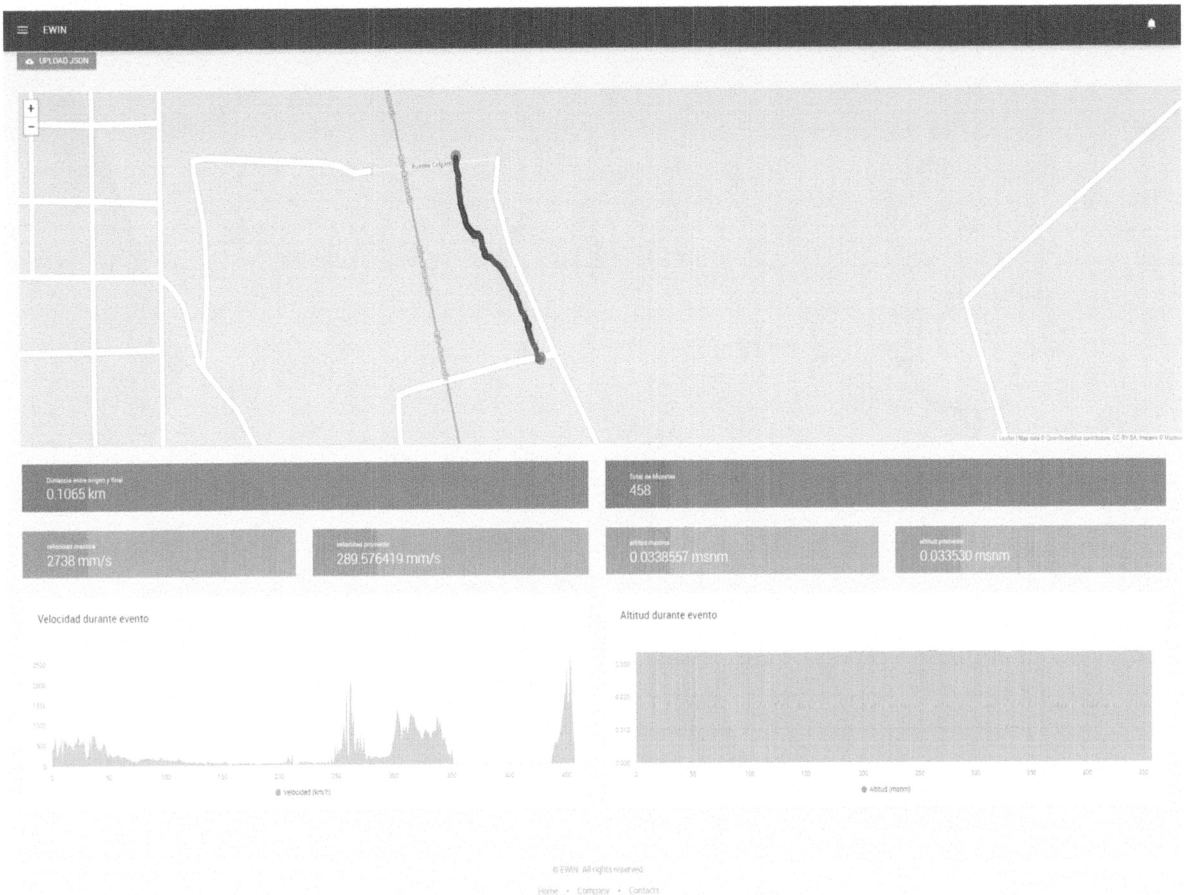

Figure 12 | Drifter tracking as seen on the web application.

Table 6 | Data retrieved from the experiment

Drifter	
Distance from the first to the last check point.	0.1065 km
Average speed	0.2895 m/s

weather stations distributed across the catchment areas collecting at least seven weather parameters and (3) dynamic supporting sensors (i.e. Drifters and RiverDrones) used to collect data during specific events. Soil moisture data are also collected using sensors connected to several nodes across the catchment.

Experiments of the network system conducted during the three main tropical storm events that occurred during the 2019 tropical cyclones season (storm Lorena 18–22 September, storm Narda September 27–October 1 and Priscila 19–20 October) demonstrated the robustness of the network to sense river levels and meteorological conditions across the city in real-time providing relevant information for flood monitoring and waning. The continuous data collection is also critical for developing hydrological and flood inundation modelling and mapping in this area where no datasets were available previously.

The work conducted throughout the project also demonstrated the importance of the multidisciplinary approach to problem-solving of flooding issues. This multidisciplinary and participatory work not only improved the outcomes but also optimised the resources used in the process. Many countries in Latin America have introduced early warning systems to minimise their flood impacts. However, most systems are not always effective due to poor robustness of the sensor's network and the high costs. Our IoT-based wireless sensor network is relatively low cost and can be replicated in areas where large budgets for flood prevention measures are not available.

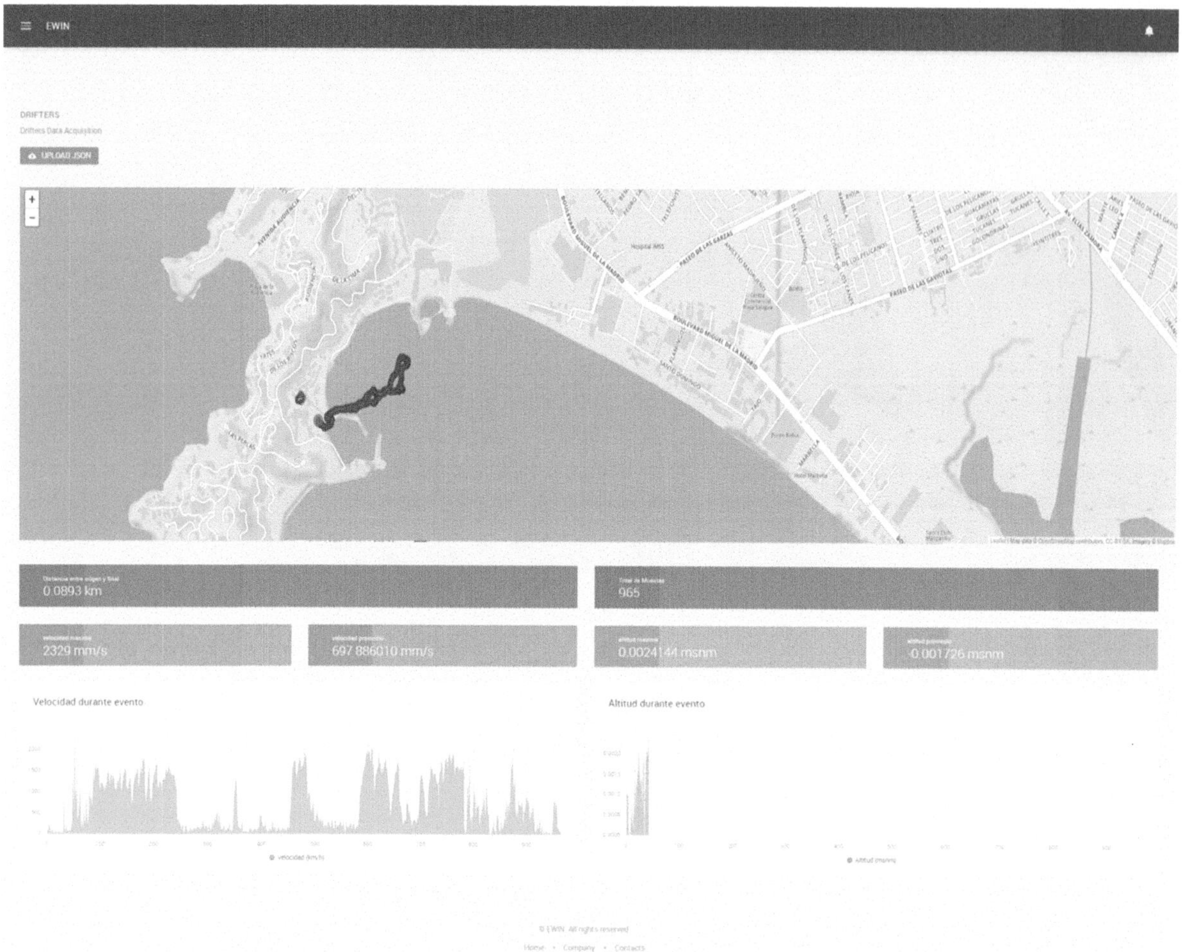

Figure 13 | Analysis MicroSD on web platform.

Figure 14 | Photos from the project January 2020 workshop.

The system proposed in this paper can provide information to the relevant flood management and emergency services and authorities to reduce risk, loss of life, damage to property and the environment; however, the system alone is not the solution to flooding problems; civil society, volunteers, organised voluntary work organisations and community organisations must be involved, in collaboration with public institutions. Women's participation is essential for effective disaster risk management. School children and youth are agents of change and must be given the space and ways to contribute to disaster risk reduction.

In terms of improvements and limitations, it is necessary to keep maintaining the system regularly and correct data issues and bugs when necessary. Additional new functionalities in the web application are needed such as easily choosing time intervals for sampling, appending the quality of the GPS signal to the datalogger. Sensor security and vandalism has been an issue although, in our network, we used higher towers and protected the sensors under bridges with a metal box. For the drifter sensors, overall, more tests are needed and it is convenient to add new functions to the current system since at the moment it does not show the distance or the intensity of the signal between the drifter and the sniffer.

ACKNOWLEDGEMENTS

This research was funded by the Engineering and Physical Sciences Research Council (EPSRC), EP/P029221/1. We thank the two anonymous reviewers for their suggestions and comments, which were very helpful in improving the paper. We thank the local water commission for providing help and support with the deployment of the sensor systems. We also thank the municipalities of Colima and Villa de Álvarez for granting permits for the installation of the sensors and support for their care.

DATA AVAILABILITY STATEMENT

All relevant data are included in the paper or its Supplementary Information.

REFERENCES

Alfieri, L., Berenguer, M., Knechtl, V., Liechti, K., Sempere-Torres, D. & Zappa, M. 2015 Flash flood forecasting based on rainfall thresholds. In: *Handbook of Hydrometeorological Ensemble Forecasting*. (Q. Duan, F. Pappenberger, J. Thielen, A. Wood, H. Cloke & J. Schaake, eds). Springer, Berlin, Heidelberg, https://doi.org/10.1007/978-3-642-40457-3_49-1.

Artiola, J. F. 2004 Monitoring surface waters, Oceans.

Bachmann, D., Eilander, D., De Leeuw, A., De Bruijn, K., Diermanse, F., Weerts, A. & Beckers, J. 2016 Prototypes of risk-based flood forecasting systems in the Netherlands and Italy. In: *3rd European Conference on Flood Risk Management (FLOODrisk 2016), E3S Web Conference*.

Camara, M., Jamil, N. R., Abdullah, A. F. B., Hashim, R. b. & Aliyu, A. G. 2020 Economic and efficiency based optimisation of water quality monitoring network for land use impact assessment. *Science of the Total Environment* 139800. doi:10.1016/j.scitotenv.2020.139800.

Cheval, S., Micu, D., Dumitrescu, A., Irimescu, A., Frighenciu, M., Ioja, C., Tudose, N. C., Davidescu, Ş. & Antonescu, B. 2020 Meteorological and ancillary data resources for climate research in urban areas. *Climate* **8** (3). doi:10.3390/cli8030037.

Constantino, T., Roberto, M., Dávila, I. & Hilda, R. 2011 Una aproximación a la vulnerabilidad y la resiliencia ante eventos hidrometeorológicos extremos en México. *Política y Cultura* **36**, 15–44. Available from: http://www.scielo.org.mx/scielo.php?script=sci_arttext&pid=S0188-77422011000200002&lng=es&tlng=es.

Davidsson, Å. 2020 Disasters as an opportunity for improved environmental conditions. *International Journal of Disaster Risk Reduction* **48**, 101590. doi:10.1016/j.ijdrr.2020.101590.

De, I., Security, H. & Find, D. 2012 Bridging the gap between New Mexico flooding events and flood coverage, 1–3.

Deng, J., Yin, H., Kong, F., Chen, J., Dronova, I. & Pu, Y. 2020 Determination of runoff response to variation in overland flow area by flow routes using UAV imagery. *Journal of Environmental Management* **265**. doi:10.1016/j.jenvman.2019.109868.

Echeverribar, I., Morales-Hernández, M., Brufau, P. & García-Navarro, P. 2019 2D numerical simulation of unsteady flows for large scale floods prediction in real time. *Advances in Water Resources* **134**, 103444. doi:10.1016/j.advwatres.2019.103444.

EM-DAT 2020 *EMDAT: The Emergency Events Database*. Available from: www.emdata.be.

Enenkel, M., Brown, M. E., Vogt, J. V., McCarty, J., Reid Bell, A., Guha-Sapir, D., Dorigo, W., Vasilaky, K., Svoboda, M., Bonifacio, R., Anderson, M., Funk, C., Osgood, D., Hain, C. & Vinck, P. 2020 Why predict climate hazards if we need to understand impacts? Putting humans back into the drought equation. *Climatic Change* **162**, 1161–1176. doi:10.1007/s10584-020-02878-0.

Ishida, T. 2017 Digital City, Smart City and Beyond.

Khalid, A. & Ferreira, C. 2020 Advancing real-time flood prediction in large estuaries: iFLOOD a fully coupled surge-wave automated web-based guidance system. *Environmental Modelling & Software* 104748. doi:10.1016/j.envsoft.2020.104748.

Khouakhi, A., Pattison, I., López-de la Cruz, J., Martinez-Diaz, T., Mendoza-Cano, O. & Martínez, M. 2019 Tropical cyclone – induced heavy rainfall and flow in Colima, Western Mexico. *International Journal of Climatology*. doi:10.1002/joc.6393.

Kundzewicz, Z. W., Kanae, S., Seneviratne, S. I., Handmer, J., Nicholls, N., Peduzzi, P., Mechler, R., Bouwer, L. M., Arnell, N., Mach, K., Muir-Wood, R., Brakenridge, G. R., Kron, W., Benito, G., Honda, Y., Takahashi, K. & Sherstyukov, B. 2014 Le risque d'inondation et les perspectives de changement climatique mondial et régional. *Hydrological Sciences Journal* **59** (1), 1–28. doi:10.1080/02626667.2013.857411.

Lee, E. A. 2008 Cyber physical systems: design challenges. In: *11th IEEE International Symposium on Object and Component-Oriented Real-Time Distributed Computing (ISORC)*, edited by IEEE, New York, pp. 363–369.

Lindenschmidt, K. E., Rokaya, P., Das, A., Li, Z. & Richard, D. 2019 A novel stochastic modelling approach for operational real-time ice-jam flood forecasting. *Journal of Hydrology* **575**, 381–394. doi:10.1016/j.jhydrol.2019.05.048.

López-Marrero, T. & Tschakert, P. 2011 From theory to practice: building more resilient communities in flood-prone areas. *Environment and Urbanization* **23** (1), 229–249. doi:10.1177/0956247810396055.

Lund, N. S. V., Falk, A. K. V., Borup, M., Madsen, H. & Steen Mikkelsen, P. 2018 Model predictive control of urban drainage systems: a review and perspective towards smart real-time water management. *Critical Reviews in Environmental Science and Technology* **48**, 279–339.

Makropoulos, C. & Sávic, D. A. 2019 Urban hydroinformatics: past, present and future. *Water* **11** (10). doi:10.3390/w11101959.

Matthews, J. H., Wickel, B. A. J. & Freeman, S. 2011 Converging currents in climate-relevant conservation: water, infrastructure, and institutions. *PLoS Biology* **9** (9). doi:10.1371/journal.pbio.1001159.

Mendoza-Cano, O., López-de la Cruz, J., Ian, P., Martinez-Preciado, M. A., Manuel, U. R. J., Edwards, R. M., Ivan, R.-L. C., Pedro, R. A. & Velazco-Cruz Jorge, A. 2019 Disaster risk resilience in Colima-Villa de Alvarez, Mexico: application of the resilience index to flash flooding events. *International Journal of Environmental Research and Public Health* **16** (12), 2128. doi:10.3390/ijerph16122128.

Miller, J. D. & Hutchins, M. 2017 The impacts of urbanisation and climate change on urban flooding and urban water quality: a review of the evidence concerning the United Kingdom. *Journal of Hydrology* **12**, 345–362. doi:10.1016/j.ejrh.2017.06.006.

Nguyen, K. V. & James, H. 2018 Measuring household resilience to floods: a case study in the Vietnamese Mekong river delta. *Ecology and Society* **18** (3). doi:10.5751/ES-05427-180313.

Noymanee, J. & Theeramunkong, T. 2019 Flood forecasting with machine learning technique on hydrological modeling. *Procedia Computer Science* **156**, 377–386.

Olya, H. G. T. & Alipour, H. 2015 Risk assessment of precipitation and the tourism climate index. *Tourism Management* **50**, 73–80. doi:10.1016/j.tourman.2015.01.010.

Pattison, I., Lane, S. N., Hardy, R. J. & Reaney, S. M. 2014 The role of tributary relative timing and sequencing in controlling large floods. *Water Resour. Res.* **50**, 5444–5458, doi:10.1002/2013WR014067.

Plümper, T., Quiroz Flores, A. & Neumayer, E. 2017 The double-edged sword of learning from disasters: mortality in the Tohoku tsunami. *Global Environmental Change* **44**, 49–56. doi:10.1016/j.gloenvcha.2017.03.002.

Ritter, J., Berenguer, M., Corral, C., Park, S. & Sempere-Torres, D. 2020 ReAFFIRM: real-time assessment of flash flood impacts – a regional high-resolution method. *Environment International* **136**, 105375. doi:10.1016/j.envint.2019.105375.

Ryan, B. 2018 Establishing information seeking pathways in slow and flash floods. *International Journal of Disaster Risk Reduction* **31**, 9–19. doi:10.1016/j.ijdrr.2018.04.005.

Tanoue, M., Hirabayashi, Y. & Ikeuchi, H. 2016 Global-scale river flood vulnerability in the last 50 years. *Scientific Reports* **6**, 1–9. doi:10.1038/srep36021.

Uteuov, A., Kalyuzhnaya, A. & Boukhanovsky, A. 2019 The cities weather forecasting by crowdsourced atmospheric data. *Procedia Computer Science* **156**, 347–356.

Zhang, X., Liu, P., Xu, C. Y., Gong, Y., Cheng, L. & He, S. 2019 Real-time reservoir flood control operation for cascade reservoirs using a two-stage flood risk analysis method. *Journal of Hydrology* **577**, 123954. doi:10.1016/j.jhydrol.2019.123954.

First received 10 August 2020; accepted in revised form 8 March 2021. Available online 26 April 2021

© 2021 The Authors

Capturing high-resolution water demand data in commercial buildings

Peter Melville-Shreeve, Sarah Cotterill and David Butler

ABSTRACT

Water demand measurements have historically been conducted manually, from meter readings less than once per month. Leading water service providers have begun to deploy smart meters to collect high-resolution data. A low-cost flush counter was developed and connected to a real-time monitoring platform for 119 ultra-low flush toilets in 7 buildings on a university campus to explore how building users influence water demand. Toilet use followed a typical weekly pattern in which weekday use was 92% ± 4 higher than weekend use. Toilet demand was higher during term time and showed a strong, positive relationship with the number of building occupants. Mixed-use buildings tended to have greater variation in toilet use between term time and holidays than office-use buildings. The findings suggest that the flush sensor methodology is a reliable method for further consideration. Supplementary data from the study's datasets will enable practitioners to use captured data for (i) forecast models to inform water resource plans; (ii) alarm systems to automate maintenance scheduling; (iii) dynamic cleaning schedules; (iv) monitoring of building usage rates; (v) design of smart rainwater harvesting to meet demand from real-time data; and (vi) exploring dynamic water pricing models, to incentivise optimal on-site water storage strategies.

Key words | low-cost water sensors, smart water meters, ultra-low flush toilet, water demand management

Peter Melville-Shreeve† (corresponding author)
David Butler
Centre for Water Systems, College of Engineering, Mathematics and Physical Sciences,
University of Exeter,
Exeter EX4 4QF,
UK
E-mail: *p.melville-shreeve2@exeter.ac.uk*

Sarah Cotterill†
School of Civil Engineering,
University College Dublin,
Belfield,
Dublin 4,
Ireland

†These authors contributed equally to this work.

HIGHLIGHTS

- A novel, low-cost, high-resolution water demand sensing strategy was tested, by deploying flush counters across seven large campus buildings.
- Making such real-time data available could deliver value in improving: water demand forecasts; maintenance strategies; cleaning strategies; building user insights; and optimal water system design.
- Water demand varied and was linked to occupancy metrics.

INTRODUCTION

Early studies on residential and commercial water use were driven by the need to quantify network demand, improve the design of water distribution systems, develop short- and long-term demand forecasts, and establish appropriate pricing structures (Buchberger & Wells 1996). Water demand at commercial premises such as offices and educational facilities has historically been measured using analogue water meters. The majority of these meters are read monthly or quarterly to coincide with the local water service provider's billing cycle (Thames Water 2020a).

This is an Open Access article distributed under the terms of the Creative Commons Attribution Licence (CC BY 4.0), which permits copying, adaptation and redistribution, provided the original work is properly cited (http://creativecommons.org/licenses/by/4.0/).

doi: 10.2166/hydro.2021.103

Where more granular water use data is needed, sub-metering systems are implemented to enable the costs of water to be suitably allocated to internal clients and other on-site users. Even where additional data are captured through sub-metering, the analogue and sparse temporal nature of the data collection prevents supply patterns from being derived. Although monthly usage data is typically sufficient for billing purposes, it has little value in terms of insight into leakage; user practices; seasonality; and responses to extreme events. In an ideal world, water resources managers might hope to access novel tools based on big data solutions (Chen & Han 2016) supported by high-performance computing (Morales-Hernández *et al.* 2020) and receive highly granular, near-real-time forecasts for water demands that could, in turn, help them to optimally manage the operation of water distribution networks upstream of a customer.

Water demand models can be built from statistical information on users, end-uses, frequency, intensity, and duration of use, as well as time of water use (Blokker *et al.* 2010, 2011). These models rely on statistical information of water-using appliances, and those that use them, instead of actual water use measurements (Blokker *et al.* 2010) and can enable a realistic estimation of demand where costs of a conventional metering approach are prohibitive. Simulations, based on a variety of non-residential buildings, have shown that water demand is a function of the way in which water is used within that building type. Water use in office buildings is primarily determined by the flushing of toilets, while water use in hotels is primarily determined by the total use within customers' rooms (e.g. WC, shower, bathroom tap, bathtub, etc.) (Blokker *et al.* 2011). Therefore, for an office building, reliable information on toilet use is of primary concern; while for a hotel, reliable information is required both on the number and occupancy of the rooms and the water-using appliances in the rooms (Blokker *et al.* 2011). Furthermore, users are the key to understanding and predicting water demand in buildings. The number of people present in a non-residential building, such as an office, will vary throughout the day, and across the week (Blokker *et al.* 2011). Pieterse-Quirijns *et al.* (2013) suggest that with a proper estimation of the number of users and appliances within a building, such models can create a realistic diurnal demand pattern. However, while some non-residential buildings, for example, offices, hotels, etc., have a defined number of users at a given time, that can be reliably estimated from the number of employees or the number of guests, other buildings, such as those on a university campus are less static and contain a large number of transient users.

In the past two decades, data-logging, storage, and low-power wide-area network (LPWAN) communications technologies have developed apace, resulting in the wide-scale application of (IoT) technologies (Gurung *et al.* 2014; Cherukutota & Jadhav 2016; Peña Queralta *et al.* 2019). This has seen the overall value, and thus, the prevalence of smart water meter technologies reach an inflection point wherein data add the sufficient value to warrant wide-scale deployments in residential and commercial premises. A study by Cominola *et al.* (2015) explored a range of sensing technologies such as accelerometers, flow meters, ultrasonics, and pressure sensors, noting that typical smart meters rely on pressure or flow measurements. More recently, studies which explore low-cost (~\$1) sensor technologies such as vibration and temperature sensors have successfully illustrated that IoT solutions can be harnessed to capture event data within water networks without the need to rely upon high-quality/high-cost (~\$50) water meters (Thomson *et al.* 2012; Martini *et al.* 2014; Pirow *et al.* 2018).

With the progress in IoT technologies, smart water meter installation programmes have been implemented by many of the UK's Water Service Providers (WSPs). One leading effort was initiated by Thames Water in January 2016 (ThamesWater 2020b). Within 2 years, they achieved significant demand reduction benefits, for example in the identification of customer-side leakage (WWT 2018). By June 2016, Thames Water had installed 240,000 smart meters providing 5.8 million data points per day at an hourly time-step. This helped them to identify 11 million litres of leakage, but perhaps more importantly, is enabling Thames Water to undertake more accurate analysis of water user behaviour within each district metered area.

A typical practice adopted by WSPs in the UK sees their smart water meter networks target hourly data capture (Hackett 2018; Thames Water 2020b). This is the result of an economic trade off wherein WSPs must balance the need for water meters to have adequate battery life (i.e. 5–10 years); an appropriate near-real-time data uplink frequency; while securing a manageable total number of data points; against the desire to have the highest resolution dataset possible. Higher-resolution, 15-min data can be captured from Anglian Water's systems (Hackett 2018). The highest rate of data collection (i.e. sub-minutely) is typically only valued when practitioners aim to implement micro-component analyses. In contrast to Thames Water, Anglian Water's 15-min data quadruples the overall data storage requirements, while presenting little value in the context of current operational approaches. However, they recognize that there could be unseen research benefits from the data, including: fixing bursts and leaks more quickly; increased accuracy of water demand forecasts; increased understanding of water use patterns; and improved customer service interfaces (Anglian Water 2020).

Researchers capturing the highest resolution data have successfully developed methodologies that enable disaggregation of customer usage (Cominola *et al.* 2015; Nguyen *et al.* 2015; Pastor-Jabaloyes *et al.* 2018). Creaco *et al.* (2016) showed how 1-min resolution data from smart water meters can add value to the development of predictive models and called for practitioners to make use of the finest resolution possible when undertaking data acquisition. As cost-effective cloud data storage has become available, capturing higher-resolution water demand data has increased. For example, micro-component analyses have been performed to identify how water meter data can be linked to each household appliance. Wills *et al.* (2018) demonstrated that demand signals for various appliances, such as dishwashers, can be successfully identified from high-resolution pulse data on smart meters.

Recent developments in reliable, low-cost telemetry systems associated with Industry 4.0 technologies have enabled local data collection to be achieved using a wider set of novel sensors. In the future, reliable battery-powered IoT technologies are forecast to provide fully connected smart cities, with data being available from any and all infrastructure within the urban realm (Lom *et al.* 2016; Rathore *et al.* 2016).

Repurposing or upgrading traditional water auditing methods represents a field of opportunity wherein analogue monitoring methods can be upgraded using IoT technologies. Analogue flush counters can be deployed to measure the frequency of use at toilets by measuring the frequency of cistern refills. These simple technologies can help estates managers understand toilet usage and thus explore in-cistern leakage rates when compared with water meter data (Waterwise 2020). They can also be deployed for short-term monitoring studies to demonstrate the economic value of undertaking appliance upgrades, for example, by switching to a waterless urinal or upgrading old toilets with oversized cisterns. Analogue devices such as these are used during water audits, however, they provide coarse resolution data (i.e. daily reads) and although they are low cost to purchase, they are expensive to operate, as the process of taking manual readings is highly laborious. Float switch sensors which include reed switches have been successfully deployed in rainwater harvesting tanks to support the efficient operation of pump controls. Studies in that field have also seen them deployed as event counters in low cost, data-logging pilots for a series of alternative water resource studies (Melville-Shreeve *et al.* 2016). Such flush counters can be easily mounted (and removed) to capture the events associated with a change in water level in a toilet cistern (i.e. flushes). Combining these event counting technologies with contemporary micro-computers and associated bespoke software permits data to be captured from many water closets (WC) and shared in raw format for further analysis.

Digital turbine meters have increased in quality and reduced in price ($\$ < 50$). Again, when combined with a micro-computer, these sensors can enable practitioners and building designers to incorporate a network of water demand monitoring within commercial premises and thus derive accurate demand management datasets. Implementation of such approaches is likely to become a new normal in the years ahead as IoT technologies fall in price and sensors can be incorporated within infrastructure (or appliances themselves) at with very low capital cost. Opportunities to correlate and link high-resolution water demand data (e.g. from washrooms within a campus) to other building data such as floor space and occupancy could pose value to building designers who, for the first time will be able to design new water systems while getting real-time feedback once the site is operational.

At a city-scale, high-resolution data from meters within water networks have been used to train forecasting algorithms, such as artificial neural networks, to predict future water demands. There remains, however, a gap in the literature as component-scale water demand data has hitherto been unavailable and existing forecast studies have been demonstrated at a district metered area scale (Antunes *et al.* 2018).

As part of a wider study providing a living laboratory for water demand management techniques, a research opportunity was identified to develop a novel dataset of high-resolution water demand across a zone of a university campus. In this paper, we investigate data from a high-resolution smart water metering platform installed as part of a large-scale water demand management programme at the University of Exeter, UK. We seek to generate a set of high-resolution water demand profiles that can be made available to researchers to enable further investigations and to support practitioners when designing commercial water supply systems. The research described herein explores the deployment, management, and interpretation of data from a series of washrooms within a campus setting. High-resolution flush counters were installed to 119 WCs in 7 buildings. In addition, digital turbine flow meters were deployed within one building to capture micro-component washroom data. Floor plans, building user data, and local water meter data were obtained from the estates team to enable patterns in water usage to be explored. Outputs from the study could present value to the global water systems design community as design tables and heuristic methods deployed at the design stage are infrequently retro-tested against operational datasets.

To inform campus managers, building designers, and water resource planners, the authors have sought to explore to following research questions associated with the dataset:

1. What is the diurnal pattern of toilet usage on a campus?
2. Are there significant variations in water demand during high and low use periods?
3. Can toilet demand be linked to building occupancy metrics?
4. Who could benefit from high-resolution real-time water demand data?

METHODS

Project development

Data were collected in a large-scale deployment of water metering technologies across seven buildings on the University of Exeter's Streatham Campus (Supplementary Annex A, Figure A1). The monitoring was deployed as part of a package to investigate, ultra-low flush (1.5 litres) toilets (ULFT) installed by Propelair® as part of a Horizon 2020 project (European Commission 2020). The trial was designed to qualitatively and quantitatively investigate the extent to which high-density deployments of the ULFT affect downstream drainage infrastructure. The project involved the replacement of traditional toilets (with 6–9 litre cisterns) with ULFT that use only 1.5 litres per flush. The new ULFT systems are electrically connected via a mains power connection. With mains power supplies available behind each ULFT, unlike typical water metering studies, the team was unconstrained by the need to operate monitoring systems using a battery. Mains powered monitoring and telemetry systems were deployed within each washroom, and high-resolution data were captured.

Early phases of project implementation were conducted between 27 June 2018 and 8 July 2018 to establish the total number of toilets within Streatham Campus. A bespoke survey application was developed within the Device Magic environment to capture washroom data (Device Magic 2020). For this scoping phase, 55 buildings and 668 toilets were surveyed across the campus, with a shortlist of 7 buildings selected for monitoring and ULFT upgrades. Following a further phase of engagement with stakeholders, a series of electrical and plumbing surveys were conducted to shortlist units for replacement.

Water demand data collection

Selected washrooms and toilets were re-surveyed to shortlist units for replacement. Installation of ULFT toilets on Streatham Campus was conducted between 1 November 2018 and 31 March 2019. Monitoring systems were installed throughout March 2019 and formally commissioned in April 2019. Initial data from April was reviewed and issues with flush sensors rectified throughout May to improve data completeness throughout the later phases of the study. For the purposes of this study, we report on a 6-month window during which robust data recording was maintained between 1 July 2019 and 31 December 2019.

The monitoring platform was installed to capture flush data at all new ULFTs in 7 buildings with a total of 38 washrooms and 119 toilets campus. Only standard washrooms received ULFTs and monitoring equipment. For operational reasons, disabled washrooms were not altered or monitored. The buildings covered by the network were (i) Amory Building; (ii) Harrison Building; (iii) Innovation Building 1 and 2 (grouped); (iv) Kay Building; (v) Laver Building; and (vi) Streatham Court.

Each washroom was fitted with a suite of sensors and a data-logging and communications hub, providing both real-time data-logging to a server and local backups (Figure 1). The washrooms and data-logging systems in the network were issued identification codes to aid in data handling and analysis. The identification code is an alphanumeric sequence which contains information on the Building, Floor, Gender, and Location of the washroom. The alphanumeric sequence is given in Supplementary Annex A, Table A1. The real-time monitoring consisted of a flush cycle monitoring of each of the ULFTs. These digital flush counters were based on a bespoke float switch mounted in the cistern of each ULFT. The sensor control units captured the time at which the cistern was refilled (i.e. flushed). Water consumption could, thus, be extrapolated from the flush count by using the product's specified flush volume of each toilet (1.5 litres). In addition, where it was feasible to retrofit within the existing plumbing, flow sensors were installed within the Harrison Building. Water Regulations Advisory Scheme (WRAS) approved that digital turbine flow sensors were fitted to monitor flows on the pipework feeding the toilets, enabling verification of the flush count data. In principle, these high-grade meters provide excellent accuracy, however, as with any retrofit plumbing, it was not always practical for the plumbers to ensure straight inlet and outlet runs to comply with the installation specifications. A standard volumetric water meter was installed on the final leg of pipework feeding each set of ULFTs within Harrison Building. These analogue meters were not connected to the real-time monitoring system and can be read manually to identify gaps in data/provide long-term records of performance as a fall back asset.

Throughout the trial, where data were lost or corrupted, a maintenance visit was conducted by a member of the project

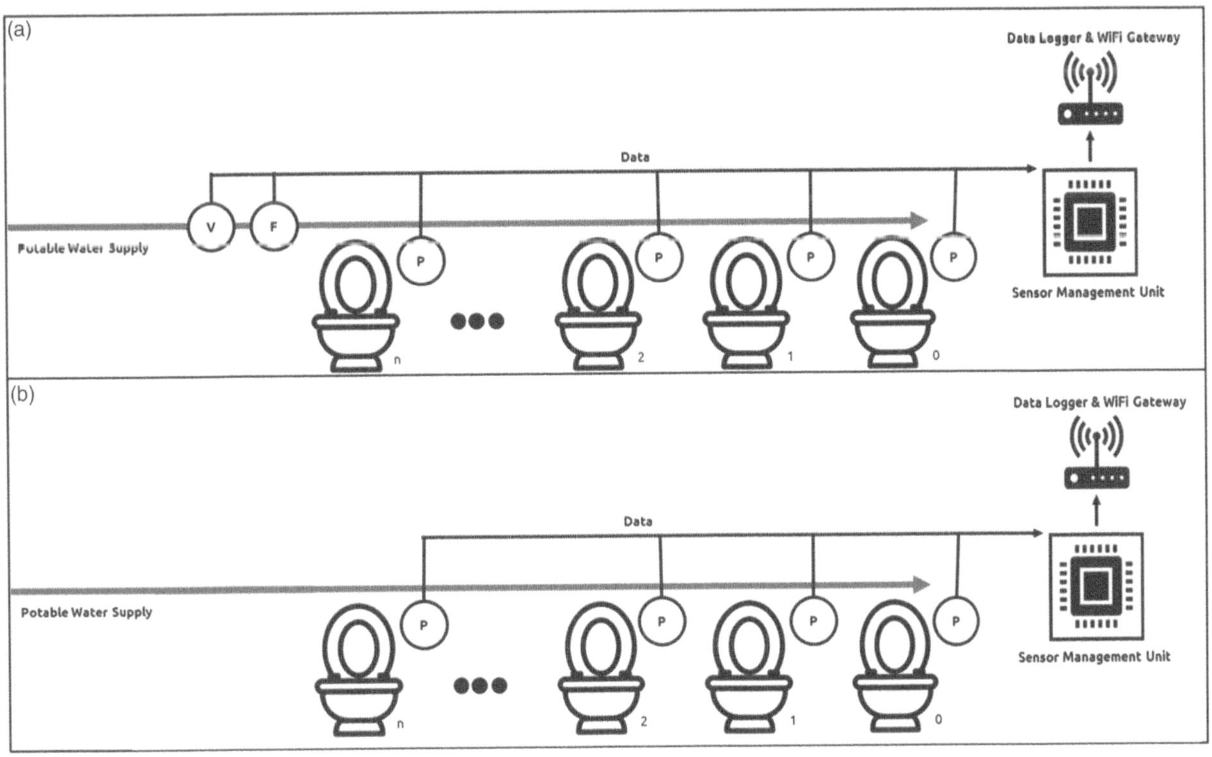

Figure 1 | (a) Schematic representation of the washroom-scale hardware/software platform for data logging used in the Propelair® trial in the Harrison Building on the University of Exeter's Streatham Campus. F = Flow Sensor, Turbine; V = Flow Sensor, Volumetric; P = Flush Counter. (b) Schematic representation of the washroom-scale hardware/software platform for data logging used in the Propelair® trial in all locations involved apart from the Harrison Building on the University of Exeter's Streatham Campus. P = Flush Counter.

team and the data recording system reset. Data completeness throughout the study is described in the Results section.

Data quality assurance: flush count data

Data collection for the 6-month period was completed and analysed for quality issues against a series of metrics. Data quality was assessed at the washroom level (Supplementary Annex B) and the building level (Table 1). Flush counts within the ULFT units can miss-record data if toilet faults occur. Such faults include continuous flush cycling or partial flush cycling, data-logging system failures, inadvertent human intervention, for example, maintenance staff adjusting float switches when undertaking maintenance.

1. Data completeness was established for each washroom and calculated as the actual number of data points collected versus the theoretical maximum for the time period. Data completeness is expressed as a percentage of the potential maximum for each monitoring unit.
2. Data integrity was calculated as the amount of valid data in the dataset, expressed as a percentage of the available dataset. In some circumstances, the data-logging system may be functioning and recording data, but this data may be blank or erroneous (e.g. if a sensor has a loose connection, etc.) Data integrity was assessed by comparing the sum of invalid data points (null or erroneous) against the total data points in the dataset.
3. Data viability was, therefore, derived by multiplying data integrity by data completeness, when compared with the maximum number of potential data points.
4. The flush count data was compared with the digital and volumetric flow meters to explore the validity of the flush count data.

Data capture and manipulation

Data were captured at a micro-computer in each washroom in a time-series format. These raw data files were shared to a remote server to enable access to the time-series data. Data manipulation tasks were coded in python to provide demand data for each washroom as plain text files with flushes and flow events captured at 1-min, 15-min, and 1-day time series. These were further manipulated to enable analysis by building and time period to facilitate the presentation of results.

Following an initial review of the available data, two 8-week subsets were selected to enable a comparison between term-time water demand and non-term-time water demand. The academic term at the University of Exeter ran from 23 September 2019 to 13 December 2019. Hence, a non-term-time window was allocated as 1 July 2019 to 31 August 2019 (low demand), while term-time water demand was assessed for the period 1 October 2019 to 30 November 2019 (high demand). The total number of flush counts per day in each washroom was summed for the two 8-week periods, giving the total number of flushes per washroom per period. These values were summed at the building level to show the total number of flushes in both the low-demand and high-demand windows campus (Supplementary Annex A, Figure A2).

Table 1 | Data quality metrics for the data collection period 01 July 2019 to 31 August 2019, including data completeness, integrity and viability, for each building

		Amory	Harrison	Innovation	Kay	Laver	Streatham
Number of	Washrooms	10	8	8	1	3	8
	Flush counters	41	28	21	2	9	18
	Flow meters	0	20	0	0	0	0
Building use	Lecturing	✓	✓	✗	✗	✓	✓
	Labs	✓	✓	✗	✗	✓	✗
	Offices	✓	✓	✓	✓	✓	✓
	Café	✓	✓	✓	✗	✗	✗
Data completeness (%)		96.4	90.7	99.6	96.9	99.8	94.9
Data integrity (%)		100	100	100	100	100	100
Data viability (%)		96.4	90.7	99.6	96.9	99.8	94.9

The mean washroom flush count over a week was calculated within each of the two 8-week subsets by averaging the number of flush counts for each day of the week (Monday to Sunday). No data points were omitted from the analysis. Any significant deviations from the mean are evaluated in the discussion.

The data from each of the 8-week subsets were summed for all of the washrooms within each of the 7 buildings. The mean building flush count over a week was then subsequently calculated, as before for the washroom, within each of the two 8-week subsets by averaging the number of flush counts for each day of the week (Monday to Sunday). This weekly profile (with error bars showing the standard deviation of each of the mean values) is shown for each of the buildings in Figure 3.

Further data were collated from the university's campus services team (Beavis 2020) to determine the capacity (i.e. maximum number of persons or occupants) and floor area (in metres squared) of each building. These metrics were used to explore variations in water demands linked to building uses. This data is shown in the tables in Figure 3. The values obtained for the 'weekday flush count' are taken as an average of the data recorded Monday to Friday, while the 'weekend flush count' is an average of the data recorded Saturday to Sunday. The total number of flushes in each 8-week period per building was divided by the capacity and by the floor area of the building (reported in Figure 3), such that the 7 buildings could be directly compared (Table 2). Data from digital flow meters was automatically converted to a minutely time-step for washrooms throughout the Harrison Building (Figures 1 and 2).

A 1-week period following the two 8-week subsets was evaluated to assess how well the mean demand profiles represented a typical week in low-demand and high-demand windows (Figure 5). The first week of September 2019 was chosen to evaluate the window of low-demand averages. The first week of December 2019 was chosen to evaluate the high-demand averages. The mean value for each building (Figure 3) is shown as a solid line; the mean flush count during the 1-week period following the 2-month average is shown as a bar chart with error bars (Figure 6).

RESULTS AND DISCUSSION

Data quality

The quality of the flush count data was evaluated for completeness, integrity, and viability. Data integrity was 100% across all buildings during the data collection period (Table 1). This demonstrates that the data recorded was not 'blank' or erroneous at any point during the 6-month collection period. Data completeness was excellent (96.4% ± 3.4) across all buildings (Table 1). The actual number of data points collected for each building ranged between 90.7% (in the Harrison Building) and 99.8% (in the Laver Building) of the theoretical maximum number of data points that could be collected for this time period.

Table 2 | Total flush count per building during the entire 6-month monitored period, the period of low demand and the period of high demand

Building	Total flushes in 6-month monitored period	Total flushes in period of low demand	Total flushes in period of high demand	Building capacity (Number of occupants)	Building floor area (m^2)
Amory	94,841	15,540	52,359	1,305	11,027
Harrison	59,752	16,143	28,748	1,334	8,968
Innovation	97,260	31,004	37,172	363	4,196
Kay	3,488	1,157	1,340	38	222
Laver	11,842	2,965	5,630	414	5,860
Streatham Court	33,358	5,478	19,044	930	3,971
Total	300,541	72,287	144,293	n/a	n/a

Building metrics including capacity (given as the maximum number of occupants) and floor area (in metres squared) are provided for each building.

Data validation

Digital water meters in the Harrison Building recorded total inflow to the ULFTs in each washroom (Figure 1). The seven Harrison building washrooms also had a volumetric water meter that was read intermittently by the maintenance team. Evidence from these seven washrooms showed that the flush count monitoring system was a robust method for recording water demand. However, a total water demand recorded on the water meters was found to be far higher than the flush count data suggested. In each case, these were found to be associated with maintenance issues such as leaky fittings, or one-off continuous flow events. Discounting these washrooms, flow meter data for the remaining five washrooms was used to verify that the average flush volume flush count data. In each case, the fit was within 1.5–1.7 litres per flush (Figure 2). However, flush counters alone fail to capture evidence of certain maintenance or high water usage events.

WC demand

Propelair® WCs were flushed 300,541 times between July and December 2019. The total number of flushes for each WC is summed at the building level for two time periods within the data collection period. Across all buildings, there were a total of 72,287 flushes in the period of low demand and 144,293 flushes in the period of high demand. The combined flush count for the two selected 8-week periods accounts for 72% of the total monitored flushes during the 6-month period. Individually, the flush count for the low-demand period and the high-demand period account for 24 and 48% of the total, respectively. The mean number of flushes per building in the period of high demand (24,048 flushes) is twice that of the period of low demand (12,047 flushes). Two of the buildings – Kay and Innovation – do not fit with this observation. This is likely due to their primary 'office use' status (Table 1). Both buildings are used by staff who operate on normal working hours contracts. Neither building has laboratories or teaching spaces, and thus, they are more representative of commercial office space. A Pearson's partial correlation was run to assess the relationship between flush count and building after adjusting for the total number of WCs monitored, given that only two WCs were monitored in Kay, while 41 WCs were monitored in Amory. Pearson's partial correlation showed that there was no statistically significant relationship when the number of WCs monitored was controlled for $r\ (9) = 0.169$, $p = 0.619$.

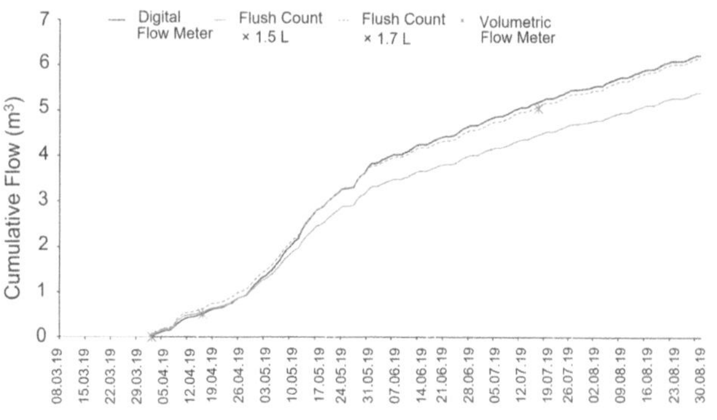

Figure 2 | Flush counter and water demand data validation for Washroom HB-00-004-M-2.

Diurnal variation in toilet usage

Across all buildings, the flush count on weekdays was considerably higher than the weekend flush count (Figure 3). The weekend flush count was 92.4% ± 4.6 lower than the weekday flush count during the low-demand period and 92.1% ± 3.9 lower than the weekday flush count during the high-demand period (Figure 3). There was little difference between the diurnal pattern of toilet use between buildings, with highest WC demand exhibited Monday to Friday and markedly reduced demand on Saturday and Sunday. In the high-demand window, Wednesday was found to have the lowest flush weekday count at the two large teaching buildings (Amory and Harrison). This could be associated with student sports activity on Wednesday afternoons and could be a useful indicator that footfall is lower. Such evidence warrants further exploration and could be useful to cleaning coordinators or café managers to understand when high footfall was experienced.

Temporal variations in water demand

Variation in water demand between the two observed periods (low and high demand) was not uniform between buildings. The Amory building, which is a mixture of

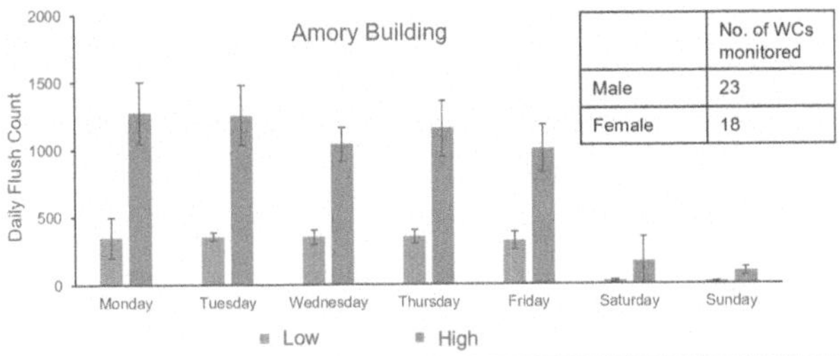

Capacity (N°. of Occupants)	Floor Area (m²)	Low Demand		High Demand	
		Weekday flush count	Weekend flush count	Weekday flush count	Weekend flush count
1305	11027	345	18	1143	128

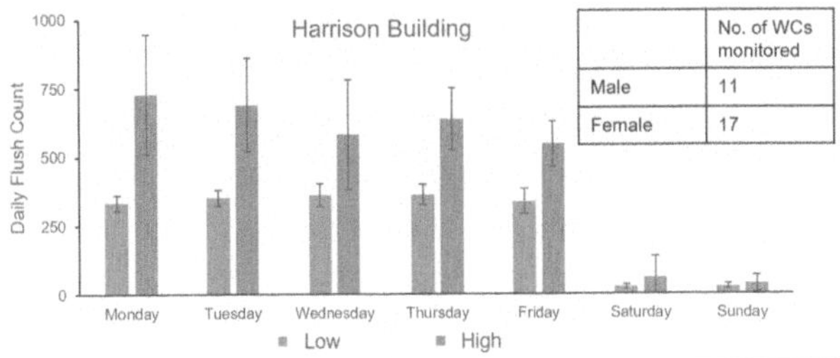

Capacity (N°. of Occupants)	Floor Area (m²)	Low Demand		High Demand	
		Weekday flush count	Weekend flush count	Weekday flush count	Weekend flush count
1334	8968	349	26	637	48

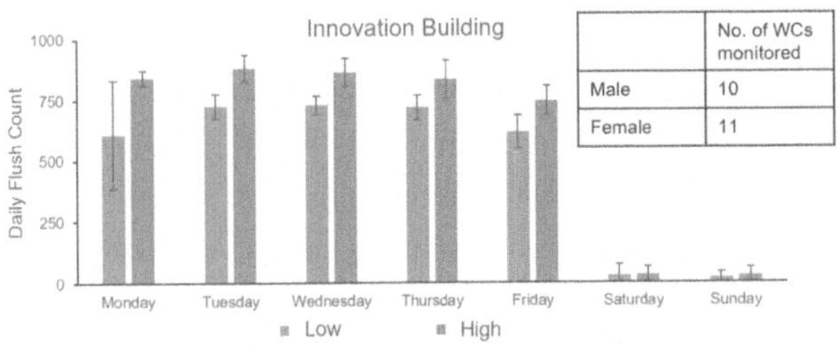

Capacity (N°. of Occupants)	Floor Area (m²)	Low Demand		High Demand	
		Weekday flush count	Weekend flush count	Weekday flush count	Weekend flush count
363	4196	680	24	833	31

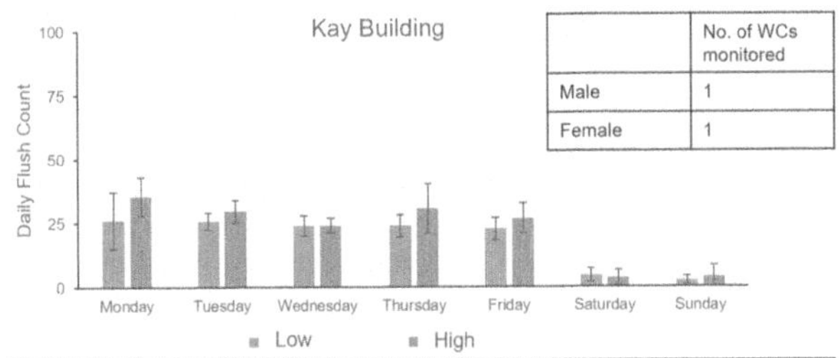

Capacity (N°. of Occupants)	Floor Area (m²)	Low Demand		High Demand	
		Weekday flush count	Weekend flush count	Weekday flush count	Weekend flush count
38	222	24	3	29	4

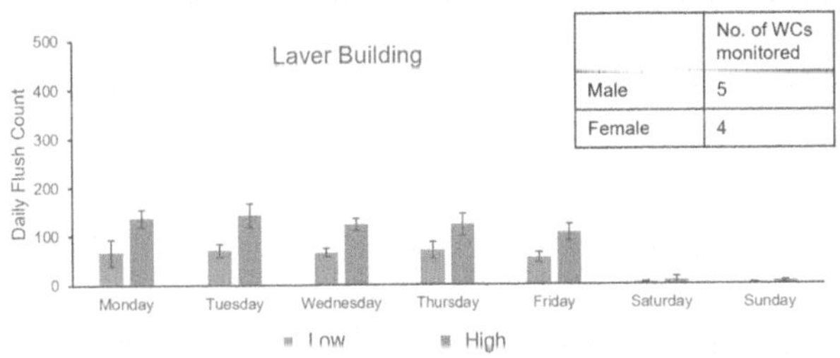

Capacity (N°. of Occupants)	Floor Area (m²)	Low Demand		High Demand	
		Weekday flush count	Weekend flush count	Weekday flush count	Weekend flush count
414	5860	65	2	126	6

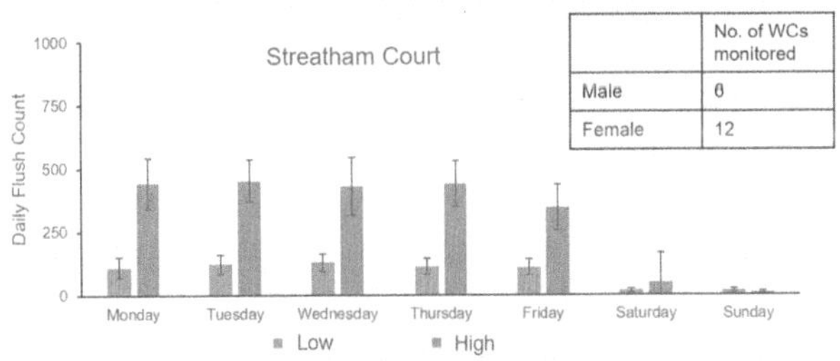

Capacity (N°. of Occupants)	Floor Area (m²)	Low Demand		High Demand	
		Weekday flush count	Weekend flush count	Weekday flush count	Weekend flush count
930	3971	116	16	422	28

Figure 3 | Mean diurnal pattern of WC use in each building during the period of low demand (01 July 2019 to 31 August 2019) and high demand (01 October 2019 to 30 November 2019). Subset tables display: (i) number of WCs monitored for each gender (top right) and (ii) building metrics including capacity and floor area (below each graph).

lecturing, teaching, office and laboratory spaces, had a weekday flush count in the period of low demand which was 69.8% lower than the period of high demand (Figure 3). On weekends, this difference was even more marked, with a low-demand period flush count 86% lower than the high-demand period. In the Innovation and Kay buildings, which are staffed with office workers and have no significant lab spaces, much smaller differences were noted. The weekday flush count in the period of low demand was only 18.4 and 17.2% lower than the period of high demand in the Innovation and Kay buildings, respectively. A similar trend was observed with weekend flush counts which were 22.6% (Innovation) and 25% lower (Kay).

A paired-samples *t*-test revealed that the mean flush count was statistically significantly higher in the period of high demand than the period of low demand as follows: Amory – a mean increase of 610 flushes (95% Confidence Interval (CI) of 518–702 flushes), $t(60) = 13.216$, $p < 0.0005$; Harrison – a mean increase of 213 flushes (95% CI of 163–262 flushes), $t(60) = 8.522$, $p < 0.0005$; Innovation

– a mean increase of 110 flushes (95% CI of 78–141 flushes), $t(60) = 6.941$, $p < 0.0005$; Kay – a mean increase of 3 flushes (95% CI of 1–5 flushes), $t(60) = 3.037$, $p = 0.004$; Laver – a mean increase of 45 flushes (95% CI of 36–54 flushes), $t(60) = 10.159$, $p < 0.0005$; and Streatham Court – a mean increase of 224 flushes (95% CI of 183–266 flushes), $t(60) = 10.851$, $p < 0.0005$.

Toilet use per building occupant

The total number of flushes across all buildings during the 6-month period totalled 300,541. This included 94,841 flushes in Amory; 59,752 flushes in Harrison; 97,260 flushes in Innovation; 3,488 flushes in Kay; 11,842 flushes in Laver; and 33,358 flushes in Streatham Court (Table 2).

The mean flush count per building occupant per day was less than 1 for all buildings except Innovation (Figure 4). Mean flush count in Innovation was 1.45 flushes/occupant/day across the 6-month period (July–December 2019). The mean flush count for the other five buildings ranged between 0.155 flushes/occupant/day (Laver) and 0.499 flushes/occupant/day (Kay). The mean flush count per building occupant was 18.5% lower in Innovation in the low-demand period (1.37 flushes/occupant/day) compared with the high-demand period (1.68 flushes/occupant/day). A similar trend was seen in Kay, where the mean flush count per building occupant was 15% lower in the low-demand period (0.491 flushes/occupant/day) compared with the high-demand period (0.578 flushes/occupant/day). In the remaining four buildings, which are all used for teaching and research as well as office use, the mean flush count per building occupant in the period of low demand was considerably lower (45–72%) than during the period of high demand.

The total flush count, for the two 8-week periods, was plotted against building capacity (maximum number of occupants) for each of the six buildings (Figure 5). Generally, as building capacity increased, so too did total flush count during both the low- and high-demand periods. The Innovation building was an outlier to the observed trend, with a much higher flush count than would be expected based on the building capacity (Figure 5). A Pearson's correlation was run to evaluate the relationship between building capacity and total flush count. The Innovation building was excluded from this analysis, as it was deemed to be an outlier because of the regular working hours of its occupants. A Pearson's correlation for flush count showed a strong, positive linear relationship with building capacity which was statistically significant for the period of low demand, $r(5) = 0.931$, $p = 0.022$ and high demand, $r(5) = 0.877$, $p = 0.05$ (Figure 5).

These analyses indicate that the data hold value in terms of providing building managers with an understanding of occupancy and footfall, especially where data streams are made available in real-time. For example, if user behaviour is assumed to be consistent throughout the year, the number of flushes per occupant could offer building

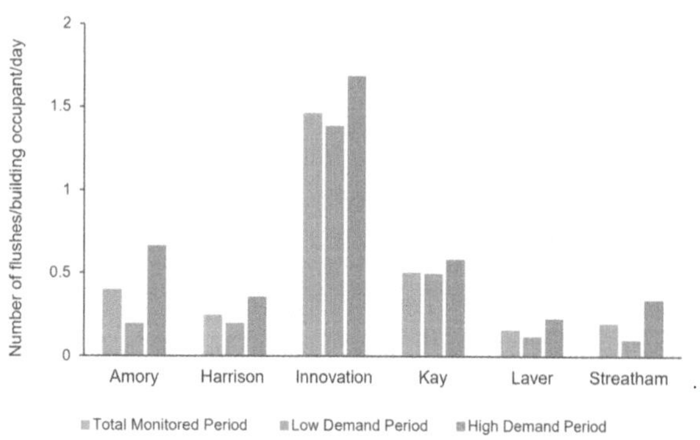

Figure 4 | Mean number of flushes per building occupant per day for the total 6-month monitoring period (July–December 2019); the period of low demand (01 July 2019 to 31 August 2019); and the period of high demand (01 October 2019 to 30 November 2019).

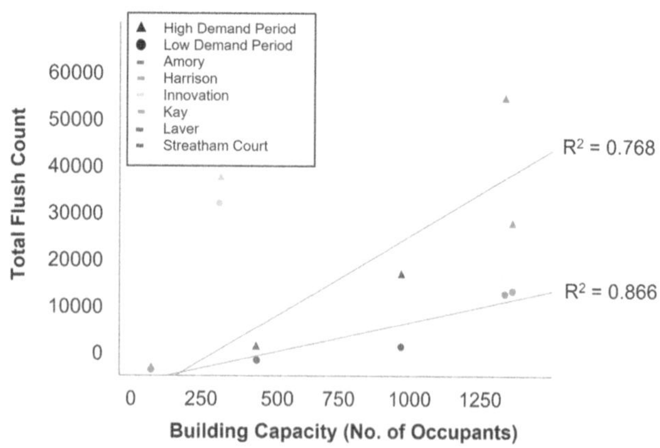

Figure 5 | Total number of flushes per number of building occupants during the period of low demand (01 July 2019 to 31 August 2019), shown as circles, and the period of high demand (01 October 2019 to 30 November 2019), shown as triangles. Each building is colour coded as follows: Amory = Blue, Harrison = Green, Innovation = Yellow, Kay = Orange, Laver = Red and Streatham Court = Purple. Please refer to the online version of this paper to see this figure in colour: https://doi.org/10.2166/hydro.2021.103.

managers a means of reflection on building use rates, and therefore, inform management practices.

Benefits of high-resolution real-time water demand data

Monthly meter readings, typically used to record water demand in commercial buildings, offer little information on daily building usage. In contrast, the highly granular, near-real-time data provided in this study have the potential to provide benefits for a wide range of users. These uses depend on data that are meaningful and of sufficient quality that the projected, demand profiles are representative of the observed water use, and that added value can be obtained from micro-component analyses which go above and beyond monthly analogue meter readings. First, we demonstrate that the demand profiles are representative, before describing a subset of potential end-users that could derive value from accessing real-time WC flush data.

The demand profiles created for the period of low demand and high demand together account for 72% of the total flushes in the 6-month monitored period. To evaluate how well the demand profiles represented a typical week in both the low-demand and high-demand windows, a 1-week period immediately succeeding the two 8-week subsets was transposed onto each profile (Figure 6).

For the period of low demand, the flush counts recorded for the first week of September fall within one standard deviation of the mean for Amory, Innovation, Kay, and Streatham Court (Figure 6). However, the total flush count in Harrison falls outside of the standard deviation on weekdays (but not for weekends) and the total flush count in Laver falls outside of the standard deviation on Wednesday only. In the period of high demand, the flush counts observed in the first week of December only fall within

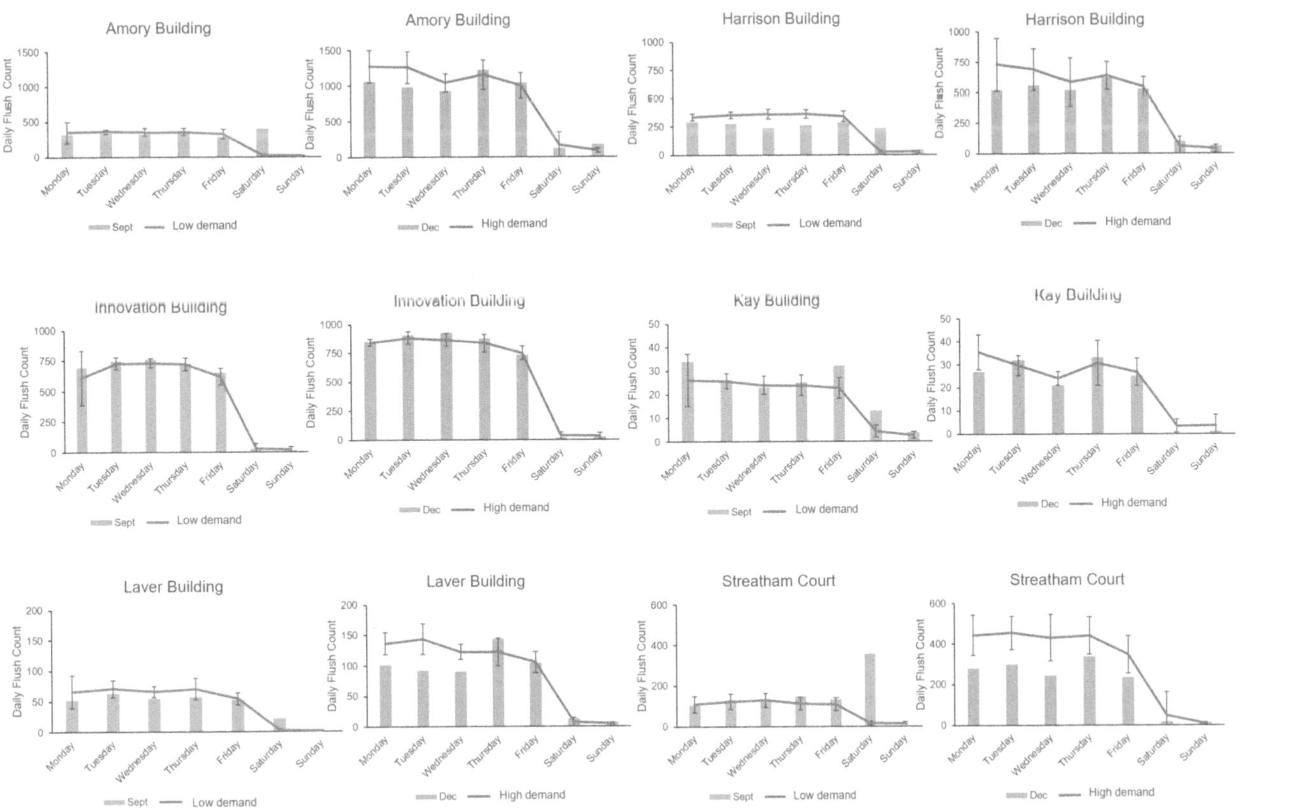

Figure 6 | Evaluating how representative the mean demand profiles are of a typical week. The first week of September 2019 (green bars) was chosen to evaluate the window of low-demand averages (green line). The first week of December 2019 (blue bars) was chosen to evaluate the high-demand averages (blue line). Error bars indicate variation in flush count over the course of the 8-week period. Please refer to the online version of this paper to see this figure in colour: https://doi.org/10.2166/hydro.2021.103.

one standard deviation of the mean demand profiles in the Harrison and Innovation buildings (Figure 6). For Amory and Kay, the total flush count falls outside of the standard deviation on only one day (Tuesday in Amory; Monday in Kay). However, in Laver and Streatham Court, the total flush count deviates from the prediction on three or more consecutive weekdays (Monday to Wednesday in Laver; Monday to Thursday in Streatham Court) (Figure 6). From this, we deduce that the demand profiles are more representative in predominantly office-use buildings (where users exhibit more steady daily behaviours) than mixed-use teaching and research buildings (where footfall is driven by the scheduling of lectures) on campus.

The design and implementation of the flush monitoring system was found to be robust, providing >90% viable data for each washroom (90.7–99.8%). The project was primarily implemented to provide evidence of water demand savings associated with the large-scale deployment of novel ULFTs. Based on data captured in this study, the ULFTs were flushed 300,541 times in 6 months. Over a year, when compared with toilets with a typical 6-litre flush volume, we estimate that this could equate to 2,585 m^3 of water savings (based on a minimum of 4.3 litre saving per flush). Assuming commercial water charges as provided by the campus estates team of approximately £5.50 per m^3, this would equate to a monetary saving in excess of £14,000 per annum, or £119.00 per toilet from reduced water costs alone.

Following initial analyses herein, we anticipate that the dataset can be explored further to understand how it can generate further value for end-users. We anticipate that the data could meet the needs of a variety of users including facilities and estates managers, potable water service, and wastewater network providers, as well as informing security and lone working protocols and maintenance management. It is important to note that there may be a much larger pool of users, than those described here, that may benefit from using such data. We anticipate that the supplementary data is well suited for use in future hackathon-style events. The following explores a series of initial concepts for further investigation:

1. *Water resource managers*: With incentives in place from the Water Services Regulation Authority for England and Wales, Ofwat, each water service provider in the UK has a team of staff providing water efficiency programmes that seek to reduce water demand using technologies and behaviour change. The campus-wide toilet demand data not only provide such users with the opportunity to interrogate the costs and benefits of the ULFT system's monitored herein, but they can also illustrate diurnal and seasonal patterns which could bring benefit to existing data-driven models within the water resources team. Conceptually, a smart water supply network can be framed wherein every water use is monitored in real time at a micro-component level. Algorithms could be developed which explore how best to satisfy these demands (e.g. by amending reservoir or pressure reducing valve set points in the distribution network) and other valuable network data such as water pressure could be captured at each WC in a future iteration of data collection efforts.
2. *Facilities managers*: The maintenance staff involved in the project raised the possibility of using the flush data to flag maintenance issues. For example, every WC could be set to automatically flush at midnight. Where a toilet fails to flush (i.e. requires maintenance), a task can be raised and the WC fixed before users report an issue. The flush data can also demonstrate which washrooms are most highly trafficked. This could enable cleaning staff to be provided with dynamic cleaning schedules. For example, their cleaning tasks could be based on usage rates, rather than a fixed regime where underused washrooms are cleaned as frequently as heavily trafficked ones. Further investigation could explore how such a revised strategy could be set up to more effectively deploy cleaning resources, which in addition to generating potential economic benefits, could also help reduce the environmental footprint (e.g. through using less cleaning products).
3. *Estates manager*: Building occupancy within campus settings is an important metric to ensure spaces are being used to their maximum capacity. Understanding the dynamic nature of usage within buildings can enable the space management team to understand departmental usage rates for hot-desks and office spaces. Campus managers could use flush data to support business cases that

promote hot-desking in locations where permanent office space is infrequently used.

4. *Water system/supply manager*: Water use patterns derived for specific building types pose opportunities to help designers to appropriately retrofit novel technologies. These can, in turn, support more efficient use of water resources. With accurate demand data, technologies such as smart rainwater harvesting can be designed with confidence to meet toilet flush demand where accurate data has been obtained. Furthermore, commercial water tanks that feed water supplies in large buildings (either by gravity or from booster sets) can be optimally sized and operated using novel control regimes. This, in turn, enables dynamic water pricing models to be explored. For example, a water service provider might conceivably provide water during periods of low demand (e.g. winter/overnight) at a lower charge rate than during periods of high demand (e.g. diurnal peaks and summer). With accurate data captured for the buildings studied, models can be developed and tested which explore novel economic incentives as a way to reduce peak water demand and more optimally incentivize good practice within customer's water behaviours. Further development of the sensor network and data capture systems at the case study site offers the potential to investigate a wider set of benchmarks and metrics related to high-resolution water demand data.

CONCLUSIONS

A systematic regime for monitoring toilet usage at a 1-min time-step was conceived and implemented alongside the retrofit of 119 ULFTs at University of Exeter's Streatham Campus. The water consumption data gathered at 119 WCs show that water demand has significant variation in term time versus non-term time periods. The results support the following conclusions:

1. Toilet use events within buildings in a campus setting can be monitored using low-cost sensors (~\$1) and were found to follow a typical weekly pattern in which weekday use is significantly higher than weekend use.
2. Toilet demand is considerably higher during term time than during holiday periods, with flush counts 46% lower on weekdays (and 48% lower on weekends) in holidays.
3. Generally, toilet use showed a strong, positive relationship ($r = 0.931$ low demand; $r = 0.877$ high demand) with building metrics such as the number of occupants in office and teaching spaces. This could offer a means of reflection on building use rates and inform management practices or the scheduling of events.
4. Mixed-use buildings, such as those used for teaching and laboratories, tended to have greater variation in WC use between term time and holidays than office-use buildings. This is likely due to the high ratio of students to staff using these buildings. Office-use buildings had a more steady WC water demand throughout the year.
5. Real-time toilet flush data can provide accurate high-resolution water demand patterns. The data captured has been provided in Supplementary Annex C to enable research practitioners to explore concepts and benefits associated with the data captured in this study, for example through hackathon-style events or data science projects.

Additional benefits were identified for further exploration as follows: (i) water resource managers can measure costs and benefits of novel water-saving technologies and could, in principle, develop building-level water supply models to inform water resource plans; (ii) facilities managers could develop alarm systems to automate maintenance reports for faulty WCs; (iii) cleaning staff can be provided dynamic cleaning schedules, wherein the highest used washrooms are cleaned most frequently; (iv) campus managers can establish building usage rates based on flush count data; (v) internet-enabled technologies such as smart rainwater harvesting or building header tanks can be optimally designed and operated based on real-time demand data; and (vi) researchers can explore opportunities for dynamic water pricing, wherein water charges are reduced when demand on the system is low to incentivize optimal use of on-site water storage tanks. Opportunities to develop a wider set of open-access water demand data at the case study site are now being explored by introducing further sensors. In turn, key stakeholders have the potential to look beyond siloed working practices to co-create value propositions that derive benefit from the data described herein.

ACKNOWLEDGEMENTS

The authors are grateful for the support of Alex Newman (Propelair), Sam Dickinson (OTA), Andy Seaman and Simon Tailford (University of Exeter estates). The installation of monitoring was funded by Horizon 2020 Grant agreement ID: 778742 and the analysis conducted through a TWENTY65 leverage grant EPSRC Reference: EP/N010124/1.

CONFLICTS OF INTEREST

During the data collection phase of the study (2018–2019), Peter Melville-Shreeve was a director of Over The Air Analytics Limited (OTA). OTA was contracted to provide project management, installation and maintenance services relating to the data capture systems used in this study.

DATA AVAILABILITY STATEMENT

All relevant data are included in the paper or its Supplementary Information.

REFERENCES

Anglian Water 2020 *Upgraded Meters*. Available from: https://www.anglianwater.co.uk/services/switch-to-a-water-meter/upgraded-meters/ (accessed 26 February 2020).

Antunes, A., Andrade-Campos, A., Sardinha-Lourenço, A. & Oliveira, M. S. 2018 Short-term water demand forecasting using machine learning techniques. *Journal of Hydroinformatics* **20** (6), 1343–1366. https://doi.org/10.2166/hydro.2018.163.

Beavis, D. 2020 Personal Correspondence. E-mail 18 February 2020 'Building Occupancy Analyses'.

Blokker, E. J. M., Vreeburg, J. H. G. & van Dijk, J. C. 2010 Simulating residential water demand with a stochastic end-use model. *Journal of Water Resources Planning and Management* **136** (1), 19–26.

Blokker, E. J. M., Pieterse-Quirijns, E. J., Vreeburg, J. H. G. & van Dijk, J. C. 2011 Simulating nonresidential water demand with a stochastic end-use model. *Journal of Water Resources Planning and Management* **137** (6), 511–520.

Buchberger, S. G. & Wells, G. J. 1996 Intensity, duration and frequency of residential water demands. *Journal of Water Resources Planning and Management* **122** (1), 11–19.

Chen, Y. & Han, D. 2016 Big data and hydroinformatics. *Journal of Hydroinformatics* **18** (4), 599–614. https://doi.org/10.2166/hydro.2016.180.

Cherukutota, N. & Jadhav, S. 2016 Architectural framework of smart water meter reading system in IoT environment. In: *International Conference on Communication and Signal Processing (ICCSP)*, Melmaruvathur, India, pp. 0791–0794.

Cominola, A., Giuliani, M., Piga, D., Castelletti, A. & Rizzoli, A. E. 2015 Benefits and challenges of using smart meters for advancing residential water demand modeling and management: a review. *Environmental Modelling & Software* **72**, 198–214.

Creaco, E., Kossieris, P., Vamvakeridou-Lyrouda, L., Makropoulos, C., Kapelan, Z. & Savic, D. 2016 Parameterizing residential water demand pulse models through smart meter readings. *Environmental Modelling & Software* **80**, 33–40. https://doi.org/10.1016/j.envsoft.2016.02.019.

Device Magic 2020 www.devicemagic.com (accessed 4 March 2020).

European Commission 2020 *Periodic Reporting for Period 2 – Propelair*. Available from: https://cordis.europa.eu/project/id/778742/reporting (accessed 10 April 2020).

Gurung, T. R., Stewart, R. A., Beal, C. & Sharma, A. 2014 Smart meter enabled water end-use demand data: platform for the enhanced infrastructure planning of contemporary urban water supply networks. *Journal of Cleaner Production*. doi:10.1016/j.jclepro.2014.09.054.

Hackett, R. 2018 *Arqiva Delivers Further Smart Metering Fixed Network Trial for Anglian*. WWT Online. Available from: https://wwtonline.co.uk/news/arqiva-delivers-further-smart-metering-fixed-network-trial-for-anglian (accessed 10 March 2020).

Lom, M., Pribyl, O. & Svitek, M. 2016 Industry 4.0 as a part of smart cities. In: *2016 Smart Cities Symposium*, Prague, 26–27 May 2016.

Martini, A., Troncossi, M., Rivola, A. & Nascetti, D. 2014 Preliminary investigations on automatic detection of leaks in water distribution networks by means of vibration monitoring. In: *Advances in Condition Monitoring of Machinery in Non-Stationary Operations. Lecture Notes in Mechanical Engineering* (G. Dalpiaz, ed.). Springer, Berlin, Heidelberg.

Melville-Shreeve, P., Ward, S. & Butler, D. 2016 Rainwater harvesting typologies for UK houses: a multi criteria analysis of system configurations. *Water* **8**, 129.

Morales-Hernández, M., Sharif, M. B., Gangrade, S., Dullo, T. T., Kao, S. C., Kalyanapu, A., Ghafoor, S. K., Evans, K. J., Madadi-Kandjani, E. & Hodges, B. R. 2020 High-performance computing in water resources hydrodynamics. *Journal of Hydroinformatics*. https://doi.org/10.2166/hydro.2020.163.

Nguyen, K. A., Stewart, R. A., Zhang, H. & Jones, C. 2015 Intelligent autonomous system for residential water end use classification: autoflow. *Applied Soft Computing* **31**, 118–131.

Pastor-Jabaloyes, L., Arregui, F. & Cobacho, R. 2018 Water end use disaggregation based on soft computing techniques. *Water* **10** (1), 46. http://dx.doi.org/10.3390/w10010046.

Peña Queralta, J., Gia, T. N., Zou, Z., Tenhunen, H. T. & Westerlund, T. 2019 Comparative study of LPWAN technologies on unlicensed bands for M2M communication in the IoT: beyond LoRa and LoRaWAN. *Procedia Computer Science* **155**, 343–350. ISSN 1877-0509. https://doi.org/10.1016/j.procs.2019.08.049.

Pieterse-Quirijns, E. J., Blokker, E. J. M., van der Blom, E. & Vreeburg and, J. H. G. 2013 Non-residential water demand model validated with extensive measurements and surveys. *Drinking Water Engineering and Science* **6**, 99–114. https://doi.org/10.5194/dwes-6-99-2013.

Pirow, N. O., Louw, T. M. & Booysen, M. J. 2018 Non-invasive estimation of domestic hot water usage with temperature and vibration sensors. *Flow Measurement and Instrumentation* **63**, 1–7.

Rathore, M. M., Ahmad, A., Paul, A. & Rho, S. 2016 Urban planning and building smart cities based on the Internet of Things using Big Data analytics. *Computer Networks* **101**, 63–80.

Thames Water 2020a *Water Meters*. Available from: https://www.thameswater.co.uk/help/water-meters (accessed 3 March 2020).

Thames Water 2020b *Smart Meters*. Available from: https://www.thameswater.co.uk/help/water-meters/getting-a-water-meter (accessed 3 March 2020).

Thomson, P., Hope, R. & Foster, T. 2012 GSM-enabled remote monitoring of rural handpumps: a proof-of-concept study. *Journal of Hydroinformatics* **14** (4), 829–839. https://doi.org/10.2166/hydro.2012.183.

Waterwise 2020 *Leaky Loos Phase II, Water Industry Collaborative Fund Project*. Available from: https://www.waterwise.org.uk/wp-content/uploads/2018/08/Leaky-Loos-Phase-II_Final-report.pdf (accessed 4 April 2020).

Wills, P., Memon, F. A., Savic, D., Merhcant, P., Roberts, M. & Martin, J. N. A. 2018 Combining surveys and flow logger data to improve the accuracy of end-use segregation of residential water consumption. In: *1st International WDSA/CCWI 2018 Joint Conference*, July 23–25, 2018, Kingston, Ontario, Canada.

WWT (Water and Wastewater Treatment) 2018 *Is the UK Ready for a Smart Meter Revolution*. Available from: https://wwtonline.co.uk/features/smartening-up-is-uk-water-metering-ready-for-a-smart-revolution-(accessed 4 March 2020).

First received 15 June 2020; accepted in revised form 14 January 2021. Available online 4 February 2021

© 2021 The Authors

A feasibility study of uninhabited aircraft systems for rapid and cost-effective plant stress monitoring at green stormwater infrastructure facilities

Kery Prettyman, Meghna Babbar-Sebens, Christopher E. Parrish and Jeremy Matthew Babbar-Sebens

ABSTRACT

Vegetation health monitoring is key to identifying early signs of water stress, pollutant-induced toxicity, and plant diseases in green urban stormwater facilities. However, rigorous monitoring to collect accurate quantitative data is an expensive and time-consuming process. This paper examines the feasibility of using uninhabited aircraft systems (UAS), in comparison to standard ground-based methods, for monitoring biomass and primary production in two bioswale cells at an urban stormwater facility. Implementation of the UAS-based approach involved flight planning in an urban area to meet resolution requirements of bioswale imagery obtained from near-infrared and red-green-blue cameras. The resulting normalized difference vegetation index (NDVI) estimated from UAS data was tracked over a 2-month period during the transition from spring to summer, showing the spatial distribution of NDVI and the change in vegetation coverage areas over time. In comparison, ground-based measurements of the fraction of intercepted photosynthetically active radiation (PAR) presented multiple practical challenges during implementation in the field, leading to over- and underestimates of intercepted PAR. Overall, UAS-derived NDVI was found to be a valuable reflectance-based, vegetation health-monitoring methodology that can be used by utilities and cities for practical, cost-effective, and rapid assessment of vegetation stress and for long-term maintenance in green stormwater facilities.

Key words | fraction of intercepted photosynthetically active radiation, green infrastructure, NDVI, stormwater, uninhabited aircraft systems, vegetation monitoring

Kery Prettyman
Meghna Babbar-Sebens (corresponding author)
Christopher E. Parrish
School of Civil and Construction Engineering,
Oregon State University,
Corvallis, OR,
USA
E-mail: *meghna@oregonstate.edu*

Jeremy Matthew Babbar-Sebens
Knife Edge Software,
Corvallis, OR,
USA

HIGHLIGHTS

- Ground-based monitoring of green infrastructure (GI) facilities for estimation of fraction of intercepted photosynthetically active radiation poses multiple operational challenges.
- Uninhabited aircraft systems (UAS) were found to enable efficient acquisition of multi-temporal normalized difference vegetation index (NDVI) data for GI site monitoring, through the workflow developed in this study.
- Time series of UAS-derived NDVI showed the expected downward trend over a two-month period starting on June 1, 2019.
- Spatial variation in UAS-derived NDVI provides plant-specific health information that may assist in GI site management.

This is an Open Access article distributed under the terms of the Creative Commons Attribution Licence (CC BY 4.0), which permits copying, adaptation and redistribution, provided the original work is properly cited (http://creativecommons.org/licenses/by/4.0/).

doi: 10.2166/hydro.2020.195

INTRODUCTION

Green infrastructure (GI) for stormwater management is an attractive solution for urban municipalities, serving to reduce urban flooding and treat contaminated stormwater, while simultaneously providing aesthetically pleasing spaces in urban areas. Plants play a vital role in GI facilities by helping reduce the flow velocity of stormwater, providing nutrients to the microbial communities in the soil, maintaining soil permeability, and taking up contaminants that infiltrate the soil in stormwater runoff (USEPA 1999; Akpor & Muchie 2010). To support function and performance of plants in GI, vegetation health-monitoring practices must be established. Monitoring the health of vegetation in GI can reveal signs of water stress, pollutant-induced toxicity, and disease. Hence, early detection of stress in vegetation is critical for long-term sustainability and effectiveness of GI, especially when there is a potential for buildup of sediments and contaminants in media and vegetation that may reduce treatment efficiency (Lucke & Nichols 2015). Methods for monitoring GI hydrology have been extensively investigated in the literature. However, methods for monitoring the health of urban GI vegetation are not well studied. Previous work on monitoring vegetation in GI in practice has primarily been limited to the identification, location, and description of vegetation density and height, with no detailed methodology included beyond a visual inspection of these plant-based elements at GI facilities such as bioswales, bioretention systems, raingardens, and green roofs (USEPA 2002, 2009).

Multiple potential methods exist that could be applied toward long-term monitoring of vegetation in GI facilities, including both ground-based and remote sensing methods. Remote sensing technology holds the potential for efficient, recurrent monitoring of large-scale green spaces for vegetation health and productivity. Recent studies have used vegetation indices and vegetation density metrics derived from aerial imagery acquired using aircraft and from satellites to explore canopy cover and urban greenness. These studies have contributed to planning for design and maintenance and valuing urban green areas (Franco & Macdonald 2018; Gadi *et al.* 2018). Studies have also explored the cooling effects of large urban green areas where various types of remote sensing data have been utilized including multispectral, hyperspectral, and light detection and ranging (LiDAR) data (Bartesaghi-Koc *et al.* 2017, 2019). Additionally, remote sensing imagery has been used in studies for relating urban GI and green spaces to effectiveness of GI in mitigating flood events (Kim *et al.* 2017; Lee 2018).

Despite the usefulness of remote sensing methods in urban green spaces, limited research exists on the application of uninhabited aircraft systems (UAS; also defined as 'unmanned aircraft systems' by Federal Aviation Administration; FAA 2016a) for monitoring and management of stormwater GI facilities and urban water infrastructure management (Hill & Babbar-Sebens 2019; McDonald 2019). For example, the use of UAS for monitoring of vegetation health in stormwater GI using vegetation indices, such as the normalized difference vegetation index (NDVI), has been limited to countable studies, such as those conducted recently by Dimitrov *et al.* (2018), in which NDVI derived from UAS waveband data was used to detect health and disease of vegetation in an urban green space of Bulgaria. However, this study did not explore the usefulness of their approach for temporal monitoring of vegetation classes and did not provide any insight on how the NDVI data could be compared with their *in situ* data for potential trade-offs in monitoring approaches. In contrast, UAS-derived NDVI has been more popular in the agriculture industry for low-cost assessment of crop health and overall field monitoring (Sankaran *et al.* 2015; Shi *et al.* 2016).

In addition to remote monitoring, individual plant growth can also be monitored using ground-based methods. Destructive methods of measuring plant growth and health are labor intensive, involving harvesting plants at the ground level, drying and weighing them to determine biomass (Prabhakaraa *et al.* 2015). Various hand-held sensors can measure parameters of plant health, including chlorophyll meters (Muñoz-Huerta *et al.* 2013) that measure vegetation greenness, chlorophyll fluorometers (Mohammed *et al.* 1995) that measure chlorophyll fluorescence, leaf porometer instruments (Toro *et al.* 2019) that measure stomatal conductance, and ceptometer instruments (Bréda 2003) that measure the fraction of intercepted photosynthetically active radiation (fIPAR). Chlorophyll meters and porometers provide measures of vegetation system functions at the leaf level, requiring large samples due to variations between leaves (Bauerle *et al.* 2004; Möller *et al.* 2007; Cordon *et al.* 2016). However, ceptometer instruments measure intercepted radiation at the plant level. The fIPAR is the amount of solar radiation in the spectral range of 400–700 nm (i.e., the visible portion of the electromagnetic spectrum) that is intercepted by vegetation for the purpose of vegetation growth. The fIPAR has been shown to have a strong correlation with dry biomass production, or net primary productivity and is used in models of dry matter production efficiency and radiation use efficiency (Monteith 1972; Rosati & Dejong 2003). The fIPAR is also considered an essential plant parameter for defining functional groups used in process-based models. It has been previously used to define functional groups for wetlands in several representative wetland regions in the USA (Williams *et al.* 2017). Additionally, studies show that the fIPAR and NDVI have a linear relationship when applied to row crops and prior to the onset of senescence in sampled vegetation (Hatfield *et al.* 1984; Serrano *et al.* 2000). This indicates that NDVI may provide an estimate of fIPAR and biophysical parameters related to fIPAR.

While the use of NDVI and fIPAR for monitoring or detecting change in crops and natural systems is a well-studied topic, there is a lack of knowledge about the effectiveness of monitoring these parameters as indicators of stress and health of vegetation in engineered systems, specifically stormwater GI systems (McDonald 2019). The implementation of remote technology to monitor GI vegetation growth and health may provide insights into vegetation-dependent performance while informing targeted operational and policy decisions. Fast detection of physiological changes in vegetation may provide indications of exposure to new or altered contaminant loads, signs of senescence and disease, water stress, or other stressors. In turn, this information can guide decisions related to adaptive mitigation or removal/disposal and replacement of vegetation and possibly soil media in GI facilities.

The goal of this study is to evaluate methods for rapid and cost-effective monitoring of GI vegetation health, with a focus on UAS platforms for remotely monitoring bioswale facility vegetation health via the estimation of NDVI. UAS are of interest as for GI health monitoring for two primary reasons. First, the spatial extents of typical stormwater GI facilities in urban areas (generally less than a few km^2) are well suited to UAS data collection. (Larger areas would be more suitable to data collection from conventional manned aircraft or satellites, whereas smaller sites could be effectively studied using solely ground-based methods.) Second, the relative low cost and logistical simplicity of UAS data collection compared with data acquisition from conventional aircraft makes repeat, high-resolution data collection possible. To compare the UAS-based approach with a ground-based vegetation health-monitoring method, this study collected data on fIPAR of the vegetation canopy, a measure that has previously been reported to be proportional to biomass production. These two measures (fIPAR and the NDVI) were then compared to explore potential relationships between each other and to determine if the NDVI can act as an estimate of fIPAR and the physiological parameters derived from fIPAR at the stormwater bioswales facility. Finally, temporal trend analysis using the UAS-derived NDVI was also evaluated for iterative monitoring of stress and vegetation health in the bioswale facility.

METHODOLOGY

Site description

The study was conducted at the OSU-Benton County Green Stormwater Infrastructure Research (OGSIR) Facility in Corvallis, Oregon. OGSIR is a field research facility for testing GI installations that captures runoff from approximately 9,300 m^2 of catchment area in the county property, as shown in Figure 1. The county property is a maintenance and service facility that is representative of many industrial sites used for parking, storage, and refueling county vehicles and storage of construction materials and equipment. Stormwater generated in the catchment is routed via underground stormwater pipes into a 5,700 liters in-line underground storage tank, and then pumped into a sedimentation bay (or, sediment bay) via an effluent pump. The stormwater in the sedimentation bay then overflows via 45°, 4 in, V-notch weirs into each of the three bioswale cells of OGSIR facility for treatment. The three cells are each 28.4 m long, 3.2 m wide, and 0.9 m deep, except at the middle of each cell where there is an additional 0.3 m deep longitudinal trench containing a perforated underdrain running along the length of the cell. While, during the study period, Cell 1 contained no vegetation, bioswale Cell 2, and Cell 3 contained vegetation

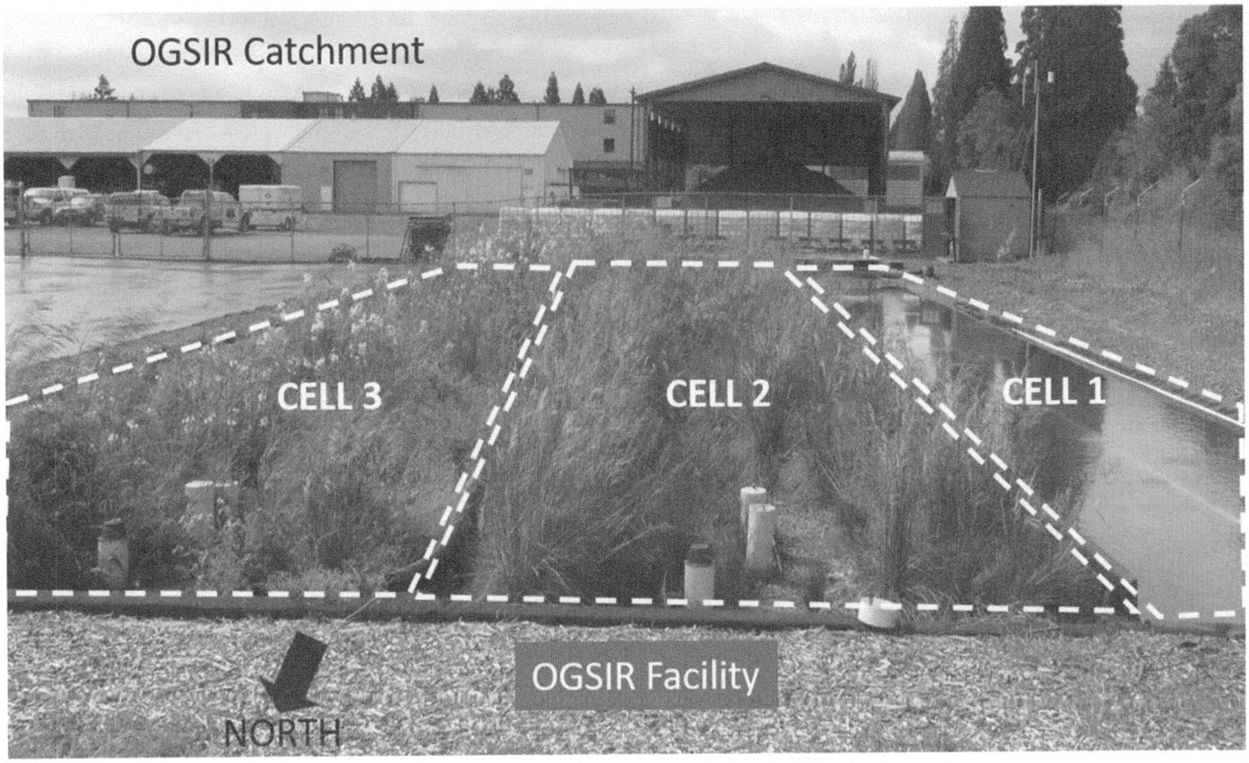

Figure 1 | OSU-Benton County Green Stormwater Infrastructure Research (OGSIR) facility and its catchment at far end.

native to the local region and resilient to local climate. Cell 2 was originally planted primarily with sedges, rushes, and grasses, and Cell 3 was planted with a mixture of grass and broad leaf vegetation. In both cells, vegetation die-off occurred between the planting date in the summer of 2014 and the data collection dates in the summer of 2019. Furthermore, the two cells have heterogenous canopies due to the irregular plant locations and variations in plant health, species, and heights (<0.3–1.2 m), with bare areas dominated by weeds during the spring and dry grass during the summer. Hence, in this study, the measurement of vegetation parameters was focused on specific areas in the two bioswale cells that covered a range of originally planted species.

Ground-based monitoring approach

Ground-based measurements of fIPAR for estimating vegetation health were made using the Accupar LP-80 ceptometer. Measurements with the ceptometer were made within 2 h of local noon, as shown in Figure 2. Table 1 lists

Figure 2 | Accupar LP-80 ceptometer in use at the OSU-Benton County Green Stormwater Infrastructure Research (OGSIR) facility and the resulting AOI and wand orientation beneath a plant canopy.

Table 1 | Dates when data were collected from the different cells at OSU-Benton County Green Stormwater Infrastructure Research (OGSIR)

Date	fIPAR collected	UAS imagery collected
June 1, 2019	–	All cells
June 15, 2019	Cell 3	All cells
June 16, 2019	Cell 2	–
June 29, 2019	Cell 2	All cells
June 30, 2019	Cell 3	–
July 27, 2019	–	All cells

that dates when fIPAR data were collected, once in the middle of June and then again at the end of June, in 2019. The mid-June sampling event involved the collection of fIPAR data on June 15, 2019 in Cell 3. However, because of the extensive human labor involved in using ceptometer and the need for the data collection to occur within 2 hours of noon, the end-of-June sampling event was split over 2 days. Cell 2 fIPAR data were collected on June 29, 2019, and Cell 3 fIPAR data were collected on June 30, 2019. Since, both days (June 29 and 30) had similar weather, the data from two cells were combined to represent end-of-June (June 29, 2019) readings that coincided with UAS scanning on June 29, 2019 (discussed in detail in the next sub-section).

PAR readings from the 80 sensors in the probe were grouped into eight segments along the 81.3 cm (32 in) length of wand. A Leica GS14 Global Navigation Satellite System (GNSS) receiver, receiving master auxiliary concept (MAC) network corrections from the Oregon Real-time GNSS Network (ORGN), was additionally used to collect coordinates at the center of each area of interest (AOI), following procedures that previous studies have shown capable of yielding accuracies on the order of 2 cm horizontally and 3 cm vertically (Allahyari 2016; Allahyari *et al.* 2018). AOIs represented physical canopy space of individual plants that were randomly selected along the length of the bioswale cells and across plant groups (e.g., sedges, rushes, grasses, and broad leaf plants).

At each AOI, the GNSS antenna pole was centered over the point, and PAR measurements were taken above the canopy and below the canopy, carefully avoiding shading the sensors during measurements. Measurements were taken using a variation on the Plant Method described by Johnson *et al.* (2010). To reduce the effects of directionality and canopy heterogeneity on measurements, measurements were taken in four directions (N-S, E-W, NE-SW, and NW-SE) with each measurement centered in the middle of the AOI, as shown in Figure 2. A total of 32 measurements were acquired both above and below the canopy at each AOI. The average of above and below canopy measurements were used to calculate the fIPAR (Serrano *et al.* 2000), as shown in the following equation:

$$\mathrm{fIPAR} = 1 - \frac{T}{S} \tag{1}$$

where T is the PAR transmitted through the canopy (i.e., the below canopy PAR measurement), and S is the incident PAR at the canopy (i.e., or the above canopy PAR measurement).

UAS-based monitoring approach

Flight mission planning

All flights in this study were conducted under 14 CFR Part 107, with the UAS operated by a Remote Pilot in Command (RPIC) holding a Remote Pilot certificate with a Small UAS rating. When deemed necessary, a visual observer was used to assist the RPIC in maintaining safe flight operations and ensuring visual line of sight to the remote aircraft at all times.

While creation of orthomosaics is well understood for large agricultural operations, capture of small urban targets at high resolution presents some unique challenges. There are two primary regulatory constraints to operating UAS in urban areas. The first is that urban areas – particularly those served by major airports – often contain controlled airspace below 400 ft, sometimes extending down to the ground surface. Fortunately, the FAA's Low Altitude Authorization and Notification Capability (LAANC) has greatly facilitated the process of applying for and obtaining an airspace authorization. In some cases, the more restrictive regulation related to operating in urban areas is the stipulation that: 'No person may operate a small unmanned aircraft over a human being unless that human being is: (a) Directly participating in the operation of the remote aircraft or (b) Located under a covered structure or inside a stationary vehicle…' (FAA 2016b). Because urban areas tend to contain large numbers of pedestrians and moving (i.e., nonstationary) vehicles, small UAS (sUAS) operation under Part 107 is significantly more challenging than in rural areas. One option is to apply for a Part 107.39 waiver. Depending on the area and the flexibility in scheduling, it may also be possible to schedule sUAS operations for times when pedestrian and vehicle traffic is restricted, such as when an area is blocked off for construction. It is recommended that municipalities adopting UAS technology for stormwater management facility monitoring maintain a formal UAS program with documented operational procedures and responsibility for managing waivers and establishing safety plans.

Another challenge is that there may be obstacles (e.g., trees, buildings, and poles) within or near the scan area. The RPIC must identify these and determine their height to establish a safe altitude for operations. Once a safe flight altitude is established, it is necessary to verify that the desired ground sample distance (GSD) can be obtained from that altitude or higher. For GI site monitoring, the limiting factor in establishing the GSD is typically the need to be able to identify individual plants in the imagery. The required altitude to achieve this GSD is a function of the physical pixel size on the camera chip and the focal length of the camera lens. Because of the complex and diverse geometry of GI vegetation, it is important to capture with a large image overlap setting in order to maximize the number of tie points. For this research, 95% overlap (endlap and sidelap) was used. For flight at low altitude with tight turnarounds, a low flight speed (1.5–3 m/s) is appropriate. This limits motion blur and allows smoother turns.

Once programmed in the mission planning software, the UAS is generally able to automatically execute the preplanned flight plan, with no direct input from the RPIC. However, the RPIC is required to ensure that the aircraft is ready for flight in all respects and to be prepared to handle any unexpected incidents aloft. In particular, it is important to ensure that the UAS compass is well calibrated on site. Failure to do so may lead to skewed images and unsteady flight. The RPIC must be prepared to immediately assume direct control if they judge that the vehicle is at risk – many UAS obstacle avoidance systems are disabled during autonomous flight.

Data acquisition and processing

Near-Infrared (NIR) and Red-Green-Blue (RGB) imagery were simultaneously collected with drones on three occasions (see Table 1), on June 1, 2019, June 29, 2019, and July 27, 2019. The flight on each day occurred within 2 hours of noon. The imagery collected on June 1, 2019 was acquired simultaneously with the built-in camera on a DJI Mavic Pro UAS and a MAPIR Survey3N NIR camera mounted externally on the Mavic Pro (DJI 2017a; MAPIR 2017). The image sets on June 29, 2019 and July 27, 2019 were acquired similarly with a DJI Phantom 4 Pro UAS with the Survey3N NIR camera mounted externally on the Phantom 4 Pro (DJI 2017b). In addition to the dates above, RGB imagery was also collected using the Mavic Pro on June 15, 2019, so that acquisition of ground-based fIPAR data coincided with dates and times of drone flights.

Imagery was acquired at an approximate altitude of 21 m above the rain garden at 1.8 m/s with 95% endlap and sidelap in a lawn mower pattern. These flight parameters enabled acquisition of high-resolution imagery with minimal motion blur and greater than the minimum amount of overlap required for structure from motion (SfM) photogrammetry, which is generally in the range of 75% front overlap ('endlap') and 60% side overlap ('sidelap') (Pix4D 2020). For safety purposes, before setting the operational flying height, the remote aircraft was first operated at a greater altitude to observe nearby obstacle heights. The highest relevant obstacle (in this case, a nearby tree) was found to be approximately 15.2 m tall, to which a safety factor of 1.4 was then applied to conclude a minimum operating altitude of 21 m. A low speed of 1.78 m/s was selected to minimize motion blur and to allow for clean turnarounds between the tightly spaced passes in the urban environment.

Tests were performed with passes aligned with the long and short dimensions of the survey area, with no observed difference in quality. The collected imagery was then processed using WebOpenDroneMap (Zecevic *et al.* 2017), an open-source SfM photogrammetry software package, to create orthomosaics with a GSD of 1 cm, with the resolution selected to enable individual plant-level analysis. The reason for investigating WebOpenDroneMap, as opposed to commercial SfM software, is that the methods of this study were deemed more likely to be implemented in practice, if the costs and relatively steep learning curves of commercial off the shelf (COTS) software could be avoided through the use of free, open-source, web-based options.

Radiometric calibration

In retrievals of NDVI from uncalibrated consumer-grade cameras on UAS, extraneous variables related to the imaging geometry and camera acquisition parameters (e.g., ISO, aperture, and shutter speed) can impact the results, if not appropriately accounted for (Berra *et al.* 2017). Although NDVI inherently performs a type of normalization (in fact, the 'N' in the acronym stands for 'normalized'), it is insufficient to account for large changes in camera and environmental parameters. This is particularly true when the RGB (red, green, and blue) and NIR imagery are acquired from separate cameras, with independently modified acquisition settings, and when acquisition is performed on different days, as done in this study. By performing radiometric calibration, it is possible to reduce the impacts of these extraneous variables and obtain pixel values that are more closely related to true surface reflectance in different image bands and, hence, NDVI values that can be meaningfully compared across time and space. The next two sub-sections provide an overview of how radiometric calibration was performed in this study.

Spectral data acquisition. Reflectance spectra were acquired with the OceanOptics Flame-S spectrometer and the OceanView software (Ocean Optics 2013, 2015). The instrument covers the VIS and NIR spectral range (350–1,000 nm), has an optical resolution (full width at half maximum) of 1.34 nm, and a sampling interval of 0.37 nm. An optical fiber with a 25° field of view transfers light into the spectrometer.

Calibration targets were selected to meet the requirements described in Smith & Milton (1999) and Haghighattalab *et al.* (2016). Specifically, matte ethylene-vinyl acetate (EVA) foam panels were used. These are commonly used calibration targets, since they are inexpensive and act as diffuse reflectors with highly Lambertian behavior and horizontal homogeneity (Jeong *et al.* 2018; Padro *et al.* 2018).

The calibration procedure was performed within 2 hours of noon on a clear, sunny day at the study site using 24 cm × 24 cm white, gray, and black matte EVA foam calibration panels mounted on black canvas and separated by several centimeters to reduce adjacency and background effects (Padro *et al.* 2018). These panels are many times larger than the pixel sizes (1 cm/pixel) of the imagery used for calibration. Reflectance measurements were taken from 5 to 8 cm above each panel at a 45° angle (Milton *et al.* 2009; Pe'eri *et al.* 2013). Prior to measurements of each panel reflectance, dark measurements and diffuse white reflectance standard measurements were taken. Reflectance measurements were normalized with the white reference standard, or the light incident on the calibration panel surface.

Seven to eight measurements were collected from different angles above each panel between 11:10 am and 11:20 am local on a clear, sunny day on June 15, 2019. This reflectance data set was used to calibrate imagery from three dates June 1, June 29, and July 27, 2019 with the assumption that no changes to the calibration panels occurred through this time span that would affect reflectance. To shield the instrument from large temperature changes, which could potentially impact the acquired spectral measurements, the instrument was placed in an insulated housing during data collection. The temperature inside the insulated housing was recorded immediately preceding the start of spectra collection for the white calibration panel, between the collection of spectra for each panel, and after all spectra were collected. The panels were stored indoors in a dry, climate-controlled location between the field acquisition dates.

Empirical line method. To perform the empirical line method (ELM), the mean reflectance value corresponding to the spectral ranges of the UAS camera image bands was found and plotted for each calibration target (Wang & Myint 2015). Since manufacturer-provided spectral sensitivity curves for the built-in RGB cameras on the UAS were not available, typical spectral ranges for complementary metal–oxide–semiconductor (CMOS) image sensors were assumed: 580–670 nm for the red band, 480–610 nm for the green band, and 400–520 nm for the blue band (Hunt *et al.* 2013). For high-accuracy applications, future work could include a step of empirically determining the spectral sensitivity curves of the camera bands, but, for purposes of this study, the impacts of assuming default (typical) spectral ranges were deemed negligible; a difference of up to ±30 nm in the assumed spectral range of the red band was calculated to produce only a 0.52–3.29% difference in the reflectance output from the ELM. In contrast to the RGB camera, the center wavelength corresponding to the

MAPIR Survey3N NIR camera is published: 850 nm (Ramseyer 2017).

The mean digital numbers (DNs) were extracted from the calibration panels in the orthophotos for each of the four bands. Only the central pixel values of each calibration target were extracted from the orthophotos to avoid boundary effects (Smith & Milton 1999; Wang & Myint 2015; Haghighattalab *et al.* 2016). The mean DN of each target in the orthomosaics was plotted against the mean reflectance value obtained from the spectrometer measurements for each camera waveband, where each point in the plot represents a different calibration target (Wang & Myint 2015). The resulting regression models were used to determine the calibration parameters (the *y*-intercepts) for each waveband (Wang & Myint 2015). The equations were then applied to the individual bands of the orthophotos.

Normalized difference vegetation index

The NIR and RGB orthomosaics generated in the previous step were georeferenced using control points at the four corners of the bioswale facility, and the bands were combined to create NDVI maps. The NDVI, Equation (2), is the normalized difference between the NIR and the visible red band. This equation can either be formulated in terms of the raw DNs in the red and NIR image bands, or in terms of surface reflectance, ρ, computed from the raw DNs, as discussed in Section 3.2.1 (Gamon *et al.* 1995):

$$\mathrm{NDVI} = \frac{\rho_{\mathrm{NIR}} - \rho_{\mathrm{Red}}}{\rho_{\mathrm{NIR}} + \rho_{\mathrm{Red}}} \tag{2}$$

Healthy leaves produce an abundance of chlorophyll pigments, which absorb highly in the red part of the spectrum (Knipling 1970). However, chlorophyll pigments have no effect on the absorption or reflectance in the NIR. Instead, the structure of the leaves, specifically the spongy mesophyll and air cavities, is responsible for the high reflectance in the NIR part of the spectrum, where NIR light is scattered and reflected (Knipling 1970).

NDVI values are floating point numbers ranging from −1 to 1, where negative values indicate the presence of water or possibly wet soil, values close to zero indicate soil or dead vegetation, values in the low positive range indicate stressed vegetation, and values in the high positive range represent healthy vegetation (Weler & Herring 2000).

Data analysis

Regression analysis

The relationship between the NDVI derived from the UAS imagery and the ground-based measurements of the fIPAR was explored by overlaying the orthomosaic and the corresponding fIPAR measurements acquired at the end of June. The mean of NDVI pixel values falling within each AOI was extracted, with each AOI of approximate area 0.5 m^2 containing between 4,000 and 5,000 pixels in the three NDVI images. Figure 3 shows the location of 18 AOIs that were sampled on June 15 in bioswale Cell 3 and the location of 41 AOIs that were sampled on June 29–30 in the two bioswale cells – Cell 2 and Cell 3. Since NIR imagery was not collected on June 15, 2019, a regression analysis was performed on the data collected at AOI locations only on June 29–30, 2019 to explore the correlation between fIPAR and NDVI for the two cells in the bioswale facility. The data points were grouped by the cell they were located inside, as well as based on which sedge, rush, or broad leaf vegetation groups the data point represented. A regression analysis was then performed separately on each group.

Spatial analysis

NDVI and fIPAR values at different locations were examined for spatial trends related to proximity of locations to the bioswale cell inlets (overflow sediment bay in Figure 3), the primary source of water during the spring growing season. As water is conveyed down the length of the cells (toward the north direction), it infiltrates and enters the underdrain. Though drip irrigation is used during the summer as a vegetation maintenance intervention in these bioswale cells, the plants in the far north end of the cells generally receive a lower annual amount of stormwater in comparison to the end closer to inlets in sediment bay. To determine if the proximity to the inlet is related to the health of vegetation, the values of NDVI and fIPAR were related to the distance from the inlet. The relationship

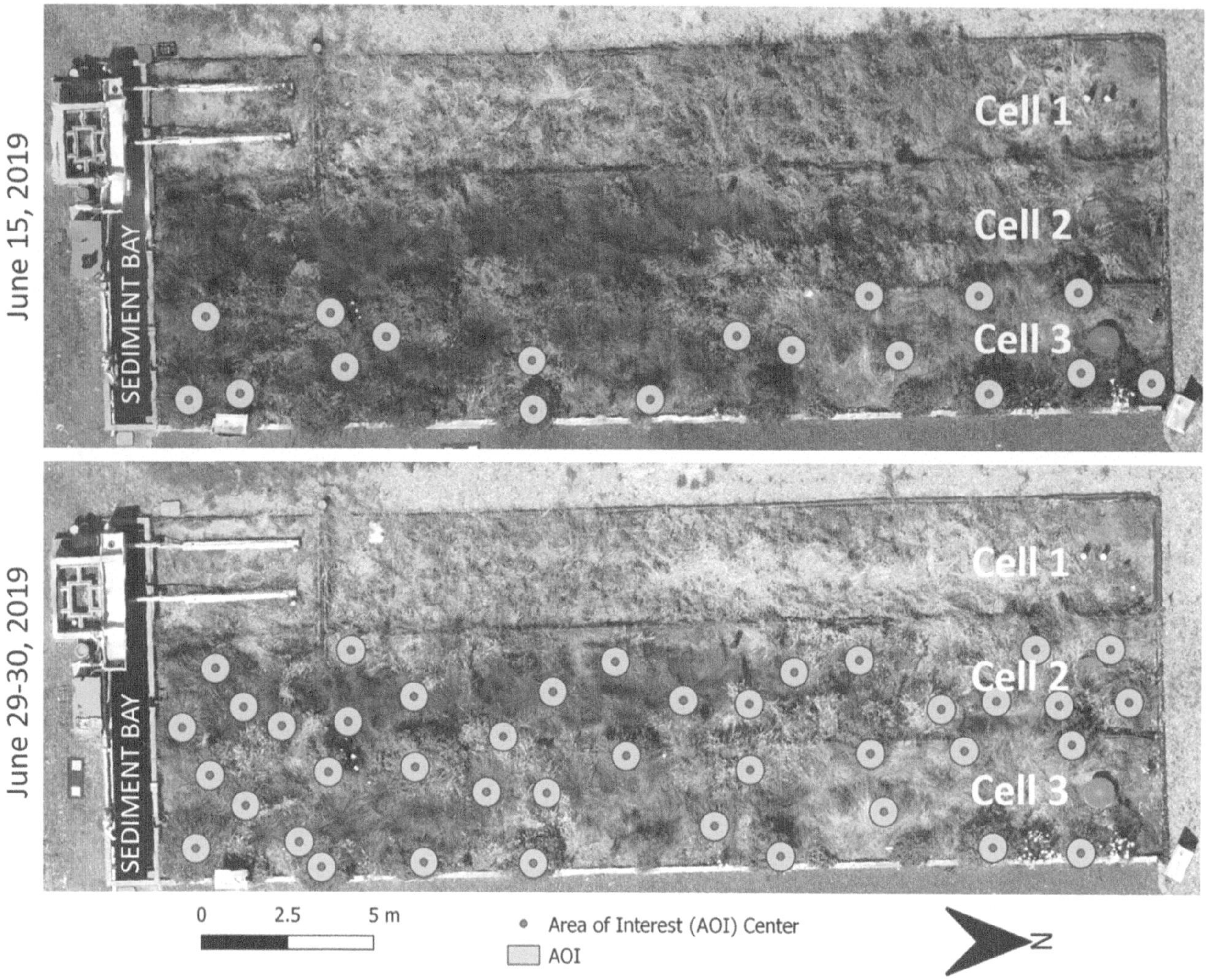

Figure 3 | Locations of data points and AOIs in bioswale Cells 2 and 3 overlaid on UAS-derived RGB imagery collected on June 15, 2019 (top) and June 29, 2019 (bottom).

between fIPAR and the distance from the inlet was examined for all sampling locations.

Temporal trend analysis

The temporal trend analysis examines NDVI data from the summer of 2019, with images acquired monthly at the end of May, June, and July. NDVI data within set ranges of values were extracted from each of the three bioswale cells to examine how the area of coverage changes for each NDVI range over the course of the 2 months. The ranges of NDVI values explored were 0.6–1.0, representing healthy vegetation, 0.2–0.6, representing unhealthy or stressed vegetation, −0.2 to 0.2, representing soil and dead vegetation, and −1.0 to −0.2, representing a complete absence of vegetation. These ranges were empirically selected to represent vegetation health status and the presence of soil or lack of vegetation for this particular study site.

RESULTS AND DISCUSSION

Ground-based monitoring of fIPAR

Spatial analysis

Figure 4 shows the classified fIPAR values overlaid on RGB orthomosaics for one cell (Cell 3) of the bioswale facility on

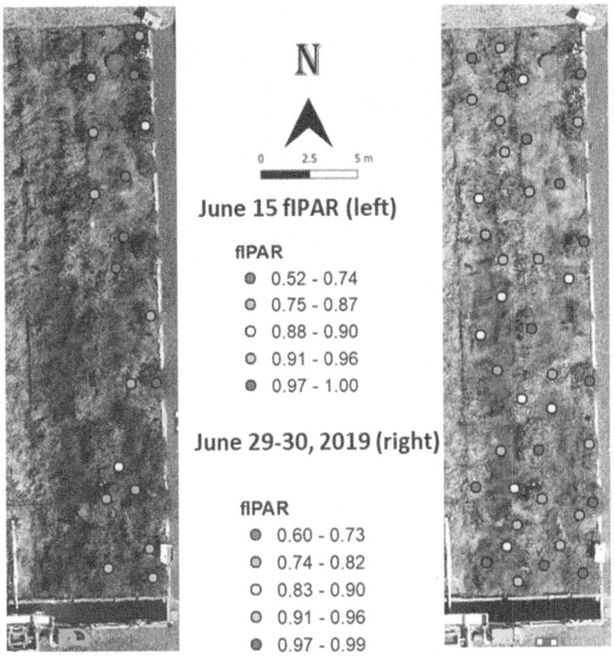

Figure 4 | Classified fIPAR at 18 AOIs for June 15, 2019 and 41 AOIs for June 29–30, 2019 overlaid on RGB orthomosaics from UAS-derived imagery obtained on June 15 (left) and June 29 (right).

June 15, 2019 and two cells (Cells 2 and 3) on June 29–30, 2019. The GNSS data for Cell 2 gathered on June 15 were unreliable (post analysis revealed that the receiver was likely not initialized) and, therefore, were not included in the fIPAR analysis. The Cell 3 sampling locations were refined on June 29 to maximize the coverage of plant groups. Figure 5 graphs these fIPAR values in relation to the distance from the inlet to each cell. When examining the fIPAR in relation to the distance from the inlet, there is very little change as the distance between the sample points and the inlet increases. The range of fIPAR values was found to be above 0.50, or more than 50% of light is intercepted by vegetation, on both dates, with no values being lower than 0.60 on June 29–30, 2019 and no values lower than 0.52 on June 15, 2019. Furthermore, the fIPAR values were found to not be consistently high for vegetation that visibly appeared healthy or low for vegetation that visibly appeared unhealthy. Although vegetation visually appears unhealthy, many species of grasses are dormant in the summer months, indicating that these species may retain density during dormancy. Hence, without additional differentiation between unhealthy and dormant vegetation and without an understanding of how the dormancy of vegetation species present in these facilities can affect vegetation density, we found that fIPAR values cannot be reliably alone used as an indicator of vegetation health. Other studies have also reported challenges in using fIPAR for sampling mixed vegetation in various growth stages and can have very different values that are dependent on the plant species (Hatfield *et al.* 1984; Serrano *et al.* 2000).

Sedges and rushes are common species in GI and are present in the OGSIR facility. The mean fIPAR value obtained on June 29–30, 2019 for a sample of eight sedge plants (including slough and dense sedge species) in the OGSIR facility, as shown in Figure 6(a), was found to be 0.92. The mean fIPAR value for seven rush plants (including hardstem bulrush and spreading rush species) sampled in the facility at the same time, as shown in Figure 6(b),

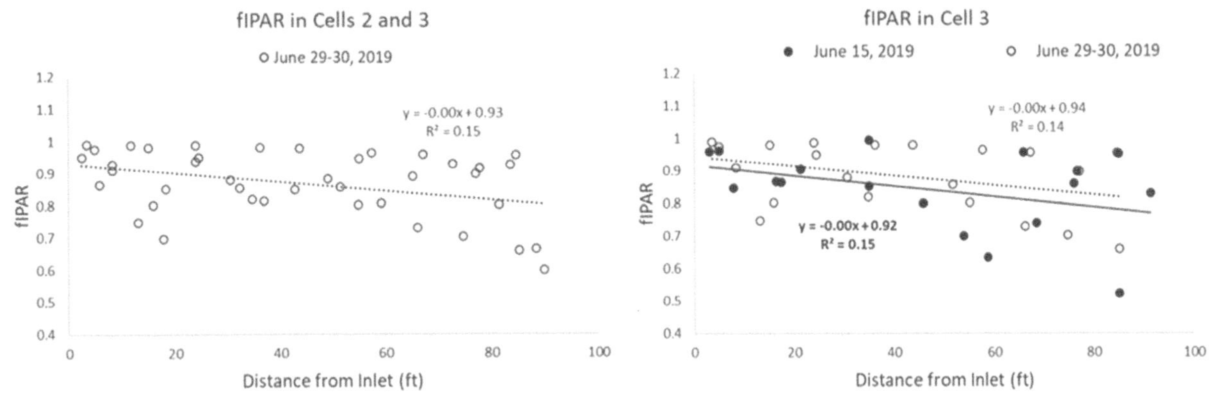

Figure 5 | fIPAR values related to the distance from the inlet, the primary water source.

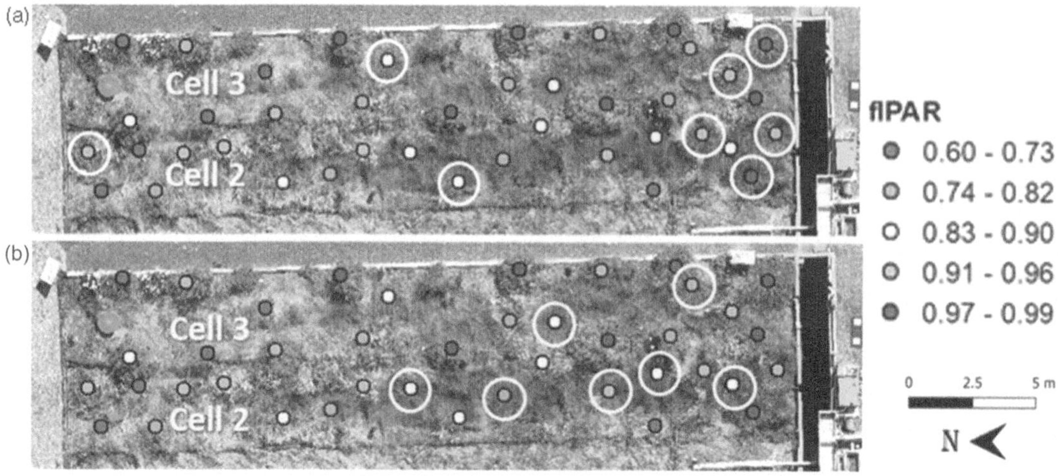

Figure 6 | Locations of sampled sedge vegetation shown circled in white in (a), and sampled rush vegetation shown circled in yellow in (b) in the OGSIR facility and associated fIPAR values collected on June 29–30, 2019. Please refer to the online version of this paper to see this figure in colour: http://dx.doi.org/10.2166/hydro.2020.195.

was found to be 0.85. Previously reported fIPAR values in the literature for slough sedge, hardstem bulrush and poverty rush include those that were sampled from natural stands in four different wetlands in the United States. These values for two wetlands in Texas, one in Delaware, and one in North Dakota were found to be 0.67, 0.32, and 0.47, respectively (Williams *et al.* 2017). Note that these values obtained in the natural wetlands are much lower than those obtained in OGSIR's engineered environment. This indicates that sedge and rush vegetation could be denser in engineered systems than in natural systems. Since fIPAR values of wetland species or species commonly used in urban GI are seldom reported in the literature, the use of values reported for natural wetlands may lead to underestimation of plant growth parameters.

Method evaluation

There are many practical challenges in obtaining reliable PAR results in bioswale cells, potentially resulting in biased fIPAR measurements. At the OGSIR facility, the cross-section geometry of the topography of bioswale cells is trapezoidal with side slopes of 4:1 (H:V). Obtaining level PAR measurements in four directions beneath the canopy of shrubs requires compensating for this side slope by raising the Accupar wand and obtaining measurements within instead of entirely beneath the canopy. This also occurs close to the bioswale cell walls, where the height of the walls requires raising the wand to acquire level measurements. This results in an underestimate of fIPAR, where the Accupar sensors are not able to detect the radiation intercepted by the entire canopy, but only part of the canopy. The opposite situation can occur when obtaining measurements next to wells and other obstructions in the bioswale cells. In this case, some AOI measurements include shadows from these obstructions, creating an overestimate of the fIPAR at the AOI.

Challenges also occur from structural differences between species and from the use of a GNSS unit to obtain coordinates at AOI center points. The rushes and sedges form dense clumps at the ground level instead of woody stems common below the canopies of shrubs. Pushing the Accupar wand through the clump at the base of sedges and rushes disturbs the plant, changing the structure of the plant and the resulting PAR measurements. Placing the GNSS antenna pole into the center of plants selected for AOIs also disturbs the plant canopy and acts as an obstruction, casting a shadow during measurement. While care was taken to minimize disruption to the vegetation when obtaining measurements, some areas of the facility have very dense vegetation where it is necessary to disturb vegetation in order to place the instruments, potentially resulting in biased measurements.

To collect data for 41 AOIs in this small GI facility on June 29 took approximately 4 hours of field labor with one additional hour of processing time. This leads to approximately 6 min of field labor per plant. The larger the GI facility, the more manpower and more instruments would be necessary to acquire sufficient data to use this method for monitoring GI vegetation health. The combined cost of purchasing the Accupar LP-80 and GNSS smart antenna and controller for a small facility, and the manpower and additional cost of more instruments required to scale this method to larger GI facilities reduces the feasibility of using this approach for routine maintenance and monitoring operations.

UAS-based monitoring approach

Figure 7 shows the orthomosaics generated with WebODM from the imagery acquired on June 1, 2019, June 29, 2019, and July 27, 2019. The June 1 and June 29 image sets were acquired in full sun, while the July 27 image set was acquired in overcast conditions. The images have a resolution of 1 cm/pixel, with minimal visible distortion. The camera settings were constant between the flights with the same UAS configuration, but the light conditions were different on each date resulting in the dissimilar brightness of orthomosaics at the sampling dates.

Radiometric calibration

Reflectance spectra. Eight reflectance spectra were recorded for each white, gray, and black calibration panel on June 15, 2019 at the OGSIR site. To reduce noise, the means of the eight observed reflectance spectra were computed for each panel, and the resultant mean reflectance spectra (Figure 8) were used in all subsequent steps. To evaluate the consistency between the observed spectra for each panel, the standard deviations were also computed and found to be low: ±3.0, ±1.2, and ±0.2 percentage points of reflectance for the white, gray, and black calibration panels, respectively.

Image DNs. The standard deviation of the DN values ranges from 0.36 to 3.77, or 0 to 1.48% of the 8-bit DN value range (0–255). Jeong *et al.* (2018) tested commercial calibration panels and found that the standard deviation of DN values ranged from 0 to 12.66, or 0 to 4.96% of the 8-bit DN value range. The DN values for these panels on all dates are well within this range. The low standard deviation of calibration panel reflectance data and DN values in the two image bands confirms that the panels exhibit Lambertian properties.

Empirical line method. Figure 9 shows the scatter plots and the two linear regression models examined for converting image DNs in the red and NIR wavebands to reflectance for the June 1, June 29, and July 27, 2019 data set. The relationship between surface reflectance and image DNs is not always linear, as shown by Wang & Myint (2015). However, higher-order polynomials can lead to overfitting, and with only three calibration targets in this study, a linear fit was deemed most appropriate. The relationship between

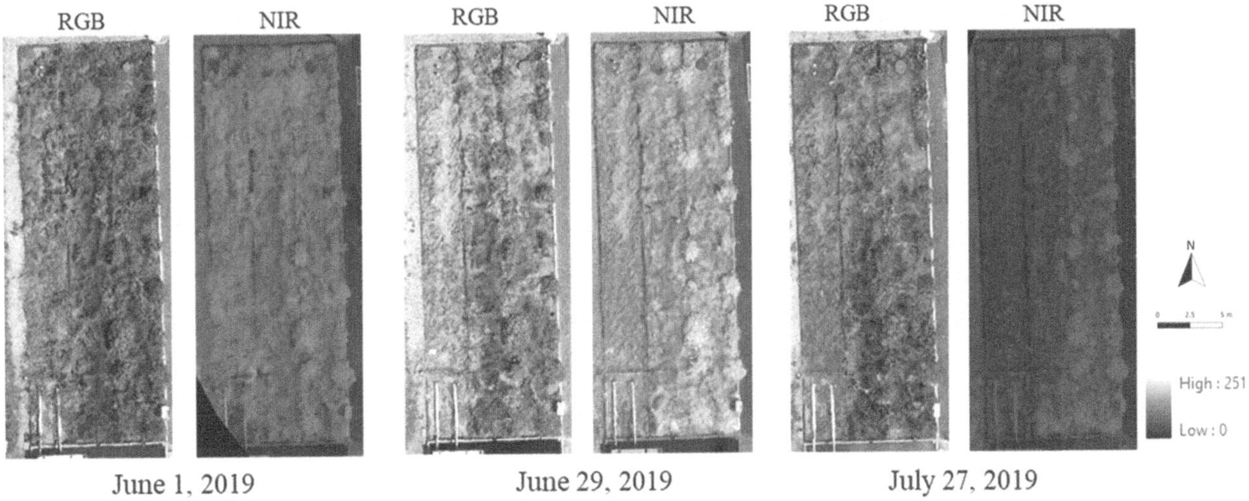

Figure 7 | RGB and NIR orthomosaics generated from image sets acquired on three dates showing the difference in coloration and brightness between the two cameras (Mavic Pro RGB on June 1, 2019 and Phantom 4 Pro RGB on June 29 and July 27, 2019) and with the same NIR camera with different environmental conditions.

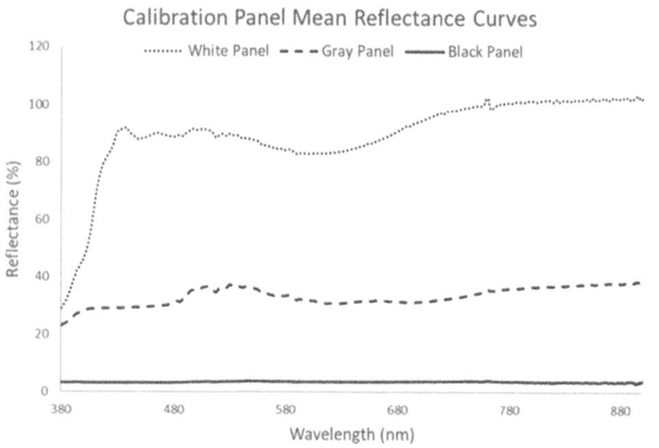

Figure 8 | Mean reflectance curves for the white, gray, and black calibration panels.

the image DNs and the surface reflectance of the calibration panels is given in the following equation and illustrated in Figure 9:

$$R_i = b\mathrm{DN}_i + a \tag{3}$$

where R_i is the converted reflectance of band i pixels, DN_i is the uncalibrated DN of band i, and a and b (the offset and scale factor, respectively) are the parameters determined through the linear regression (Wang & Myint 2015; Jeong *et al.* 2018).

Each of these calibration equations in Figure 9 has a negative y-intercept. To avoid nonsensical negative reflectance values, the red-band DNs below 54.61, 75.48, and 78.67 and the NIR-band DNs below 9.03, 6.85, and 4.45 for the June 1, 2019, June 29, 2019, and July 27, 2019 data sets, respectively, were all set to the lowest possible value of zero reflectance.

Method evaluation

Acquiring UAS imagery in the field involved minimal time and effort, with each flight mission for scanning all bioswales at the OGSIR facility being completed under 1 h. Each mission included setup time and break-down time, and acquisition of one to two backup data sets. The two UAS used for the three flights produced good quality data,

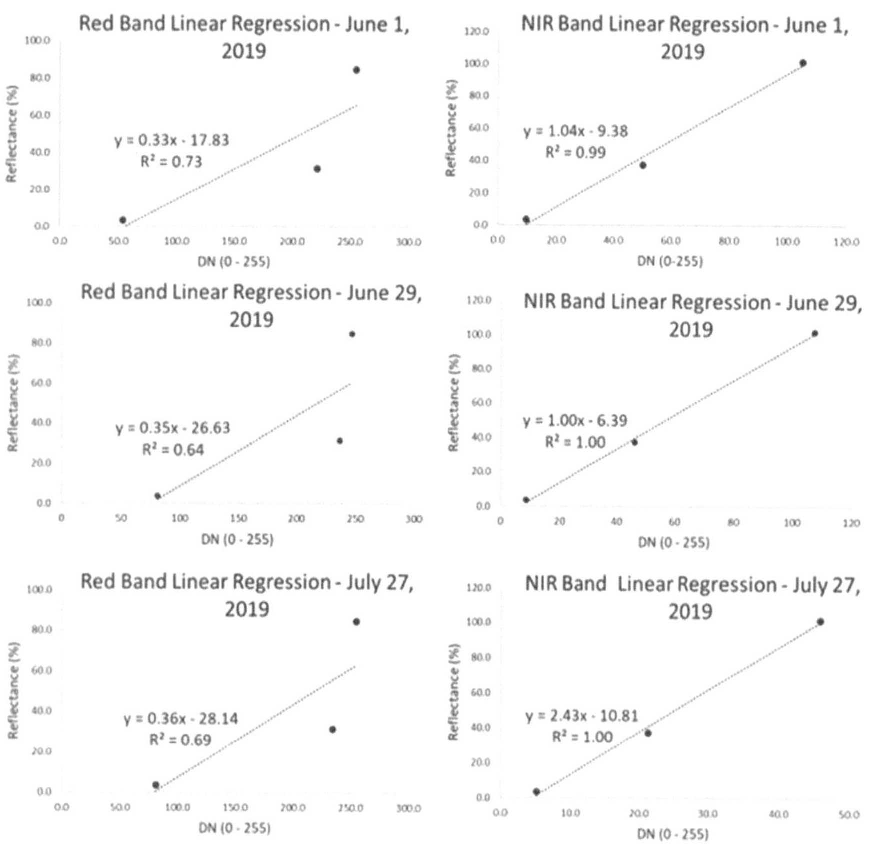

Figure 9 | The linear relationship between calibration panel surface reflectance and image DN for the three flight dates, June 1, June 29, and July 27, 2019.

with minimal distortion in the final orthomosaics obtained from both the DJI Mavic Pro and the DJI Phantom 4 Pro data sets. Data processing times with the structure-from-motion software WebODM varied by camera, with the NIR data sets processing requiring up to 1–2 h and the RGB data sets processing requiring 12–24 h. Data were processed with a 2.20 GHz AMD A8-7410 APU. Obtaining a validation set of EVA foam panels is inexpensive with EVA foam panels frequently sold in packages with multiple panels of various colors.

Normalized difference vegetation index

Figure 10 shows classified NDVI maps of the OGSIR bioswale facility for the three dates with the same 10 classes, where there are five classes from −1.0 to 0.0 and five from 0.0 to 1.0. The spatial distribution of each class is shown in the figure. The shades of green represent healthy vegetation, yellow represents senescent or fading vegetation, orange and brown represent soil and dead vegetation, and red represents no vegetation. The values close to −1 are in the gravel/rock on the side of the rain garden. Cell 1 contains only weeds, showing the effect they have on the NDVI in all cells in the data sets. Once these weeds die, the remaining green areas in Cells 2 and 3 represent the original vegetation planted in the facility in 2014. The largest concentration of green values, or healthy vegetation on June 29 and July 27, is close to the inlet on the south end of the facility, with the coverage of green values decreasing at one-third of the length of the facility and continuing to decrease moving away from the inlet.

Spatial analysis

Figure 11 shows the NDVI change in relation to the distance from the inlet. The drop in the NDVI values for the entire

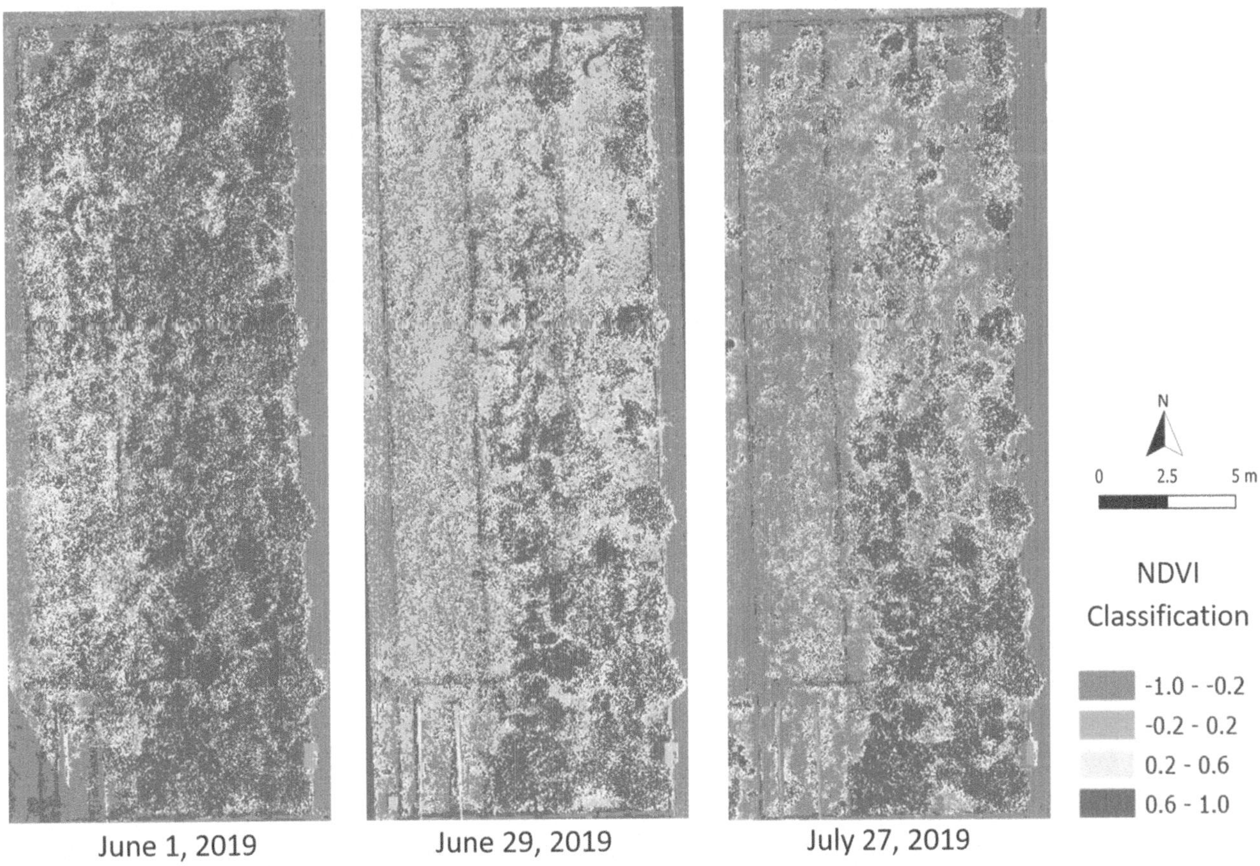

Figure 10 | Classified NDVI data for the three dates showing the spatial distribution of NDVI values in bioswale cells. Please refer to the online version of this paper to see this figure in colour: http://dx.doi.org/10.2166/hydro.2020.195.

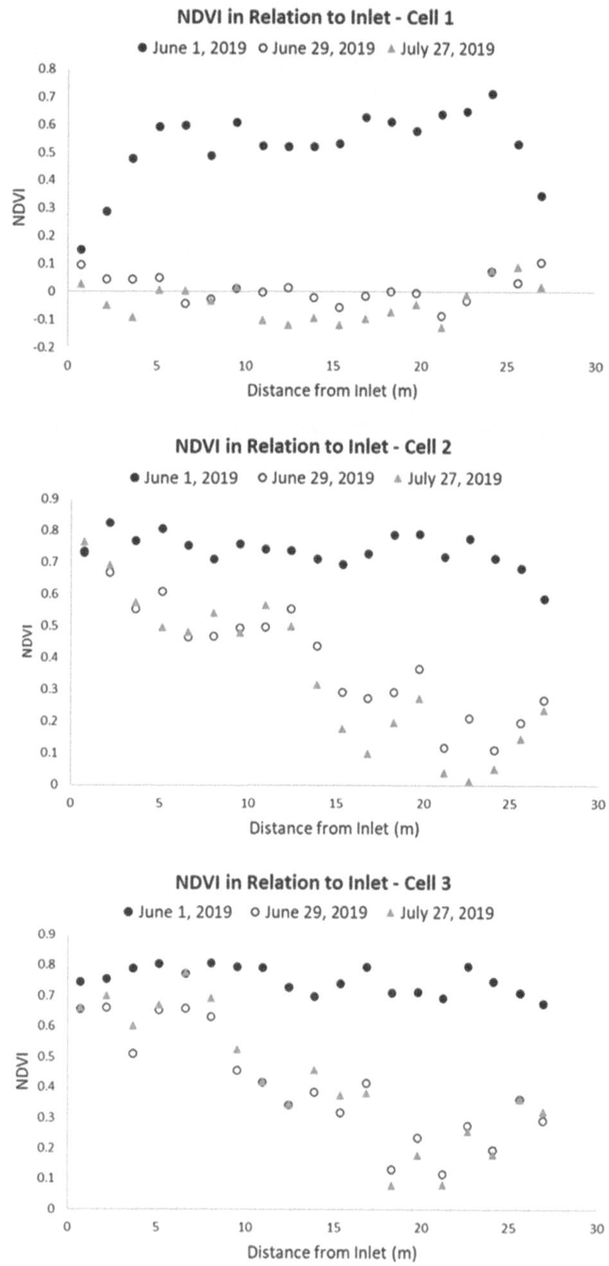

Figure 11 | NDVI related to distance from the inlet for Cells 1, 2, and 3.

area of Cell 1 represents the change in weed coverage as water availability and the temperature changed from June 1 to later in summer (June 29 and July 27). In addition to Cell 1, the high NDVI values in Cell 2 and Cell 3 on June 1, 2019 also coincide with the extensive coverage of weeds. Once the weeds died in Cell 2 and Cell 3 later in the summer, the NDVI values in these cells also experienced a downward trend, especially at locations further away from the inlet. NDVI values in Cell 2 stayed between 0.45 and 0.75 within 12 m from the Cell 2 inlet, with values falling below 0.4 at distances larger than 15 m from the inlet. NDVI values stayed between 0.5 and 0.8 within 9 m from the Cell 3 inlet, with values falling below 0.4 at distances larger than 18 m from the inlet.

Note that eight of the 41 AOIs that were used for ground-based monitoring were at sedge locations, while seven were at rush locations (Figure 6(a)). The mean NDVI of the eight sedges estimated using the UAS-based monitoring approach was found to be 0.82, 0.78, and 0.82 on June 1, June 29, and July 27, respectively, while the mean NDVI for the rush locations was 0.67, 0.79, and 0.69 on the same three dates. Though we recommend additional soil and plant-tissue sampling in future studies to explore this relationship for GI systems, strong relationships between NDVI, biomass, soil, and water have been reported for natural wetlands (e.g., Boelman *et al.* 2003; Goswami *et al.* 2015). Considering these previous studies, the high NDVI values indicate dense vegetation and a correspondingly high biomass in the OGSIR facility, especially close to the inlet. Changes in NDVI-based biomass estimates over time can indicate the presence of stressors and spatial variations in these estimates can indicate stress in vegetation caused by spatial variability in the plant's environment. Because NDVI is a function of not only environmental stressors (e.g., drought) but also plant type, it is important to monitor NDVI for multiple species within a stormwater facility over time in order to best inform management decisions.

Temporal trend analysis

Figure 10 also indicates a visible change in healthy vegetation coverage from the beginning of June through the end of July. To examine temporal trends in vegetation health, we calculated the area covered by healthy vegetation, unhealthy or stressed vegetation, soil and dead vegetation, and a complete absence of vegetation. NDVI values ranging from 0.6 to 1.0 in Cell 2 covered 60 m^2 of the facility on June 1 and decreased by 44% by June 29. NDVI values close to zero and below zero covered 6 m^2 on June 1 and increased by 83% by June 29. The areas covered by these NDVI values in Cell 2 remained close to constant during July with changes in the area of 10% or less in all classes. NDVI values ranging from 0.6 to 1.0 in Cell 3 covered close to 74 m^2 on June 1 and declined by 48% by June 29. NDVI values indicating soil, dead vegetation, and no vegetation in Cell 3 covered 6 m^2 on June 1 and increased by 84% by June 29. Changes in the area of 11% or less occurred in Cell 3 during July in all classes. NDVI values ranging from 0.6 to 1.0 in Cell 1, which contained only weeds, covered an area of 32 m^2 on June 1 and decreased by 80% by June 29. NDVI values close to zero and below zero covered an area of 14 m^2 on June 1 and increased by 81% by June 29. The area covered by NDVI classes in Cell 1 changed by 21–28% during July, with classes representing stressed vegetation or no vegetation showing an increase in area and classes representing healthy vegetation and dying vegetation showing a decrease in area.

Method evaluation

The NDVI data generated from the calibrated UAS image sets proved useful in detecting changes in the area covered by vegetation classes over time and in exploring the spatial distribution of NDVI values. However, an important limitation to be aware of is that NDVI generated from different sensors cannot be directly compared without examining the spectral sensitivity regions and band peaks included in the bandpass filters of each CMOS sensor (Miura *et al.* 2006). The CMOS sensors in the Mavic Pro and Phantom 4 Pro likely have differing spectral ranges in the red spectral response curve and differing red peaks. Unless the differences are minor, they will affect the resulting NDVI values. It is important to be aware that many manufacturers, including DJI, do not make the spectral response curves available to the public. Finally, while we attempted to account for this limitation in this study by using the radiometric calibration panels and the ELM to convert to surface reflectance, additional studies may provide insight on the impacts of using different cameras without the radiometric calibration step.

Zhang *et al.* (2015) found that shadows cause an overestimate of NDVI, but when computing vegetation parameters from NDVI, such as vegetation fractional coverage, the results were not strongly affected by the presence of shadows. A side-by-side comparison of the June 29 RGB orthomosaic and the June 29 classified NDVI map indicated that part of the area covered by the highest NDVI values is shaded. While some of the areas may correspond to healthy vegetation cast in shadows and some represent shadows cast by obstructions in the facility in areas of dry or dead vegetation, these areas have very high NDVI values indicating an overestimate of NDVI.

There are many methods for detecting shadows in imagery, but histogram thresholding is the simplest method and assumes that shadow and sunlit areas have different histogram levels (Adeline *et al.* 2013). Increased accuracy can be reached by applying a shadow-detection methodology to the image bands prior to generating the NDVI data. However, UAS flights should be planned to reduce the presence of shadows by performing flights close to noon and considering the effect of seasonality on shadow size. Even without removing shaded areas from the analysis, this method allows for tracking changes in facility vegetation once a baseline of health is established. Deviations from this baseline should trigger a maintenance plan of action.

This is a promising method that could be used toward long-term periodic monitoring of facilities. With access to image analysis software, like ESRI ArcMap, the NDVI is easy to compute and analysis provides useful information on the health of vegetation in GI. The method can be improved using higher quality sensors designed for NDVI generation and using shadow-detection methods. Consistently using the same sensors for each flight ensures that resulting NDVI data will be comparable across a temporal data set, eliminating the need to generate cross-sensor models to translate reflectance from one sensor to match reflectance from another (Miura *et al.* 2006). Using a sensor that has both red and NIR bands also reduces the need for radiometric calibration using the ELM. However, the NDVI results obtained from inexpensive sensors, calibration methodology, and NDVI classification performed in this study provide a good idea of vegetation health with the potential to inform long-term maintenance decisions.

Comparison of ground-based and UAS-based monitoring of bioswales

Figure 12 shows the scatter plots and regression models to determine the NDVI relationship to fIPAR for sedge (Figure 12(a)), rush (Figure 12(b)), and broad leaf (Figure 12(c))

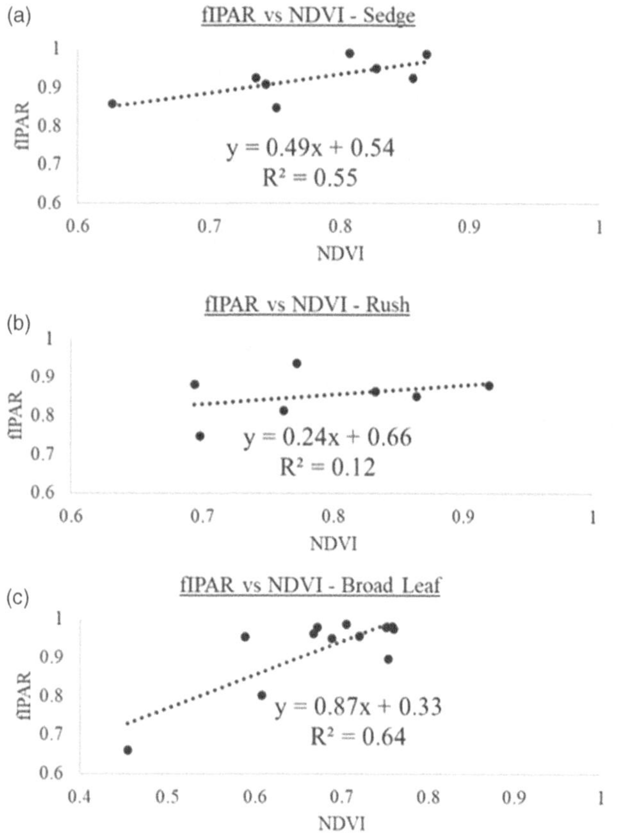

Figure 12 | Scatter plots and regression models of the NDVI relationship to fIPAR of sedge, rush, and broad leaf vegetation groups including vegetation from both cells of the OGSIR bioswale facility for data collected on June 29–30, 2019.

vegetation groups in the bioswale cells, using the data collected on June 29, 2019. It appears that the broad leaf and sedge NDVI-fIPAR relationship is stronger with coefficients of determination (R^2) of 0.64 and 0.55, with a very weak relationship for the rush group ($R^2 = 0.12$). Since the vegetation groups have small sample sizes, there is a potential for outliers to affect the regression model. For example, in Figure 12(c), the low point (bottom left of the figure) appears to have both high influence and leverage on the regression analysis. This outlier may be a function of the small sample size. Future studies are recommended to take this issue into consideration when collecting additional fIPAR samples of each vegetation group, so that the relationship for different types of stormwater GI vegetation species can be determined more accurately.

Previous studies have shown a strong relationship between radiometer-derived NDVI and ground-based measurements of fIPAR in wheat crops during the growth phase with coefficients of determination of 0.81 and 0.97 (Hatfield *et al.* 1984; Serrano *et al.* 2000). Strong relationships were also found between UAS-derived NDVI and ground-based measurements of fIPAR in peach and citrus orchards with coefficients of determination of 0.88 and 0.85, respectively (Guillen-Clement *et al.* 2012). However, these relationships were found for agricultural study sites with the relationship examined for a single crop. In contrast, the bioswale facility at the OGSIR facility contains mixed vegetation and presents unique challenges related to accessing the sampling point and taking consistent measurements in densely planted areas and varying canopy types.

CONCLUSIONS AND FUTURE WORK

This study examined the feasibility of using UAS for efficient, multitemporal monitoring of vegetation in urban bioswales. UAS imagery data sets were acquired with two remote aircrafts, and the data were processed in an open-access, web-based structure from motion (SfM) software utility to generate 1-cm resolution orthomosaics. These orthomosaics were then radiometrically calibrated to surface reflectance measurements to reduce any inaccuracies caused by the camera parameters and illumination conditions and to ensure that the multiple sensors represent the actual reflectance. This is especially important when comparing and combining data from multiple sensors and when comparing data from multiple dates. Calibration was performed via the ELM using spectral reflectance data of three calibration panels using a field spectrometer.

NDVI maps were generated from the calibrated red and NIR bands of the orthomosaics. Three NDVI maps were examined for spatial distribution and temporal changes. Cell 1 contained only weeds during the study time period, while Cells 2 and 3 contained vegetation originally planted in the facility and seasonal weeds. High NDVI values dominated all three cells on the early June NDVI map, representing the abundance of healthy weeds at the end of spring. At the end of June and July, the seasonal weeds had died and the high NDVI values corresponded to only the healthy bioswale vegetation originally planted in the facility.

Spatially, mean NDVI values in Cells 2 and 3 were found to be highest in the first 9–12 m measured down the length of the facility from the inlet, or the third of the facility closest to the inlet. Cell 3 maintained higher NDVI values in this area of the facility than Cell 2, indicating that mixed vegetation remains healthier when close to the primary water source than when vegetation is limited to the sedges, rushes, and grasses of Cell 2. Most of the mean NDVI values in Cell 3 in the 9 m closest to the inlet were above 0.6, representing very healthy vegetation in this region. When monitoring the NDVI over time and spatially, knowledge of the facility growth patterns increases the ability to establish a baseline of vegetation health. Deviations in the baseline can trigger site visits by GI facility managers and technicians, further data acquisition, maintenance, and analysis of potential causes of the deviation. Overall, NDVI maps can prove valuable for long-term monitoring of GI vegetation, especially in large GI facilities. Furthermore, this study concluded that UAS represents an efficient and cost-effective method of obtaining the NDVI data.

A second component to the study involved comparing the UAS-derived NDVI data against fIPAR measurements collected using a hand-held ceptometer to see if the former could be used to predict the latter. fIPAR measurements were obtained at multiple sampling locations in Cells 2 and 3 of the OGSIR bioswale facility. Results from this portion of the study revealed no statistically strong relationship between fIPAR and UAS-based NDVI. The practical challenges in obtaining large samples of consistent measurements using the fIPAR methods likely influenced the exploration of a relationship between the two parameters. The bioswale facility is inherently different from the agricultural study sites investigated in the previous literature, with an irregular planting arrangement, mixed species in the canopy make-up, and unregulated growth and health. Agricultural sites, where fIPAR and NDVI have been previously explored, generally have regular rows of a single crop maintained for consistent growth and health. The structure of agricultural sites promotes consistency in data collection methods with fairly controlled conditions. A lack of a relationship between NDVI and fIPAR precludes the use of NDVI results as a proxy for fIPAR in biomass productivity models of bioswale and similar urban GI facilities. However, the NDVI derived from calibrated imagery is a valuable reflectance-based vegetation health parameter and as a stand-alone indicator of stress in vegetation.

In future studies, it may be of interest to test different methods of field measurement of fIPAR (specifically, methods that enable more consistent measurements across varying species, plant densities, and canopy types) and to then reassess the relationship between field-measured fIPAR and UAS-derived NDVI. More consistent biomass measurements can be obtained using destructive methods where vegetation samples cut at ground level are dried, weighed, and used to represent the biomass of all facility vegetation (Prabhakaraa *et al.* 2015). Additionally, allometric equations can be used to determine above-ground biomass. Allometric equations for bioswale vegetation can be developed from biometric variables unique to vegetation species including plant height, diameter, crown area, and crown shape (Ali *et al.* 2015).

Another recommended extension of the methods of this study is to further investigate UAS mission planning to reduce shadows and the associated impacts on NDVI. With flights planned for the same time of day using the same flight path for individual facilities, the effect of shadows becomes dependent on seasonal changes in the sun angle. Future research is recommended to investigate how integrated UAS with additional cameras and machine learning could assist with adaptive scanning of areas of interest in bioswales. For large-scale implementation of UAS-based monitoring of urban GI facilities, additional work could also focus on operational efficiency and effectiveness, especially improving cycle times, reducing pilot workloads, and managing evolving regulations. A final recommendation for future work is to utilize additional calibration panels in the radiometric calibration procedure, to improve the regression models.

ACKNOWLEDGEMENTS

We would like to convey gratitude to students in the Hydroinformatics Research Group, with special thanks to Mr Azad Dazaea, for their assistance with collection of data. Part of this research was financially supported by Award ER18-C3-1230 funded by Strategic Environmental Research and Development Program (SERDP) at U.S. Department of Defense.

DATA AVAILABILITY STATEMENT

All relevant data are included in the paper or its Supplementary Information.

REFERENCES

Adeline, K. R. M., Chen, M., Briottet, X., Pang, S. K. & Paparoditis, N. 2013 Shadow detection in very high spatial resolution aerial imagers: a comparative study. *ISPRS Journal of Photogrammetry and Remote Sensing* **80**, 21–38. doi:10.1016/j.isprsjprs.2013.02.003.

Akpor, O. B. & Muchie, M. 2010 Remediation of heavy metals in drinking water and wastewater treatment systems: Processes and applications. *International Journal of the Physical Sciences* **5** (12), 1807–1817.

Ali, A., Xu, M. S., Zhao, Y. T., Zhang, Q. Q., Zhou, L. L., Yang, X. D. & Yan, E. R. 2015 Allometric biomass equations for shrub and small tree species in subtropical China. *Silva Fennica* **49** (4), 10. Article id 1275. doi:10.14214/sf.1275.

Allahyari, M. 2016 *Accuracy Evaluation of Real-Time GNSS Survey Observations. **M.S. Thesis***, Department of Civil and Construction Engineering, Oregon State University, Corvallis, OR.

Allahyari, M., Olsen, M. J., Gillins, D. T. & Dennis, M. L. 2018 Tale of two RTNs: rigorous evaluation of real-time network GNSS observations. *Journal of Surveying Engineering* **144** (2), 05018001.

Bartesaghi-Koc, C., Osmond, P., Peters, A. & Matthias, I. 2017 A methodological framework to assess the thermal performance of green infrastructure through airborne remote sensing. *Procedia Engineering* **180**, 1306–1315. doi:10.1016/j.proeng.2017.04.293.

Bartesaghi-Koc, C., Osmond, P. & Peters, A. 2019 Mapping and classifying green infrastructure typologies for climate-related studies based on remote sensing data. *Urban Forestry & Urban Greening* **37**, 154–167. doi:10.1016/j.ufug.2018.11.008.

Bauerle, W. L., Weston, D. J., Bowden, J. D., Dudley, J. B. & Toler, J. E. 2004 Leaf absorptance of photosynthetically active radiation in relation to chlorophyll meter estimates among woody plant species. *Scientia Horticulturae* **101**, 169–178. doi:10.1016/j.scienta.2003.09.010.

Berra, E. F., Gaulton, R. & Barr, S. 2017 Commercial off-the-shelf digital cameras on unmanned aerial vehicles for multitemporal monitoring of vegetation reflectance and NDVI. *IEEE Transactions on Geoscience and Remote Sensing* **55** (9), 4878–4886. doi:10.1109/TGRS.2017.2655365.

Boelman, N. T., Stieglitz, M., Rueth, H. M., Sommerkorn, M., Griffin, K. L., Shaver, G. R. & Gamon, J. A. 2003 Response of NDVI, biomass, and ecosystem gas exchange to long-term warming and fertilization in wet sedge tundra. *Oecologia* **135** (3), 414–421. doi:10.1007/s00442-003-1198-3.

Bréda, N. J. J. 2003 Ground-based measurements of leaf area index: a review of methods, instruments and current controversies. *Journal of Experimental Botany* **54** (392), 2403–2417. https://doi.org/10.1093/jxb/erg263.

Cordon, G., Lagorio, M. G. & Paruelo, J. M. 2016 Chlorophyll fluorescence, photochemical reflective index andnormalized difference vegetative index during plant senescence. *Journal of Plant Physiology* **199**, 100–110. doi:10.1016/j.jplph.2016.05.010.

Dimitrov, S., Georgiev, G., Georgieva, M., Gluschkova, M., Chepisheva, V., Mirchev, P. & Zhiyanski, M. 2018 Integrated assessment of urban green infrastructure condition in Karvolo urban area by in-situ observations and remote sensing. *One Ecosystem* **3**, e21610. doi:10.3897/oneeco.3.e21610.

DJI. 2017a *Mavic Pro User Manual (V2)*. DJI. Available from: https://www.dji.com/mavic/info.

DJI. 2017b *Phantom 4 Pro/Pro+ User Manual*. DJI. Available from: https://www.dji.com/phantom-4-pro/info.

Federal Aviation Administration. 2016a *14 CFR Parts 21, 43, 61, 91, 101, 107, 119, 133, and 183: Operation and Certification of Small Unmanned Aircraft Systems*. U.S. Department of Transportation.

Federal Aviation Administration. 2016b *Summary of Small Unmanned Aircraft Rule (Part 107)*. Available from: https://www.faa.gov/uas/media/Part_107_Summary.pdf (accessed 25 May 2020).

Franco, S. F. & Macdonald, J. L. 2018 Measurement and valuation of urban greenness: remote sensing and hedonic application to Lisbon, Portugal. *Regional Science and Urban Economics* **72**, 156–180. doi:10.1016/j.regsciurbeco.2017.03.002.

Gadi, V. K., Garg, A., Prakash, S., Wei, L. & Andriyas, S. 2018 A non-intrusive image analysis technique for measurement of heterogeneity in grass species around tree vicinity in a green infrastructure. *Measurement* **114**, 132–143. doi:10.1016/j.measurement.2017.09.010.

Gamon, J. A., Field, C. B., Goulden, M. L., Griffin, K. L., Hartley, A. E., Joel, G., Penuelas, J. & Valentini, R. 1995 Relationships between NDVI, canopy structure, and photosynthesis in three California vegetation types. *Ecological Applications* **5** (1), 28–41.

Goswami, S., Gamon, J. A., Vargas, S. & Tweedie, C. E. 2015 Relationships of NDVI, biomass, and leaf area index (LAI) for six key plant species in Barrow, Alaska. *PeerJ PrePrints*. doi:10.7287/peerj.preprints.913v1.

Guillen-Clement, M. L., Zarco-Tejada, P. J., Berni, J. A. J., North, P. R. J. & Villalobos, F. J. 2012 Mapping radiation interception in row-structured orchards using 3D simulation and high-resolution airborne imagery acquired from a UAV. *Precision Agriculture* **13**, 473–500. doi:10.1007/s11119-012-9263-8.

Haghighattalab, A., Perez, L. G., Mondal, S., Singh, D., Schinstock, D., Rutoski, J., Ortiz-Monasterio, I., Singh, R. P., Goodin, D. & Poland, J. 2016 Application of unmanned aerial systems for high throughput phenotyping of large wheat breeding nurseries. *Plant Methods* **12** (35), 1–15. doi:10.1186/s13007-016-0134-6.

Hatfield, J. L., Asrar, G. & Kanemasu, E. T. 1984 Intercepted photosynthetically active radiation estimated by spectral reflectance. *Remote Sensing of Environment* **14**, 65–75. doi:10.1016/0034-4257(84)90008-7.

Hill, D. J. & Babbar-Sebens, M. 2019 Promise of UAV-assisted adaptive management of water resources systems. *Journal of Water Resources Planning and Management* **145**, 2017–2020. doi:10.1061/(asce)wr.1943-5452.0001081.

Hunt, E. R., Doraiswamy, P. C., McMurtrey, J. E., Daughtry, C. S. T., Perry, E. M. & Akhmedov, B. 2013 A visible band index for remote sensing leaf chlorophyll content at the canopy scale. *International Journal of Applied Earth Observation and Geoinformation* **21**, 103–112. doi:10.1016/j.jag.2012.07.020.

Jeong, Y., Yu, J., Wang, L., Shin, H., Koh, S.-M. & Park, G. 2018 Cost-effective reflectance calibration method for small UAV images. *International Journal of Remote Sensing* **39** (21), 7225–7250. doi:10.1080/01431161.2018.1516307.

Johnson, M.-V. V., Kiniri, J. R. & Burson, B. L. 2010 Ceptometer deployment method affects measurement of fraction of intercepted photosynthetically active radiation. *Agronomy Journal* **102** (4), 1132–1137. doi:10.2134/agronj2009.0478.

Kim, H. W., Kim, J.-H., Li, W., Yang, P. & Cao, Y. 2017 Exploring the impact of green space health on runoff reduction using NDVI. *Urban Forestry & Urban Greening* **28**, 81–87. doi:10.1016/j.ufug.2017.10.010.

Knipling, E. B. 1970 Physical and physiological basis for the reflectance of visible and near-infrared radiation from vegetation. *Remote Sensing of the Environment* **1**, 155–159. doi:10.1016/S0034-4257(70)80021-9.

Lee, H. K. 2018 A panel data analysis of a spatial measurement of green infrastructure and its potential effectiveness on peak streamflow. *Environment, Development and Sustainability*. doi:10.1007/s10668-018-0210-3.

Lucke, T. & Nichols, R. W. B. 2015 The pollution removal and stormwater reduction performance of street-side bioretention basins after ten years in operation. *Science of the Total Environment* **536**, 784–792. doi:10.1016/j.scitotenv.2015.07.142.

MAPIR. 2017 *MAPIR Survey3 Quick Start Guide*. Available from: https://cdn.shopify.com/s/files/1/0878/4732/files/MAPIR_Survey3_Manual_v1.pdf?12365009533491002433 (accessed 25 May 2020).

McDonald, W. 2019 Drones in urban stormwater management: a review and future perspectives. *Urban Water Journal*. https://doi.org/10.1080/1573062X.2019.1687745.

Milton, E. J., Schaepman, M. E., Anderson, K., Kneubühler, M. & Fox, N. 2009 Progress in field spectroscopy. *Remote Sensing of Environment* **113**, S92–S109.

Miura, T., Huete, A. & Yoshioka, H. 2006 An empirical investigation of cross-sensor relationships of NDVI and red/near-infrared reflectance using EO-1 Hyperion data. *Remote Sensing of Environment* **100**, 223–236. doi:10.1016/j.rse.2005.10.010.

Mohammed, G. H., Binder, W. D. & Gillies, S. L. 1995 Chlorophyll fluorescence: a review of its practical forestry applications and instrumentation. *Scandinavian Journal of Forest Research* **10** (1–4), 383–410. doi:10.1080/02827589509382904.

Monteith, J. L. 1972 Solar radiation and productivity in tropical ecosystems. *Journal of Applied Ecology* **9**, 747–766. doi:10.2307/2401901.

Muñoz-Huerta, R. F., Guevara-Gonzalez, R. G., Contreras-Medina, L. M., Torres-Pacheco, I., Prado-Olivarez, J. & Ocampo-Velazquez, R. V. 2013 A review of methods for sensing the nitrogen status in plants: advantages, disadvantages and recent advances. *Sensors* **13**, 10823–10843.

Möller, M., Alchanatis, V., Cohen, Y., Meron, M., Tsipris, J., Naor, A., Ostrovsky, V., Sprintsin, M. & Cohen, S. 2007 Use of thermal and visible imagery for estimating crop water status of irrigated grapevine. *Journal of Experimental Botany* **58** (4), 827–838. doi:10.1093/jxb/erl115.

OceanOptics. 2013 *OceanView Installation and Operation Manual*. Ocean Optics, Inc. Available from: https://oceanoptics.com/product/oceanview-downloads/ (accessed 25 May 2020).

OceanOptics. 2015 *Flame Miniature Spectrometer User Manual*. Ocean Optics, Inc. Available from: https://oceanoptics.com/product/flame-spectrometer/ (accessed 25 May 2020).

Padro, J.-C., Munoz, F.-J., Avila, L. A., Pequer, L. & Pons, X. 2018 Radiometric correction of Landsat-8 and Sentinel-2A scenes using drone imagery in synergy with field spectroradiometry. *Remote Sensing* **10**, 1687. doi:10.3390/rs10111687.

Pe'eri, S., McLeod, A., Lavoie, P., Ackerman, S., Gardner, J. & Parrish, C. 2013 Field calibration and validation of remote sensing surveys. *International Journal of Remote Sensing* **34** (18), 6423–6436.

Pix4D. 2020 *Pix4D Documentation*. Available from: https://support.pix4d.com/hc/en-us/articles/203756125-How-to-verify-that-there-is-enough-overlap-between-the-images.

Prabhakaraa, K., Hively, W. D. & McCarty, G. W. 2015 Evaluating the relationship between biomass, percent groundcover and remote sensing indices across six winter cover crop fields in Maryland, United States. *International Journal of Applied Earth Observation and Geoinformation* **39**, 88–102. doi:10.1016/j.jag.2015.03.002.

Ramseyer, N. 2017 Pre-Process Survey3 Images in MAPIR Camera Control (MCC). *MAPIR Camera*. Available from: https://www.mapir.camera/blogs/guide/pre-process-survey3-images-in-mapir-camera-control-mcc (accessed 25 May 2020)

Rosati, A. & Dejong, T. M. 2003 Estimating photosynthetic radiation use efficiency using incident light and photosynthesis of individual leaves. *Annals of Botany* **91**, 869–877. doi:10.1093/aob/mcg094.

Sankaran, S., Khot, L. R., Espinoza, C. Z., Jarolmasjed, S., Sathuvalli, V. R., Vandemark, G. J., Miklas, P. N., Carter, A. H., Pumphrey, M. O., Knowles, N. R. & Pavek, M. J. 2015 Low-altitude, high-resolution aerial imaging systems for row and field crop phenotyping: a review. *European Journal of Agronomy* **70**, 112–123. https://doi.org/10.1016/j.eja.2015.07.004.

Serrano, L., Filella, I. & Penuelas, J. 2000 Remote sensing of biomass and yield of winter wheat under different nitrogren supplies. *Crop Science* **40**, 723–731. doi:10.2135/cropsci2000.403723x.

Shi, Y., Thomasson, J. A., Murray, S. C., Pugh, N. A., Rooney, W. L., Shafian, S., Rajan, N., Rouze, G., Morgan, C. L. S., Neely, H. L., Rana, A., Bagavathiannan, M. V., Henrickson, J., Bowden, E., Valasek, J., Olsenholler, J., Bishop, M. P., Sheridan, R., Putman, E. B., Popescu, S., Burks, T., Cope, D., Ibrahim, A., McCutchen, B. F., Baltensperger, D. D., Avant Jr, R. V., Vidrine, M. & Yang, C. 2016 Unmanned aerial vehicles for high-throughput phenotyping and agronomic research. *PLoS One*. https://doi.org/10.1371/journal.pone.0159781.

Smith, G. & Milton, E. 1999 The use of the empirical line method to calibrate remotely sensed data to reflectance. *International Journal of Remote Sensing* **20** (13), 2653–2662. doi:10.1080/014311699211994.

Toro, G., Flexas, J. & Escalona, J. M. 2019 Contrasting leaf porometer and infra-red gas analyser methodologies: an old paradigm about the stomatal conductance measurement. *Theoretical and Experimental Plant Physiology* **31**, 483–492. https://doi.org/10.1007/s40626-019-00161-x.

United States Environmental Protection Agency (USEPA), Office of Water. 1999 *Preliminary Data Summary of Urban Storm Water Best Management Practices*. United States Environmental Protection Agency, Office of Water, Washington, D.C

United States Environmental Protection Agency (USEPA), Office of Water. 2002 *Urban Stormwater BMP Performance Monitoring: A Guidance Manual for Meeting the National Stormwater BMP Database Requirements*. Available from: https://nepis.epa.gov.

United States Environmental Protection Agency (USEPA), Office of Water. 2009 *Urban Stormwater BMP Performance Monitoring Manual*. Available from: http://bmpdatabase.org/Docs/2009%20Stormwater%20BMP%20Monitoring%20Manual.pdf.

Wang, C. & Myint, S. 2015 A simplified empirical line method of radiometric calibration for small unmanned aircraft systems-Based remote sensing. *IEEE Journal of Selected Topics in Applied Earth Observations and Remote Sensing* **8** (5), 1876–1885. doi:10.1109/JSTARS.2015.2422716.

Weler, J. & Herring, D. 2000 Measuring Vegetation (NDVI & EVI). *NASA: Earth Observatory*. Available from: https://www.earthobservatory.nasa.gov/features/MeasuringVegetation.

Williams, A. S., Kiniri, J. R., Mushet, D., Smith, L. M., McMurry, S., Attebury, K., Land, M., McCarty, G. W., Shaffer, J. A., Effland, W. R. & Johnson, M.-V. V. 2017 Model parameters for representative wetland plant functional groups. *Ecosphere* **8** (10), e01958. doi:10.1002/ecs2.1958.

Zecevic, Z., Popovic, T. & Krstajic, B. 2017 Cloud based solution for automatic image mosaicking and georeferencing. In: *Proceedings of 22nd International Scientific-Professional Conference Information Technology*. Available from: https://researchgate.net/publication/315839453.

Zhang, L., Sun, X., Wu, T. & Zhang, H. 2015 An analysis of shadow effects on spectral vegetation indexes using a ground-based imaging spectrometer. *IEEE Geoscience and Remote Sensing Letters* **12** (11), 2188–2192. doi:10.1109/LGRS.2015.2450218.

First received 28 May 2020; accepted in revised form 13 November 2020. Available online 2 December 2020

 © 2021 The Authors

Data assimilation in hydrodynamic models for system-wide soft sensing and sensor validation for urban drainage tunnels

Rocco Palmitessa, Peter Steen Mikkelsen, Adrian W. K. Law and Morten Borup

ABSTRACT

Tunnels are increasingly used worldwide to expand the capacity of urban drainage systems, but they are difficult to monitor with sensors alone. This study enables soft sensing of urban drainage tunnels by assimilating water level observations into an ensemble of hydrodynamic models. Ensemble-based data assimilation is suitable for non-linear models and provides useful uncertainty estimates. To limit the computational cost, our proposed scheme restricts the assimilation and ensemble implementation to the tunnel and represents the surrounding drainage system deterministically. We applied the scheme to a combined sewer overflow tunnel in Copenhagen, Denmark, with two sensors 3.4 km apart. The downstream observations were assimilated, while those upstream were used for validation. The scheme was tuned using a high-intensity event and validated with a low-intensity one. In a third event, the scheme was able to provide soft sensing as well as identify errors in the upstream sensor with high confidence.

Key words | combined overflow tunnel, data assimilation, distributed urban drainage models, model-based sensor validation, soft sensing, urban drainage tunnel

Rocco Palmitessa (corresponding author)
Peter Steen Mikkelsen
Morten Borup
Department of Environmental Engineering,
Technical University of Denmark,
Lyngby,
Denmark
E-mail: *rocp@env.dtu.dk*

Rocco Palmitessa
Adrian W. K. Law
School of Civil and Environmental Engineering,
Nanyang Technological University,
Singapore 639798,
Singapore

HIGHLIGHTS

- We propose a data assimilation scheme tailor-made for urban drainage tunnels that can efficiently assimilate observations into an ensemble of 1D hydrodynamic models.
- We tested and validated our methodology with a real case study.
- The results support our hypothesis that the scheme is capable of promoting the hydrodynamic model to a soft sensing tool, which can be further used for validating physical sensors.

INTRODUCTION

Tunnels are increasingly integrated in urban drainage systems (UDSs). Notable examples include the pioneering Tunnel and Reservoir Plan in Chicago (USA), the Thames Tideway Tunnel in London (UK) for overflow interception and the Deep Tunnel Sewerage System in Singapore for the centralized collection of used water. Despite the variety of uses and designs, their defining feature is the ability to expand the capacity of UDSs while conveying water to final treatment and disposal/re-use (Palmitessa *et al.* 2018). The depth and scale of urban drainage tunnels also allow UDSs to minimize disruption at the surface level and achieve attractive cost-benefit ratios.

To take full advantage of their capacity, tunnels need to be optimally controlled. A control system includes

This is an Open Access article distributed under the terms of the Creative Commons Attribution Licence (CC BY 4.0), which permits copying, adaptation and redistribution, provided the original work is properly cited (http://creativecommons.org/licenses/by/4.0/).

doi: 10.2166/hydro.2020.074

monitoring equipment to measure the current state of the system, actuators to change the dynamics of the system and a control strategy aimed at reaching defined objectives. It is essential to have as much knowledge of the state of the system as possible to take the right control decisions. The monitoring and control of tunnel systems primarily relies on sensor data. Water level and flow sensors measure the hydraulic state in strategically chosen locations. Additional information can be indirectly derived knowing the state of the control actuators, e.g. if gates are open or pumps are activated. However, the sensor data is limited by the instrument accuracy and has mostly a point-wise validity. Models can augment the monitoring of drainage tunnels in real time by providing a system-wide estimate of the hydraulic variables as compared to the point-wise observations from the sensors. The model estimates could also ideally be used for validating observations.

Models incorporate our knowledge of the governing hydrological and hydraulic processes for the UDSs. The degree of detail and accuracy varies widely from conceptual models to Computation Fluid Dynamics (CFD) models, and each serves specific purposes. Distributed Urban Drainage Models (DUDMs) built in state-of-the-art software such as Mike Urban, SWMM and InfoWorks, balance a sufficient level of detail with a feasible computational cost and are commonly used for both design and optimization of UDSs. DUDMs rely on a large number of parameters, related to geometry, materials, etc., to be able to simulate the complex and non-linear behaviour of UDSs. Many of these parameters can be derived from the asset information, e.g. diameter and length of the pipes, others need to be calibrated. Deletic *et al.* (2012) identify the calibration process as a key source of uncertainty in urban drainage modelling together with the model input, e.g. rainfall data and dry weather inflows, and the model structure. Moreover, the calibration is sensitive to the quantity and quality of available measurements (Tscheikner-Gratl *et al.* 2016), which are typically limited in UDSs.

Data assimilation (DA) is capable of improving the accuracy of model predictions by correcting the model states and/or parameters dynamically based on information from the sensors, thus preventing the model from drifting far from reality (Hutton *et al.* 2011). The correction or 'update' can be either restricted to the locations where the observations are available, or propagated in space depending on the physical relationship among the states. DA algorithms can be broadly classified as deterministic, which correct the model variables based on some prior assumptions on their uncertainty, and statistical, which dynamically assess and compare the uncertainty of model predictions and observations, and update the model with a weighted average of the two, with more weight being given to the less uncertain (Borup 2014). In ensemble-based methods, the uncertainty of model predictions is computed by running an ensemble of the same model, where each ensemble member is forced with a differently perturbed input (e.g. initial conditions, boundary conditions and model parameters).

The use of DA in rural hydrologic forecasting is well-established (Liu *et al.* 2012) but has yet to reach the same level of adoption in the urban context. Borup *et al.* (2014) suggest that only recently the need for controlling UDSs in real time and the availability of communication and computing resources have prompted the application of DA in urban drainage modelling. Previous studies have explored the challenges and benefits of both deterministic and statistical approaches. Hansen *et al.* (2014) updated a DUDM by inserting or extracting water at every computational time step at the location with available observations. Direct correction requires limited computational resources but does not provide a measure of uncertainty and can only propagate the update downstream. Borup *et al.* (2011) assimilated downstream flow measurements to indirectly correct the (unknown) infiltrating inflow upstream, thus propagating the update across the entire system. Similarly, Hutton *et al.* (2014) used the downstream residual errors to update the upstream states. While the deterministic approaches have proven to perform well with gradually changing hydraulic conditions, they cannot produce uncertainty estimates and cannot easily assimilate ambiguous information from multiple sensors. For these reasons, ensemble-based methods are to be preferred over the deterministic methods if the required computational resources are available (Borup *et al.* 2014). Lund *et al.* (2019) limited the computational cost of running large ensembles by replacing the DUDM with a lumped surrogate model, at the expense of the spatial resolution of the model estimates.

The quantity and quality of sensor data in UDSs is not always reliable enough for monitoring purposes. This is even more so in large drainage tunnels where sensors are difficult to maintain, due to their location deep underground in a dangerous and harsh environment. If a detailed hydrodynamic model correctly updated in real time is capable of producing reliable, probabilistic estimates throughout the entire system, it can serve as a soft (software) sensing tool to complement the hard (hardware) physical sensors. Independent observations can then be automatically validated against the model estimate, given our degree of confidence in the sensor and the updated model.

The aim of this study is to enable soft sensing and sensor validation of urban drainage tunnels by developing and validating a DA scheme tailor-made for tunnel models that can efficiently assimilate observations into an ensemble of models. Our proposed scheme restricts the assimilation only to a subsystem of the UDS, which includes the tunnel and its connected structures. By doing so, we significantly decrease the computational cost of the assimilation, while retaining the spatial resolution and level of detail of the fully distributed model.

METHODOLOGY

Assimilation scheme

The application of ensemble-based DA to DUDMs requires running in parallel several instances of the same model. To limit the computational overhead, we develop a scheme that isolates an 'inner' model of the tunnel from the 'outer' DUDM, where the output from the outer model is used as the boundary condition for the inner model, and it is only this model that is updated. The inner model retains the same level of detail as the outer model and includes the tunnel itself and its inlet and outlet structures (e.g. overflows, dropshafts and pumping stations).

Setting the inner model boundaries is a critical step in the implementation of the scheme. It is a case-specific procedure but should satisfy some general criteria:

(i) the inner model should include the smallest number of structures besides the tunnel itself, to minimize the computational cost;
(ii) the inner model should include all locations with sensors whose data is relevant for the assimilation;
(iii) at the tunnel inlets, the boundaries should be set at locations where no backwater effects from the tunnel are expected, e.g. overflows;
(iv) the outlet should include only the structures necessary to simulate the emptying of the tunnel, e.g. pumping stations and safety overflows.

To optimize for the first and second criteria, the boundaries are ideally set at locations where the sensors are placed. In principle, the boundary conditions could be either water levels or discharges. However, level gauges are preferred because they are usually more common and accurate than flow meters.

The scheme is implemented in three stages, as seen in Figure 1. In an offline setting, the stages are sequential, so that one stage is completed for the entire simulation period before moving to the next. In an online scenario, the sequence should be repeated at each time step. In the first stage, a single instance of the drainage model receives as input rainfall observations. From its output, the simulated states at the tunnel boundaries are read. The boundary conditions are perturbed according to a user-defined perturbation model in the second stage, to account for the uncertainty of the drainage model. In the final stage, each set of perturbed boundary conditions is linked to an identical copy of the tunnel model. Therefore, each member of the tunnel model ensemble returns a different prediction of the tunnel states, to represent the uncertainty of the model. Given the spread of the predicted tunnel states and the uncertainty assigned to the corresponding observations, the DA algorithm computes the updated states. The process is repeated at each time step and the updated states serve as the initial condition for the prediction at the following time step. If the observations are missing at a given point in time, the assimilation step is bypassed, and the predicted tunnel states are used directly as the initial condition for the following prediction.

1D Hydrodynamic models

DUDMs in state-of-the-art software such as Mike Urban (DHI 2019) include a lumped-conceptual surface module, that converts precipitation data into stormwater inflows, and a hydrodynamic module, that simulates the storage and conveyance of water in the system. 1D hydrodynamic models

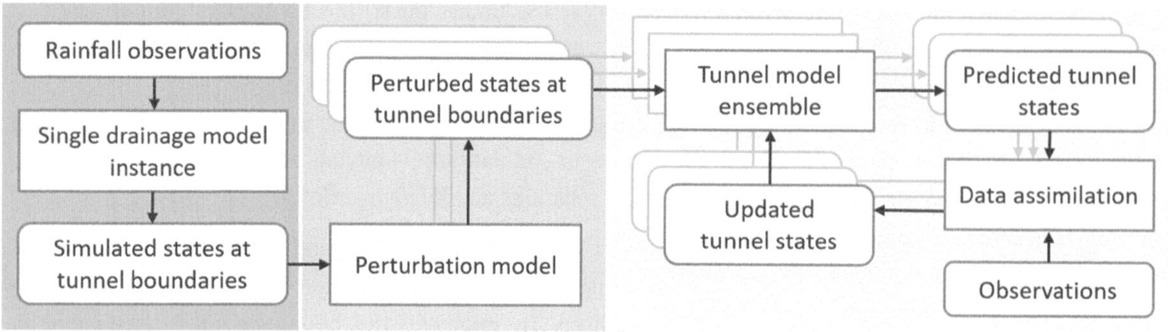

Figure 1 | Data flow of a time step in the proposed data assimilation scheme: (i) a single instance of the urban drainage model is forced with rainfall observations and the states at the tunnel boundaries are read from its output; (ii) the boundary states are perturbed to account for the model uncertainty and (iii) the states predicted by the tunnel model ensemble are compared to the available observations to compute the update, which serves as the initial condition for the following prediction.

compute the change over time and space of the hydraulic variables by solving the continuity and momentum equations, also known as the St. Venant equations. The transition between the free-surface flow and pressurized flow is numerically dealt with by introducing a narrow virtual opening with insignificant volume longitudinally at the top of all pipes, called the Preissmann slot, which makes it possible to use the St. Venant equations, intended for free-surface conditions, for pressurized flows as well. Water levels h and discharges Q are computed at pre-defined time and space intervals. The computational grid is generated automatically by the software, or with a user-specified number of grid points (Figure 2).

Ensemble-based data assimilation

The Kalman Filter is a sequential assimilation method, which means that the model propagates in time its internal variables or state vector x and is updated whenever measurements d are available. At time k, the state vector x_k is given by the model M as:

$$x_k = M(x_{k-1}, u_{k-1}, e_{k-1}) \tag{1}$$

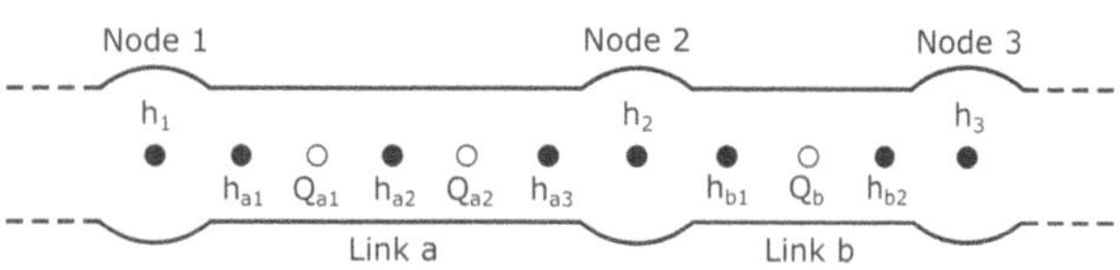

Figure 2 | Example of computational grid in 1D hydrodynamic models; h and Q indicate, respectively, water level and discharge computation nodes (adapted from https://manuals.mikepoweredbydhi.help/2020/Cities/MOUSEPipeFlowReference.pdf).

where u is the model input and e is the system noise, which represents the uncertainty of the model forecast. For hydrodynamic models, the vector x usually includes both water levels and discharges, i.e. all the h and Q variables shown in Figure 2.

If one or more observations are available, the analysis or updated states x^a are computed as a weighted average of the model forecast or background states x^f and the observations using the analysis equation:

$$x^a = x^f + K(d - Hx^f) \tag{2}$$

where H is a matrix mapping the model states to the observations and K is the Kalman gain matrix. The elements of K act as weights between the vectors x^f and d. The higher the value of the gain, the closer the state is updated to the observation. The Kalman gain is calculated as:

$$K = P^f H^T (HP^f H^T + R)^{-1} \tag{3}$$

where P^f is the forecast error covariance matrix and R is the observation error covariance matrix. If n is the size of the state vector, then P has the dimensions n^2. The Kalman Filter requires P^f to be propagated in time analytically, which is too costly an operation for large models, and is only valid for linear models.

To overcome these two limitations, Evensen (1994) introduced the Ensemble Kalman Filter (EnKF), which uses an ensemble of model instances to propagate in time the error statistics. In the EnKF, the error covariance matrix is replaced by the ensemble covariance, which is calculated as:

$$P = \frac{1}{m-1}\sum_{i=1}^{m}(x_i - \bar{X})(x_i - \bar{X})^T = \frac{1}{m-1}AA^T \tag{4}$$

where $X = [x_i, \ldots, x_m]$ is the ensemble of model states, m is the ensemble size, $\bar{X}$ is the ensemble mean, and $A = [A_i, \ldots, A_m]$ is the ensemble of anomalies $A_i = (x_i - \bar{X})$. Therefore, the dimensions of P are here reduced to $m*n$. The gain computed with the error covariance is then used to update the ensemble members individually. To prevent the ensemble spread reducing too rapidly, the EnKF requires the use of perturbed observations, which means that each member is associated with a different realization of the observation noise consistent with the observation error covariance. However, this introduces sampling error, especially for small ensembles.

We choose to use the Deterministic Ensemble Kalman Filter (DEnKF), which eliminates the need for perturbing the observations (Sakov & Oke 2008). It does so by (i) applying the analysis to the mean ensemble forecast $\overline{X^f}$ instead of each ensemble member, (ii) updating the ensemble anomalies using only half the Kalman gain:

$$A^a = A^f - \frac{1}{2}KHA^f \tag{5}$$

and (iii) calculating the analysed ensemble by adding the updated ensemble mean to the updated anomalies:

$$X^a = A^a + [\overline{X^a}, \ldots, \overline{X^a}] \tag{6}$$

Prediction metrics

The user-defined perturbation model and the chosen DA technique will likely require some tuning parameters. To compare the performance of the scheme for different sets of parameters and events and assess the efficacy of the DA, the quality of the ensemble predictions needs to be quantified. Quality refers to the correspondence between the predictions and the matching observations (e.g. Bennett *et al.* 2013) and can be broken down in several aspects or attributes, e.g. accuracy, reliability and sharpness (Murphy 1993). The accuracy is related to the difference between individual predictions and observations. In case of a probabilistic prediction, the accuracy can either be computed from the mean or the full distribution function. A prediction is deemed to be reliable if there is an agreement between the prediction probability and mean observed frequency, i.e. if on average x% of the observations fall within the x% prediction interval. Therefore, predictions that consistently over- or underestimate the frequency interval of the observations are considered unreliable. The sharpness is related to how the prediction probability approaches zero and one, so does not depend on the observation. A sharp prediction tends to have probabilities clustered around the mean, as opposed to values near 0 or 1. Prediction metrics typically focus on one or more attributes of the prediction quality. Therefore, the choice of metrics depends on the objective of the verification and generally a combination of metrics is needed for a thorough analysis (Anctil & Ramos 2019). We focus on the predicted water levels and identified six metrics suitable to compare the ensemble predictions against the sensor observations (Table 1). Some of these are designed for deterministic forecasts, so they are applied to the ensemble mean.

The mean absolute error (MAE) and the root mean square error (RMSE) are commonly used measures of

Table 1 | Comparison of prediction metrics (the measuring unit refers to water level predictions)

Verification metric	Unit	Range	Best score	Quality attribute	Type of prediction
Mean absolute error (MAE)	m	0 to ∞	0	Accuracy	Deterministic
Root mean square error (RMSE)	m	0 to ∞	0		
Nash–Sutcliffe efficiency (NSE)	–	−∞ to 1	1		
Coverage bias (CB)	–	−1 to 1	0	Reliability	Probabilistic
Average band width (ABW)	m	0 to ∞	0	Sharpness	
Continuous Rank Probability Score (CRPS)	m	0 to ∞	0	Mix of the above	

accuracy. They are calculated as:

$$\mathrm{MAE} = \frac{1}{N}\sum_{k=1}^{N} |\bar{X}_k - d_k| \tag{7}$$

$$\mathrm{RMSE} = \left[\frac{1}{N}\sum_{k=1}^{N} (\bar{X}_k - d_k)^2\right]^{1/2} \tag{8}$$

where $\bar{X}_k$ is the mean across ensemble members at the k time step, d_k is the corresponding observation and N is the total number of time steps in the observation period. Both metrics retain the same unit of the prediction, but the RMSE penalizes large errors, which in urban drainage are known to occur especially for peak flows and water levels that can be significantly affected by timing errors due to uncertain rain inputs.

The Nash–Sutcliffe efficiency (NSE) is calculated as:

$$\mathrm{NSE} = 1 - \frac{\sum_{k=1}^{N} (\bar{X}_k - d_k)^2}{\sum_{k=1}^{N} (\bar{d} - d_k)^2} \tag{9}$$

where $\bar{d}$ is the mean value of the observations over the observation period. NSE ranges between 1 and $-\infty$, with the highest value being a perfect fit. Negative values indicate the ensemble mean is a worse predictor than the mean of the observations. NSE is a normalized metric, so it is useful to compare predictions across different events.

The coverage bias (CB) can be used to quantify the reliability of a probabilistic prediction. For a given quantile level β, the prediction interval is defined as:

$$I_k(\beta) = [X_k^l;\ X_k^u] \tag{10}$$

where X_k^l and X_k^u are, respectively, the lower and upper limits of the ensemble prediction at levels $\beta/2$ and $1-\beta/2$. The observed coverage $n(\beta)$ is the fraction of observations falling within $I_k(\beta)$ during the observation period and is expected to match the nominal coverage $1-\beta$. CB for a given quantile β is defined as the discrepancy between the nominal and observed coverage

$$\mathrm{CB}(\beta) = 1 - \beta - n(\beta) \tag{11}$$

and ranges between -1 and 1 (Thordarson *et al.* 2012). If the bias is negative, the observed coverage is larger than the nominal one and the prediction is overestimating the uncertainty. The opposite is true if the bias is positive. Therefore, the CB can be used to quantify the reliability of a probabilistic prediction, since the ideal reliability corresponds to $\mathrm{CB}(\beta) = 0$ for any $\beta \in [0, 1]$.

The average band width (ABW) for a given quantile level β is calculated as:

$$\mathrm{ABW}(\beta) = \sum_{k=1}^{N} I_k(\beta) \tag{12}$$

and is an indicator of the prediction sharpness, although not sufficient to characterize it (Gneiting *et al.* 2007). In this study, it is computed to assess how the perturbation and assimilation affect the spread of the ensemble prediction.

The continuous ranked probability score (CRPS) for a given kth pair of observation-ensemble prediction is calculated as:

$$\mathrm{CRPS}(k) = \int_{-\infty}^{\infty} [F_k(x) - F_k^d(x)]^2 dx \tag{13}$$

where x is the predicted variable, $F_k(x)$ and $F_k^d(x)$ are, respectively, the cumulative distribution function of the kth probabilistic prediction and of the corresponding observation (Hersbach 2000). When the observation is a unique value, its distribution function is given by:

$$F_k^d(x) = \begin{cases} 0 & \text{for } x < d_k \\ 1 & \text{for } x \geq d_k \end{cases} \tag{14}$$

which is also known as the Heaviside function. To remove the time dependency, the continuous ranked probability score is averaged over the observation period in this study (simply referred to as CRPS). For a deterministic forecast, $F_k(x)$ can also be approximated with a step-wise function and the CRPS is reduced to the mean absolute error. Similar to the MAE, the CRPS retains the measuring unit of the predicted variable and ranges between 0 and ∞, with 0 equal to a perfect score. Since the CRPS measures the quadratic distance between the cumulative distribution of the prediction and the observation, it mainly quantifies the accuracy of the prediction while also carrying information about its reliability and sharpness.

Model-based sensor validation

One of the potential benefits of DA is the model-based cross-validation of sensors, i.e. using the updated model to identify malfunctioning sensors. This requires that the sensor under analysis is not used for the model updating, in order to ensure that the model prediction and the sensor observation are independent. All sensor observations carry a degree of uncertainty. Assuming for simplicity that (i) the observation error is normally distributed around the measured quantity with variance σ^2_{obs}; (ii) the ensemble is regarded as Gaussian with variance σ^2_{ens} and (iii) the model ensemble provides a correct representation of the distribution of the measured quantity, then the sensor values should be distributed around the ensemble mean with a variance of $\sigma^2_{obs,\ ens} = \sigma^2_{obs} + \sigma^2_{ens}$. This implies that the standardized residuals

$$r_{std} = \frac{(d - \bar{X})}{\sigma_{obs,ens}} \tag{15}$$

should be normally distributed with zero mean and a variance of $\sigma^2_{obs\ ens}$, and that about 95% of the values should fall within the [−2,2] interval. Therefore, the standardized residuals could be used to test the validity of a sensor by checking if the mean of r_{std} over a given time period is statistically different from zero and if its value exceeds a confidence interval of [−2,2] for prolonged periods in time. In the current article, we are only using r_{std} for visual identification of suspicious sensor data. However, the validation process could be automated and the user could in a real-time setting be alerted when the sensor data does not pass the quality check.

EXPERIMENTAL SETUP

Case study

The Damhus combined sewer overflow tunnel in Copenhagen, Denmark, was used as a case study. The tunnel is integrated within a large combined sewer system that drains to the local wastewater treatment plant. We were provided with an up-to-date and highly detailed MIKE URBAN model of the urban drainage system (Figure 3(a)). This DUDM was forced with rainfall intensity series from 10 gauges distributed in and around the catchment perimeter and included all dry weather inflow patterns (both domestic and industrial). MIKE URBAN automatically assigns to each sub-catchment the rainfall intensity from the nearest gauge to simulate the spatial distribution of the precipitation. The large size of the model, which counts more than 15,000 links (equivalent to 800 km of pipes) and a similar number of nodes, translates into a high computational cost, making this an interesting case study for testing the proposed DA scheme.

Several overflow structures were originally constructed in the catchment to discharge excess combined sewage directly into the local stream in case of intense rain events. Recent environmental regulations have limited the permitted number of overflows per year, with the aim of protecting the water quality at the stream and the nearby harbour. In response, the Damhus tunnel was built along the path of the stream to intercept the outflow of the overflow structures connected to the eight tunnel dropshafts (Figure 3(b)). These range in diameter between 1.5 and 12.5 m and part of their volume contributes to the tunnel storage capacity. Ultrasonic water level sensors placed at the most upstream and downstream dropshafts are used to monitor the filling degree of the tunnel. With a length of 3.4 km and a constant diameter of 3 m, the tunnel system can store up to 29,000 m^3 of water. The invert level of the tunnel varies between 13.7 and 15.7 m below the ground level. Main dimensions are reported in Figure 3(e) as relative to the Danish Vertical Reference 1990 (DVR90).

Each dropshaft receives inflow from one or more inlet pipes connecting the pre-existing overflow structures. Most of the overflows are passive, so they are activated when the water level upstream exceeds a fixed crest level. Others are regulated with gates and valves to control the timing and magnitude of the discharge. As shown in Figure 3(c) for the most upstream dropshaft, some inlet pipes receive inflow from a single overflow, others branch off to connect multiple overflows. The 17 boundaries of the tunnel model are set in agreement with the assimilation scheme criteria, and mostly coincide with the overflow structures. Three of the four boundary locations upstream are also equipped

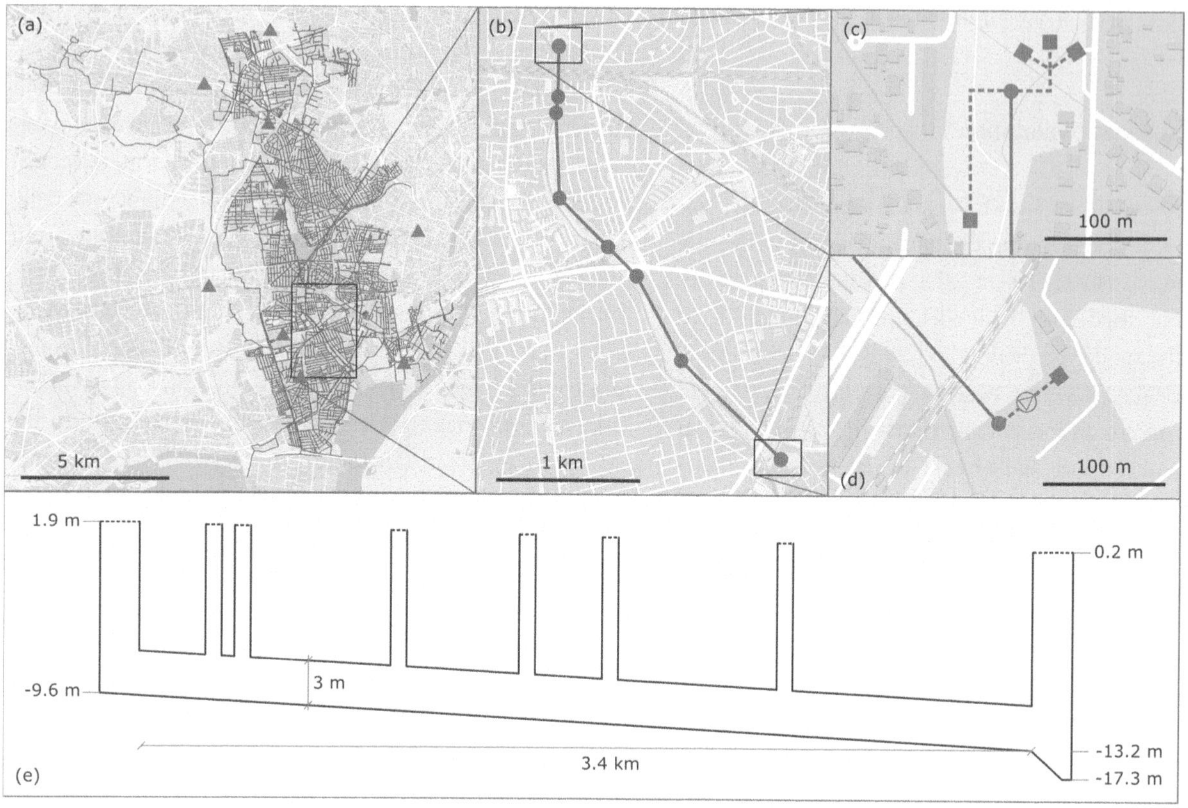

Figure 3 | (a) Hydrodynamic model of the studied urban drainage system, links (blue lines) and location of rainfall gauges (red triangles); (b) Aerial view of the tunnel system, pipes (blue line), location of dropshafts (circles) and water level gauges (red circles); (c) Detail of tunnel inlets (dashed lines) at the upstream dropshaft, location of tunnel model boundaries (squares) and water level gauges (red square); (d) Detail of tunnel outlet at the downstream dropshaft, pumps and gauged tunnel model boundary (red square) and (e) Longitudinal profile of tunnel system with main dimensions (not in scale). Please refer to the online version of this paper to see this figure in colour: https://doi.org/10.2166/hydro.2020.074.

with water level sensors, which provide valuable information on the filling of the tunnel, given that in most cases more than 50% of the tunnel inflow comes from the upstream inlets.

At the downstream dropshaft, the water level builds up gradually as the tunnel fills up until there is no more overflow into the tunnel or the max storage capacity is met. To prevent water from backing up upstream, a safety overflow releases excess water to the stream when the tunnel has reached its full capacity. Five large pumps lift the stored water from a cavity at the bottom of the downstream dropshaft back to the combined sewer. The pumps are controlled by the water level measured in the tunnel and at the pumps outlet. To ensure a realistic simulation of the tunnel emptying, the pumps and their outlet are retained in the tunnel model, with the outlet serving as a boundary (Figure 3(d)).

Observations

Since the tunnel receives overflows from the combined sewer, it is only employed during high-intensity rainfall events. Once the tunnel content has been pumped back to the sewer, sediments are removed by flushing. An 'event' here refers to a cycle of partial/complete filling, emptying and flushing. Eight events occurred in the period between August and December 2018. Among them, three characteristic events (A, B and C) were selected to cover various seasons (respectively, late summer, autumn and early winter) and a wide range of tunnel filling degrees (Figure 4). The sensor data from these three independent events were used in this study.

Rainfall intensity observations from the 10 gauges linked to the drainage model were obtained at 1-min resolution from a national archive hosted by the Danish

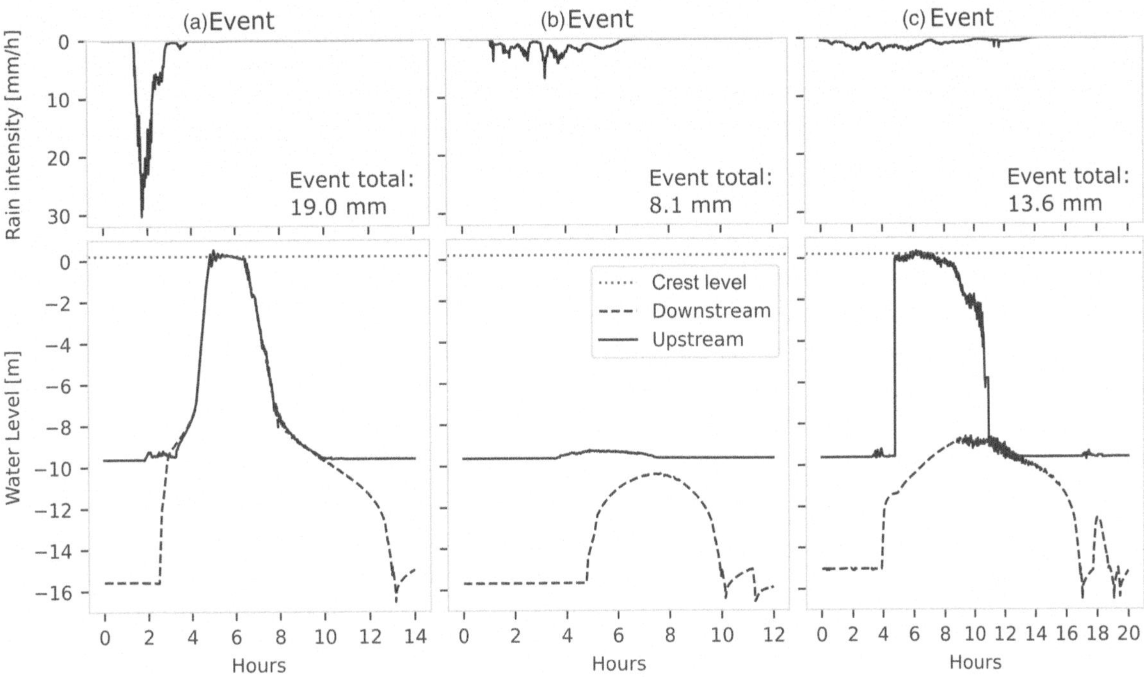

Figure 4 | Top row, arithmetic mean of the 10 available rainfall intensity series; bottom row, water level observed at the upstream and downstream tunnel dropshafts, starting from the empty tunnel (all values are relative to the Danish Vertical Reference 1990); each column represents one of the three events investigated (a, b and c).

Meteorological Institute (Jørgensen *et al.* 1998). For Event A, the average of the 10 rainfall series showed a peak intensity of 30.2 mm/h (8.4 μm/s) and a total duration of about 2.5 h. This first event was recorded at the beginning of August and had the characteristics of a summer storm, or cloudburst. On the other hand, Events B and C, which happened in October and December, showed a moderate peak intensity of 6.8 and 2.5 mm/h (1.9 and 0.7 μm/s), respectively, but extended for durations of 5.5 and 14 h, respectively.

Water level observations from six ultrasonic sensors were obtained at 1-min resolution from the local utility company. They covered the most upstream and downstream dropshafts, three of the upstream boundaries and the pumps outlet for the three events. During Event A, large volumes of combined sewage were rapidly discharged into the tunnel, which filled up entirely in about 2.5 h. At this point, the water level rose above the crest of the safety overflow at 0.2 m and the excess water was discharged into the receiving water body (see a flat plateau in the middle, bottom-left panel of Figure 4). Once the water level at the downstream outlet dropped below the control set-point, the pumps were activated to return the water to the combined sewer for disposal at the nearby treatment plant. When the tunnel pipes were completely filled, which occurs at −6.6 m, the observed water level upstream almost matched the level downstream, with differences ranging between a few centimetres and 0.5 m. The lower rain intensity of Event B led only to the partial filling of the tunnel. All the combined sewer overflow was stored in the tunnel and pumped back without significantly building up at the upstream dropshaft, where the maximum recorded water level was about 0.4 m above the invert as opposed to about 10 m in the first event. Given its peak intensity, Event C was expected to behave similarly to Event B. While this was mostly the case for the downstream water level, sudden high water levels were recorded upstream, thus suggesting a possible malfunctioning of the upstream sensor. The incident was later explained by the tunnel operator. The upstream dropshaft includes a separate flushing chamber as tall as the dropshaft itself. At the end of each event, the tunnel is flushed by opening a valve at the bottom of the chamber. During Event C, the flushing chamber was accidentally filled beyond its capacity and the overflowing water spilled into the detection zone of the sensor located in the main chamber, thus causing a false echo.

DA framework

The assimilation of water level observations in the ensemble of tunnel models was carried out with the DA framework developed by the Danish Hydraulic Institute (DHI) and embedded in the beta version of MIKE URBAN 2019. Earlier versions of this framework were previously used with ground water models (Zhang *et al.* 2015) and urban drainage models (Borup *et al.* 2018), the advantage being the complete automation of the assimilation process, which removes the need to directly access the model states at each time step. Setting up the framework still includes, though, some manual steps: (i) generate an ensemble of size m by cloning the MIKE URBAN model m times; (ii) introduce the model error by perturbing the input (in our case, the boundary conditions); (iii) link the observation series to the corresponding structures in the model; (iv) assign the uncertainty of the observations; (v) define the spatial extent of the assimilation for each observation location, a process known as localization; and (vi) choose the assimilation algorithm for the update (with the option of running the framework as an open-loop simulation without assimilation). The software then automates the parallel running of the ensemble members and, at each time step, propagates the model states forward, updates them based on the available observations and replaces the predicted states with the updated ones. The process is repeated until the end of the simulation period.

Implementation

The inner model was set to consist of the tunnel and the nearest connecting structures only (the dark-blue and red parts in Figure 3(b)–3(d)) with a total of 28 links. The inner model can simulate a 2-day event in 10 s that takes the model of the full system 2 h to simulate, corresponding to a ~700 times reduction in the computational time. This implied that the main computational effort of the DA scheme in the specific implementation was to run the outer model, when using an ensemble size of 100 and updating the ensemble to observations once per minute.

The construction of the perturbation model is crucial for the success of the assimilation scheme and is case-specific. For our case study, the uncertainty of the simulated water levels at the boundaries of the tunnel model can be formulated in terms of time delay and peak value. To capture these two features, the boundary conditions were perturbed with a time displacement uniformly distributed within the interval $[-\tau_{\max}, \tau_{\max}]$ and a normal distributed multiplication factor with zero mean and standard deviation σ_p. Both parameters were assumed constant in time but varied across locations and ensemble members. This is a rather simplistic approach that ignores the spatial and temporal correlation of the model errors at the tunnel boundaries. However, we find this sufficient for the current study, and our choice of perturbation scheme facilitates the tuning process and allows us to regulate independently the time delay and the peak values of the model boundaries.

A total of five series of water level observations were assimilated in the tunnel model ensemble, 1 from the downstream dropshaft and 4 from the surrounding boundaries. The sensor at the downstream dropshaft also measures water levels in the pump sump, where large variations in level correspond to small changes in volume. For simplicity, the cavity was not included in the model and all observations lower than the lowest tunnel invert were set equal to it. The sensor data from the upstream dropshaft was only used for cross-validation and to test the ability of the update to propagate across the entire tunnel. To account for the different ranges of water level at the tunnel dropshafts and in the combined sewer, and how they can affect the uncertainty of the measurement, we set separate values of the standard deviation at the downstream tunnel dropshaft $\sigma_{o,t}$ and at the boundary locations $\sigma_{o,b}$. Given the different hydraulic conditions inside and outside the tunnel, we also adjusted the localization settings separately, limiting the assimilation of the observation at the downstream dropshaft only to the states inside the tunnel and performing a point-wise update at the boundary location. The latter effectively reduces the uncertainty of the simulated states at some of the tunnel boundaries based on the information provided by the sensors, with the benefit of limiting unrealistically perturbed states.

Due to the arbitrary nature and correlation of the perturbations, some of the model realizations are likely to simulate extreme hydrodynamic conditions, which can cause the failure of the computational engine. To alleviate this phenomenon, we enlarged the Preissmann slot to 10% of the pipe diameter. The enlarged width slightly changed the volume of the system and therefore contributed to the overall uncertainty of the model, but helped in simulating the pressure waves induced in the tunnel by extreme inflows, thus stabilizing the simulations.

Tuning of ensemble and DA parameters

In the proposed implementation, a total of five parameters (τ_{max} and σ_p in the perturbation model, $\sigma_{o,t}$, $\sigma_{o,b}$ and m in the DA framework) need to be tuned by the user to optimize the performance of the scheme. To better understand the individual effect of the parameters, each was treated as an independent variable and tested with an initial guess and two alternative options, while the remaining parameters were fixed. The preferred value among the three options was selected based on the response of the scheme quantified with the proposed prediction metrics and left unchanged in the following tests. The scheme was tuned with Event A, and the chosen parameters were subsequently applied to Events B and C.

RESULTS AND DISCUSSION

Tuning (Event A)

The scheme parameters were individually tuned by trial and error for Event A. In each trial, the error of the updated ensemble at the downstream dropshaft was quantified using the selected prediction metrics (Table 1). CRPS was used as reference metric, given its ability to capture various attributes of the prediction quality. MAE, RMSE and NSE were used for comparison and added interpretability but are expected to behave consistently with the reference metric. CB and ABW were both computed for the 90% quantile interval and used to gain insight into the reliability and sharpness of the ensemble prediction. The tuning started from the parameters of the perturbation model (τ_{max} and σ_p), then moved to the parameters of the assimilation scheme ($\sigma_{o,t}$ and $\sigma_{o,b}$) and ended with the ensemble size m (Table 2).

The results of the tuning are discussed in detail in the Supplementary material. The process is highly specific to the case study and the event analysed, but leads to some general insights: (i) the value of the perturbation parameters needs to be large enough to represent the uncertainty of the model while avoiding the simulation of unrealistic flow conditions; (ii) the standard deviation of the observations is the single most sensitive parameter, given that it directly affects the extent of the update; (iii) the size of the ensemble only marginally affects the result, but a larger ensemble will return a better description of the states distribution. The final selection of parameters (corresponding to the 5th trial) returns the lowest CRPS of all trials but cannot be

Table 2 | Prediction metrics of assimilated, ensemble water level prediction at the downstream dropshaft during Event A given by different sets of scheme parameters

	Parameters					Verification metrics					
Trial	τ_{max} (min)	σ_p (–)	$\sigma_{o,t}$ (m)	$\sigma_{o,b}$ (m)	m (–)	MAE (m)	RMSE (m)	NSE (–)	CB_{90} (–)	ABW_{90} (m)	CRPS (m)
1	5	0.01	0.5	0.05	100	0.55	0.90	0.96	0.62	0.23	0.43
2	10	–	–	–	–	0.58	0.99	0.95	0.56	0.37	0.45
3	30	–	–	–	–	0.42	0.72	0.97	0.45	0.39	0.30
4	–	0.02	–	–	–	0.49	0.79	0.97	0.33	0.88	0.31
5	–	0.04	–	–	–	0.14	0.20	1.00	0.20	0.51	0.09
6	–	–	1	–	–	0.24	0.34	0.99	0.17	0.81	0.18
7	–	–	3	–	–	0.48	0.72	0.97	0.08	1.76	0.56
8	–	–	0.5	0.1	–	0.16	0.23	1.00	0.24	0.54	0.10
9	–	–	–	0.3	–	0.20	0.29	1.00	0.25	0.61	0.11
10	–	–	–	0.05	50	0.19	0.28	1.00	0.34	0.45	0.11
11	–	–	–	–	25	0.20	0.30	1.00	0.38	0.44	0.12
Selection (5)	**30**	**0.04**	**0.5**	**0.05**	**100**	0.14	0.20	1.00	0.20	0.51	0.09

assumed to be the global optimum. Figure 5 (left column) shows the observed, perturbed and updated water levels at the downstream and upstream locations (upper and middle row).

Scheme verification (Events B and C)

The tuned set of parameters was tested with two independent events (B and C) of lower intensity (Figure 5, middle and right columns). Despite the heterogeneity of the hydraulic conditions, the updated ensemble predictions at the downstream dropshaft showed a similar error in all three events, with CRPS ranging between 0.09 and 0.13 m and NSE always higher than 0.97 (Table 3). These values represent a significant improvement compared to the open-loop prediction without assimilation, where the CRPS ranges between 1.06 and 2.06 m and the NSE assumes negative values. The large discrepancy between the observed water level and the mean ensemble prediction prior to the assimilation can be attributed to the error in the simulated inflow and outflow to and from the tunnel. While the tunnel inflow is governed by the water level at the boundaries read from the output of the outer model, the tunnel outflow depends on the control rules of the pumps, which

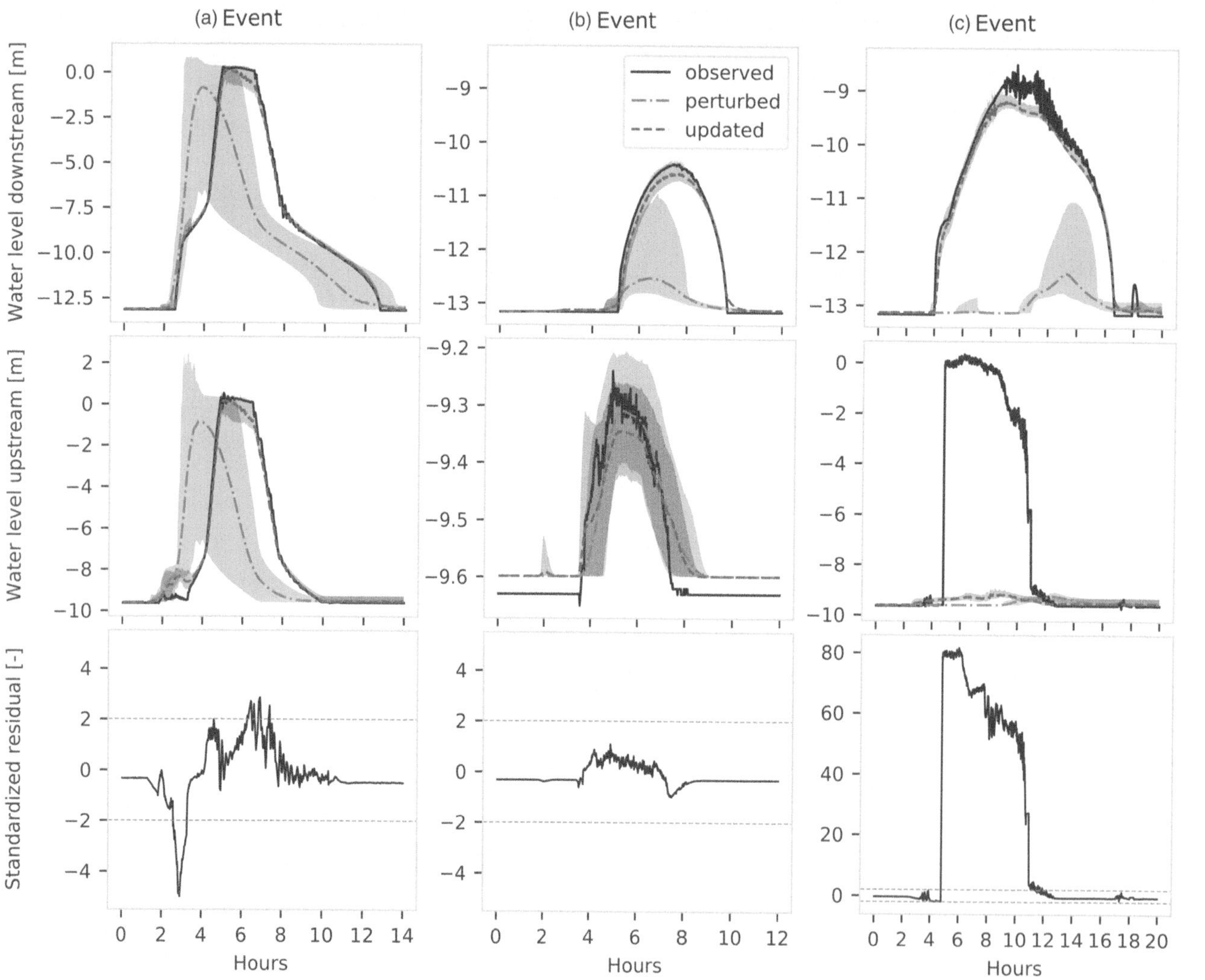

Figure 5 | Water levels for Events a, b and c at the downstream (top row) and upstream dropshaft (middle row): observations in solid black line, and model ensembles (bands delimited by 5 and 95% quantiles, mean shown as dashed lines) without (green) and with (purple) data assimilation; Standardized residual (bottom row) of updated water level at upstream dropshaft; each column represents one of the three events investigated (A, B and C). Please refer to the online version of this paper to see this figure in colour: https://doi.org/10.2166/hydro.2020.074.

Table 3 | Prediction metrics of ensemble water level prediction at the downstream and upstream dropshafts, without and with the assimilation of the observations at the downstream dropshaft and as some of the tunnel boundaries

		Downstream metrics						Upstream metrics					
Event	DA	MAE (m)	RMSE (m)	NSE (–)	CB_{90} (–)	ABW_{90} (m)	CRPS (m)	MAE (m)	RMSE (m)	NSE (–)	CB_{90} (–)	ABW_{90} (m)	CRPS (m)
A	No	2.76	3.61	0.35	0.53	4.12	1.92	3.37	4.28	− 0.22	0.65	4.35	2.46
	Yes	0.14	0.20	1.00	0.20	0.51	0.09	0.31	0.42	0.99	0.18	0.74	0.15
B	No	1.34	1.58	− 1.35	0.89	0.73	1.06	0.05	0.06	0.80	0.16	0.21	0.09
	Yes	0.14	0.16	0.97	0.06	0.34	0.09	0.05	0.06	0.82	0.20	0.18	0.09
C	No	2.33	2.75	− 2.16	0.89	0.44	2.06	3.57	5.57	− 0.66	0.80	0.20	3.42
	Yes	0.20	0.26	0.97	0.46	0.28	0.13	3.00	5.01	− 0.43	0.77	0.27	2.87

are activated based on a comparison between the observed water levels and pre-defined set-points. When the control scheme is replicated in the hydrodynamic model, the observed levels are replaced with the simulated ones. Therefore, an error in the simulated states can cause the pumps to behave differently from reality and the model drifts further from the observations. Meanwhile, the assimilation reduces the discrepancy between the observed and simulated water levels, and in doing so indirectly corrects the timing and magnitude of the pumped outflow.

Sensor validation

Once the quality of the updated predictions has been verified, they can be used to validate independent observations that are not assimilated in the model ensemble. This was the case for the observed water level at the upstream dropshaft. To quantify the discrepancy between the observed and expected value at the upstream location, we computed the standardized residual using Equation (15) assuming a standard deviation of observation error of $\sigma_{\text{obs}} = 0.1\,\text{m}$. During Event A, the tunnel was filled to its maximum capacity, at which point the difference in water level at the upstream and downstream dropshaft is expected to be minimal. The observed values are in agreement with this assumption. Also, for most of Event A, the updated ensemble prediction upstream shows a similar accuracy to the downstream prediction, with CRPS of 0.15 m and NSE of 0.99 (Table 3), which is a significant improvement over the open-loop prediction.

The filling degree was much lower during Event B, and the water level at the upstream dropshaft raised little more than 0.3 m above the invert as opposed to almost 3 m at the downstream dropshaft. It can be assumed that the inflow moved rapidly downstream and accumulated only in the downstream section of the tunnel. As observed, the water levels upstream and downstream are expected to exhibit different behaviours and be mostly uncorrelated. This is further confirmed by the observed delay between the peak values. This circumstance likely explains the only marginal improvement in the ensemble prediction after the assimilation of the downstream observation. Since the update is driven by the Kalman gain, which in turn is a function of the covariance of the model states, only the states correlated with the downstream water level are expected to be affected by the update. Nonetheless, both open-loop and updated ensemble prediction upstream showed similar accuracy to the downstream water level with CRPS equal to 0.09 m.

Similar to Event B, Event C was driven by a low-intensity precipitation event of long duration, resulting in a partial filling of the tunnel as shown by the downstream observation. This is in agreement with the ensemble prediction upstream, both before and after the assimilation. However, the observation from the upstream sensor suggests that a large volume of water entered the tunnel and filled it up completely. However, while r_{std} mostly stayed within the confidence interval [−2,2] during Events A and B, it raised to above 50 for several hours during Event C, clearly indicating that the sensor was malfunctioning for this period in time (Figure 5, bottom row).

CONCLUSIONS

We developed a DA scheme tailor-made for system-wide soft sensing and sensor validation for urban drainage tunnels. DA has the potential to equip hydrodynamic models with soft sensing capabilities, by improving their accuracy and expressing their uncertainty. Ensemble-based DA methods are particularly suitable for DUDMs but have high computational costs. By separating the 'inner' tunnel model from the 'outer' drainage model and updating only the inner model, our proposed scheme retains the benefits of distributed models and ensemble-based DA, while restraining the computational overhead. We applied the assimilation scheme to a 1D hydrodynamic model of a 3.4 km-long combined overflow tunnel, whose outer model counts over 15,000 links (equivalent to 800 km of pipes) and a similar number of nodes. The inner model covered only the tunnel and connecting structures with a total of 28 links and was roughly 700 times faster than the outer model. When using an ensemble size of 100 and updating the ensemble to observations once per minute, the main computational effort was related to running the outer model. The performance of the scheme was observed to be sensitive to the choice and tuning of the perturbation model that simulated the uncertainty of the tunnel input. A similar sensitivity was observed in the tuning of the standard deviation of the observations. We note that the tuning process is highly case-specific and requires balancing the degrees of confidence in the model and the sensors, while staying within realistic flow conditions.

In all three events tested, the updated water levels at the downstream dropshaft scored better when compared to observations than the ensemble predictions without assimilation. The update was also correctly propagated across the tunnel as confirmed by an independent observation at the most upstream dropshaft. Thus, the model generated an accurate, system-wide estimate of the hydraulic conditions. In the third event, the large discrepancy between updated and observed states enabled the automatic detection of a false echo in the upstream sensor. In this case, the updated ensemble was capable of reconstructing the likely water level upstream, acting as a soft sensing tool. Further research is needed to promote the proposed scheme into a reliable, operational tool, by for example introducing an automatic assessment of both observation and model error.

ACKNOWLEDGEMENTS

We thank Lone Bo Jørgensen and Anders Breinholt from the Greater Copenhagen Utility (HOFOR) and the Danish Meteorological Institute (DMI) for providing the model and data for the case study. We also thank DHI for granting access to an early version of MIKE URBAN's DA framework. This work is part of a joint PhD programme between the Technical University of Denmark and the Nanyang Technological University, Singapore.

DATA AVAILABILITY STATEMENT

Data cannot be made publicly available; readers should contact the corresponding author for details.

REFERENCES

Anctil, F. & Ramos, M. 2019 Verification metrics for hydrological ensemble forecasts. In: *Handbook of Hydrometeorological Ensemble Forecasting*. Springer Berlin Heidelberg, Berlin, Germany, pp. 893–922. doi:10.1007/978-3-642-39925-1.

Bennett, N. D., Croke, B. F. W., Guariso, G., Guillaume, J. H. A., Hamilton, S. H., Jakeman, A. J., Marsili-Libelli, S., Newham, L. T. H., Norton, J. P., Perrin, C., Pierce, S. A., Robson, B., Seppelt, R., Voinov, A. A., Fath, B. D. & Andreassian, V. 2013 Characterising performance of environmental models. *Environmental Modelling and Software* **40** (February 2017), 1–20. http://dx.doi.org/10.1016/j.envsoft.2012.09.011.

Borup, M. 2014 *Real Time Updating in Distributed Urban Rainfall Runoff Modelling*. PhD Thesis, Department of Environmental Engineering, Technical University of Denmark, Kgs. Lyngby, Denmark.

Borup, M., Grum, M. & Mikkelsen, P. S. 2011 Real time adjustment of slow changing flow components in distributed urban runoff models. In: *Proceedings of 12th International Conference on Urban Drainage*, pp. 11–16. Available from: http://orbit.dtu.dk/ws/files/6373440/prod11321622323249.PAP005261.pdf (accessed 14 January 2019).

Borup, M., Grum, M., Madsen, H. & Mikkelsen, P. S. 2014 Updating distributed hydrodynamic urban drainage models. In *Proceedings of the 13th International Conference on Urban Drainage*, pp. 1–8. Available from: http://orbit.dtu.dk/ws/files/101102837/BorupArticle.pdf (accessed 14 January 2019). doi:10.13140/2.1.1682.1766.

Borup, M., Madsen, H., Grum, M. & Mikkelsen, P. 2018 Technical note on the dynamic changes in Kalman gain when updating hydrodynamic urban drainage models. *Geosciences* **8** (11), 416. doi:10.3390/geosciences8110416.

Deletic, A., Dotto, C. B. S., McCarthy, D. T., Kleidorfer, M., Freni, G., Mannina, G., Uhl, M., Henrichs, M., Fletcher, T. D., Rauch, W., Bertrand-Krajewski, J. L. & Tait, S. 2012 Assessing uncertainties in urban drainage models. *Physics and Chemistry of the Earth* **42–44**, 3–10. doi:10.1016/j.pce.2011.04.007.

DHI 2019 *MIKE URBAN*. Collection System, Hørsholm, Denmark.

Evensen, G. 1994 Sequential data assimilation with a nonlinear quasi-geostrophic model using Monte Carlo methods to forecast error statistics. *Journal of Geophysical Research* **99** (C5). doi:10.1029/94jc00572.

Gneiting, T., Balabdaoui, F. & Raftery, A. E. 2007 Probabilistic forecasts, calibration and sharpness. *Journal of the Royal Statistical Society: Series B (Statistical Methodology)* **69** (2), 243–268. http://doi.wiley.com/10.1111/j.1467-9868.2007.00587.x.

Hansen, L. S., Borup, M., Møller, A. & Mikkelsen, P. S. 2014 Flow forecasting using deterministic updating of water levels in distributed hydrodynamic urban drainage models. *Water (Switzerland)* **6** (8), 2195–2211. doi:10.3390/w6082195.

Hersbach, H. 2000 Decomposition of the continuous ranked probability score for ensemble prediction systems. *Weather and Forecasting* **15** (5), 559–570. doi:10.1175/1520-0434(2000)015 < 0559:DOTCRP > 2.0.CO;2.

Hutton, C. J., Vamvakeridou-Lyroudia, C., Kapelan, Z. & Savić, D. 2011 Real-time modelling and data assimilation techniques for improving the accuracy of model predictions. In *PREPARED 2011.010 Deliverable D3.6.2*, p. 83.

Hutton, C. J., Kapelan, Z., Vamvakeridou-Lyroudia, L. & Savić, D. 2014 Real-time data assimilation in urban rainfall-runoff models. *Procedia Engineering* **70**, 843–852. doi:10.1016/j.proeng.2014.02.092.

Jørgensen, H. K., Rosenørn, S., Madsen, H. & Mikkelsen, P. S. 1998 Quality control of rain data used for urban runoff systems. *Water Science and Technology* **37** (11), 113–120. doi:10.2166/wst.1998.0448.

Liu, Y., Weerts, A. H., Clark, M., Hendricks Franssen, H.-J., Kumar, S., Moradkhani, H., Seo, D.-J., Schwanenberg, D., Smith, P., van Dijk, A. I. J. M., van Velzen, N., He, M., Lee, H., Noh, S. J., Rakovec, O. & Restrepo, P. 2012 Advancing data assimilation in operational hydrologic forecasting: progresses, challenges, and emerging opportunities. *Hydrology and Earth System Sciences* **16** (10), 3863–3887. doi:10.5194/hess-16-3863-2012.

Lund, N., Madsen, H., Mazzoleni, M., Solomatine, D. & Borup, M. 2019 Assimilating flow and level data into an urban drainage surrogate model for forecasting flows and overflows. *Journal of Environmental Management* **248** (December 2018), 109052. doi:10.1016/j.jenvman.2019.05.110.

Murphy, A. H. 1993 What is a good forecast? An essay on the nature of goodness in weather forecasting. *Weather and Forecasting* **8** (2), 281–293. doi:10.1175/1520-0434(1993)008 < 0281:WIAGFA > 2.0.CO;2.

Palmitessa, R., Borup, M. & Mikkelsen, P. S. 2018 Urban tunnel systems for conveyance and storage of storm- and wastewater: features, classification, and modelling. In: *Proceedings of the 11th International Conference on Urban Drainage Modelling* (G. Mannina, ed.). Palermo, Italy, pp. 251–254. Available from: http://orbit.dtu.dk/ws/files/154733729/UDM18_abstract_Palmitessa_2906.pdf.

Sakov, P. & Oke, P. R. 2008 A deterministic formulation of the ensemble Kalman filter: an alternative to ensemble square root filters. *Tellus, Series A: Dynamic Meteorology and Oceanography* **60A** (2), 361–371. doi:10.1111/j.1600-0870.2007.00299.x.

Thordarson, F. Ö., Breinholt, A., Møller, J. K., Mikkelsen, P. S., Grum, M. & Madsen, H. 2012 Evaluation of probabilistic flow predictions in sewer systems using grey box models and a skill score criterion. *Stochastic Environmental Research and Risk Assessment* **26** (8), 1151–1162. doi:10.1007/s00477-012-0563-3.

Tscheikner-Gratl, F., Zeisl, P., Kinzel, C., Rauch, W., Kleidorfer, M., Leimgruber, J. & Ertl, T. 2016 Lost in calibration: why people still do not calibrate their models, and why they still should – a case study from urban drainage modelling. *Water Science and Technology* **74** (10), 2337–2348. doi:10.2166/wst.2016.395.

Zhang, D., Madsen, H., Ridler, M. E., Refsgaard, J. C. & Jensen, K. H. 2015 Impact of uncertainty description on assimilating hydraulic head in the MIKE SHE distributed hydrological model. *Advances in Water Resources* **86**, 400–413. http://dx.doi.org/10.1016/j.advwatres.2015.07.018.

First received 30 April 2020; accepted in revised form 27 November 2020. Available online 15 December 2020

© 2021 The Authors

Taking water efficiency to the next level: digital tools to reduce non-revenue water

J. Cassidy IWA, B. Barbosa IWA, M. Damião, P. Ramalho IWA, A. Ganhão IWA, A. Santos IWA and J. Feliciano IWA

ABSTRACT

Efficiency optimization of urban water systems is a growing concern for water utilities worldwide. This case study aimed at evaluating the impact of using cloud-based tools on the reduction of both real (real-time network monitoring) and apparent water losses (integrated customer meters management) in two water utilities. The incorporation of smart water solutions with a methodology for the management and operation of the systems allowed us to diagnose, prioritize areas and define actions to improve efficiency. Using a real-time monitoring tool allowed us to categorize bursts and to evaluate their impact on water loss volumes and to identify operational inefficiencies regarding detection and repair times, particularly in small and medium bursts. Additionally, the implementation of an integrated customer meters management tool allowed for an optimized meter management reducing apparent losses by estimating metering errors more accurately, enabling the water utilities to replace meters based on specific lifespan. Digitalization, through the implementation of optimized algorithms and early warning systems, allowed the analysis of data in a methodical and prompt manner resulting in non-revenue water reduction up to 8% in 3 years while improving the digital organization of data and its quality (reliability and accuracy), interdepartmental organization and communication, capacity building and utilities' image among stakeholders.

Key words | apparent losses, cloud-based solutions, digital water, efficiency projects, non-revenue water, real losses

J. Cassidy IWA (corresponding author)
B. Barbosa IWA
M. Damião
P. Ramalho IWA
A. Ganhão IWA
A. Santos IWA
J. Feliciano IWA
AGS – Administração e Gestão de Sistemas de Salubridade S.A.,
Quinta da Fonte Office Park – Edifício Q54-D. José – Piso 2, 2770-203 Paço de Arcos,
Portugal
E-mail: *jcassidy@ags.pt*

HIGHLIGHTS

- Two water utilities implemented web-based tools to promote the sustainable reduction of non-revenue water.
- Implementing a real-time monitoring tool, utility A and B decreased 13 million m^3/year in real losses.
- Using an integrated customer meters management web-tool, both utilities decreased apparent losses by 3.1 million m^3/year.
- Non-revenue water reduced by 8% overall, resulting in savings of 5.8 million € in 3 years.

This is an Open Access article distributed under the terms of the Creative Commons Attribution Licence (CC BY-NC-ND 4.0), which permits copying and redistribution for non-commercial purposes with no derivatives, provided the original work is properly cited (http://creativecommons.org/licenses/by-nc-nd/4.0/).

doi: 10.2166/hydro.2020.072

INTRODUCTION

During the last century, the global demand for water has been rapidly increasing due to population growth, food and energy security policies, macro-economic processes such as trade globalization, changing consumption patterns and urbanization. Specifically, increasing urbanization causes highly localized pressures on freshwater resource availability (WWAP 2015). Water losses from urban water distribution systems can amount to one-third of the water abstracted which may imply significant revenue loss (Ananda 2019). With such water distribution losses, water utilities play a key role in guaranteeing an optimized management of the urban water cycle whilst ensuring economic and environmental sustainability. The activities needed to manage water losses can be categorized in three stages: (1) assessment and monitoring of water losses, (2) planification of cost-effective measures and (3) implementation of reactive and proactive actions to detect leakage and illegal use (Farley 2001; AL-Washali *et al.* 2016).

As such, to tackle water losses, one must first calculate the water balance which quantifies the total water volume entering the system and divides it into revenue water and non-revenue water (NRW). NRW is further divided into unbilled authorized consumption and water losses which are typically divided into real losses (leakage in mains, service connections and storage tanks) and apparent losses (customer meter inaccuracies and unauthorized consumption) (IWA 2000; Alegre *et al.* 2016).

The complex configuration of water distribution systems must be considered when selecting the real losses identification methodology to be applied. As such, dividing the system into district metered areas (DMA) to analyse the mass balance is commonly used as it results in better leakage detection, location isolation and thus, reduce the detection time (Farley 2001 and references therein; Eliades & Polycarpou 2012). The identification and localization of bursts/leaks can be performed using hardware methods or software-based methods (Li *et al.* 2015). Hardware methods include acoustic detection methods (e.g., leak correlators and leak noise loggers) and non-acoustic methods (e.g., gas injection, ground penetrating radar technology or infrared photography). These methods are increasingly more accurate but often incur high costs given the expensive equipment and the large number of human resources needed. With the increased innovation in the internet-of-things (IoT) and decrease of cost observed in such devices (e.g., online water flow and pressure meters), practitioners are more and more looking into the deployment of software-based solutions. These are less expensive to implement and provide crucial information on a more limited area where a possible leak/burst may be occurring (Li *et al.* 2015; Wu & Liu 2017). Furthermore, real-time monitoring and statistical analysis of parameters such as flow allows us to determine the minimum night flow analysis of each DMA and to close the water balance on a daily or even hourly basis and subsequently estimate the level of real loss (Thornton *et al.* 2008; Loureiro *et al.* 2016).

While real losses refer to the physically lost water, apparent losses refer to the sum of water that has been used but has not been paid for (AL-Washali *et al.* 2016). In well-managed utilities, metering errors account for a high percentage of the apparent losses (Arregui *et al.* 2018). To reduce metering errors, water utilities traditionally replace customer meters based on their age. However, the error associated with each meter can be due to meter wear and tear, incorrect installation practice, lack of maintenance or calibration, incorrect meter type and class for the application, incorrect meter sizing, demand profile or demand type problems (Criminisi *et al.* 2009). These metering errors can be minimized by implementing a continuous programme which includes field work (e.g., water meter installation and periodic auditing), laboratory work (e.g., audit meters and assessment of meter's accuracy) and management tasks (e.g., cost-benefit analysis, meter selection and meter replacement plan) (Arregui *et al.* 2012). There are several water meter types which are classified on the flow measuring mechanism (mechanic, electromagnetic and ultrasonic) which affects the measurement accuracy. In addition, the model and brand of the meters can also have an effect of such measurement even when using the same measurement technology (Karadirek 2020). As such, an optimized and utility-specific replacement plan which takes into consideration all these steps is crucial to maintain apparent losses to a minimum.

In summary, NRW reduction projects typically require high amounts of data generated from the daily operations of water utilities which need to be analysed in a quick and effective manner. As such, for utilities aiming to improve its efficiency, the development and implementation of cloud-based solutions to assist such projects in reducing real and apparent losses is a must have. Having cloud-based solutions is particularly important since having a third-party provider exempts the water utilities from having specialized hardware or IT personnel to maintain the solutions. Additionally, cloud-based solutions offer faster implementation and updates allowing its use in a short time.

On one hand, real-time monitoring tools may perform a comparison between normalized water consumption patterns and real online flows, triggering alerts for anomalous or extraordinary events such as leaks, pipe bursts, meter or transmission malfunctions or anomalous consumptions. On the other hand, the integrated customer meters management tool may compute metering errors based on diameter, consumption, age, brand and model. Subsequently, it generates a list with all the meters to be replaced, potentially stopped, with anomalous consumptions and wrongly sized. The case study presented herein aimed at evaluating the impact of using two online tools on the reduction of both real (*flowise* for network monitoring) and apparent water losses (*meterwise* for integrated management of meters).

METHODS

Main objectives

The work herein presented aimed to evaluate operational improvements and its impact on NRW reduction programmes promoted by AGS, a Portuguese water company, owned by Marubeni, and responsible for the management of 13 water utilities in Portugal and Brazil under concession agreements and public–private partnerships and for the provision of engineering services to water utilities in Europe, South America and Asia.

Specifically, the work presented focuses on two Chilean water utilities through the analysis and management of anomalous events detected by the real-time monitoring software and the reduction of apparent losses by using a cloud-based software which selects critical customer meters to be replaced. The NRW reduction programmes in each water utility are assessed in this study from 2015 (year 0) to 2018.

Detection of bursts and anomalous consumption (*flowise*)

For the reduction of real losses, AGS developed *flowise*, a real-time network monitoring system which integrates information from historic flow measurements and other hydraulic variables such as pressure, water levels, chlorine concentration or conductivity. The data are updated and continuously analysed by algorithms which have been published and validated by the scientific community (Loureiro *et al.* 2016). Data are automatically validated and normalised by detecting and correcting anomalies due to metering problems, like faulty transmission or inadequate acquisition. This normalization aims at obtaining data with a regular time step, preferentially of 15 min, where the mean flow value (or other variables) is obtained through numerical integration.

Additionally, *flowise* is structured based on existing DMA of each distribution system. DMA are first identified by an engineering team and singled out in *flowise*. These DMA correspond to geographical areas described in the existing cadastre with the specific flow being defined by a single or a set of meters. Subsequently, consumption patterns are determined and regularly updated for each specific DMA reflecting daily (weekdays, Saturday, Sunday and holidays), weekly and seasonal human behaviours. Subsequently, a comparison between the normalized consumption patterns and online flow measurements is performed. Alarms are triggered when significant differences (considered when three or eight consecutive flow measurements are above the 95th percentile, for large and medium bursts, respectively) are detected (Figure 1). These differences are due to abnormal uses and can be categorized into events such as leaks, pipe bursts, meter or transmission malfunctions and anomalous consumption. The categorization of events is done by the operational team in the field. These alarms are automatically sent to selected users, so that action can be taken to minimize or prevent the consequences of such events.

Apparent losses reduction (*meterwise*)

The methodology used to develop the web-based solution to assist in apparent loss reduction (*meterwise*) followed the principles presented by Arregui *et al.* (2006). The methodology determines the optimal lifespan of a meter considering the balance between the cost of meter

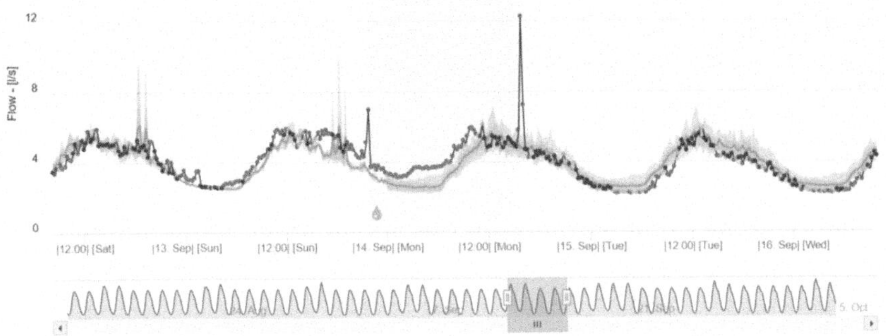

Figure 1 | Example of a burst detected by the real-time monitoring tool – *flowise*. Black line and coloured dots refer to real flow measurements (black dots – values within normal pattern; yellow dots – values above 80th percentile or below 20th percentile; red dots – values above 95th percentile or below 5th percentile). The dark grey line is the 50th percentile, the medium grey area is between 20th and 80th percentiles, and the light grey area is between 5th and 95th percentiles. Please refer to the online version of this paper to see this figure in colour: http://dx.doi.org/10.2166/hydro.2020.072.

replacement and the gains obtained from the reduction of sub measurements. Briefly, the time (year) when replacing a meter is economically favourable is determined taking into consideration the Net Present Value (NPV) of the replacement Chain (NPVC) (Equation (1)) which is calculated over a period of 30 years. The optimized lifespan corresponds to the year when $NPVC_t < NPVC_{t-1}$.

$$\mathrm{NPVC}_t = \left[\frac{(1+r')^i}{(1+r')^{i-1}} \right] * \mathrm{NPV}_t \quad (1)$$

where NPV_t is the net present value for year t and is determined by the sum of the initial cost of the meter with the generated cash-flows since the beginning of the project until year t; r' is the real discount rate which is the discount rate taking in consideration the inflation rate; and i is the years since installation. The annual cash-flow represents the revenue based on the customer meter's billing and is determined according to the following equation:

$$\mathrm{Cash-flow} = \frac{V_R * \bar{C}}{(1+r')^{(i-1)}} \quad (2)$$

where $\bar{C}$ is the average tariff and V_R is the registered volume and is calculated using Equation (3).

$$\mathrm{Registered\ volume} = \mathrm{AYC} + \mathrm{AYC} * \varepsilon_i + \mathrm{AYC} * (i-1) * a \quad (3)$$

where AYC is the average yearly consumption; ϵ_i is the initial error; and a is the degradation rate.

In order to determine the degradation rate, which is the rate of decay in metering accuracy over time (increase of metering error), and the initial error which is the customer meter's error at the time of installation, a group of customer meters from each utility was selected for laboratory bench tests which allowed us to build the error curve.

In the case study presented, the number of meters selected for testing in each utility was also dependent on the available budget of each utility. As such, in order to reduce possible inaccuracies, a compromise was obtained by selecting a range of different meters to comply with the budget. To have a representative sample, the group of meters selected for testing was selected from clusters with the highest volume consumption and was defined with the analysis of the areas, ages and models. The representative clusters were also divided by different water sources to account for different water quality. The number of meters selected for testing is shown in the Results and Discussion section.

For each meter, an error curve was obtained in order to determine its evolution in time as well as each meter global error. The weighted global error for each meter was obtained by crossing its error curve with the histogram of the average water consumption. Based on the global error of each meter and its age, the degradation rate was obtained using a linear regression.

Based on the results obtained and applying the lifespan methodology, the two utilities presented replaced customer meters with critical errors in an optimized way to reduce apparent losses.

System's design architecture and implementation

Producing data is now the easiest step towards the real-time analysis of integrated information. The management and analysis of this data are the bottleneck of organization implementing smart system. When developing such a system, it is essential to understand which data are needed to support the tools and understand the wide range of databases needed to collect all the data. As such, AGS designed the system in order to securely collect vast amounts of information from multiple sources and formats providing insight and helping decision-making (Figure 2). For network monitoring purposes, it is important to ensure that all relevant variables are being properly metered using adequate equipment and are integrated in a systematized and automatic manner. This means that flow, pressure, chlorine, pH or other data from each individual meter are sent to a database for storage. In many cases, the different metering equipment installed in the utility from different suppliers can result in data being stored in different formats and in multiple databases such as SCADA or telemetry systems which will disperse and hamper the analysis of data. This issue can be overcome by gathering all data from all platforms to a single platform as explained in this section.

Operationally, data from multiple origins (e.g., SCADA, Telemetry and other) are imported using Representational State Transfer (REST) APIs into the real-time network monitoring module (*flowise*) where it is aggregated and normalised to generate processed data and warning events (alarms). More specifically, all metered data are converted into a normalized time step and stored in a centralized database using a standard format. Subsequently, data can be analysed to perform water balances in DMA considering data from the water network input and output meters in an easy and expeditious manner, even if meters are initially associated to different systems. The monitoring system was designed to collect data from original databases with the highest frequency possible (real-time); however, some water utilities do not have data with such frequency and adopt a time step that balances the need for information and the cost of communication (e.g., a 15 min or hourly frequency).

To understand the behaviour of water distribution networks, it is also important to have information about each DMA or distribution system characteristic in terms of number of customers, pipe length, diameter and materials, number of service connections or area, to name a few. Most of these variables are discrete and stored in the geographic information system (GIS). This alphanumeric and spatial data (pipes presented as lines/polylines and DMA presented as polygons) can be directly uploaded to *flowise* as shape files.

Other data such as customer data, meter data and meter readings from a customer relationship management (CRM) database are processed using specialized algorithms by the integrated customer meter management tool (*meterwise*). *meterwise* can connect with any CRM used for commercial management. This is done either through a direct

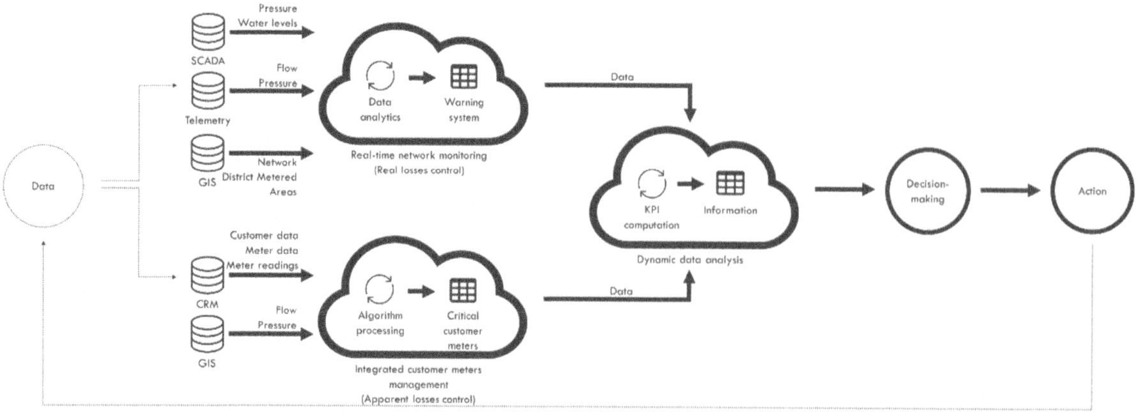

Figure 2 | System's design and architecture.

connection to the CRM or with a copy or extraction of the original database. Thus, to collect the indispensable data regarding customer meters, two REST webservices were developed to obtain: (1) customer meter's features (diameter, installation date, equipment type, brand, model or precision class) and (2) meter readings (reading date and volume). Data regarding the meter's cadastre and readings are stored in different tables and are automatically associated by using each meter's ID. Update frequency is customizable depending on the utility's practices for customer meter readings. For water utilities that do not have smart metering technology, a monthly update is generally enough.

The resulting analyses from both modules are integrated in a business intelligence solution for in-depth analysis and reporting which will assist in the definition of action plans. Such action plans might include the replacement of customer meters or the repair or replacement of a water pipe. The data resulting from these actions will be uploaded to the system and updated.

In summary, the software tools' infrastructure is secure (all integrations are IP restricted and all tenants are logically separated from each other), reliable (all data are backed up into cloud servers and have a guaranteed 99.999% service level agreement (SLA) uptime), flexible (hardware resources are available on demand with no need to install new physical components) and directly accessible by a virtual private network (VPN). In order to access the utility's databases and infrastructure, data are remotely transferred without further local interventions. The end users can access the tools using any device through their browser which allows remote assistance as shown in this study.

RESULTS AND DISCUSSION

Characterization of the water utilities

The main characteristics of the water system of each utility are presented in Table 1. The two Chilean water utilities are responsible for supplying 1.3 million inhabitants and managing 400,000 customer meters which measure an average volume of 9.8 million m^3/month. In 2015 (year 0), utility A and utility B presented NRW levels of 35.8 and 42.8%.

Table 1 | Utilities' water system characteristics (2018)

Dimension	Variable	Units	Utility A	Utility B
Water system	Customers	No.	161, 896	240, 444
	Network length	km	1,117	2,032
	District meter areas	No.	115	116
Infrastructural condition	Mains bursts	No.	781	188
	Service connection bursts	No.	5,077	5,231
Operational control	Flow meters	No.	146	140
	Pressure meters	No.	230	170

To assist in managing water loss reduction, both utilities have several IoT flow and pressure meters strategically placed in the network which communicate with *flowise* in real-time. In addition to differences in flow data, *flowise* detects differences in pressure data from pressure reduction valves (PRV) or pressure critical points (PCP). These differences can indicate and pinpoint the occurrence of bursts when a significant pressure reduction is detected.

Whilst some meters were installed before the kick-off of the project, a few more IoT meters were strategically installed throughout the project. While utility A installed 11 and 125 flow and pressure meters, respectively, utility B installed 12 and 147 flow and pressure meters, respectively. The reasoning behind this was to achieve total sectorization of the network (DMA) and monitoring via *flowise* to reduce NRW and comply with the local regulatory demands.

Technical and operational challenges

When implementing such NRW and digitization projects, several technical and operational challenges may arise. Being responsible for the management of the entire urban water cycle in their regions, the utilities, presented in this study, faced the challenge of decreasing their NRW levels while having to continuously supply over 400,000 customers which are scattered over large areas. At the start of the projects, water operations had a large dependency on manual labour and internal procedures had low levels of sophistication, control and traceability. This made assessing performance either at DMA level or utility level and auditing results a lengthy and slow process. Records of daily operations were made on paper and were not analysed unless a major problem occurred.

These procedures not only hampered the decrease of NRW but also allowed a continuous degradation of infrastructure. The existing software before year 0 was very poor as it only allowed the support of day to day operations, not allowing a comprehensive monitoring in terms of spatial coverage and data frequency. Additionally, not having the information audited and centralized in a systemic way caused constraints in terms of misrepresented information with different data being presented for the same variable depending on the employee who provided it. This caused interdepartmental communication to be a difficult and time-consuming task.

In such a sense, digitalization of water services was a key topic to be addressed in the utilities. These challenges were overcome during the implementation of the projects presented in this study.

Detection of anomalous flow events

The implementation of *flowise* allowed the generation of 2,544 and 3,592 anomalous events in utility A and utility B, respectively, from 2015 to 2018. The different anomalous events were analysed and categorized by the utility operators as (i) illegal consumption, (ii) bursts, (iii) high demand customers, (iv) maintenance works, (v) PRV adjustment, (vi) zone changes, (vii) pipe flushing and (viii) others (which includes water used for fire services, third-party bursts and tank overflows) (Figure 3).

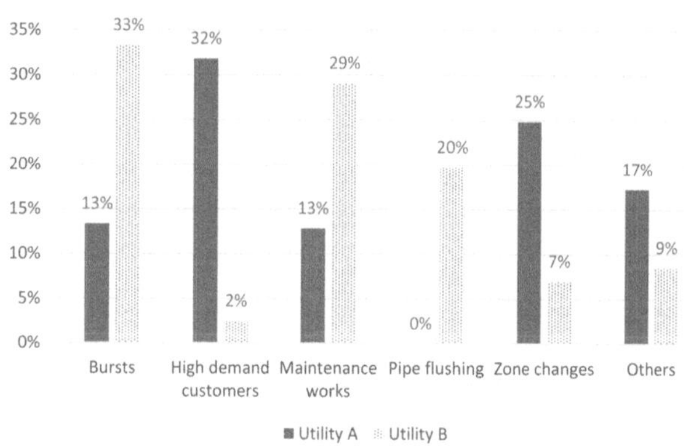

Figure 3 | Flow increase events distribution per cause in utility A and B (2015–2018).

The majority of flow increase events in utility A are due to high demand customers (32%) and zone changes (25%). Bursts come in fourth place, representing 13% of total events, indicating that the water infrastructure condition can be compromised in some locations. The high percentage of high demand customers in utility A is due to its location close to the coastal area with high consumers in the port and services. Due to seasonal water scarcity, the zone changes, which involve modifying the network sectorization, are crucial in this utility to ensure supply due to quantity and quality issues. In utility B, most events are due to bursts (33%) followed by maintenance works (29%) and pipe flushing (20%). From this analysis, one can conclude that categorizing the events is a crucial step to fully understanding the nature of the increase which in most cases is not due to real losses.

Categorization of bursts and reduction of leak runtime

The degradation of water supply infrastructure mainly due to age and operational conditions can cause issues like pipe bursts, loose joints and fittings (Eliades & Polycarpou 2012). This degradation will ultimately result in leakages and bursts which will increase the water loss in the system and consequently cause consumer problems, health risks and financial losses (Farley 2001). Depending on the nature and size of the bursts (slowly developing or abrupt), different approaches must be taken. As such, the identified bursts were categorized in three different categories: (1) small bursts or leaks – corresponding to average flows less than 5 l/s, (2) medium bursts – corresponding to average flows between 5 and 15 l/s and (3) large bursts – corresponding to average flows higher than 15 l/s.

During the period of 2015–2018 (Figure 4), utility A presented 71 small bursts, 88 medium bursts and 80 large bursts. The large bursts' flow was mostly between 21.7 and 55.7 l/s with the maximum detected flow being 231.5 l/s. On the other hand, utility B presented 323 small bursts, 266 medium bursts and 434 large bursts. In terms of flow, the largest burst reached 291.3 l/s with most large bursts having flows between 22.0 and 44.3 l/s.

Average leaks and bursts runtime include three components: awareness, location and repair time, known as the ALR concept (Lambert & Morrison 1996). This will

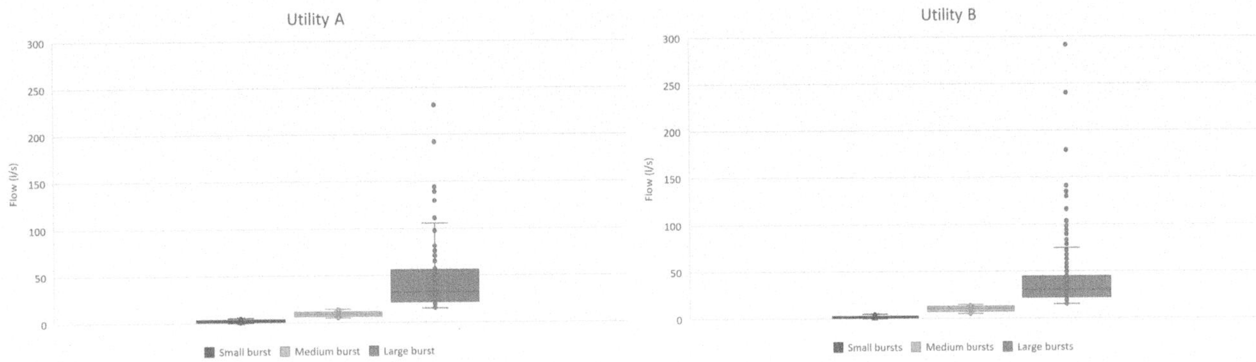

Figure 4 | Categorization of bursts detected by *flowise* in each utility (2015–2018).

allow us to determine in a more accurate way the effective real losses volume of a burst, considering both the flow rate and duration time.

Most lost volume in both utilities results from small and medium bursts (Figure 5) which is probably a result of the long time needed for ALR. In fact, most physical water losses at yearly scale can be attributed to the background leakages from the presence of leaks from joints, holes and cracks (Covelli *et al.* 2015).

In both utilities, large bursts present similar behaviour with the lowest runtime when compared to the other type of bursts. Utility A and utility B had an average runtime of 0.3 and 0.2 days, respectively.

Differences can be observed when comparing the runtime in small and medium bursts in the two utilities. In 2015, at the start of the NRW reduction project, utility A took on average 15.2 and 4.7 days to locate and repair small and medium bursts, respectively. Utility B, in the

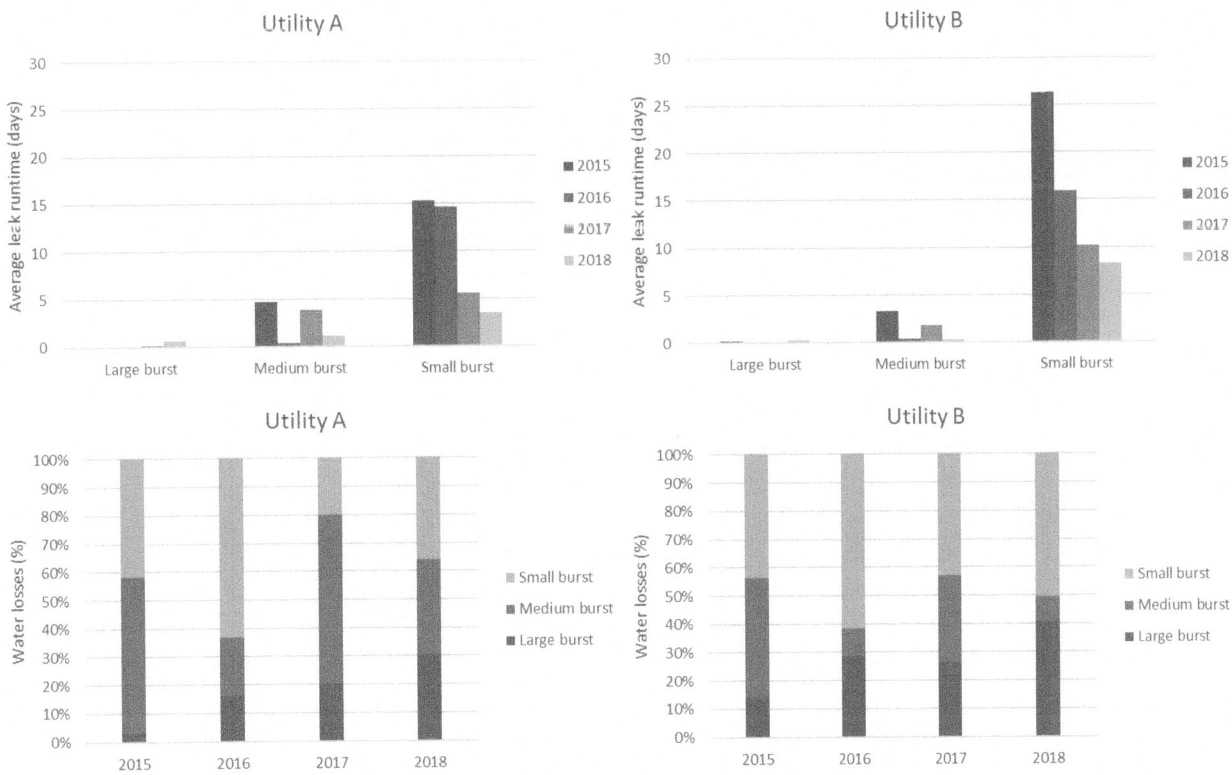

Figure 5 | Average leak runtime and water losses percentage per burst group in utilities A and B – comparison between results in 2015 (project kick-off), 2016, 2017 and 2018.

same year, had longer runtimes for small bursts, 26.3 days and shorter runtimes for medium bursts, taking on average 3.3 days. The differences observed in small bursts can be explained based on the geographical location of each utility. Utility A is located in a desert area, where it is easier to locate small bursts that can only be detected by acoustic equipment. Whereas utility B is located in an area where the groundwater level is very high, making it more difficult to locate bursts. For medium bursts, these differences are less evident since they can be visibly detected in both utilities. In addition, utility A has more human resources available and thus, more effective when developing active leakage detection activities.

These high detection and repair times had large potential for improvement as they corresponded to operational inefficiencies which could be tackled and upgraded. Based on this analysis, targets for reducing detection time and repair time in medium and small bursts were defined considering each utility's context. AGS supported local operational teams through weekly follow-up meetings focused on improving active leakage detection planning and reducing repair times. Results were continuously monitored on the real-time monitoring tool enabling a better control of systems' performance.

Results show a significant reduction of the leak runtime in small and medium bursts for water utilities A and B after implementing *flowise* (Figure 5). During the course of the projects, utility A reduced leak runtime of small bursts from 15.2 to 3.4 days while in utility B, the reduction was from 26.3 to 8.3 days. For medium bursts, the reduction of runtime was from 4.7 to 1.2 days and 3.3 to 0.3 days for utilities A and B, respectively. This reduction has also led to a reduction in the percentage of water loss related to small and medium bursts (Figure 5), contributing to a significant decrease of NRW levels, overall. The results obtained demonstrate that by implementing a real-time monitoring tool, the existing and new problems were highlighted in a methodical and systematic manner enabling the utilities to identify the small and medium bursts faster and thus, to implement oriented actions to successfully reduce runtime and consequently water loss. Based on the infrastructure problems evidenced by *flowise*, utility A and utility B successfully replaced 36 and 57 km of their network, respectively.

Customer metering inaccuracies determination

In order to determine the customer metering errors, and taking into account the considerations presented in the Methods section, a global sample of 687 meters was selected (304 meters for utility A and 383 meters for utility B) for laboratory analysis through water meters bench testing. Each water utility selected the domestic meters to be tested based on the highest volume consumption and representativity. The selected meters included DN15 and DN20 diameters of different brands, models and ages (Figure 6). Laboratory bench tests were conducted on these meters in certified laboratories.

The metrological tests conducted allowed us to determine the average error curve of the meters per age and the evolution of the weighted global error and subsequently, the degradation rate. For comparison purposes and as an example, the results obtained for one brand and model are presented in Figure 7. The degradation rate is higher in utility A (−0.8585%/year) than in utility B (−0.4174%/year) for the same brand and model. This difference is most likely due to the different water quality and meter installation

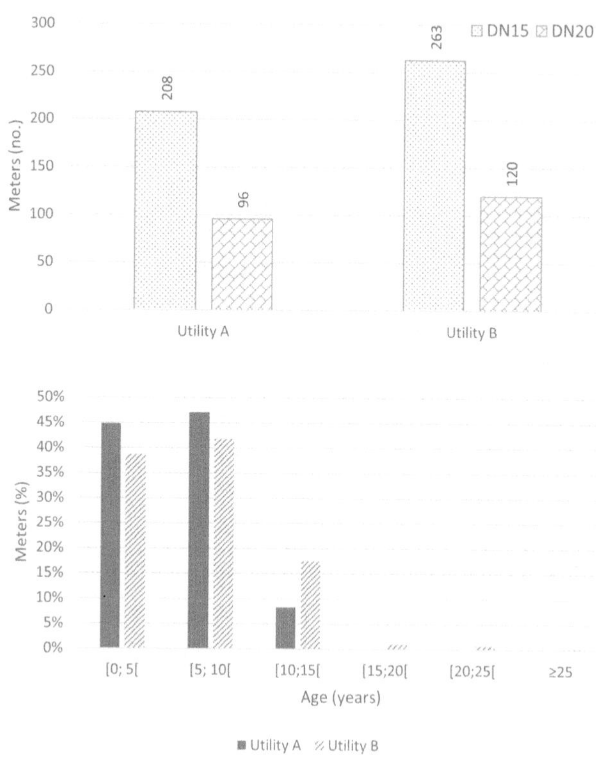

Figure 6 | Distribution of tested meters by diameter (top) and age (bottom).

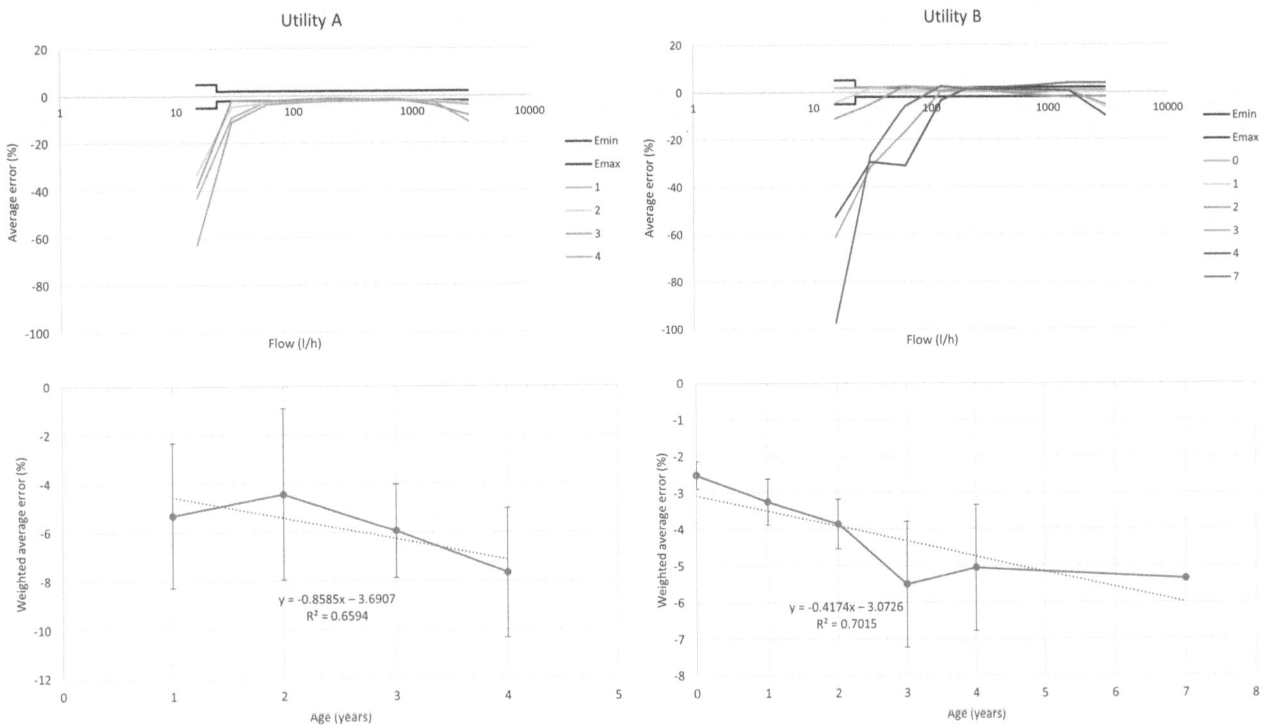

Figure 7 | Average error curve of the meters per age (top) and evolution of the weighted global error (bottom) for the same brand and model used in both utilities.

procedures which are two of the most common causes for meter deterioration (Criminisi *et al.* 2009).

A *meterwise* solution was implemented and tailor made for each utility based on the customer meter's initial errors and degradation rates. Thus, data from each utility's active meters were imported into *meterwise*. This covered a 2-year period, 180,000 customer meters for utility A and 275,000 customer meters for utility B as well as meter readings data covering, corresponding to over 4,115,000 registries. Additionally, certain specificities from each utility were considered such as the average water tariff and cost of customer meters purchase and installation.

The previous results show that initial error and degradation rate are clearly different between utilities. Thus, the meters in each utility will have different lifespans and consequently, different renewal needs. Based on the degradation rate obtained, the lifespan was calculated for different monthly consumptions (Figure 8). This estimation was performed assuming an average tariff of 1.62 and 1.29 €/m^3 and a meter replacement cost of 30 and 31.22 €/(DN15 meter), for utilities A and B, respectively. Results show

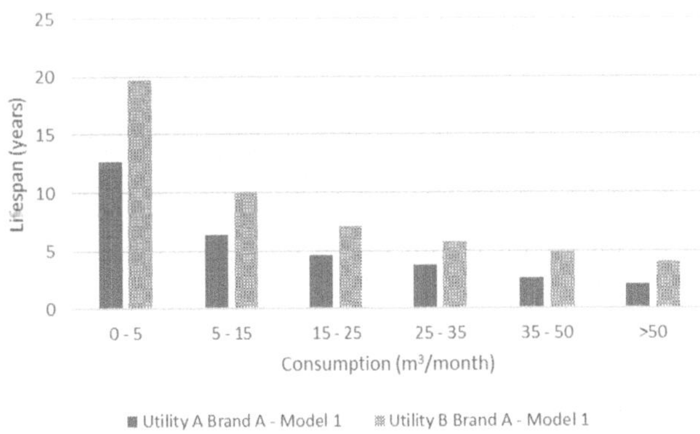

Figure 8 | Lifespan of Brand A – model 1 for each utility and for different monthly consumptions.

that for the same meter model, utility A will need to replace the meters between 2 and 7 years earlier, depending on the consumption, than utility B.

Such a result clearly indicates that the implementation of a customer meters management tool which incorporates the methodology described can translate into a significant reduction of apparent loss volumes and, thus, into significant revenue increases for a water utility.

This study confirms that the same meter type can have different ageing processes when installed in different water utilities, that different types of meters behave differently within the same water utility (Arregui *et al.* 2018) and that meter selection should be done based on several parameters such as water quality, consumption, cost and ease of installation (Karadirek 2020).

Reduction of apparent losses

Two alternatives for replacing meters were analysed. Alternative 1 considered the replacement policy at the beginning of the project, in which all meters with ages equal to or above 8 years are replaced. On the other hand, in Alternative 2, the meters are to be replaced based on the lifespan estimation presented in the previous sections and which is the outcome of *meterwise*.

Utility A, due to water scarcity, has a higher water tariff which results in high gains regardless of the approach taken (Table 2). Nevertheless, replacing the customer meters based on the lifespan (Alternative 2) would need lower initial investment and higher gains and result in a lower payback period. This advantage is more notorious in utility B. By investing 2.8 times less, the utility can increase the NPV by 1,079,765 € and the payback period from 4.1 to 1.8 years.

Therefore, in both utilities presented, Alternative 2 represents the most optimized solution for meter replacement management. The theoretical results obtained (Table 2) were in line with what was obtained in both utilities. During the project, utility A replaced close to 52,884 meters having an average payback of 1.39 years. On the other hand, utility B replaced 53,254 meters and had an average payback of 2.16 years. Thus, by implementing *meterwise*, utilities were able to increase their revenue whilst decreasing the initial investment.

Table 2 | Economic viability study

Utility	Alternative	Investment (€)	NPV (€)	Payback period (years)
A	1	1,382,478	1,431,697	1.6
	2	988,986	1,971,952	1.0
B	1	3,414,107	4,442	4.1
	2	1,236,554	1,084,206	1.8

5-year period with a 10% discount rate.

Global reduction of non-revenue water

In order to reduce NRW, AGS assisted the utilities to incorporate a methodology for the management and operation of the systems, allowing us to diagnose, prioritize areas and define actions to improve efficiency. The implemented tools and actions allowed for a significant reduction of NRW throughout the years of the projects (Figure 9).

Utility A decreased real water losses 7.1 million m^3/year, and utility B decreased 5.9 million m^3/year.

In terms of apparent losses, in 2017, both utilities adopted the methodology incorporated in *meterwise* to determine which customer meters to replace. This approach, estimating metering errors based on each meter's lifespan rather than basing decision solely on the meters' age, assisted in decreasing the apparent losses volume (Figure 9) whilst promoting the increase in utilities' revenues (Table 2). This approach led to a decrease of apparent losses of 1.0 and 2.1 million m^3/year in utility A and utility B, respectively (2015–2018).

The implementation of *flowise* and *meterwise* allowed the analysis of data in a methodical and prompt manner resulting in NRW reduction of 8.2 and 6.7% in utility A and B, respectively, corresponding to a total volume of 12.8 million m^3/year, in the course of the project (2015–2018) (Figure 9).

Digitalization of the water sector

The urban water sector has been facing complex challenges which can be summarized in being able to supply water to customers, with high quality and quantity, while keeping operations affordable. Thus, it is crucial for a water utility to implement smart water solutions to collect and analyse data in a more efficient and coherent manner. By installing IoT devices and tools to aid in data analytics, such as the tools presented in this study, the water utilities can better understand their system and, thus, optimize the operational work and teams whilst guaranteeing the service. The outputs obtained from implementing operational tools such as *flowise* and *meterwise* can be integrated in a dynamic data analysis platform (waterwise; Figure 2) to calculate key performance indicators to assess performance at operational, commercial and financial levels. This will assist water

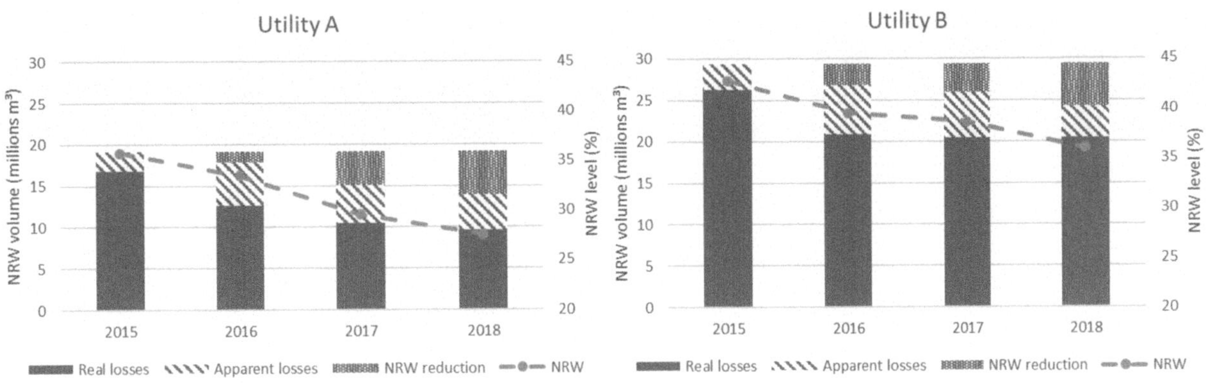

Figure 9 | Real losses, apparent losses and NRW level evolution from 2015 to 2018 in utilities A and B.

utilities in the definition of operational actions and support decisions on if, when and how to invest in equipment and activities such as network metering and sensoring, customer meters, pipe rehabilitation, active detection equipment and others. Ultimately, with the use of IoT and software tools, the water utilities can improve asset management, workforce transformation and reduce NRW in a more prompt manner which will improve customer service and reduce financial losses (Ramos *et al.* 2020).

In the last years, new technologies have emerged exponentially and have slowly been adopted by water utilities worldwide which present different digital maturity levels (Sarni *et al.* 2018). Unfortunately, digitalization of urban water systems is typically left behind when compared with other infrastructures. This is probably due to the fear of new vulnerabilities in a sector that is essential to human life. In addition, water utilities are extremely complex organizations with different and, frequently, isolated siloes. However, the fact that this is a slow moving sector can be seen as an opportunity since lessons can be learned from other sectors and risks mitigated (Moy De Vitry *et al.* 2019). Different stakeholders should continue to converge to scale and catalyse the adoption of digital solutions to minimize the risks such as water scarcity and security, increasing demand, old and undermaintained infrastructure and climate change.

CONCLUSIONS

Real-time monitoring and data analytics tools allow a more comprehensive, diverse and accurate understanding of systems' behaviour leading to a more effective control of the distribution network and water supply system. Thus, AGS developed and implemented software tools to support operational management, combining two important features, the ability to intervene in systems through events generation and the ability to explore data in a long-term period (based on historical data). This will ultimately contribute to a considerable increase of the environmental and financial efficiency level in water utilities.

The case study presented has highlighted these advantages. On one hand, the use of *flowise* for real-time monitoring allowed us to categorize bursts and to evaluate their impact on water loss volumes in order to identify operational inefficiencies regarding detection and repair times, particularly in small and medium bursts. This analysis reinforced the importance of operational work efficiency on the reduction of NRW levels, mainly regarding the increased understanding of network events and the reduction of the response time to those events, with a focus on small and medium bursts, usually responsible for the higher volumes of water loss in utilities.

On the other hand, the implementation of *meterwise* for optimized customer meter management promoted the review of water utilities' internal processes to guarantee higher metrological control and customer meters data reliability. By estimating metering errors more accurately, utilities were able to replace meters based on lifespan rather than solely on the meter's age. This further reduced the apparent losses, reducing costs and increasing revenues.

Overall, during the course of the projects, both utilities replaced 106,138 customer meters and rehabilitated 93 km of their smart water network in an optimized way. The project implemented, with the crucial support of *flowise* and *meterwise* for data analytics, reduced NRW levels by 8% which resulted in reducing 13 million m^3/year of real water loss and savings of 5.8 million Euros in 4 years which can be reallocated in other areas of water loss control. In addition, having a common goal, interdepartmental organization and communication improved, capacity building increased and image among stakeholders improved, both at customer and regulatory level. The improvement of the digitalization process in these utilities has allowed them to take water efficiency to a new dimension.

DATA AVAILABILITY STATEMENT

All relevant data are included in the paper or its Supplementary Information.

REFERENCES

Alegre, H., Baptista, J. M., Cabrera, E., Cubillo, F., Duarte, P., Hirner, W., Merkel, W. & Parena, R. 2016 *Performance Indicators for Water Supply Services*, 3rd edn. Water Intelligence Online. IWA Publishing, London, UK.

AL-Washali, T., Sharma, S. & Kennedy, M. 2016 Methods of assessment of water losses in water supply systems: a review. *Water Resources Management* **30** (14), 4985–5001.

Ananda, J. 2019 Determinants of real water losses in the Australian drinking water sector. *Urban Water Journal* **16** (8), 575–583.

Arregui, F., Cabrera Jr., E. & Cobacho, R. 2006 *Integrated Water Meter Management*. IWA, London, UK.

Arregui, F. J., Soriano, J., Cabrera, E. & Cobacho, R. 2012 Nine steps towards a better water meter management. *Water Science and Technology* **65** (7), 1273–1280.

Arregui, F. J., Gavara, F. J., Soriano, J. & Pastor-Jabaloyes, L. 2018 Performance analysis of ageing single-jet water meters for measuring residential water consumption. *Water* **10** (5), 1–18.

Covelli, C., Cozzolino, L., Cimorelli, L., Morte, R. D. & Pianese, D. 2015 A model to simulate leakage through joints in water distribution systems. *Water Science and Technology: Water Supply* **15** (4), 852–863.

Criminisi, A., Fontanazza, C. M., Freni, G. & La Loggia, G. 2009 Evaluation of the apparent losses caused by water meter under-registration in intermittent water supply. *Water Science and Technology* **60** (9), 2373–2382.

Eliades, D. G. & Polycarpou, M. M. 2012 Leakage fault detection in district metered areas of water distribution systems. *Journal of Hydroinformatics* **14** (4), 992–1005.

Farley, M. 2001 *Leakage Management and Control: A Best Practice Training Manual*. WHO, Geneva, Switzerland, pp. 1–98.

IWA 2000 *The Blue Pages: Losses From Water Supply Systems: Standard Terminology and Recommended Performance Measures*.

Karadirek, I. E. 2020 An experimental analysis on accuracy of customer water meters under various flow rates and water pressures. *Journal of Water Supply: Research and Technology – AQUA* **69** (1), 18–27.

Lambert, A. & Morrison, J. A. E. 1996 Recent developments in application of 'bursts and background estimates' concepts for leakage management. *Water and Environment Journal* **10** (2), 100–104.

Li, R., Huang, H., Xin, K. & Tao, T. 2015 A review of methods for burst/leakage detection and location in water distribution systems. *Water Science and Technology: Water Supply* **15** (3), 429–441.

Loureiro, D., Amado, C., Martins, A., Vitorino, D., Mamade, A. & Coelho, S. T. 2016 Water distribution systems flow monitoring and anomalous event detection: a practical approach. *Urban Water Journal* **13** (3), 242–252.

Moy De Vitry, M., Schneider, M. Y., Wani, O., Manny, L., Leitao, J. P. & Eggimann, S. 2019 Smart urban water systems: what could possibly go wrong? *Environmental Research Letters* **14** (8), 081001.

Ramos, H. M., McNabola, A., López-Jiménez, P. A. & Pérez-Sánchez, M. 2020 Smart water management towards future water sustainable networks. *Water* **12** (1), 1–13.

Sarni, W., White, C., Webb, R., Cross, K. & Glotzbach, R. 2018 *Digital Water: Industry Leaders Chart the Transformation Journey*. IWA.

Thornton, J., Sturm, R. & Kunkel, G. 2008 *Water Loss Control*, 2nd edn. McGraw-Hill, New York.

Wu, Y. & Liu, S. 2017 A review of data-driven approaches for burst detection in water distribution systems. *Urban Water Journal* **14** (9), 972–983.

WWAP (United Nations World Water Assessment Programme) 2015 *The United Nations World Water Development Report 2015: Water for a Sustainable World*. Paris.

First received 30 April 2020; accepted in revised form 27 November 2020. Available online 15 December 2020

© 2021 The Authors

An ethical decision-making framework with serious gaming: a smart water case study on flooding

Gregory Ewing and Ibrahim Demir IWA

ABSTRACT

Sensors and control technologies are being deployed at unprecedented levels in both urban and rural water environments. Because sensor networks and control allow for higher-resolution monitoring and decision making in both time and space, greater discretization of control will allow for an unprecedented precision of impacts, both positive and negative. Likewise, humans will continue to cede direct decision-making powers to decision-support technologies, e.g. data algorithms. Systems will have ever-greater potential to effect human lives, and yet, humans will be distanced from decisions. Combined these trends challenge water resources management decision-support tools to incorporate the concepts of ethical and normative expectations. Toward this aim, we propose the Water Ethics Web Engine $(WE)^2$, an integrated and generalized web framework to incorporate voting-based ethical and normative preferences into water resources decision support. We demonstrate this framework with a 'proof-of-concept' use case where decision models are learned and deployed to respond to flooding scenarios. Findings indicate that the framework can capture group 'wisdom' within learned models to use in decision making. The methodology and 'proof-of-concept' system presented here are a step toward building a framework to engage people with algorithmic decision making in cases where ethical preferences are considered. We share our framework and its cyber components openly with the research community.

Key words | decision support, ethics, flooding, human-centered AI, smart water systems

Gregory Ewing (corresponding author)
Ibrahim Demir IWA
Department of Civil and Environmental Engineering,
University of Iowa,
300 S. Riverside Dr., Iowa City, IA 52245, USA
E-mail: *gregory-ewing@uiowa.edu*

HIGHLIGHTS

- Water Ethics Web Engine $(WE)^2$ is a web framework to incorporate voting-based ethical preferences into water resources decision support.
- We share $(WE)^2$ openly at our project repository: https://github.com/uihilab/waterethicswebengine.
- A proof-of-concept use case is presented where decision models are learned and deployed with flooding scenarios.
- Results indicate that the framework can capture group 'wisdom' in learned models.

INTRODUCTION

Sensor networks, which is built on the back of the latest digital communication technologies, are increasingly being deployed in urban sewer networks and at the regional scales to monitor flooding and water quality of rivers (Habibi *et al.* 2017; Mullapudi *et al.* 2017; Jones *et al.* 2018; Yildirim & Demir 2019). Concurrently, control technologies are being deployed alongside sensors which allow for operators to actively manipulate these systems in places and in a

This is an Open Access article distributed under the terms of the Creative Commons Attribution Licence (CC BY 4.0), which permits copying, adaptation and redistribution, provided the original work is properly cited (http://creativecommons.org/licenses/by/4.0/).

doi: 10.2166/hydro.2021.097

way that was previously inconceivable (Kerkez *et al.* 2016; Mullapudi *et al.* 2018). Furthermore, the continued adoption of control technologies also introduces new stakeholders by which our water environment is *actively* manipulated (Rentsch 2019). These new technologies allow for higher-resolution monitoring via novel sensors (Sermet *et al.* 2019), social media (Sit *et al.* 2019), and decision making (Demir *et al.* 2018; Sermet *et al.* 2020) both in time and space. Where historically our water infrastructure, such as stormwater ponds, were designed as passive structures there is now the capability to control releases *actively* and *precisely* from a growing, distributed amount of water infrastructure, oftentimes as small as stormwater ponds.

A proliferation of new sensing and control locations means considerably more complex systems to address persistent water resources challenges such as flooding and water quality. Yet, to do so requires harnessing the unprecedented complexity of these new systems. Toward this end, there is a growing body of research on control schemes for water systems (Overloop 2006; Mollerup *et al.* 2017; Schütze *et al.* 2018). Recent studies mainly focus on the challenges of controller design to meet the objectives of physically based water *quantity* (flooding; Sadler *et al.* 2019; Mullapudi *et al.* 2020; Sun *et al.* 2020) and sometimes *quality* (total suspended solids or different 'indicator' pollutants; Muschalla *et al.* 2014; Sharior *et al.* 2019; Troutman *et al.* 2020). The collective work demonstrates the potential for improved system performance with a coordinated and increasingly automated control approach and the promise of the new 'smart' water paradigm.

However, little work exists to determine whether control schemes with physically based objectives are consistent with socially normative expectations of 'right' and 'wrong' actions. Because the primary objective of these control studies is to investigate the performance of control strategies, many experimental designs use toy networks or simplified abstractions of real networks. One consequence of this experimental design is that the social and economic variations within and across the network are not considered in the development of control schemes, nor in determining the impact of the controller performance. Yet, when considering the deployment of these distributed technologies across an entire city or region, these demographic factors become relevant. Consider the case, for example, where some flooding is an inevitable outcome within the catchment area of a stormwater network with dynamic control capabilities; how *ought* a controller act to consider the societal implications of such flooding? Or, if a control scheme consistently recommends distributing damages in poor neighborhoods and benefits in richer neighborhoods, *ought* its directives to be followed? These normative questions of *ought* – across populations, landscapes, communities, etc. – pose serious ethical and moral dilemmas, especially where negative impacts are unavoidable and uneven.

While the ethical questions of civil infrastructure are as old as the infrastructure itself, the challenge to incorporate moral and ethical preferences into the automated, algorithmic decision-making tools that water resources management will rely upon are, indeed, new. The novelty of this challenge is supported by two trends. First, because sensor networks and control systems allow for higher-resolution monitoring and decision making in both time and space, it follows that greater discretization of control will allow for an unprecedented precision of impacts, both positive and negative. A second and complimentary trend is as the complexity of systems grows, humans will continue to cede direct decision-making powers to decision-support technologies such as data algorithms (Sermet & Demir 2018, 2019). Systems will have ever-greater potential to effect human lives, and yet, humans will be insulated from these direct decisions. An important topic to explore is whether decision-support tools for water resources management can integrate socially normative expectations with physically based objectives.

Individual to institutional: ethics, morality, and machines

The challenge to consider social concepts into what are seemingly technical problems is non-trivial. The social and the technical are intertwined to make a sociotechnical problem (Jonoski 2002; Vojinović & Abbott 2012). This understanding motivates our review of both social and technical work as it relates to our efforts. Generally, we can consider ethics and morality along a spectrum that spans from the individual to the institutional. At the individual level, 'normative' ethics considers theories and schema to determine the appropriate actions of an individual or an agent. At the institutional level, studies of ethics and morals consider the power structures that shape norms, their compliance, and their impact on individuals and groups. Work in biology, philosophy, psychology, and sociology investigates questions along with this spectrum. The field of machine ethics complements the efforts from the natural and social sciences by developing methods to incorporate decision making into artificial technologies that are consistent with human's (society's) normative expectations of their behavior (Moor 2006). Here, we provide a brief background on these fields and how they relate to incorporating normative ethics into smart water systems. Though some disciplines make distinctions between ethics and morality, we use them as interchangeable in the context of this effort. For consistency within various disciplines, we defer to their respective terminology when referencing their work.

Normative ethical theories in philosophy

In the study of ethics, or moral philosophy, there are three fundamental traditions: *deontological*, *consequential*, and *virtue*. Respectively, these camps choose to inspect morality from three overlapping, but not identical, questions: 'What is the right thing to do?', 'How is the best possible state of affairs achieved?', and 'What qualities make for a good person?' (Grayling 1995). Importantly, each of these questions, and the responses of the ethical theories, are posed for *individuals*.

The *deontological approach*, or duty ethics, considers an action to be moral based upon *a set of rules* that deem an action permissible, impermissible, or obligatory (Alexander & Moore 2016). A strength of the deontological approach is the clarity of the rule to *direct* actions. However, multiple rules can require contradicting actions, which is a key weakness of a purely deontological approach.

In contrast, the *consequentialist* tradition judges the correctness of an act on its *outcome* (Sinnott-Armstrong 2019), in that the correct action is the action that leads to the best outcome by some specified *objective function*, such as maximizing happiness or minimizing costs. Egalitarianism and utilitarianism are well-known consequentialist moral theories. A concern for consequential approaches is which factors should be included in determining the normative value of an outcome. Though both consequential in the approach, utilitarian and egalitarian models can produce incompatible solutions due to which factors are given importance. This tension is referred to as the equality-efficiency or the fairness-efficiency dilemma (Binmore 1998).

In the *virtue ethics* tradition, virtues are the fundamental, irreducible unit by which to define normativity, meaning that they are derivative to neither the outcome of actions (consequential) nor duty to perform an action (deontological) (Hursthouse & Pettigrove 2018). In agent-based virtue ethics, agents' *motivations* ascribe the rightness and wrongness of an act and agents learn virtue from 'exemplars of goodness' (Zagzebski 2004). This understanding buoys the concept of learning models of normative behavior by observing the human performance of similar tasks.

Morality in the sciences

Evolutionary biology tells us that morals are adaptations to social living; when prehistoric humans began to form larger groups, survival was dependent upon the group, and therefore, what was best for it could supersede the priorities of the individual (Krebs 2008). Psychology and the disciplines of biology are interested in the mechanisms and processes of the human brain as they relate to developing and making moral judgments (Haidt 2007). As a distinction from other fields, sociology interrogates morality at scales beyond the individual. When sociology does focus on the morality of individuals, it is almost always in relation to a larger group. Contemporary sociology of morality includes both moral theorizing and experimental science to uncover moral truths (Bykov 2019). Work is not a shared substantive focus, but the recognition that moral evaluations and categorizations are an essential part of struggles in 'social fields' (Hitlin & Vaisey 2013).

When considering how to incorporate moral and normative sentiments into intelligent infrastructure, the sociological literature provides intriguing insights. The theory of 'thick' and 'thin' moral concepts contends that there are thin moral concepts that are 'methodologically tractable', through hypothetical situational tests. Conversely, thick concepts – like dignity, integrity, humanness, etc. – do not lend themselves to parsimonious description or measurement, making thick concepts more difficult to explore by the experiment (Abend 2011). Furthermore, there is a lack of evidence for how thin and thick concepts relate to each other. Thus, a holistic approach would incorporate exercises in which feedback on both thin and thick moral concepts can be collected. Furthermore, other studies show that higher social class predicts increased unethical behavior (Piff *et al.* 2012), that lower-class individuals are more likely to be compassionate to another's suffering (Stellar *et al.* 2012), that different social classes use different criteria in anticipated cost–benefit analyses (Trautmann *et al.* 2013), and that lower-class individuals are more likely to perform an unethical act if the act was to the benefit of others (Dubois *et al.* 2015). Together, these findings clarify the process and partners needed when operationalizing feedback on normative expectations of smart water systems.

Integrating moral and normative sentiments into intelligent systems

Machine learning (ML), artificial intelligence (AI), and data algorithms generally have been – and stand to be – applied in a wide range of scenarios from data augmentation (Demiray *et al.* 2021) to forecasting (Sit & Demir 2019; Xiang *et al.* 2020). There is, however, growing acknowledgement that, unexamined, these techniques can institutionalize the biases and structural prejudices that persist in the world, especially in decision-making contexts that directly affect humans such as in job hiring, evaluating credit scores, and predicting repeat offenders during parole processes (O'Neil 2016). Consequently, popular culture, industry, government, and academia have focused attention upon the intentional and ethical application of AI and created a proliferation of AI ethics frameworks (Hagendorff 2019; Jobin *et al.* 2019). Two primary concerns of deploying these technologies in a complex world are as follows: (1) their instantiated purpose and behavior may not be well understood and (2) the systems may take irrevocable acts before humans have the data to discern their error (Samuel 1960).

These concerns connect to the questions of building frameworks and governance models to responsibly integrate algorithms into systems that affect humans. Toward this aim, the agenda of 'society-in-the-loop' (SITL) proposes to build an *algorithmic social contract* where human-in-the-loop principles and general stakeholder values are integrated into an iterative development process (Rahwan 2017). In its proposal, SITL forwards the tenets of *algorithmic regulation* (O'Reilly 2013). The algorithmic regulation proposed requires a deep understanding of the desired outcome that outcomes are monitored, that the algorithm adjusts based on new data, and periodic deeper analysis on algorithm performance.

Recent work in machine ethics can be understood in the context of building methodologies and frameworks toward SITL and algorithmic regulation (Tolmeijer *et al.* 2020). A crowdsourced, voting approach has been proposed as a flexible method to incorporate moral sentiments for AI applications (Conitzer *et al.* 2015, 2017). Crowdsourced voting has already been tested on a massive scale to query preferences on resolving moral dilemmas of autonomous vehicles using a pairwise comparison experimental setup (Awad *et al.* 2018). After data collection, concepts from computational social choice (Chevaleyre *et al.* 2007) and ML classification techniques can be used together to build preference models of individuals and groups (Noothigattu *et al.* 2017). These preference models, which in theory represent the real sentiments of the participants, can then be used in decision-support algorithms. Two examples of this process are in the development of an algorithm to decide tiebreaks in a theoretical kidney exchange market (Freedman *et al.* 2019) and an algorithm to support a fair and efficient dispatch of food donations (Lee *et al.* 2019). Critically, these examples apply voting-based preference aggregation within a larger, participatory framework. We observe that such frameworks to generally be able to (a) identify relevant belief features to base decisions upon; (b) assay preference along each relevant belief feature using pairwise comparison testing; (c) learn a preference model from the preference assays; (d) use the learned preference model in experimental decision-making scenarios; (e) analyze the outcomes of the preference model and identify if/where model-driven outcomes are incongruous with stated values or objectives; and (f) iterate on and/or deploy the learned preference model. In the 'Materials and methods' section, we provide further details for this process.

Currently, open frameworks which support all or part of the above workflow are limited (if any) in the literature. Previous studies have built custom web-voting applications (step b) for their specific applications but did not share generalized source code. Other studies used proprietary web platforms to collect preference data. Furthermore, we are unaware of any framework that has shared analytical tools for post-play data processing and preference model development.

Toward this end, we propose an integrated and generalized framework to incorporate voting-based ethical and normative preferences into water resources decision-support schemes. We then demonstrate the framework with a proof-of-concept use case where decision models are learned and deployed to respond to flooding scenarios. The results indicate that the framework can capture group 'wisdom' in learned models and use this to make decisions. Furthermore, we share our generalized framework openly with the research community.

The remaining sections are organized as follows. Section 'Materials and methods' provides details on the methodology of each step of the framework listed above, and cyber components used in handling the step. Section 'Results and discussion' demonstrates the framework with a proof-of-concept use case for decision making for flood response with a discussion of the framework. Finally, the 'Conclusion' presents the larger context of incorporating normative expectations into smart water systems.

MATERIALS AND METHODS

In this section, we describe the approach and technologies used to employ the methodological framework introduced previously. First, we describe how to identify relevant belief features. Next, we describe the process employed to collect people's preferences. This subsection includes a description of the generalized web framework developed to collect voting-based preferences, its web architecture, game play, and database architecture. Finally, this section details the post-play data analysis to derive preference models, how to use them to make decisions, and the analytics toolbox developed to support this exercise.

Identifying and assaying relevant belief features

Example methodologies to identify relevant belief features for decision making (step a) can be found in Lee *et al.* (2019) and Freedman *et al.* (2019) and include survey and interview techniques. Relevant belief features are those that someone would use to make a deliberative action. For example, in the case of choosing between two flooding outcomes, relevant belief features to be considered could be public costs, private costs, injuries, deaths, and environmental impact. Once belief features have been established, different scenarios can be created that vary along with the belief features. The scenarios are presented to individuals in pairs. Participants must choose which outcome they prefer from each pair presented to them. Their choices are recorded and used later to learn a preference model.

To collect preference data from participants, we built an integrated web-based serious gaming platform. Serious gaming is used in a variety of fields for training, decision making, and education (Susi *et al.* 2007). In water resources, serious gaming is used to explore the multifaceted challenges such as multi-hazard mitigation (Carson *et al.* 2018). Web-based serious gaming offers easily accessible and user-friendly interfaces with flexible architectures for various skill levels (Xu *et al.* 2020). In many serious gaming applications, the game play offers the user an opportunity to explore real problems and strategies without the consequences of their actions impacting the real world. Instead, play informs a value system that guides behavior and action during a future, real-world event. Play is recognized as an important feature in the development of value systems and morals in humans. Furthermore, actions within the context of play (i.e. games) give rise to different value systems compared with a work context, even when considering the same topic (Bargheer 2018). These value systems result in different moral treatments of the same topic which can result from starting in the context of play or work. Gamification allows for parsimonious yet engaging descriptions of the ethical dilemmas at hand. More engaged users are more likely to play for longer and contribute more to the model development. The following subsections describe the web architecture, game play, and database architect of our serious game approach.

Water Ethics Web Engine $(WE)^2$ architecture

The Water Ethics Web Engine $(WE)^2$ is an open and integrated framework that allows a rapid deployment of web applications to investigate moral preferences via pairwise comparisons. The framework is comprised of a PHP-based application engine and use case web template and includes a database architecture on the back end (Figure 1). Full documentation of the framework and source code can be found in GitHub: https://github.com/uihilab/WaterEthics-WebEngine. Researchers provide their application and experimental information to the engine via two configuration files (site_meta.json and scenarios.json) using the JavaScript Object Notation (JSON) format. These files, along with supplied images, are stored on a Web server and accessed during the game play by the application engine. Data generated during the game play are logged in a web-accessible database. Once data are collected, researchers have access to analytics tools and a portal for data export.

Case study information described above is stored on a Web server in a unique directory. When a user navigates their browser to the specific game version, given to them by the researcher, the engine generates the Web page from the content stored in case study files (i.e. the site metadata, scenario content, and the images). Scenarios displayed to participants are chosen randomly from the total set supplied.

Game play

Users are presented with a homepage that provides a brief mission statement on the purpose of the game and a button to start the game play. During the game play, users are presented with two scenario windows displayed side-by-side. In each scenario windows are the descriptions of

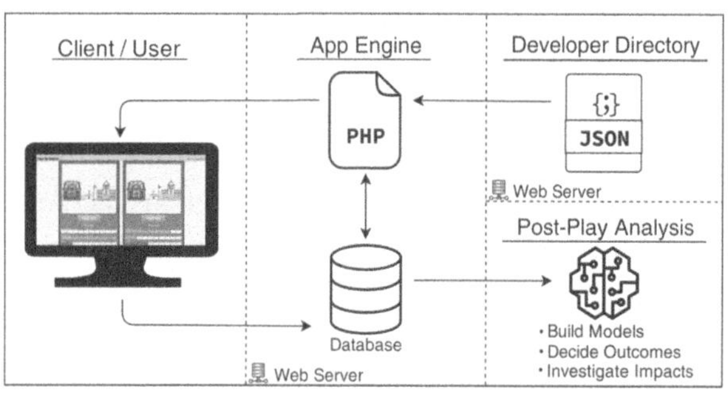

Figure 1 | System architecture of the web-based decision-making framework. The application engine, database, and researcher-supplied data are located on a Web server. When clients, or users, navigate to a project directory, the application engine builds the project using the information supplied in the static JSON files. After the game play, the researcher may use the post-play analytical tools to investigate results and learn preference models.

the event and an action button with a user-defined decision, e.g. 'Flood This' or 'Save This'. Descriptions come in some combination of three forms: an image, info bar, and written description. All description types are supplied by the researcher and are customizable. Users are instructed to use the descriptions to determine what outcome they prefer for the scenario. To choose the preferred decision, users click on the action button on which the decision is recorded, and a new scenario is displayed. Once the user has provided their preference for all scenarios, they will be guided to a results page that provides the user with a description of their aggregate preferences in relation to all others who have played the game and to the absolute possible outcomes along the belief features provided in the descriptions.

Post-play analysis

In this section, we describe the methods used in the post-play analysis. This includes learning preference models, making decisions with these models, and the iterative process of improving and deploying them. We also describe our analytic toolkit to facilitate these efforts.

Learning preference models

By using the paired comparison experimental design, parsimonious random utility models can be leveraged to learn preference models for individuals or groups (step c) (Tsukida & Gupta 2011). When a participant votes that they prefer one outcome over another, one can hypothesize that the outcome has, on average, a greater utility. It is assumed that the utility of a decision relies on weighing the tradeoffs of each option across the belief features that describe them. When participants provide decisions across a set of scenarios, they provide a classified dataset from which a preference model can be learned via classification techniques used in machine learning.

There are numerous methods of using pairwise comparison data to build preference models (Conitzer *et al.* 2015; Noothigattu *et al.* 2017; Lee *et al.* 2019; Freedman *et al.* 2019). Here, we employ the Thurstone's Law of Comparative Judgement Case V model to convert paired comparisons into group quality scores (Tsukida & Gupta 2011). For a group of participants who have made judgments on the same scenarios, an estimate of the mean quality difference between option A and B, μ_{AB}, is calculated for each scenario as follows:

$$\mu_{AB} = \Phi^{-1}\left(\frac{C_{A,B}}{C_{A,B} + C_{B,A}}\right) \quad (1)$$

where $\Phi^{-1}(x)$ is the inverse cumulative distribution function (CDF) of the standard normal distribution, and $C_{A,B}$, $C_{B,A}$ are the number of votes for each option received. For example, $C_{A,B} = 27$ would mean that 27 people prefer outcome A over outcome B.

Mean differences can be related to scenario belief features as:

$$\mu_{AB} = \beta^T(X_A - X_B) \quad (2)$$

where X_A, X_B are the vectors of the belief features for option A and option B, respectively. The learned preference model of the group is the estimate of the belief feature weighting factors of β^T and can be found using linear regression, or other classification techniques, given that:

$$\begin{bmatrix} \mu_{AB,1} \\ \mu_{AB,2} \\ \vdots \\ \mu_{AB,n} \end{bmatrix} = \beta^T \begin{bmatrix} (X_A - X_B)_1 \\ (X_A - X_B)_2 \\ \vdots \\ (X_A - X_B)_n \end{bmatrix} \quad (3)$$

where $\mu_{AB,i}$ is the estimated mean utility difference of scenario i of the n shared scenarios and $(X_A - X_B)_i$ is the difference in belief features of i of the n shared scenarios.

After a preference model β^T is learned, we can investigate its behavior. To do so, we need to calculate μ_{AB} in Equation (2) using the belief features of new scenarios. If μ_{AB} is positive, then choice A is selected. If μ_{AB} is negative, then choice B is selected. Decisions generated using the preference model can then be compared against various benchmarks, such as the historical performance of a system or against some specified definition of fairness and efficiency. Committing to this activity will require a reflection on whether the algorithm is meeting its purpose. After interrogating the behavior of the learned preference model, the researcher has a clear choice to continue to refine the preference model iteratively or incorporate the preference model into their operation. As a practical matter, one can choose to pursue both continuously and simultaneously.

To facilitate the workflow of data retrieval, learning preference models, and using them to make decisions on new scenarios, we developed a data analytics toolkit using the Python programming language. The toolkit includes a data service interface with a PostgreSQL database, data structures and methods to streamline the workflow. The toolkit and documentation of workflow are provided in the project repository.

RESULTS AND DISCUSSION

In this section, the web framework and methodology are applied in a proof-of-concept use case with a group of undergraduate engineering students at the University of Iowa ($n = 409$). The use case explores the preferences of individuals to flood scenarios and the result of using their preference models to make decisions.

Between 1980 and 2018, Iowa experienced 26 flood disasters where damages exceeded \$1 billion (Immerman & Immerman 2018). Most recently, damages due to the spring 2019 flooding events in Iowa are estimated at \$1.6 billion (Hardy & Cannon 2019). Furthermore, between 1988 and 2016, there were a total of 951 flood-related presidential disaster declarations in the state (Eller 2018). In total, the sum of damages over the last 40 years is estimated at \$41 billion, or a little more than \$1 billion per year. In response, the State of Iowa has supported flood mitigation and flood preparedness through the Iowa Flood Center (IFC) and the Iowa Watershed Approach (Weber *et al.* 2018). Yet still, when floods occur, the response requires lay people and decision makers alike to make fraught decisions that take on ethical and moral dimensions (Bosman *et al.* 2019; Kelley 2019; Marso 2019). Furthermore, actions taken in the lead up to and during a flood event can result in litigation from damaged parties which contribute to a fractured response. As such, decision-support recommendations from a voting-based framework could increase confidence and coordination in a community's flood response.

The primary goal of this use case was to demonstrate how voting-based preferences of specific flooding scenarios can be used to decide responses for a collection of flood scenarios. A secondary goal is to analyze the cumulative effect of these responses. To design flooding scenarios to use with the web-based ethics framework, we first determined relevant belief features. Belief features were chosen in an exploratory fashion based upon stated priorities of stakeholders as detailed in news publications and personal experience. Relevant belief features used to describe the impact of various flooding outcomes, such as *public costs*, *private costs*, *injuries*, *deaths*, and *environmental damage*. Given the proof-of-concept nature of our application, the process of identifying and clarifying relevant belief features is beyond the scope of this paper. Next, 17 different scenarios were randomly generated. Scenarios present two options of flooding outcomes and consist of two assets with varying descriptors (e.g. flood national chain grocer in low-income neighborhood or flood middle-income multi-family home). For each asset, unitless values were assigned for the impact of flooding along with each belief feature. Furthermore, each asset description includes an illustration and a text (Figure 2).

To collect preference data, we had freshmen vote on a five-scenario subset of the 17 scenarios during an ethics module in the University of Iowa, College of Engineering's Introduction to Engineering Problem Solving course. To build class section-specific preference models, all students

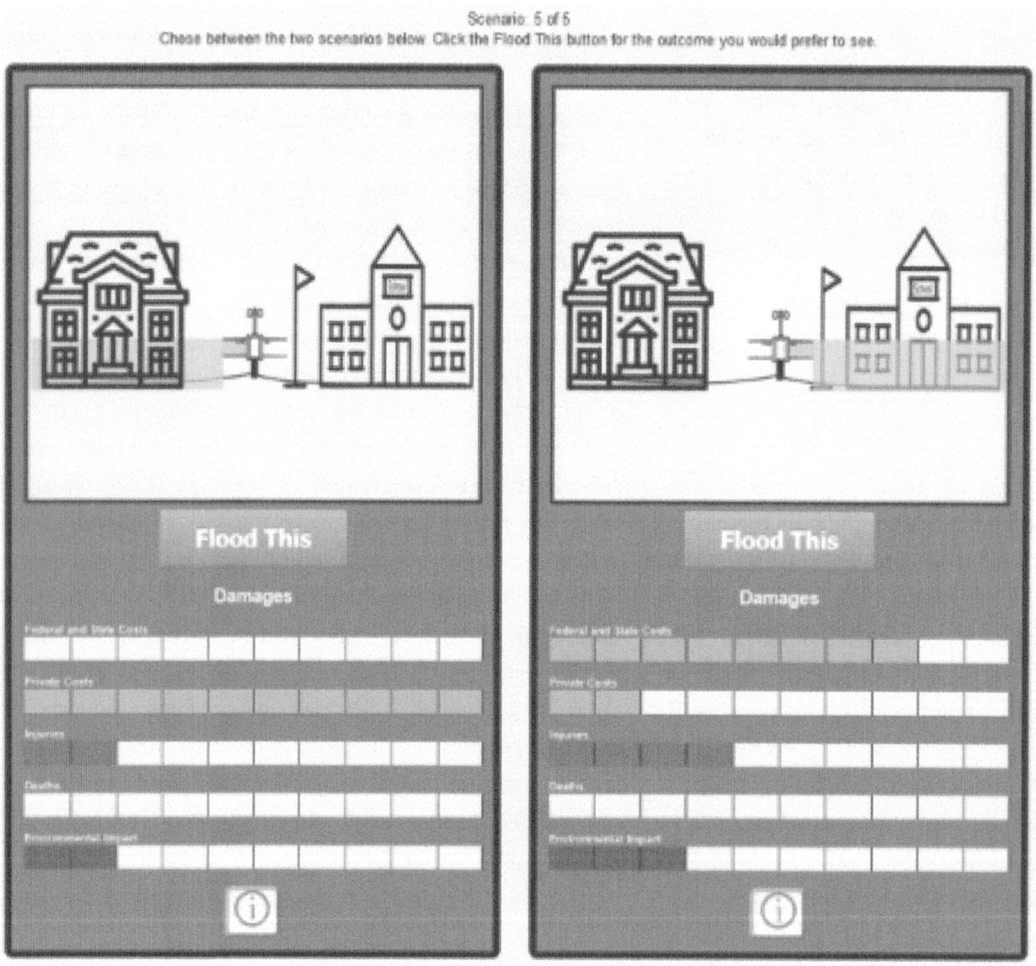

Figure 2 | Example scenario and game play provided by the (WE)² framework. Participants are served multiple scenarios where they are provided two outcomes and their descriptions as photos, info bars, and text (displayed when information button, button is click). Upon clicking the action button, e.g. 'Flood This', the response is logged in the database and the participant is served another scenario.

within a section were served the same five scenarios (Table 1). All responses were recorded in a web-accessible database and related to unique user ids and the class section keyword name.

To learn models from the collected data, votes were aggregated by class section. Participants who did not provide a vote for all five scenarios were not included in model generation. Preference models were learned using multiple linear regression. Five models were learned including one for each class section and one model that is the average of the four class sections. Five model parameters (β values) describe each model, one parameter for each belief feature provided in scenarios (Figure 3 and Table 2). A negative model parameter indicates the model's preference to minimize the impact along the given dimension. A positive model parameter indicates the preference to maximize the impact along the given dimension. This is due to

Table 1 | Each class section is referenced by a letter, section name

Section name	N	Scenarios
a	103	2, 3, 10, 15, 16
b	111	2, 3, 4, 5, 6
c	106	8, 9, 10, 11, 12
d	89	13, 14, 15, 16, 17

N details that how many students provided preferences through the game play. Scenarios column lists which five scenarios were served to each class section.

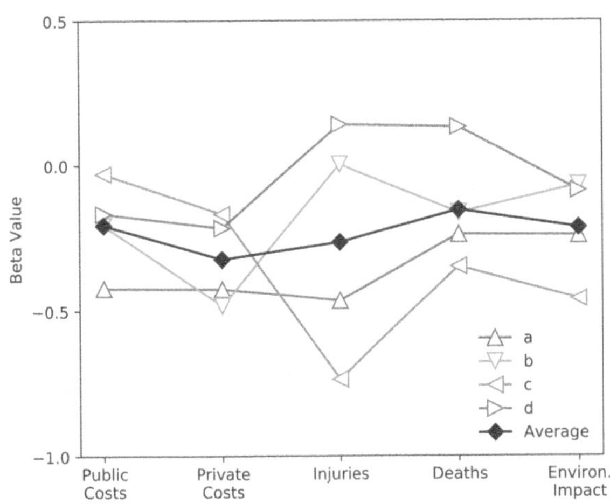

Figure 3 | Learned preference models for each of the four class sections and their average. Negative beta values indicate a preference to minimize the impact for a category.

the experimental design and our definition of marginal utility. For example, if it is preferred to minimize public costs, then the difference in public cost between the preferred option and the undesired option would be negative; or, that the preferred option results in fewer public costs. This negative value is multiplied by a negative β value to result in a positive contribution to the utility of that choice. All learned models minimize public costs to some degree; model a gives the most priority to public cost minimization and model c gives the least. However, no model gives minimizing public cost the highest utility against the other belief features. Models b, d, and Average give the highest utility to private cost minimization. Models a and c give the highest utility to the minimization of injuries. However, model a maximizes the minimization of public costs, private costs, and injuries almost equally.

To analyze the impact of the learned models, we simulated decisions on all 17 scenarios using the learned models. To add reference outcomes, we also simulated decisions of five models that, respectively, only prioritized one of the belief features. In effect, these reference models made decisions based on a single criterion instead of some combination of the five criteria of the learned models. Decisions, realizing the flood of the left or right scenario, for all 17 scenarios were made (Figure 4). The procedure is as follows: the dot product was taken between the vector $(X_A - X_B)$ and each of the ten β^T models. The result, the mean utility difference between option A (left) and option B (right) determines how each model would decide between the two outcomes; for a negative difference, chose option A (left), and for a positive difference, chose option B (right). Ties, which only occurred with the reference models, were noted and interpreted as that outcome could go either way. In general, learned models voted similarly, showing the agreement of vote on 11 out of 17 scenarios.

To compare how the outcomes differed between the models, we calculated the total impact each model *avoided* along with each belief feature and normalized them against the minimum and maximum possible avoidable damage (Figure 5(a)). A score of one indicates that the maximum damage was avoided, while a score of zero indicates that the minimum damage was avoided. All learned models avoided high levels of death and environmental impacts,

Table 2 | Beta values for each preference model used to make decisions

Experimental model	Beta values Public Costs	Private Costs	Injuries	Deaths	Environmental Impact
a	– 0.42	– 0.42	– 0.46	– 0.24	– 0.24
b	– 0.21	– 0.48	0.00	– 0.16	– 0.07
c	– 0.03	– 0.17	– 0.74	– 0.35	– 0.46
d	– 0.17	– 0.22	0.14	0.13	– 0.09
Average	– 0.21	– 0.32	– 0.26	– 0.15	– 0.21
Minimize Public Costs	– 1	–	–	–	–
Minimize Private Costs	–	– 1	–	–	–
Minimize Injuries	–	–	– 1	–	–
Minimize Deaths	–	–	–	– 1	–
Minimize Environmental Damages	–	–	–	–	– 1

Models a, b, c, and d are the learned models from each class section, and Average is the average of each of these learned models. These models weigh multi-dimensional effects to determine a decision. These models contrast with the reference models, which only consider the impact along with a single category.

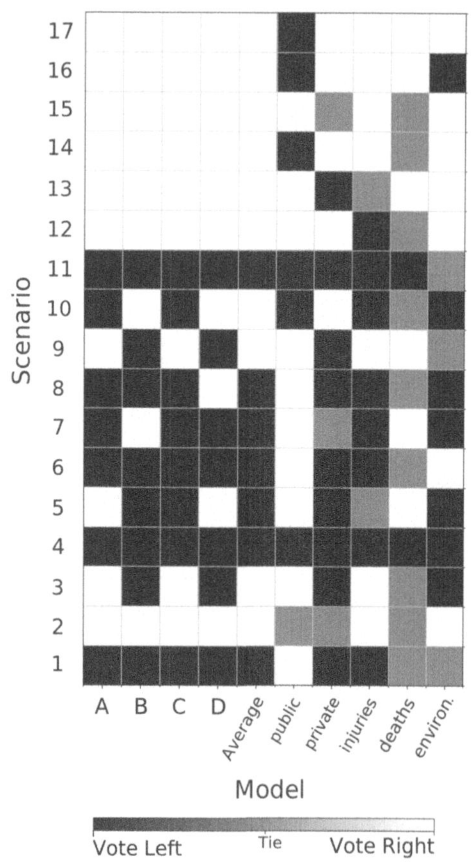

Figure 4 | Voting results from each of the five learned models and five reference models. Each model was used to vote, left or right, on the outcome of all scenarios generated for the flooding use case.

while showing less agreement and less ability to avoid public costs. The same exercise was performed for the reference models (Figure 5(b)). Although each model scored maximally in their respective categories, it does not suggest with these example scenarios that it is a productive strategy to prioritize only one category over all others.

Because each of the 17 scenarios has only two options, flood left or flood right, there are 2^{17} – or 131,072 – unique decision combinations. We normalized and ranked all outcomes along with each category, allowing us to build percentile curves (Figure 6). Only model a score above the 50th percentile in all categories. The remaining four learned models performed above the 50th percentile in all but one category, public costs. Across the five categories and five models, 17 out of the 25 outcome scores ranked above the 90th percentile, which indicates strong performance for all learned models.

To compute an overall outcome score for each of the 2^{17} voting possibilities, we performed a simple summation of the normalized outcome scores of each category for each model. These overall scores, too, were ranked and percentile scores calculated (Table 3 and Figure 5(c)). Shockingly, model a achieved the highest possible overall score of 3.922 (out of a maximum of 5) for rank 1 of 2^{17}. Model c achieved a cumulative outcome score of 3.849 for a rank of 6 of 2^{17}. All learned models achieved scores that put them in the 98th percentile of all outcomes in avoiding

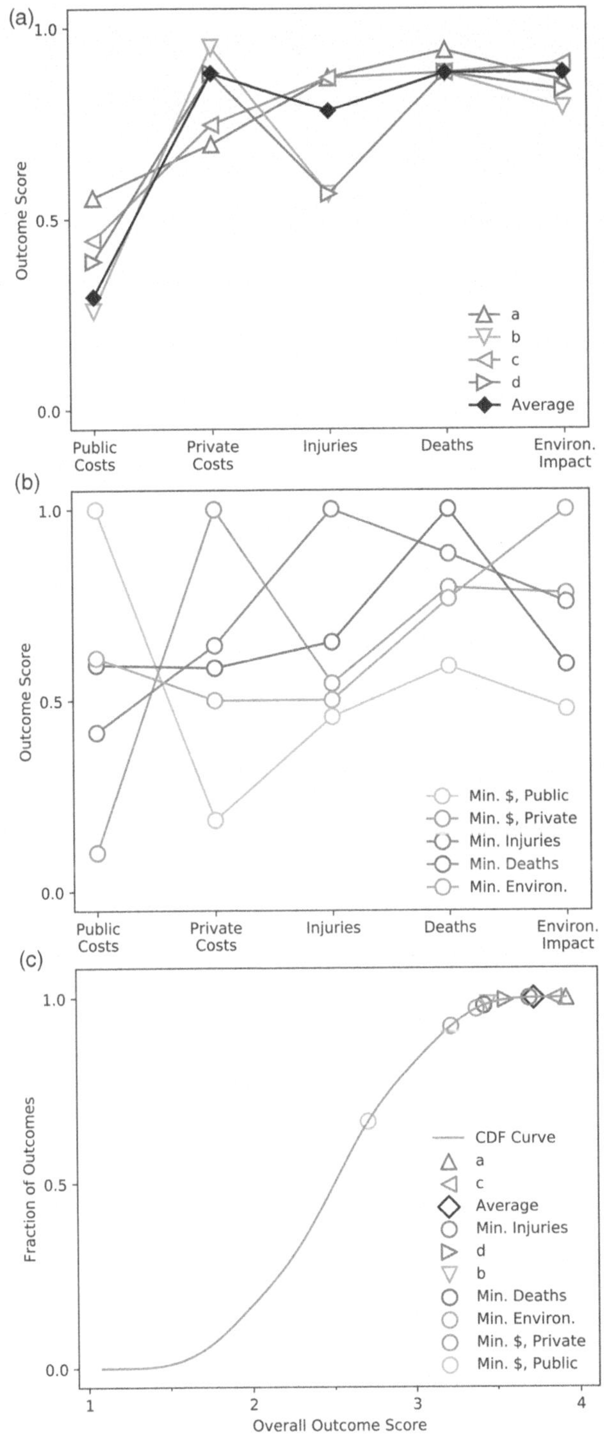

Figure 5 | Outcome scores are calculated as the avoided damage along a belief feature normalized by the maximum possible avoidance. Outcome scores were calculated for the learned models (a) and reference models (b). A model's cumulative outcome score is calculated as the sum of all outcome scores. Each model's cumulative score is ranked against all possible cumulative outcomes (c). Model a performed achieved the highest possible overall outcome score of 3.92 and a rank of 1 in 2^{17}.

total damages. Interestingly, only one of the single objectives, reference models performed better than any of the learned models.

DISCUSSION

The framework presented here demonstrates the ability to use a voting-based system to aggregate human preferences to ethical decisions in smart water systems. Collected data can be used to learn models of preferred behavior which can then be used to make decisions on new scenarios. This data-driven approach is novel in helping researchers *learn* the utility function from a large, potentially very large cohort of people and not assume an understanding of the utility function to be used to judge outcomes *a priori*. To this end, water professionals and researchers can investigate how algorithmic components of smart water systems or disaster response perform in relation to people's normative expectations of right and wrong.

The framework follows a consequentialist ethical theory, as the definitions of utility and performance rankings are based upon the *outcomes* of each scenario. However, learning utility functions without any *a priori* knowledge mitigates the limitations of traditional consequentialist approaches. As discussed earlier, consequentialist approaches come with limitations such as what factors (belief features) to use to describe outcomes and the fairness-efficiency tradeoffs between the following different moral theories (utilitarian vs. egalitarian strategies). One strength of this approach is that belief features may also be crowdsourced, which allows many people to define an inclusive list of important belief features. Another strength is that by learning the models of action from people's decision, it is unnecessary to proclaim beforehand what fairness-efficiency tradeoffs should be made. Instead, the fairness-efficiency priorities are captured within the learned models. However, these learned models may not agree with the institutional understanding of rights or justice. Further effort may be required to integrate learned preference models with our ideals and aspirational sense of justice and fairness. Finally, this approach currently does not address the difficulty of decision making under uncertainty. This is an area of future work. Overall, we identify the

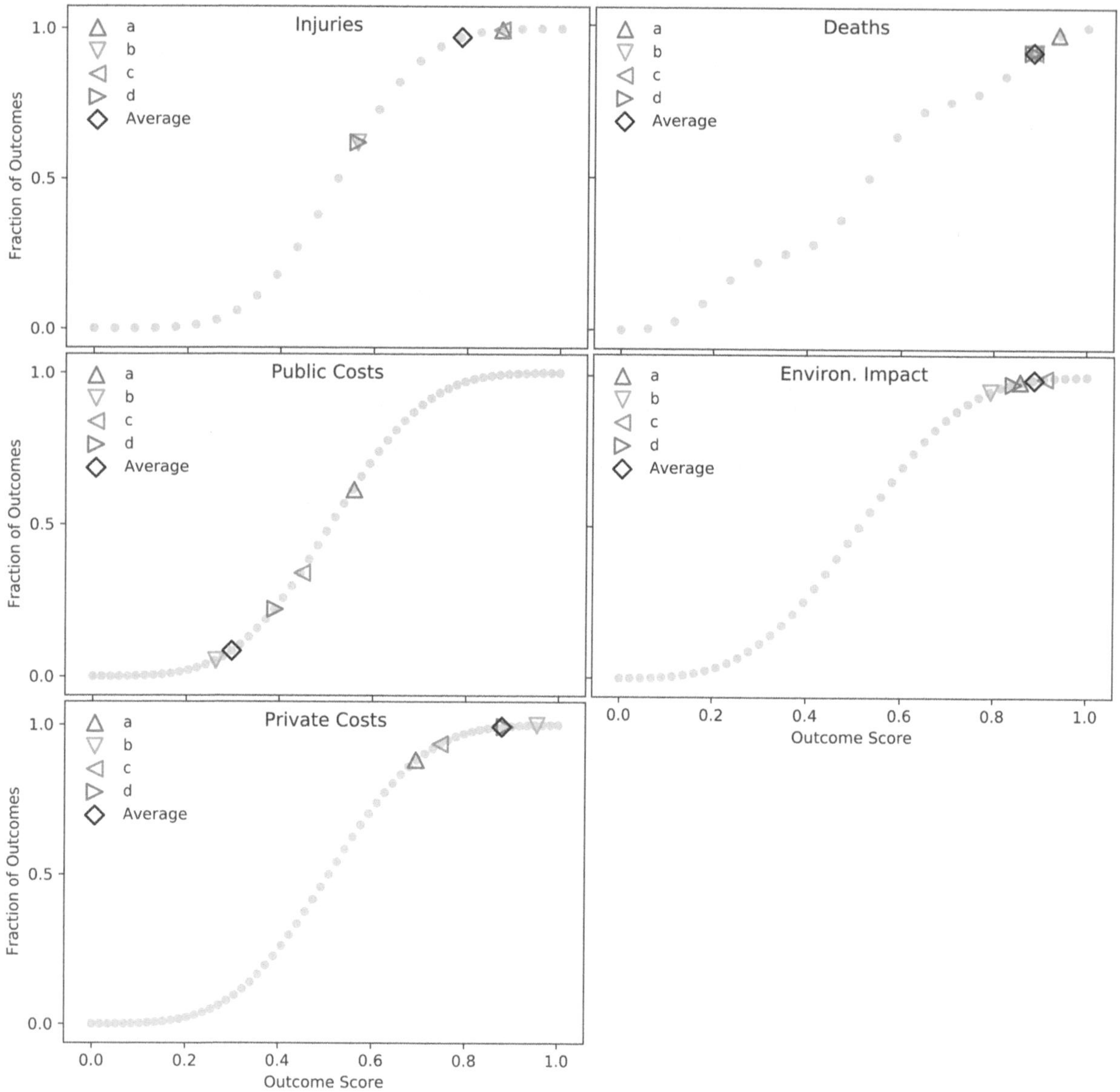

Figure 6 | Ranking curves for each of the five learned models along each of the five relevant belief features. Overall, the learned models performed well. Model a, which had the highest cumulative score when considering minimization of damages, scored highest among the models in three categories: deaths, injuries, and public costs. Yet, it also performed the worst in minimizing private costs and was third in environmental impact minimization.

inability of a single person, namely the researcher, to insert bias into the calculations as a positive outcome of using this framework.

Results from the experiment show the remarkable possibility for the models to choose outcomes that rank highly when considering the cumulative outcome. Models a and c appear to achieve their astonishingly high ranking because they chose outcomes that did not favor private cost minimization at the expense of public costs. This observation is supported by the β values for public and private costs that are equal for model a. These findings are relevant only within the context of our theoretical proof-of-concept. More robust studies are required to make that claim generally. Furthermore, the cumulative scores and ranking method used here communicate only a single concept of 'success'. It is reasonable to justify other performance metrics beyond summing the normalized results of each category. Rather, the cumulative outcome score analysis can be

Table 3 | Cumulative impact scores for each learned and reference model in the rank order

Experimental model	Cumulative outcome	Percentile	Rank
a	3.922	0.99999	1
c	3.849	0.99995	6
Average	3.726	0.99944	73
Minimize Injuries	3.699	0.99911	117
d	3.555	0.99406	779
b	3.447	0.98280	2,254
Minimize Deaths	3.423	0.97894	2,760
Minimize Environmental Damages	3.376	0.96988	3,948
Minimize Private Costs	3.219	0.92303	10,089
Minimize Public Costs	2.708	0.66605	43,772

A total of 2^{17} possible outcomes.

instructive as a first step toward an application-specific evaluation technique.

Critiques of paired comparison for modeling moral preferences follow two forms: misplaced moral subject and lack of meaning of dilemmas descriptors. The first form states that the subject, or the moral actor, in the dilemma should move from the individual to the institutional. Consider the Trolley Problem, a classic philosophical dilemma in which the subject, the trolley driver, must choose between two bad decisions (Thomson 1985). One could ask: How did society fail to enforce safety standards such that an individual must intervene in a life and death scenario? Or, why has society allowed public infrastructure to be so underfunded that it poses the risk of catastrophic failure? Broadly, this critique emphasizes that moral and ethical dilemmas should be interrogated from an institutional or societal level and not exclusively at the individual. As such, a goal of smart water research and institutional design may ask: how can organizations be structured such that moral decisions made by an autonomous agent are minimized?

The second criticism is that the characteristic classifiers used in the moral dilemmas under describe scenarios and, in doing so, elicit not a users' moral ideas but their biases along the dimension of the classifiers (Everett Jaques 2019). Put another way, if an experimental setup provides only information on the age or race of a candidate to receive medical treatment what is the experiment doing but forcing participants to display their ageist and racist biases to make healthcare decisions? And, by extension, if we use these data to learn a model of 'ethical' decision making, are we not simply training a model to be biased just like us?

These criticisms are addressed in the methodological design. Contextually, pairwise comparison preference testing should not be understood to possess deep ethical or normative meaning independently of all the exercises in the stated framework. Next, because it is not well understood how thin moral concepts relate to thick moral concepts (Abend 2011), we cannot *a priori* assume that these thin normative preferences collected in pairwise comparison tests translate to thicker concepts. As such, thicker moral and ethical concepts can be incorporated via activities beyond pairwise comparison testing, such as the numerous rounds of interviews with stakeholders (Lee *et al.* 2019). Finally, strong reaction to an action or a method derived from the framework is itself a measure of normative values and can be integrated via the iterative process; collect data and act upon both thin and thick concepts of morality to improve system performance.

In practice, it is unlikely that a strictly consequentialist framework would be operational in real-world scenarios. Instead, a hybrid decision-making process would be employed. For example, a system could trigger automatic human oversight if a decision is anticipated to reach a specified damage threshold. Likewise, because the learned models predict whether one outcome is better than another, a mean utility difference between two outcomes close to zero suggests that there is a very weak preference. Thus, in these cases, a human review could also be triggered. These rule-based heuristics can set guardrails on the strictly consequentialist models, while also providing further opportunity for society-in-the-loop principles. Furthermore, a critical step for future work will be to explore how decisions derived from $(WE)^2$ preference models impact system outcomes via their integration with hydraulic models.

CONCLUSION

Societal values are embedded into our built world – water systems are no exception – but these values are rarely inspected as part of the scoping of technical solutions.

Instead, values are treated as priors – immutable, unstated, and implicit as they relate to the objectives of infrastructure. Yet, our infrastructure itself evolves. Increasing resolution in sensing and control of our water environment will allow for unprecedented precision of impacts, both positive and negative. At the same time, humans will continue to cede direct decision-making powers to decision-support technologies such as data algorithms. This new paradigm of smart and autonomous water systems will create new operational capabilities and new opportunities to [re]evaluate these values and explicitly incorporate them into operations.

The methodology and 'proof-of-concept' presented here are a first step toward building a framework for engaging people in algorithmic decision making in cases where normative and ethical preferences are considered. We developed the web-based $(WE)^2$, which is a generalized framework with serious gaming to collect normative preferences through paired comparison testing. Although our framework was designed for water applications, the framework is generalizable and can be used for any paired comparison exercise in any field. Preferences collected using $(WE)^2$ can then be used with our data analytics toolbox to build decision-support preference models and investigate their behavior. These resources, including documentation and tutorials, are shared openly and can be found in the project repository.

We observe that the strength of this framework is that it can prime conversations on values and system expectations at every step of the process, forwarding an iterative process. By doing so, practitioners can work to unobscure AI, ML, and data-driven techniques from behind jargon and demystify the 'black box' processes of decision-support algorithms. We anticipate benefits in deploying our integrated framework in education, operational, and outreach contexts.

Efforts toward incorporating ethics and norms into smart systems must be considered in a sociotechnical context. Importantly, this means that the solution to the development of a technology that is faithful to a society's values may not necessarily be technical in nature at all. Instead, findings from studies could support a structural, social solution as opposed to a solution reliant upon a technological artifice. Though aspects of the work can be technological, it should not preclude results finding that a structural or institutional solution is preferred. The application of AI, ML, and data-driven techniques to water sector problems does not alone make a system 'smart'. Instead, 'smart' water should be conceived as the use of these tools to forward an explicitly recognized objective of the society.

DATA AVAILABILITY STATEMENT

All relevant data are available from an online repository or repositories (https://github.com/uihilab/waterethicswebengine).

REFERENCES

Abend, G. 2011 Thick concepts and the moral brain. *Archives Europeennes de Sociologie* **52** (1), 143–172. https://doi.org/10.1017/S0003975611000051.

Alexander, L. & Moore, M. 2016 Deontological ethics. In: *The Stanford Encyclopedia of Philosophy* (E. N. Zalta, ed.). Winter 201. Metaphysics Research Lab, Stanford University.

Awad, E., Dsouza, S., Kim, R., Schulz, J., Henrich, J., Shariff, A., Bonnefon, J. F. & Rahwan, I. 2018 The moral machine experiment. *Nature* **563** (7729), 59–64. https://doi.org/10.1038/s41586-018-0637-6.

Bargheer, S. 2018 A pragmatist theory of morality. In: *Moral Entanglements: Conserving Birds in Britain and Germany*. University of Chicago Press, Chicago, p. 336.

Binmore, K. 1998 Egalitarianism versus utilitarianism. *Utilitas* **10** (3), 353–367. https://doi.org/10.1017/S0953820800006269.

Bosman, J., Turkewitz, J. & Williams, T. 2019 In the Midwest, relentless floods dredge up 'shadow' of 1993. *New York Times*, June 3, 2019. Available from: https://www.nytimes.com/2019/06/03/us/midwest-floods.html.

Bykov, A. 2019 Rediscovering the moral: the 'old' and 'new' sociology of morality in the context of the behavioural sciences. *Sociology* **53** (1), 192–207. https://doi.org/10.1177/0038038518783967.

Carson, A., Windsor, M., Hill, H., Haigh, T., Wall, N., Smith, J., Olsen, R., Bathke, D., Demir, I. & Muste, M. 2018 Serious gaming for participatory planning of multi-hazard mitigation. *International Journal of River Basin Management*. https://doi.org/10.1080/15715124.2018.1481079.

Chevaleyre, Y., Endriss, U., Lang, J. & Maudet, N. 2007 A short introduction to computational social choice, pp. 51–69. https://doi.org/10.1007/978-3-540-69507-3_4.

Conitzer, V., Brill, M. & Freeman, R. 2015 Crowdsourcing societal tradeoffs. In *Proceedings of the 2015 International Conference on Autonomous Agents and Multiagent Systems*, pp. 1213–1217. Available from: http://dl.acm.org/citation.cfm?id=2773305.

Conitzer, V., Sinnott-Armstrong, W., Borg, J. S., Deng, Y. & Kramer, M. 2017 Moral decision making frameworks for artificial intelligence. In *AAAI Conference on Artificial Intelligence*, pp. 4831–4835. https://doi.org/10.1111/j.1467-8276.2009.01357.x.

Demir, I., Yildirim, E., Sermet, Y. & Sit, M. A. 2018 FLOODSS: Iowa flood information system as a generalized flood cyberinfrastructure. *International Journal of River Basin Management*. https://doi.org/10.1080/15715124.2017.1411927.

Demiray, B. Z., Sit, M. & Demir, I. 2021 D-SRGAN: DEM super-resolution with generative adversarial networks. *SN Computer Science* **2**, 48. https://doi.org/10.1007/s42979-020-00442-2.

Dubois, D., Rucker, D. D. & Galinsky, A. D. 2015 Social class, power, and selfishness: when and why upper and lower class individuals behave unethically. *Journal of Personality and Social Psychology*. https://doi.org/10.1037/pspi0000008.

Eller, D. 2018 Flooding has slammed every Iowa county since 1988, some as many as 17 times. *Des Moines Register*, April 29, 2018. Available from: https://www.desmoinesregister.com/story/money/agriculture/2018/04/29/iowa-flood-center-ranks-disaster-damages-billions-wapsipinicon-river/422336002/.

Everett Jaques, A. 2019 Why the moral machine is a monster. In *University of Miami Law School: We Robot Conference*, Miami, p. 10. Available from: https://robots.law.miami.edu/2019/wp-content/uploads/2019/03/MoralMachineMonster.pdf.

Freedman, R., Borg, J. S., Sinnott-Armstrong, W., Dickerson, J. P. & Conitzer, V. 2019 Adapting a kidney exchange algorithm to align with human values, p. 115. https://doi.org/10.1145/3278721.3278727.

Grayling, A. C. 1995 *Philosophy: A Guide Through the Subject*. Oxford University Press. Available from: https://books.google.com/books?id=YLR8ngEACAAJ.

Habibi, H., Nazari, B., Norouzi, A., Noh, S. J., Seo, D. J., Sinha, S., Yu, X., Bartos, M., Kerkez, B., Lakshman, L., Zink, M., Lyons, E., Philips, B. J., Jangyodsuk, P. & Gao, J. 2017 Integrated sensing and prediction of flash floods for the Dallas-Fort Worth Metroplex (DFW). In *97th American Meteorological Society Annual Meeting*.

Hagendorff, T. 2019 *The Ethics of AI Ethics – An Evaluation of Guidelines, No. 2018 (February)*. Available from: http://arxiv.org/abs/1903.03425.

Haidt, J. 2007 The new synthesis in moral psychology. *Science*. https://doi.org/10.1126/science.1137651.

Hardy, K. & Cannon, A. 2019 Iowa flooding: damage from floodwaters reaches $1.6B, Gov. Kim Reynolds estimates. *Des Moines Register*, March 22, 2019. Available from: https://www.desmoinesregister.com/story/news/2019/03/22/iowa-flooding-damage-estimate-governor-kim-reynolds-federal-flood-relief-trump-nebraska-missouri/3232934002/.

Hitlin, S. & Vaisey, S. 2013 The new sociology of morality. *Annual Review of Sociology* **39** (1), 51–68. https://doi.org/10.1146/annurev-soc-071312-145628.

Hursthouse, R. & Pettigrove, G. 2018 Virtue ethics. In: *The Stanford Encyclopedia of Philosophy* (E. N. Zalta, ed.), Winter 201. Metaphysics Research Lab, Stanford University.

Immerman, M. & Immerman, E. 2018 Getting from Push to Pool – An Alternative Starting Point in the Water Quality Debate. Des Moines, IA. Available from: http://regionalstrategic.com/assets/rsq_2018_summer_essay3.pdf.

Jobin, A., Ienca, M. & Vayena, E. 2019 The global landscape of AI ethics guidelines. *Nature Machine Intelligence* **1** (9), 389–399. https://doi.org/10.1038/s42256-019-0088-2.

Jones, C. S., Davis, C. A., Drake, C. W., Schilling, K. E., Debionne, S. H. P., Gilles, D. W., Demir, I. & Weber, L. J. 2018 Iowa statewide stream nitrate load calculated using in situ sensor network. *Journal of the American Water Resources Association*. https://doi.org/10.1111/1752-1688.12618.

Jonoski, A. 2002 *Hydroinformatics as Sociotechnology. Ihe Dissertation*, Taylor & Francis, p. 27. Available from: https://books.google.com/books?id=IVD-FVDxbdAC.

Kelley, T. J. 2019 The fight to tame a swelling river with dams that may be outmatched by climate change. *New York Times2*, March 21, 2019. Available from: https://nyti.ms/2TLVuKa.

Kerkez, B., Gruden, C., Lewis, M., Montestruque, L., Quigley, M., Wong, B., Bedig, A., Kertesz, R., Braun, T., Cadwalader, O., Poresky, A. & Pak, C. 2016 Smarter stormwater systems. *Environmental Science and Technology*. https://doi.org/10.1021/acs.est.5b05870.

Krebs, D. L. 2008 Morality: an evolutionary account. *Perspectives on Psychological Science* **3** (3), 149–172. https://doi.org/10.1111/j.1745-6924.2008.00072.x.

Lee, M., Kusbit, D., Kahng, A., Kim, J. T., Yuan, X., Chan, A., See, D., Noothigattu, R., Lee, S., Psomas, A. & Procaccia, A. D. 2019 WeBuildAI: participatory Framework for Algorithmic Governance. *Proc. ACM Hum.-Comput. Interact.* **3** (CSCW). https://doi.org/10.1145/3359283.

Marso, A. 2019 Act of god? Or government? In fight over flood, farmers blame army corps of engineers. *The Kansas City Star*, March 24, 2019. Available from: https://www.kansascity.com/news/state/missouri/article228237519.html.

Mollerup, A. L., Mikkelsen, P. S., Thornberg, D. & Sin, G. 2017 Controlling sewer systems – a critical review based on systems in three EU cities. *Urban Water Journal* **14** (4), 435–442. https://doi.org/10.1080/1573062X.2016.1148183.

Moor, J. H. 2006 The nature, importance, and difficulty of machine ethics. *IEEE Intelligent Systems* **21** (4), 18–21. https://doi.org/10.1109/MIS.2006.80.

Mullapudi, A., Wong, B. P. & Kerkez, B. 2017 Emerging investigators series: building a theory for smart stormwater systems. *Environmental Science: Water Research and Technology* **3** (1), 66–77. https://doi.org/10.1039/c6ew00211 k.

Mullapudi, A., Bartos, M., Wong, B. & Kerkez, B. 2018 Shaping streamflow using a real-time stormwater control network. *Sensors (Switzerland)* **18** (7), 1–11. https://doi.org/10.3390/s18072259.

Mullapudi, A., Lewis, M. J., Gruden, C. L. & Kerkez, B. 2020 Deep reinforcement learning for the real time control of stormwater systems. *Advances in Water Resources* **140**, 103600. https://doi.org/10.1016/j.advwatres.2020.103600.

Muschalla, D., Vallet, B., Anctil, F., Lessard, P., Pelletier, G. & Vanrolleghem, P. A. 2014 Ecohydraulic-driven real-time control of stormwater basins. *Journal of Hydrology* **511**. https://doi.org/10.1016/j.jhydrol.2014.01.002.

Noothigattu, R., Gaikwad, S. '. S., Awad, E., Dsouza, S., Rahwan, I., Ravikumar, P. & Procaccia, A. D. 2017 A voting-based system for ethical decision making, pp. 1587–1594. Available from: http://arxiv.org/abs/1709.06692.

O'Neil, C. 2016 *Weapons of Math Destruction: How Big Data Increases Inequality and Threatens Democracy*. Crown, New York, NY. https://www.tandfonline.com/doi/full/10.1080/01972243.2017.1354593.

O'Reilly, T. 2013 Open data and algorithmic regulation. *Beyond Transparency: Open Data and the Future of Civic Innovation* **21**, 289–300.

Overloop, P.-J. 2006 *Model Predictive Control on Open Water Systems*. IOS Press, Amsterdam.

Piff, P. K., Stancato, D. M., Coteb́, S., Mendoza-Denton, R. & Keltner, D. 2012 Higher social class predicts increased unethical behavior. *Proceedings of the National Academy of Sciences of the United States of America*. https://doi.org/10.1073/pnas.1118373109.

Rahwan, I. 2018 Society-in-the-loop: programming the algorithmic social contract. *Ethics Inf Technol* **20**, 5–14. https://doi.org/10.1007/s10676-017-9430-8.

Rentsch, J. 2019 Can a retention pond be high-tech? Walmart and Maryland officials hope so. *Delmarva Now*, November 20, 2019. Available from: https://www.delmarvanow.com/story/news/2019/11/20/walmart-state-maryland-retrofit-water-retention-ponds-bay-health/4239417002/.

Sadler, J. M., Goodall, J. L., Behl, M., Morsy, M. M., Culver, T. B. & Bowes, B. D. 2019 Leveraging open source software and parallel computing for model predictive control of urban drainage systems using EPA-SWMM5. *Environmental Modelling and Software*. https://doi.org/10.1016/j.envsoft.2019.07.009.

Samuel, A. L. 1960 Some moral and technical consequences of automation – a refutation. *Science* **132** (3429), 741–742. https://doi.org/10.1126/science.132.3429.741.

Schütze, M., Lange, M., Pabst, M. & Haas, U. 2018 Astlingen – a benchmark for real time control (RTC). *Water Science and Technology* **2017** (2), 552–560. https://doi.org/10.2166/wst.2018.172.

Sermet, Y. & Demir, I. 2018 An intelligent system on knowledge generation and communication about flooding. *Environmental Modelling and Software*. https://doi.org/10.1016/j.envsoft.2018.06.003.

Sermet, Y. & Demir, I. 2019 Towards an information centric flood ontology for information management and communication. *Earth Science Informatics*. https://doi.org/10.1007/s12145-019-00398-9.

Sermet, Y., Villanueva, P., Sit, M. A. & Demir, I. 2019 Crowdsourced approaches for stage measurements at ungauged locations using smartphones. *Hydrological Sciences Journal*. https://doi.org/10.1080/02626667.2019.1659508.

Sermet, Y., Demir, I. & Muste, M. 2020 A serious gaming framework for decision support on hydrological hazards. *Science of the Total Environment*. https://doi.org/10.1016/j.scitotenv.2020.138895.

Sharior, S., McDonald, W. & Parolari, A. J. 2019 Improved reliability of stormwater detention basin performance through water quality data-informed real-time control. *Journal of Hydrology*. https://doi.org/10.1016/j.jhydrol.2019.03.012.

Sinnott-Armstrong, W. 2019 Consequentialism. In: *The Stanford Encyclopedia of Philosophy* (E. N. Zalta, ed.), Summer 201. Metaphysics Research Lab, Stanford University.

Sit, M. & Demir, I. 2019 Decentralized flood forecasting using deep neural networks. *arXiv Preprint* arXiv:1902.02308.

Sit, M. A., Koylu, C. & Demir, I. 2019 Identifying disaster-related tweets and their semantic, spatial and temporal context using deep learning, natural language processing and spatial analysis: a case study of Hurricane Irma. *International Journal of Digital Earth*. https://doi.org/10.1080/17538947.2018.1563219.

Stellar, J. E., Manzo, V. M., Kraus, M. W. & Keltner, D. 2012 Class and compassion: socioeconomic factors predict responses to suffering. *Emotion*. https://doi.org/10.1037/a0026508.

Sun, C., Svensen, J. L., Borup, M., Puig, V., Cembrano, G. & Vezzaro, L. 2020 An MPC-enabled SWMM implementation of the Astlingen RTC benchmarking network. *Water* **12** (4), 1034. https://doi.org/10.3390/w12041034.

Susi, T., Johannesson, M. & Backlund, P. 2007 *Serious Games – An Overview* (Technical Report). *Elearning*.

Thomson, J. J. 1985 The trolley problem. *The Yale Law Journal* **94** (6), 1395. https://doi.org/10.2307/796133.

Tolmeijer, S., Kneer, M., Sarasua, C., Christen, M. & Bernstein, A. 2020 Implementations in machine ethics: a survey. Available from: http://arxiv.org/abs/2001.07573.

Trautmann, S. T., van de Kuilen, G. & Zeckhauser, R. J. 2013 Social class and (un)ethical behavior: a framework, with evidence from a large population sample. *Perspectives on Psychological Science*. https://doi.org/10.1177/1745691613491272.

Troutman, S. C., Love, N. G. & Kerkez, B. 2020 Balancing water quality and flows in combined sewer systems using real-time control. *Environmental Science: Water Research & Technology* 1357–1369. https://doi.org/10.1039/c9ew00882a.

Tsukida, K. & Gupta, M. R. 2011 *How to Analyze Paired Comparison Data*. UWEE Technical Report.

Vojinović, Z. & Abbott, M. B. 2012 *Flood Risk and Social Justice: From Quantitative to Qualitative Flood Risk Assessment and Mitigation*. *Water Intelligence Online*. https://doi.org/10.2166/9781780400822.

Weber, L. J., Muste, M., Bradley, A. A., Amado, A. A., Demir, I, Drake, C. W., Krajewski, W. F., Loeser, T. J., Politano, M. S., Shea, B. R. & Thomas, N. W. 2018 The Iowa watersheds project: Iowa's prototype for engaging communities and professionals in watershed hazard mitigation. *International Journal of River Basin Management*. https://doi.org/10.1080/15715124.2017.1387127.

Xiang, Z., Yan, J. & Demir, I. 2020 A rainfall-runoff model with LSTM-based sequence-to-sequence learning. *Water Resources Research* **56** (1), e2019WR025326.

Xu, H., Windsor, M., Muste, M. & Demir, I. 2020 A web-based decision support system for collaborative mitigation of multiple water-related hazards using serious gaming. *Journal of Environmental Management* **255**, 109887. https://doi.org/https://doi.org/10.1016/j.jenvman.2019.109887.

Yildirim, E. & Demir, I. 2019 An integrated web framework for HAZUS-MH flood loss estimation analysis. *Natural Hazards*. https://doi.org/10.1007/s11069-019-03738-6.

Zagzebski, L. T. 2004 *Divine Motivation Theory*. https://doi.org/10.1017/CBO9780511606823.

First received 8 June 2020; accepted in revised form 4 March 2021. Available online 18 March 2021

© 2021 The Authors

Interactive decision support methodology for near real-time response to failure events in a water distribution network

E. Nikoloudi, M. Romano, F. A. Memon and Z. Kapelan

ABSTRACT

The present study proposes a new interactive methodology and an interactive tool for the response to water network failure events facilitating near real-time decision-making. The proposed methodology considers: (i) a structured yet flexible approach supporting and guiding the operator throughout the entire response process to water network failure events, while allowing the operator to have a final say; (ii) a novel interaction with the operator in near real time via the proposed tool (e.g. allowing operators to propose different 'what-if' scenarios without being hydraulic experts); (iii) the provision of automatically generated advice (e.g. optimal response solutions and assessed end-impacts) – although optimal response solutions not identified in near real time yet; and (iv) improved impact assessment using realistic impact indicators that cover different aspects of the event – which are consistently calculated for every proposed response solution (to facilitate easy comparison between different response solutions). The new methodology was applied on a semi-real case study. The results obtained demonstrated the potential of the new response methodology and its application through the interactive tool to improve water utilities' current practice. This was accomplished through supporting/guiding operators in the identification of effective response solutions with low end-impact on the consumers and low cost for the utility.

Key words | decision-support tool, event management, impact assessment, near real-time response, water distribution network failure events

E. Nikoloudi (corresponding author)
F. A. Memon
Z. Kapelan
Centre of Water Systems, Department of Engineering, Mathematics and Physical Sciences,
University of Exeter, Centre of Water Systems,
North Park Road, Exeter EX4 4QJ,
UK
E-mail: *en273@exeter.ac.uk*

M. Romano
United Utilities Group PLC,
Lingley Green Avenue, Warrington WA5 3LP,
UK

Z. Kapelan
Department of Water Management,
Delft University of Technology,
Stevinweg 1, 2628CN Delft,
The Netherlands

HIGHLIGHTS

- Novel structured overall response methodology supporting operators throughout the whole response process.
- Improved impact assessment.
- Realistic selection of operational interventions (based on operational costs, the availability of different types of interventions, etc.).
- Novel near real-time interaction with the control room operator (manual 'what-if' scenarios and automatic optimal solutions).

This is an Open Access article distributed under the terms of the Creative Commons Attribution Licence (CC BY 4.0), which permits copying, adaptation and redistribution, provided the original work is properly cited (http://creativecommons.org/licenses/by/4.0/).

doi: 10.2166/hydro.2020.101

INTRODUCTION

The water industry in the UK and worldwide faces considerable challenges in making effective use of sensor and other data that are collected in water distribution systems (WDSs) in near real time (typically every 15–30 mins). These data are still not used much in a water utility's control room, especially when it comes to identifying a suitable strategy to respond to failure events in near real time, i.e. events such as major pipe bursts, equipment failure or water treatment work (WTW) shutdowns. Relevant academic work has not adequately addressed this challenge mainly due to the focus on specific stages (i.e. isolation, impact assessment or intervention) rather than the overall response process. Furthermore, for an effective near real-time response, there is still a need to develop: (1) improved impact assessment methods that are based on realistic metrics used in the water industry and that are also used in a consistent manner to facilitate easy comparison between different response solutions (i.e. a response intervention or a set of interventions), (2) more realistic selection of response interventions to be implemented (e.g. based on operational costs, the availability of different types of interventions, etc.) and (3) effective interaction with the control room operators that takes into account their expert judgement, preferences and experience.

In this study, a novel response methodology that aims to fulfil the above needs is proposed. The new response methodology is implemented via an interactive decision-support tool entitled the Interactive Response Planning Tool (IRPT). The IRPT is used to guide/support operators in identifying an effective response solution in near real time (i.e. usually required up to 1 h after the event detection/localisation). The main aim of this study is to show the potential of the IRPT to improve utilities' current practice by supporting/guiding operators in the identification of low end-impact (i.e. the total impact after the implementation of the response solution) and low-cost response solutions.

This paper is organised as follows. Firstly, background information relevant for the present study is presented. Subsequently, the new response methodology (including its concept, details of the indicators used in the impact assessment and information regarding the optimisation of the response interventions) is described. Later, results from a semi-real case study are presented and discussed. Finally, conclusions from the application of the new methodology and of the interactive tool are drawn.

BACKGROUND

An 'event' is denoted here as any failure that has a negative impact on a WDS's performance in terms of the water utility's temporary inability to deliver a regular service. An efficient event management process in WDSs can be divided into three principal stages: event detection, event localisation and event response (Vamvakeridou-Lyroudia *et al.* 2010; Romano *et al.* 2011; Jung *et al.* 2015; Kapelan *et al.* 2017). The first two stages involve detecting and localising the event in the network and raising the relevant alarm. The third stage is associated with the decisions and actions required to reduce and, ultimately, eliminate the negative impact of the event on the consumers (Jeong *et al.* 2006; Bicik *et al.* 2009; Nayak & Turnquist 2016). The first two stages have been researched extensively in the literature (Bicik *et al.* 2011; Romano *et al.* 2011; Casillas Ponce *et al.* 2014; Romano *et al.* 2014; Jung *et al.* 2015; Okeya *et al.* 2015; Laucelli *et al.* 2016; Romano 2019; Zhou *et al.* 2019). Hence, the focus of this paper is on the event response stage.

The event response stage typically includes two sub-stages, namely isolation and recovery (Vamvakeridou-Lyroudia *et al.* 2010, Mahmoud *et al.* 2018). The isolation sub-stage aims to minimise the negative initial impact of an event and prepare the affected part of the network for follow-on repairs. This sub-stage has been thoroughly studied in the past by several authors such as Jun & Loganathan (2007) and Giustolisi & Savic (2010). Hence, it is not the subject of the present work. The recovery sub-stage, on the other hand, involves impact assessment of the event and selection of the best response solution. This sub-stage is the focus of this study.

Several methods for event impact assessment, such as Kapelan *et al.* (2006), Kao & Li (2007), Giustolisi *et al.*

(2008), Bicik *et al.* (2009) and Qi *et al.* (2018), have been proposed in the literature. These methods are all based on rather theoretical impact assessment metrics. This aspect is, therefore, further investigated in this work by developing and using a wider range of improved impact assessment metrics that are based on real-life indicators used by water utilities.

The selection of the best response solution to implement is strongly dependent on the preceding steps of the event management process. An effective event management process should be based on an integrated methodology that takes into account all the preceding stages. However, in the literature, the recovery sub-stage has been approached mainly through (1) proposing theoretical mitigating methods for the minimisation of the consequences of a physical attack (e.g. Jeong *et al.* 2006; Jeong & Abraham 2009; Turner *et al.* 2012) and (2) proposing generic decision-support systems (DSSs). For example, Bicik *et al.* (2008) proposed a general risk-based DSS methodology for supporting operators in decision-making against a failure. Vamvakeridou-Lyroudia *et al.* (2010) proposed an integrated Intervention Management Model, in the context of a general DSS methodology for operational WDS management. Finally, Mahmoud *et al.* (2018) proposed an integrated methodology for the near real-time response to pipe burst events in WDSs. All these methods, however, are rather academic in nature as they do not take into account the complexities of real-world response problems. This aspect is accounted for in this work by making use of the control room operators' extensive knowledge and experience via their interaction with the IRPT.

Paez *et al.* (2020), in their summary paper, presented several methods for the response to WDS events after an earthquake disaster (i.e. they considered real-life and complex pipe network failures). Later, Zhang *et al.* (2020) proposed an optimisation-based framework to maximise resilience of a WDS after a disaster-type event (e.g. earthquake). However, both these studies identified the optimum set of response interventions that includes pipe repair or replacement only (i.e. without proposing different types of response interventions). This limitation is circumvented in this work by developing a methodology that utilises multiple intervention types (e.g. rezoning, water injection) and that also enables identifying the best time for their implementation in the field in order to restore supply.

RESPONSE METHODOLOGY

Current practice response methodology

Different utilities deal with events in a different way and use more or less structured approaches. This section briefly describes a response methodology mainly based on *ad hoc* response interventions that can be considered typical for the UK water sector. In this methodology, the response interventions are largely based on the experience and expert judgement of control room operators, despite various systems being used by the operators to support their decisions.

The detection of an event in a water utility is nowadays usually done in two possible ways: (a) through customer calls (i.e. reporting no water/low pressure/discolouration, etc.) and/or (b) through an automated detection system (i.e. alarms generated based on flow and/or pressure data). Once the detected event is confirmed and approximately localised (e.g. roughly based on customer calls' addresses and/or using other semi-automated means), the utility typically mobilises some available water trucks, called Alternative Supply Vehicles (ASVs). This is done as an immediate restoration measure after an initial impact assessment usually carried out manually and/or with limited hydraulic model support. Here, an assessment involving the calculation of the water volume required to be supplied per hour (and hence the number of ASVs required per hour) based on the affected district metered areas (DMAs) normal water demand may also be carried out. At the same time, in the control room, after further manual (e.g. by checking service reservoirs' levels using online systems) and/or hydraulic model-supported initial impact assessment, operators request isolation of the event. Isolation is then carried out either as soon as possible (e.g. if the service reservoirs' levels are quickly dropping or there is significant third-party damage) or later in the day, depending on severity/time of the event and other factors. There are also occasions where the repair can be conducted without isolating the failure (i.e. under pressure). If isolation is required, the isolation valves are usually identified manually – as the closest operable valves to the event. With some ASVs already on site (or not), the control room operators then attempt to identify the most suitable response solution (e.g. how many more ASVs should be sent to the site, a suitable rezoning plan, overland bypasses, etc.) to be implemented while the repair is being carried out. Online map systems, offline connectivity maps, calculation sheets and hydraulic models can be used by the operators for this purpose.

Bearing in mind the above, it is worth stressing that despite using hydraulic models for some of the aforementioned activities can be considered as common practice, hydraulic analysis is not always carried out thoroughly due to limitations in terms of the time that can be dedicated to this activity, the skills required to run hydraulic simulations, the ability to only test a few scenarios and the difficulty to consistently assess their end-impact.

New response methodology

The concept

The new response methodology proposed in this paper consists of the following main steps: Step (1) initial impact assessment, Step (2) identification of the isolation plan, Step (3) manual identification of a response solution proposed by an operator, Step (4) automatic identification of a response solution generated using optimisation and Step (5) identification of the response solution to be implemented in the field. Note that these five steps do not need to be necessarily carried out in a sequential manner as presented here.

The implementation of the new methodology within the IRPT is conducted through the following three-stage routine in each step: Stage (1) involves obtaining the operators' inputs, Stage (2) involves carrying out hydraulic simulations to assess the end-impact/cost for each solution and Stage (3) involves visualising the calculated end-impact of each solution. The new response methodology's steps are described in more detail in the remainder of this section and are also shown as a flowchart in Figure 1.

Step 1. Following the confirmation (i.e. detection and localisation) of an event, an initial impact assessment is performed assuming the 'do nothing' scenario. At this point, the operators are asked if isolation needs to take place or if it can be carried out.

a. If yes, the operators create an event isolation plan in Step 2.
b. If no, they move to Step 3 to propose a manual solution.

Step 2. For the identification of the isolation plan, the best (i.e. closest to the event) set of valves is automatically provided to the operators by the IRPT and the operators are then asked if they are satisfied with this set (e.g. if the identified valves can been localised and are operable). If they are not satisfied, they ask the IRPT to automatically provide the next best set of isolation valves. As soon as the best set of isolation valves is selected, the operators input into the IRPT the isolation duration and different potential isolation start times. Then, the IRPT automatically calculates the end-impact of the different isolation start times, and these are presented to the operators. In view of the calculated end-impacts, the operators can then select a desired start time of isolation. Once the isolation plan is finalised, the operators are asked if they consider the resulting end-impact low.

a. If yes, they proceed with the implementation of the 'isolation only' final solution without applying any further intervention. The operators then move to Step 6.
b. If no (or isolation is not possible), they then identify the more comprehensive response solution as follows. The operators are asked if they want the IRPT to automatically generate an optimal solution.
    i. If yes, they move to Step 4.
    ii. If no, they proceed by proposing a manual solution in Step 3.

Step 3. In this step, the operators are able to propose a manual solution by interacting with the IRPT. Here, the IRPT firstly enables the operators to input their desired and available (e.g. accessible/operable rezoning valves) intervention(s) and the start time of this (these) intervention(s). It then provides decision support to the operators by assessing and visualising the end-impact/cost of the proposed manual solution. Then the operators are asked if they are satisfied with the end-impact/cost of their proposed manual solution.

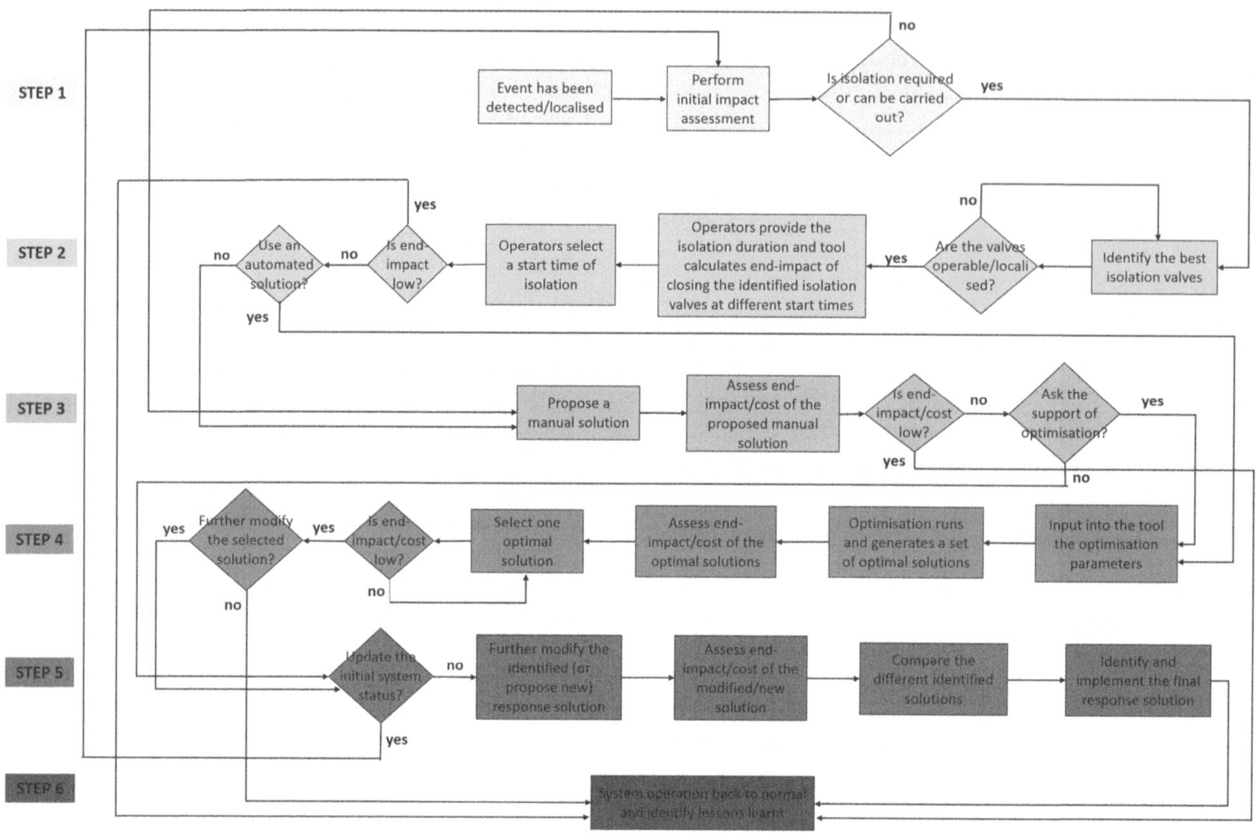

Figure 1 | New response methodology's steps.

a. If yes, a final solution has been found and the operators move to Step 6.
b. If no, the operators are allowed to propose alternative manual solutions and compare their end-impact/cost by moving to Step 5 or ask the IRPT to automatically generate an optimal solution in Step 4.

Step 4. Operators input into the IRPT all the desired and available interventions, as well as a time range in which the various interventions could start. Then, the optimisation runs and optimal solutions are automatically generated and assessed by the IRPT. Operators are then able to select one (or more) optimal solution(s) on the Pareto front (depending on whether or not the end-impact/cost is low). Finally, they are asked if they wish to further modify this (these) solution(s) manually.

a. If yes, they move to Step 5.
b. If no, a final solution has been found and they move to Step 6.

Step 5. In this step, the operators are first asked if they want to go back all the way to Step 1, with the modified system state used as a starting point. This is done to account for the fact that the situation may have changed in the meantime. If not, they can further modify a solution from Step 3 or Step 4, as well as proposing new solution. Here, the IRPT enables operators to compare all of the identified solutions consistently (i.e. consistent impact metrics) and with support of effective visualisations (e.g. multiple maps in a single window) of the end-impacts and costs. All this enables the operators to select the final solution they wish to implement.

Step 6. Once the system operation is back to normal, the operators identify the lessons learned.

In the IRPT, the hydraulic simulations are carried out by using EPANET2 (Rossman 2000) and pressure-driven network modelling based on methodology developed by Paez *et al.* (2018). The demand-driven analysis conducted by EPANET2 accurately estimates the nodal demands in normal operating conditions but not in the pressure-deficient ones that occur during various system failures (e.g. during a pipe burst or some equipment failure). Hence, in this study, the original EPANET2 hydraulic model is modified by using the approach proposed by Paez *et al.* (2018). This method works by adding suitably chosen dummy elements to the original EPANET2 model. This creates the pressure-driven model that is able to simulate hydraulic conditions in the network under both normal- and pressure-deficient conditions. Paez *et al.*'s pressure-driven method has been selected here as it was thoroughly tested, validated and demonstrated to work effectively on real-sized networks producing accurate hydraulic results (Paez *et al.* 2018). The use of this model provides additional computational burden for the calculation of different impact indicators but then so does any other pressure-driven method – this is simply the price to pay for being able to simulate pressure-driven conditions in the pipe network. Finally, note that the selection of a pressure-driven model is not the focus of this study, i.e. any other reliable and accurate pressure-driven model can be used instead within the response methodology presented in this paper.

The IRPT also links to the Quantum Geographic Information System (QGIS) software to visualise the spatial distribution of end-impact on a suitable map of the analysed water system.

The key novelties of the new response methodology proposed are as follows: (i) structured yet flexible approach supporting and guiding the operator throughout the entire response process (from detection and localisation of a failure event to the implementation of the identified response solution in the field) while allowing the operator to have a final say, (ii) novel interaction with the operator in near real time (i.e. up to 1 h) via the IRPT (e.g. 'what-if' scenarios) without hydraulic expertise requirements, (iii) provision of automatically generated advices (e.g. optimal response interventions and assessed impacts) – although the optimal response interventions are not yet provided in near real time due to the long (e.g. several hours) optimisation time currently required (longer than the time typically available in a control room for identifying a response), (iv) improved impact assessment (based on realistic impact indicators) that covers different aspects of the event – which are consistently calculated for every proposed response intervention (to facilitate easy comparison between different response solutions) and (v) more realistic selection of operational interventions (based on operational costs, the availability of different types of interventions, etc.).

Impact assessment

The IRPT provides to operators the capability to automatically assess the end-impact of a proposed solution (i.e. in Stage 2 of each step of methodology) based on realistic metrics. In the IRPT, a consistent framework for end-impact assessment (i.e. same impact metrics calculated for every proposed response solution) is implemented. This facilitates the comparison of different response solutions (in Step 5 of the methodology) and enables more informed decision-making. Furthermore, the IRPT allows the operators to perform this comparison without the need for them to be hydraulic model experts.

The impact indicators proposed in this paper have been developed bearing in mind the UK water industry practice as well as previous relevant literature (e.g. Bicik *et al.* 2009). Most of these indicators have not been used before in this context (at least in the published literature). The following aspects of end-impact are considered here: water supply interruption, low pressure impact and discolouration risk increase (DRI) impact. More specifically, the following indicators are used: (1) customer minutes lost (CML), (2) average minutes low pressure (AMLP), (3) unaccounted for water (UW) and (4) DRI. AMLP and UW are calculated for different customer types, namely residential, industrial and sensitive (i.e. schools and hospitals). The impact horizon in the new response methodology is the period of time for which the end-impact is assessed. It starts from the detection/localisation time of an event and lasts until the repair is completed (i.e. time period over which restoration interventions can be implemented).

CML is defined as the mean duration customers are without water supply (i.e. equivalent to pressure ≤ 3 m in the main) in a given reporting year. CML is a real-life indicator used in water utilities nowadays and is calculated for every discrete pressure area (DPA; i.e. discrete areas within a DMA). It is measured in minutes per customer (mins/cust). In this study, CML is found as follows:

$$\text{CML [mins/cust]} = \sum \frac{\text{Cust}_{\text{SI}} \times \text{Dur}_{\text{SI}}}{\text{Cust}} \tag{1}$$

where Cust_{SI} is the total number of customers in each DPA affected by supply interruption at least during one time-step (i.e. 15 mins) over the impact horizon; Dur_{SI} is defined as the length of time for which properties are without a continuous supply of water in mins – only events with duration ≥3 h are taken into account; Cust is the total connected customers at year end (fixed number for each utility).

The AMLP indicator is defined as follows:

$$\text{AMLP [mins/cust]} = \frac{\text{Cust}_{\text{LP}} \times \text{Dur}_{\text{LP}}}{\text{Cust}} \tag{2}$$

where Cust_{LP} is the total number of customers affected by low pressure (i.e. minimum_pressure < pressure < required_pressure, where minimum_pressure is usually considered in UK utilities as equal to 3 m and required_pressure as equal to 15 m) at least during one time-step (i.e. 15 mins) over the impact horizon; Dur_{LP} is the average low pressure impact duration over the impact horizon in mins; Cust is the total number of connected customers at year end (fixed number for each utility).

The UW indicator is calculated as follows (Bicik *et al.* 2009):

$$\text{UW [m}^3] = \frac{900}{1000} \times \sum_{t=1}^{T} \sum_{i=1}^{N} \times \begin{cases} \dfrac{(D_{i,\text{req}}(t) - D_i(t)) \times \text{Cust}_i}{\text{Cust}_{i,\text{count}}} & \text{if } P_i(t) < P_{\text{req}} \\ 0 & \text{otherwise} \end{cases} \tag{3}$$

where T is the impact horizon (h), t is the simulation time (with assumed time-step of 15 mins = 900 s), $D_{i,\text{req}}(t)$ is the requested demand at node i and time t in l/s, $D_i(t)$ is the delivered demand at node i and time t in l/s, $\text{Cust}_{i,\text{count}}$ is the number of customers supplied from demand node i. Note that the requested water demand may be undelivered due to either complete interruption or low pressure (i.e. pressure < required_pressure).

The DRI is estimated based on a combination of the methods found in Beuken *et al.* (2008) and Bicik *et al.* (2009). Beuken *et al.* (2008) suggest to calculate discolouration risk based on minimum and maximum velocities and maximum flow rates. Here, the minimum and maximum flow velocities in an average demand day are calculated with a hydraulic model for each pipe. The same model is used to estimate the largest flow rate for each pipe under the same demand conditions. Then, a score is assigned to each pipe for each discolouration risk type (i.e. based on velocities and on flow rates). Hence, a score of 1 means low, a score of 2 means moderate and a score of 3 means high discolouration risk. The discolouration risk for every pipe is calculated as the sum of the scores based on both velocity and flow rate. The resulting discolouration risk scores are grouped in five severity categories (Beuken *et al.* 2008), i.e. 'VERY LOW' with a total score of 2, 'LOW' with a total score of 3, 'MODERATE' with a total score of 4, 'HIGH' with a total score of 5 and 'VERY HIGH' with a total score of 6.

Once the discolouration risk score for every pipe has been found, the DRI for every pipe can be calculated as the difference between the discolouration risk score under 'failure' and normal conditions. The 'failure' condition is defined here as the WDS condition after the occurrence of the event and/or the implementation of the intervention(s). The DRI is then ranked based on the total score increase, i.e. 'NO RISK' with a total score increase equal to 0, 'LOW INCREASE' with a total score increase equal to 1, 'MODERATE INCREASE' with a total score increase equal to 2 or 3 and 'HIGH INCREASE' with a total score increase equal to 4.

Following the calculation of the DRI for every pipe in the network, the number of pipes with at least 'LOW INCREASE' (i.e. with total score increase equal to 1 or higher) is used to estimate the DRI (based on a modification from the equation in Bicik *et al.* 2009):

$$\text{DiscRiskIncrease} = \text{count}_1^{N_\text{p}} \times \begin{Bmatrix} \text{Disc}_{j,\text{failure}} - \text{Disc}_{j,\text{norm}} \geq 1 & \text{if Disc}_{j,\text{failure}} > \text{Disc}_{\text{j,norm}} \\ 0 & \text{otherwise} \end{Bmatrix} \tag{4}$$

where N_p is the number of pipes in the network and $\text{Disc}_{j,\text{norm}}$ and $\text{Disc}_{j,\text{failure}}$ are the total discolouration risk of pipe j under normal and failure conditions, respectively.

Interventions optimisation

The IRPT provides to operators the capability to automatically identify a number of optimal solutions (i.e. in Step 4 of the methodology) by solving a two-objective optimisation problem. The two objectives are the minimisation of total end-impact (of a response solution) and the minimisation of the total cost associated with this solution.

The total (i.e. aggregated) end-impact is estimated by normalising and then adding up the values of the individual impact indicators defined in the previous section. Before aggregating, the normalised indicators are multiplied with specified weights based on the priority/preferences of the operators as follows:

$$\text{Total\_impact} = \sum_{i=1}^{4} (w_i f_i) \tag{5}$$

where i is the index of each impact indicator with $i \in [1, 4]$; f_i is the normalised impact indicator i and w_i is the weight of impact indicator i with $\sum w_i = 1$.

The impact indicators are normalised in the range [0, 1] as follows:

$$x_{\text{new}} = \frac{x - x_{\min}}{x_{\max} - x_{\min}} \tag{6}$$

where x_{new} is the normalised impact indicator value; x is the non-normalised impact indicator value; $x_{\min}$ is the minimum impact indicator value and $x_{\max}$ is the maximum impact indicator value.

The second objective function is the total cost of the selected response solution, calculated as follows:

$$\text{Total\_cost} = c_{\text{rez}} d_{\text{rez}} N_{\text{rez}} + c_{\text{ASV}} h_{\text{ASV}} \tag{7}$$

where c_{rez} is the cost (£) per hour of manipulating (i.e. opening or closing) a single rezoning valve; d_{rez} is the time it takes to open and close a single rezoning valve (in h); N_{rez} is the number of rezoning valves to open/close in the specific response solution; c_{ASV} is the cost (£) per hour of ASV injection and h_{ASV} is the total time of ASV injection (i.e. hours of injection from all the ASVs sent to site). The above hourly costs (i.e. c_{rez}, c_{ASV}) can be calculated from the total employee rechargeable (i.e. equal to the inflated value of the average employee costs) divided by the number of working days per annum. It is stressed that the cost function presented here does not aim to calculate the precise cost of a specific solution. It mainly aims to point out the cost difference between different solutions and to identify those solutions that reduce significantly the end-impact with least increase in cost, as these solutions are likely to be selected by the decision-makers.

The decision variables of the optimisation problem are (1) the operational interventions used and (2) the start times of their implementation. The operational interventions considered in this methodology are (1) rezoning by valve manipulations (i.e. opening of initially closed boundary valves), (2) water injection at different network locations and (3) combination of these. Water injection, which is a novel type of intervention considered in this study, is carried out through the ASVs. In this study, an ASV is modelled as a tank linked to the injection point through a pump (to manage the pressure pumped into the network) and a valve (to allow water flow from the tank to the system). Usually, utilities dispatch three ASVs of 30 m^3 to every injection point in order to guarantee continuous supply to the affected node/customers. In this study, to simplify the coding required in the IRPT, one artificial ASV with volume equal to 90 m^3 (i.e. 3×30 m^3) is modelled at each injection point.

It is important to highlight here that apart from the type of interventions, the start time of each intervention is a decision variable too. Likewise, it is important to stress that rezoning is assumed to last until the repair is complete (i.e. as in utility's general practice) and, hence, its duration is not considered as a decision variable. ASV injection, on the other hand, is carried out until the tank (modelled at each injection point, see above) gets empty. This may happen before the repair is complete, depending on the water demand (under normal conditions) of the affected area.

In view of the above, each identified optimal solution takes the form of an action plan, as it was also done in Sophocleous *et al.* (2018). The Non-Dominated Sorting Genetic Algorithm II or NSGA II (Deb *et al.* 2002) is used in this paper to solve the optimisation problem. This method has already been proved to be appropriate for solving a similar optimisation problem (Mahmoud *et al.* 2018). The mathematical description of the present multi-objective problem, as well as the optimisation constraints considered in this study, can be found in the Supplementary Material.

It is stressed here that optimising for minimum end-impact and cost has multiple benefits for a utility. The most important benefit is reducing the impact on the customers which can be costly in many ways (financially but also in terms of reputation, etc.). A couple of other examples related to costs include: (1) operational savings in the long term as many events may occur each year – although the cost of a single response solution may be small (e.g. hundreds of pounds) and (2) less time spent on site for opening valves or injecting water – this could benefit utilities in terms of more efficient scheduling of the technicians' activities.

CASE STUDY

The present case study aims to illustrate the benefit of a response solution identified through interaction with the IRPT (hereafter referred to as the 'New methodology response') by comparing it to a response solution based on utilities' current practice (hereafter referred to as the 'Current practice response'). For this purpose, a semi-real case study (described hereunder) was considered. Then, the IRPT's steps are implemented for the case study's event in order to identify the 'New methodology response'. The 'New methodology response' is ultimately compared with the 'Current practice response', in order to demonstrate the benefit resulting from the operator's interaction with the IRPT.

Description of the semi-real case study

The case study used here is based on the following real system and event. On Saturday 2nd November 2019 at 14:00, a WTW that serves approximately 100,000 customers located in the North West of England shut down following observation of high turbidity levels. This event was due to a burst on a main within the WTW. The shutdown resulted in intermittent supply and low pressure to some customers. The WTW remained shut until the quality of the water leaving the WTW could be assured to meet the required standards. The utility mobilised ASVs to the area and implemented network changes (i.e. rezoning) in order to minimise customer end-impact. Bottled water was delivered directly to priority services and sensitive customers. The repair was completed 24 h after the shutdown.

In the IRPT, the shutdown is modelled by closing the pipe downstream the service reservoir directly fed by the WTW (the WTW feeds this service reservoir only), in order to facilitate the hydraulic simulations. As far as the actual utility's response actions are concerned, a number of simplifications and assumptions were made to simplify the coding required in the IRPT. For example, ASV injection at each point is carried out by using a single artificial ASV (equivalent to the 3×30 m^3 ASVs usually sent to every injection point). However, in reality ASVs supplied water intermittently at some injection points (i.e. started at different times during the event and did not inject water consecutively) and with more than three ASVs used in some cases. Additionally, the rezoning valves considered in the IRPT's simulations do not necessarily coincide with the rezoning valves actually used by the utility during the event. This is due to the fact that the hydraulic model used did not precisely reflect the real valves' layout. Finally, the actual start times of the interventions have been rounded to the next hour (e.g. if an intervention started at 19:30 in real-life, then in the IRPT it assumed to start at 20:00).

Because of all of the above, we refer to the case study under scrutiny as 'semi-real' (i.e. based on a real system and event, but with several simplifications and assumptions). Bearing in mind the typical response strategy described earlier, we refer to the response actions shown in Table 1 as the 'Current practice response' although they only approximate (i.e. in terms of total end-impact, start time of impact, affected areas, etc.) the actual real-life response. This said, it is also important to stress here that many factors may have influenced the actual response actions taken by the utility. These factors have not been accounted for in this study and, hence, the term 'Current practice response' should be construed accordingly.

New methodology response

In this section, the identification of the response solution through the IRPT's steps is presented. It is worth stressing at this point that Step 2 of the IRPT methodology is not applied

Table 1 | Semi-real case study's event timeline and 'Current practice response' interventions

Saturday Nov 2nd 2019	
14:00	WTW shut down
20:00	ASV injection in DMA 009 (one injection point)
21:00	ASV injection in DMA 004 (one injection point)
22:00	ASV injection in DMA 010 (one injection point)
Sunday Nov 3rd 2019	
00:00	Rezoning from DMA 001 to DMA 005 (five rezoning valves)
00:00	Rezoning from DMA 005 to DMA 006 (five rezoning valves)
01:00	ASV injection in DMA 007 (two injection points)
09:00	ASV injection in DMA 003 (one injection point)
09:00	Rezoning from DMA 003 to DMA 008 (two rezoning valves)
14:00	Repair is completed

here because the event is considered to be the shutdown (i.e. not the burst). Hence, in this case, operators do not need to ask the support of the IRPT for the identification of the best isolation start time and of the best isolation valves to close.

Initial impact assessment (methodology Step 1)

The first step of the methodology is to apply the initial impact assessment. Here, operators aim at assessing the initial end-impact over the impact horizon (i.e. until the repair is completed). Therefore, for the purposes of this work, they input into the IRPT the repair completion time as 24 h. This is because, despite the fact that the completion time can only roughly be estimated before the actual repair commences, 24 h is considered to be a reasonable period over which the repair of a major burst is likely to be carried out. Figure 2(a) shows the location of the considered service

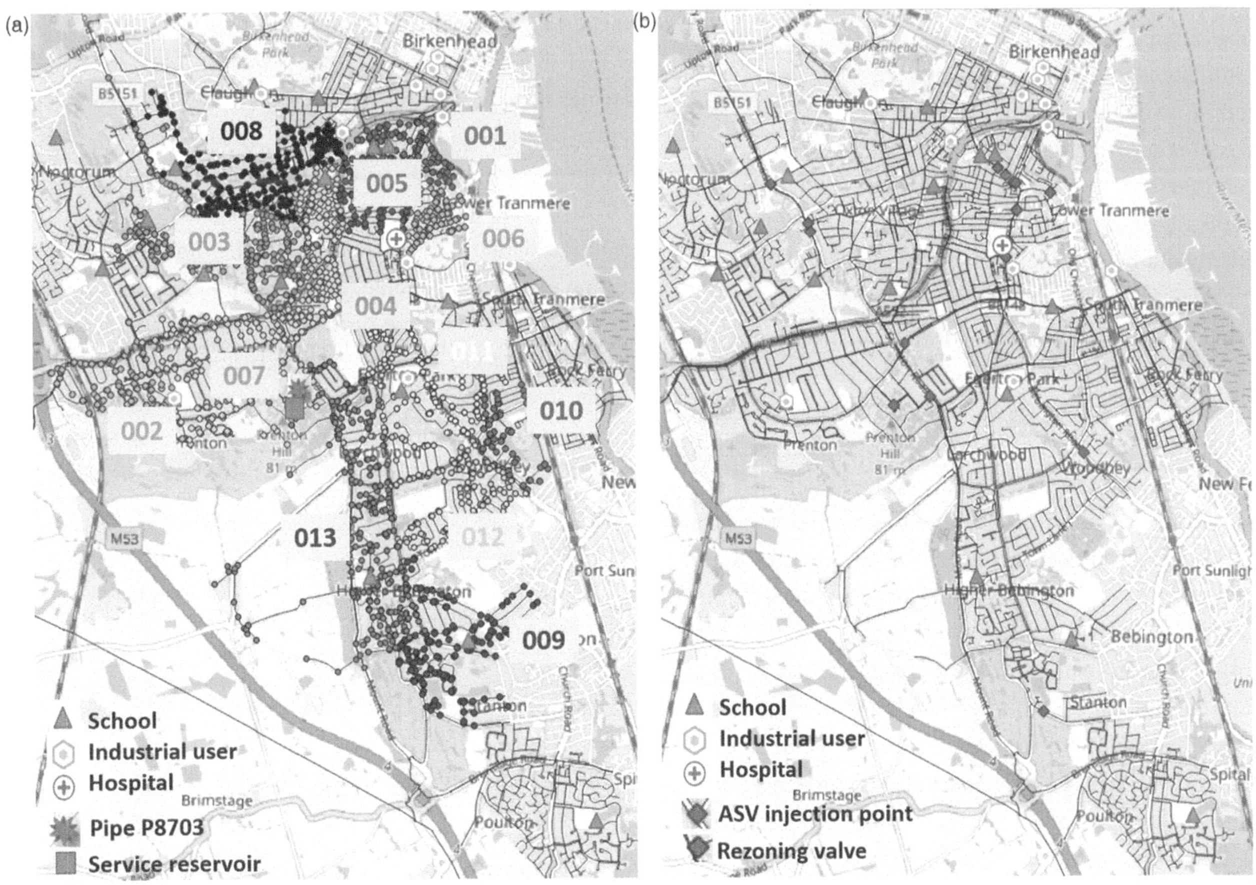

Figure 2 | (a) Location of the considered service reservoir (fed by the WTW), simulated closed pipe P8703, schools, industrial users, hospital and DMAs; and (b) location of the selected available interventions (i.e. rezoning valves and ASV injection points).

reservoir and the downstream pipe that was closed (i.e. pipe P8703) for modelling the shutdown, as well as the location of the industrial users, schools, hospital and the network model's DMAs (each DMA represented with different colouration).

The values of the four impact indicators, for the initial condition of the system (denoted hereafter as 'No response'), are calculated in the IRPT. Considering that the total number of customers registered in the utility is equal to 3,293,080 (value obtained by the utility), P_{min} is equal to 3 m and P_{req} is equal to 15 m (as applied in water utilities' practice), then CML is equal to 4.0 mins/cust, AMLP is equal to 3.6 mins/cust, UW is equal to 3,330 m^3 and DRI is equal to 14 pipes (out of the 8,950 pipes in this section of network). This results in a total initial end-impact as equal to 11.1%. The location of the affected customers with supply interruption (SI) for more than 3 h is shown in Figure 5(a) (with purple-gradient colouration), in sets of 3 h (i.e. 3–6, 6–9, 9–12, >12 h). The above impact values (computed over the 24-h impact horizon) highlight the significance of this event that affected a wide area comprising different DMAs. The affected area also includes two schools and one industrial node and the hospital (all purple-gradient coloured depending on the SI duration). However, the risk of discolouration (or DRI) could be considered low (i.e. only 14 pipes are at high risk).

Using the IRPT, operators are also able to check the pressure over the impact horizon and, hence, get a view of when the aforementioned affected customers start getting end-impact. At the bottom of Figure 5(a), the pressure graph for DMA 005 (selected as an example here because the hospital is located in that DMA) is presented. It can be noticed that DMA 005 (and, hence, the hospital too) starts being affected approximately 5 h after the shutdown, if nothing is done. The IRPT also provides the capability to visualise the other aspects of the impact, such as the low pressure duration at each node, the volume of undelivered water at each node and the DRI at each pipe, as well as for different DMAs, in a similar way as shown in Figure 5(a). All this is a significant advantage over what done/available as part of current practices. This step takes approximately 3 mins to be completed on the PC used in this study (Intel processor, Core i5-6200U CPU at 2.30 GHz and 64-bit Windows 7), i.e. as long as a single impact evaluation takes for the present complex network.

Manually proposed solution (methodology Step 3)

For the purposes of this work, a fictional operator proposes a realistic (i.e. that could have potentially be identified in the utility's control room) manual solution (denoted hereafter as 'New response – manual') in Step 3 of the methodology, after having carried out the initial end-impact assessment. The available interventions are shown in Figure 2(b). Looking at the initial end-impact (in Figure 5(a)), the fictional operator decides to inject water into the affected DMAs 003, 004, 010 and 007 and rezone-affected DMAs 005, 006 and 008 by opening all the available rezoning valves. This is because the fictional operator wants to intervene into all affected DMAs where available interventions exist. Then, looking at the pressure graphs of the DMAs where ASV injection is available (i.e. DMAs 003, 004, 010 and 007), presented in Figure 3 for the 'No response' scenario, he/she decides to start injecting into these DMAs 5 h after the shutdown (i.e. when the impact starts in the horizon). This is because he/she wants to allow plenty of time to mobilise the ASVs and also allow injection to start at 19:00 when a peak in demand is expected. He/she finally decides to rezone as soon as possible (here assuming 2 h after the shutdown to allow plenty of time for technicians to get to site), because rezoning for longer periods is expected to significantly reduce end-impact without increasing cost (i.e. rezoning duration does not affect cost, see Equation (7)). Bearing in mind the above, it is worth stressing that the

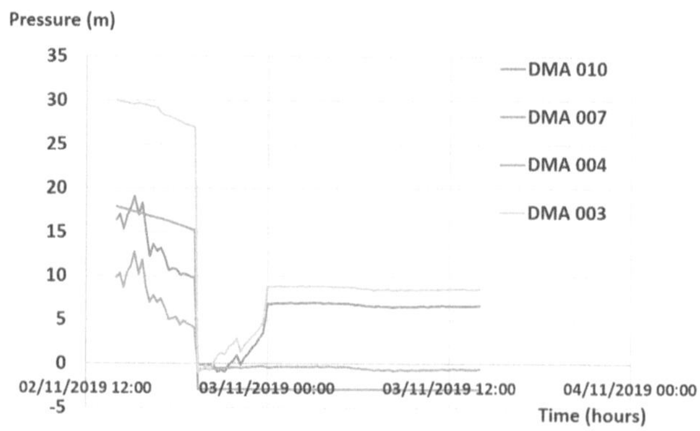

Figure 3 | Pressure vs. time of different DMAs for the 'No response' case.

IRPT supports operators with modelling different 'what-if' scenarios in an easy way. For example, the same fictional operator could have also tried opening fewer valves to rezone those DMAs just by modifying the information he/she did input in one field of the IRPT's graphical user interface.

Assuming that c_{rez} is equal to £27, c_{ASV} is equal to £32 (both provided by the utility to make this manual solution more realistic) and $d_{rez} = 2$ h (i.e. 1 h for technicians to open one rezoning valve and another 1 h to close it), then CML is equal to 1.2 mins/cust, AMLP equal to 1.7 mins/cust, UW equal to 1,235 m^3 and DRI is equal to 316 pipes in the network. The total end-impact is equal to 4.5% and the cost is equal to £813. The location of the affected customers with SI of more than 3 h and the pressure graph of DMA 005 after applying the 'New response – manual' solution are shown in Figure 5(b). This step takes approximately 3 mins to be completed on the PC used in this study, i.e. as long as a single impact/cost evaluation of one manual solution takes for the present complex network.

Optimal solution (methodology Step 4)

After having assessed the end-impact/cost of their manually proposed solution in Step 3, the fictional operator asks the support of the optimisation in Step 4. Details of the assumptions/simplifications applied in the optimisation carried out here are as follows: (1) each rezoning valve can open at different times in the horizon in the range of 1 h (i.e. the earliest time rezoning can usually start) and 23 h (i.e. the latest time rezoning can start in the present 24-h hydraulic simulation) after the shutdown, (2) injection at each ASV point can start at a different time in the range of 1 h (i.e. the earliest time injection can usually start) and 23 h (i.e. the latest time rezoning can start in the present 24-h hydraulic simulation) after the shutdown (the time ranges in points 1 and 2 can be easily modified by the operators) and (3) the operators decide to give priority to reducing CML and UW as in utilities' practice these indicators are usually considered important. Despite its importance, it was chosen not to give higher priority to DRI in this optimisation problem, because minimisation of the second objective (i.e. cost) is expected to substantially reduce the risk of discolouration (due to minimisation of the number of valve manipulations and ASV injections). Based on the above, the following indicator weights are used (see Equation (5)): $w_{CML} = 0.4$, $w_{AMLP} = 0.15$, $w_{UW} = 0.3$ and $w_{DRI} = 0.15$. These are indicative weight factors selected here to illustrate the desired priority, but can be easily changed in the IRPT by the decision-maker.

For the calculation of the total end-impact, the maximum values of the impact indicators (i.e. for normalisation, see Equation (6)) are calculated as follows. For the CML and AMLP, the total number of customers in the section of network under scrutiny is equal to 46,545, and the simulation duration is equal to 24 h = 1,440 mins. The maximum value of UW is equal to the total volume of water required to supply the whole section of network under normal operation (equal to 175,530 m^3). The maximum value of DRI is equal to the total number of pipes in the section of network (equal to 8,750 pipes). Hence, $CML_{max} = 20.35$ mins/cust, $AMLP_{max} = 20.35$ mins/cust, $UW_{max} = 175{,}530$ m^3 and DRI = 8,750 pipes. The minimum values of all impact indicators are equal to 0.

In the present optimisation problem, the number of decision variables is equal to 36 (i.e. 18 time variables and 18 valve status manipulation variables). The population size was set to 15 and the number of maximum generations to 100 (although 45 generations were required for the identification of the final Pareto front). In Figure 4, the NSGA II Pareto front of optimal solutions (i.e. black colour), the 'New response – manual' solution (blue colour) and the 'No response' case (i.e. yellow colour) are presented. The present optimisation problem was completed in a couple

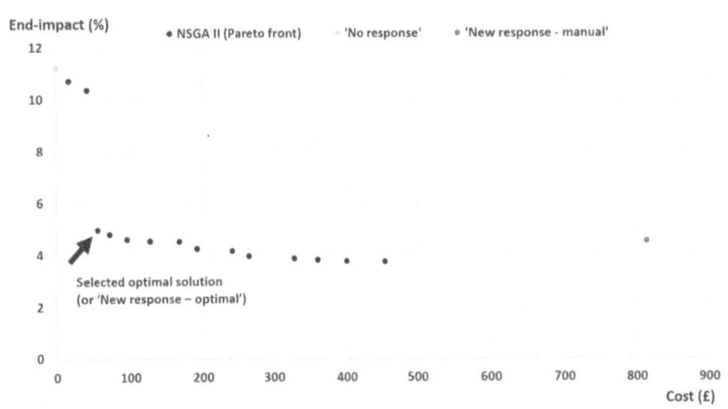

Figure 4 | End-impact (%) vs. cost (£) of optimal (NSGA II) solutions, the 'New response – manual' and the 'No response'.

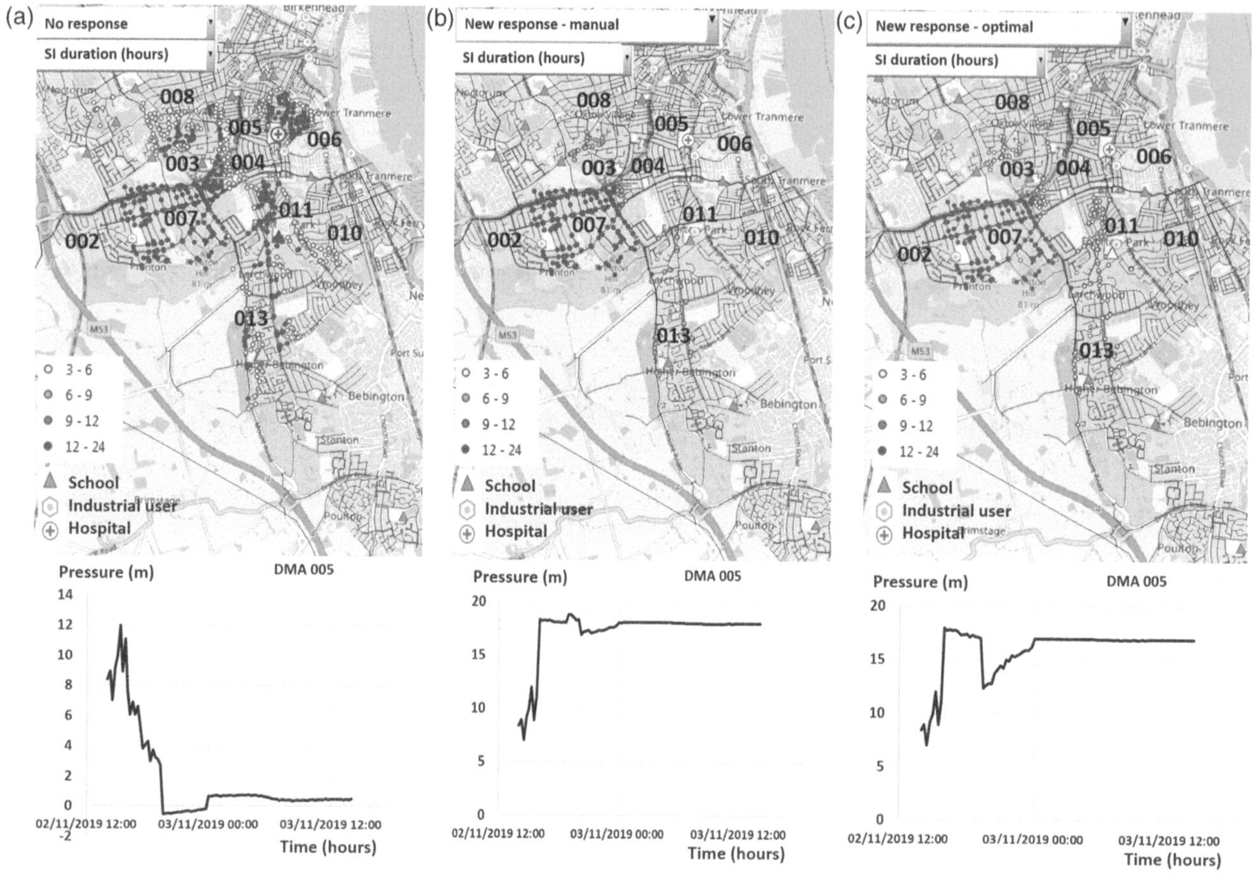

Figure 5 | Customers affected with SI and pressure graph for the (a) 'No response'; (b) 'New response – manual' and (c) 'New response – optimal'.

of days, i.e. not in real time. This is considered as a drawback of the present methodology, which is addressed here by developing a heuristic-based optimisation methodology (i.e. this will be presented in future work as the focus of this paper is the new response methodology).

Identification of the final response plan (methodology Step 5)

After the optimisation (Step 4 of the methodology) is completed, in Step 5, the fictional operator decides to compare the identified optimal solutions with the 'New response – manual' solution in order to identify the best response plan. Here, for illustration reasons, he/she selects one solution from the Pareto front in order to compare it with the 'New response – manual' solution. The selected optimal solution (denoted hereafter as 'New response – optimal', pointed with a black arrow in Figure 4) is a solution with significantly reduced end-impact for a small cost increase compared with the rest of the optimal solutions with less cost and bigger impact (i.e. solutions found at the left side of the selected one) on the Pareto front. Such a solution is quite likely to be selected by a decision-maker.

The values of the impact indicators of the 'New response – optimal' solution (as well as those of the 'New response – manual' and 'No response') are shown in Table 2. As can be seen from this table, the values of CML and UW are significantly reduced in the 'New response – optimal' compared with the 'No response' case. This implies that the considered weight factors were effective in this problem. It is also observed that the 'New response – manual' solution obtains smaller impact values for the CML, AMLP and UW compared with the 'New response – optimal' solution. However, cost and DRI are significantly reduced in the 'New

Table 2 | Total end-impact and cost of 'No response', 'Current practice response', 'New response – manual' and 'New response – optimal' (or 'New methodology response')

	CML (mins/cust)	AMLP (mins/cust)	UW (m^3)	DRI (–)	Cost (£)	Total end-impact (%)
No response	4	3.6	3,330	14	0	11.1
Current practice response	2.1	2	1,825	273	894	6.5
New response – manual	1.2	1.7	1,235	316	813	4.5
New response – optimal (or New methodology response)	1.6	2	1,475	92	55	5

response – optimal' solution compared with the 'New response – manual' solution due to the optimisation enforcing cost minimisation. As expected, minimisation of cost function has also reduced the number of rezoning valves to open and the injection time, resulting in reduced disturbance in pipe flows, and consequently reduced risk of discolouration.

The 'New response – optimal' solution also suggests only one intervention, i.e. opening one rezoning valve which feeds DMA 005, starting 2 h after shutdown. No injection from ASVs is suggested, which explains the minimised cost (i.e. £55) of this solution. The significantly reduced total end-impact of the 'New response – optimal' solution (i.e. 5%), compared with the 11.1% of 'No response', is a consequence of starting the rezoning very early in the simulation (although only one valve is opened). It is also observed that the total end-impact of the 'New response – manual' solution (i.e. 4.5%) is not significantly lower than the total end-impact of the 'New response – optimal' solution (i.e. 5%). However, the cost of the 'New response – optimal' solution (i.e. £55) is much lower than the cost of the 'New response – manual' solution (i.e. £813).

Furthermore, Figure 5(c) shows that in the 'New response – optimal' solution, the number of affected customers with SI has been reduced when compared with the 'No response'. However, in the 'New response – manual' solution, the affected area is smaller than the affected area in the 'New response – optimal' solution (e.g. DMA 011 is not affected with CML when applying the 'New response – manual' solution, but there is CML impact when applying the 'New response – optimal' solution). In both solutions, the hospital is not affected anymore (see also the pressure graphs in Figure 5(b) and 5(c)) when compared with the 'No response' case. If applying the 'New response – manual' solution, the initially (i.e. 'No response') affected schools (in DMAs 011 and 013) and the industrial node (in DMA 011) are not affected anymore. However, in the 'New response – optimal' solution, the school in DMA 011 is still affected, although for fewer hours (i.e. between 3 and 6 h) compared with the 'No response' case. Through Figure 5(b) and 5(c) operators are also informed that DMA 007 still has high SI impact (i.e. almost the whole DMA is affected with SI >12 h, although the total CML is low) in both 'New response – manual' and 'New response – optimal' solutions. It is reminded at this point that the only available intervention in DMA 007 is injection from two ASV points (see Figure 2(b)). Hence, the IRPT also serves the purpose of informing the operators that they should look into more available ASV points and/or available rezoning (e.g. from adjacent unaffected DMA 002) in DMA 007.

At this point in time (i.e. in Step 5 of methodology), based on the information obtained by using the IRPT, the fictional operator has to make the following decision: (1) apply the 'New response – manual' solution due to the reduced CML, AMLP and UW when compared with the 'No response', (2) apply the 'New response – optimal' solution where CML, AMLP and UW are also reduced when compared with the 'No response' (although higher than those in the 'New response – manual' solution) but with cost and DRI impact much lower than the cost and DRI impact of the 'New response – manual' solution, (3) test/assess a different manual solution (i.e. 'what-if' scenario) and compare it with the other identified solutions or (4) select a different optimal solution from the Pareto front in Figure 4 and compare it with the other identified solutions. For the purpose of this work, the different 'what-if' scenarios and the different optimal solutions are discounted due to space limitation. Hence, based on the results obtained, it is assumed that the fictional operator is more likely to select the 'New response – optimal' solution because of the minimum DRI (and small cost), as well as relatively low values of all the other impact indicators. The 'New response – optimal' solution is therefore considered to be the 'New methodology response' in the remainder of this paper. This step can be completed in a number of minutes in a water utility control room, when a limited number of comparisons takes place. In this case study, operators carried out a single comparison between the 'New response – manual' and the 'New response – optimal'; hence, it is assumed that this step was completed in approximately 5 mins (as an example). From the above, it is shown that the whole response methodology can be implemented within 1 h, required for the near real-time response decision-making.

Comparison between 'Current practice response' and 'New methodology response'

In Table 2, CML, AMLP, UW, DRI, cost and total end-impact calculated by the IRPT for the 'Current practice response' described in Table 1 are shown. It can be noticed that even though CML, AMLP and UW are reduced when compared with the 'No response' scenario, the 'New methodology response' (i.e. the 'New response – optimal' solution) offers further improvements. Indeed, the 'New methodology response' further reduced all impact indicators (especially DRI and cost), except AMLP which remained the same. The 'New methodology response' also suggested fewer interventions to implement (i.e. opening of only one rezoning valve compared with opening of 12 valves and injecting from 5 points in the 'Current practice response'), justifying the significant improvement in DRI and cost.

In light of the above, it can be concluded that the 'New methodology response' identified through interaction with the IRPT outperforms the 'Current practice response'.

CONCLUSIONS

This paper presents a novel overall response methodology that aims to support/guide water utility operators in making better informed decisions regarding water network failures. The new response methodology considers (1) improved impact assessment methods (based on realistic metrics used in the water industry), (2) consistent impact assessment (i.e. impact metrics consistently calculated for every response solution to facilitate easy comparison), (3) provision of automatically generated advices (e.g. optimal response interventions and assessed end-impacts), (4) more realistic selection of operational interventions (based on operational costs, the availability of different types of interventions, etc.) and (5) novel near real-time (i.e. identification of a solution within 1 h after event detection/localisation) interaction with the control room operator that takes into account their expert judgement/experience (e.g. proposing 'what-if' scenarios) without hydraulic model expertise requirements. The new methodology is implemented via an interactive decision-support tool aiming to support operators in making better informed decisions.

The application of the new methodology on a semi-real case study showed that the tool enabled operators to identify a more effective response solution (i.e. reduced end-impact and cost) compared with the 'Current practice response'. This is because the tool allowed operators to compare alternative response strategies (i.e. manually created by the operator and automatically generated by the IRPT through optimisation). This comparison was facilitated by the consistent impact assessment (i.e. same metrics assessed for each solution) used in the tool, as well as by the comprehensible impact metrics (i.e. well-known metrics in utilities), impact coverage (shown in maps) and cost of different solutions (shown in graphs). Hence, this application showed the potential of the IRPT to be used by utilities to make better and more informed decisions. All the methodology's steps apart from the optimisation step can currently be conducted in near real time (i.e. to identify a solution within 1 h after event detection/localisation).

Future work on further improvement of the proposed methodology and tool will focus on: (1) the improvement of the DRI index (e.g. to take into account the flow direction into the pipe network, as well as the pipe material), (2) the improvement of ASV injection modelling (i.e. injection in one point at different times in the day should be modelled) and (3) the improvement of the optimisation methodology in order to identify optimum solutions faster (i.e. in near real time). The latter point is of crucial importance to effectively enable optimal solutions to be used in the near real-time response framework presented in this paper.

ACKNOWLEDGEMENTS

This work was funded as part of the Water Informatics Science and Engineering Centre for Doctoral Training (WISE CDT) under a grant from the Engineering and Physical Sciences Research Council (EPSRC) (grant no. EP/L016214/1), as well as from United Utilities. The present work has made use of resources, data and information provided by United Utilities.

DATA AVAILABILITY STATEMENT

All relevant data are included in the paper or its Supplementary Information.

REFERENCES

Beuken, R. H. S., van den Boomen, M., Blaauwgeers, H. G. P. & van Daal, H. A. 2008 Feasibility study on quantitative risk analysis of drinking water networks. In: *Water Distribution Systems Analysis*, Proceedings of 10th Annual Water Distribution Systems Analysis Conference WDSA2008 (K. Van Zyl, ed.). American Society of Civil Engineers, Kruger National Park, South Africa, pp. 1–11.

Bicik, J., Makropoulos, C., Joksimović, D., Kapelan, Z., Morley, M. S. & Savić, D. A. 2008 Conceptual risk-based decision support methodology for improved near real-time response to WDS failures. In: *Water Distribution Systems Analysis*, Proceedings of 10th Annual Water Distribution Systems Analysis Conference WDSA2008 (K. Van Zyl, ed.). American Society of Civil Engineers, Kruger National Park, South Africa, pp. 1–10.

Bicik, J., Kapelan, Z. & Savić, D. A. 2009 Operational perspective of the impact of failures in water distribution systems. In: *World Environmental and Water Resources Congress 2009: Great Rivers*, Proceedings of World Environmental and Water Resources Congress 2009 (S. Starrett, ed.). American Society of Civil Engineers, Kansas City, MO, USA, pp. 1–10.

Bicik, J., Kapelan, Z., Makropoulos, C. & Savić, D. A. 2011 Pipe burst diagnostics using evidence theory. *Journal of Hydroinformatics* **13** (4), 596–608.

Casillas Ponce, M. V., Garza Castañón, L. E. & Cayuela, V. P. 2014 Model-based leak detection and location in water distribution networks considering an extended-horizon analysis of pressure sensitivities. *Journal of Hydroinformatics* **16** (3), 649–670.

Deb, K., Pratap, A., Agarwal, S. & Meyarivan, T. A. M. T. 2002 A fast and elitist multiobjective genetic algorithm: NSGA-II. *IEEE Transactions on Evolutionary Computation* **6** (2), 182–197.

Giustolisi, O. & Savic, D. 2010 Identification of segments and optimal isolation valve system design in water distribution networks. *Urban Water Journal* **7** (1), 1–15.

Giustolisi, O., Kapelan, Z. & Savic, D. 2008 Algorithm for automatic detection of topological changes in water distribution networks. *Journal of Hydraulic Engineering* **134** (4), 435–446.

Jeong, H. S. & Abraham, D. M. 2009 Water rationing model for consequence minimization of water infrastructure destruction. *Journal of Water Resources Planning and Management* **135** (2), 80–89.

Jeong, H. S., Qiao, J., Abraham, D. M., Lawley, M., Richard, J. P. & Yih, Y. 2006 Minimizing the consequences of intentional attack on water infrastructure. *Computer-Aided Civil and Infrastructure Engineering* **21** (2), 79–92.

Jun, H. & Loganathan, G. V. 2007 Valve-controlled segments in water distribution systems. *Journal of Water Resources Planning and Management* **133** (2), 145–155.

Jung, D., Kang, D., Liu, J. & Lansey, K. 2015 Improving the rapidity of responses to pipe burst in water distribution systems: a comparison of statistical process control methods. *Journal of Hydroinformatics* **17** (2), 307–328.

Kao, J. J. & Li, P. H. 2007 A segment-based optimization model for water pipeline replacement. *Journal of American Water Works Association* **99** (7), 83–95.

Kapelan, Z., Giustolisi, O. & Savić, D. A. 2006 Risk assessment of water supply interruptions due to mechanical pipe failures. In *Proceedings of the Combined International Conference of Computing and Control for the Water Industry, CCWI2007 and Sustainable Urban Water Management, SUWM2007*, 2003, pp. 1–5.

Kapelan, Z., Savic, D. & Mahmoud, H. 2017 A response methodology for reducing impacts of failure events in water distribution networks. *Procedia Engineering* **186**, 218–227.

Laucelli, D., Romano, M., Savić, D. & Giustolisi, O. 2016 Detecting anomalies in water distribution networks using EPR modelling paradigm. *Journal of Hydroinformatics* **18** (3), 409–427.

Mahmoud, H. A., Kapelan, Z. & Savić, D. 2018 Real-time operational response methodology for reducing failure impacts in water distribution systems. *Journal of Water Resources Planning and Management* **144** (7), 04018029.

Nayak, M. A. & Turnquist, M. A. 2016 Optimal recovery from disruptions in water distribution networks. *Computer-Aided Civil and Infrastructure Engineering* **31** (8), 566–579.

Okeya, I., Hutton, C. & Kapelan, Z. 2015 Locating Pipe Bursts in A District Metered Area via Online Hydraulic Modelling. *Procedia Engineering* **119**, 101–110.

Paez, D., Suribabu, C. R. & Filion, Y. 2018 Method for extended period simulation of water distribution networks with pressure driven demands. *Water Resources Management* **32** (8), 2837–2846.

Paez, D., Filion, Y., Castro-Gama, Quintiliani, C., Santopietro, S., Butler, D., Farmani, R., Fu, G., Meng, F., Sweetapple, C., Diao, K., Huang, Y., Ulanicki, B., Zhang, Q., Zheng, F., Abraham, E., Deuerlein, J., Gilbert, D., Piller, O., Balut, A., Brodziak, R., Bylka, J., Zakrzewski, P., Cai, J., Chenhao, O., Jinliang, G., Shiyuan, H., Yuanzhe, L., Mahmoud, H. A., Nikoloudi, E., Romano, M., Sophocleous, S., Woodward, K., Creaco, E., Di Nardo, A., Di Natale, M., Santonastaso, G. F., Bibok, A., Aguilar, A., Cuero, P., Gonzalez, S., Muñoz, S., Pérez, J., Posada, A., Robles, J., Salcedo, C., Vargas, K., Franchini, M., Galelli, S., Hoon Kim, J., Iglesias-Rey, P., Kapelan, Z., Saldarriaga, J., Savic, D. & Walski, T. 2020 The battle of post-disaster response and restoration (BPDRR). *Journal of Water Resources Planning and Management (ASCE)* **146** (8), 04020067.

Qi, Z., Zheng, F., Guo, D., Zhang, T., Shao, Y., Yu, T. & Maier, H. R. 2018 A comprehensive framework to evaluate hydraulic and water quality impacts of pipe breaks on water distribution systems. *Water Resources Research* **54** (10), 8174–8195.

Romano, M. 2019 Review of Techniques for Optimal Placement of Pressure and Flow Sensors for Leak/Burst Detection and Localisation in Water Distribution Systems. *ICT for Smart Water Systems: Measurements and Data Science. The Handbook of Environmental Chemistry, vol 102* (A. Scozzari, S. Mounce, D. Han, F. Soldovieri & D. Solomatine, eds). Springer, Cham, Switzerland.

Romano, M., Kapelan, Z. & Savić, D. A. 2011 Burst detection and location in water distribution systems. In: *World Environmental and Water Resources Congress 2011: Bearing Knowledge for Sustainability*, Proceedings of World Environmental and Water Resources Congress 2011 (R. E. Beighley II & M. W. Killgore eds), pp. 1–10. American Society of Civil Engineers, Palm Springs, CA, USA.

Romano, M., Kapelan, Z. & Savić, D. A. 2014 Automated detection of pipe bursts and other events in water distribution systems. *Journal of Water Resources Planning and Management* **140** (4), 457–467.

Rossman, L. A. 2000 *EPANET 2: Users Manual.* USEPA, Cincinnati, OH, USA.

Sophocleous, S., Nikoloudi, E., Mahmoud, H., Woodward, K. & Romano, M. 2018 Simulation-based framework for the restoration of earthquake-damaged water distribution networks using a genetic algorithm. In *WDSA/CCWI Joint Conference Proceedings (Vol. 1).*

Turner, J. P., Qiao, J., Lawley, M., Richard, J. P. & Abraham, D. M. 2012 Mitigating shortage and distribution costs in damaged water networks. *Socio-Economic Planning Sciences* **46** (4), 315–326.

Vamvakeridou-Lyroudia, L. S., Bicik, J., Morley, M., Savic, D. & Kapelan, Z. 2010 A real-time intervention management model for reducing impacts due to pipe isolation in water distribution systems. In: *Water Distribution Systems Analysis 2010*, Proceedings of 12th Annual Water Distribution Systems Analysis Conference WDSA2010 (K. E. Lansey, C. Y. Choi, A. Ostfeld & I. L. Pepper, eds). American Society of Civil Engineers, Tucson, AZ, USA, pp. 209–221.

Zhang, Q., Zheng, F., Chen, Q., Kapelan, Z., Diao, K., Zhang, K. & Huang, Y. 2020 Improving the resilience of postdisaster water distribution systems using dynamic optimization framework. *Journal of Water Resources Planning and Management* **146** (2), 04019075.

Zhou, X., Tang, Z., Xu, W., Meng, F., Chu, X., Xin, K. & Fu, G. 2019 Deep learning identifies accurate burst locations in water distribution networks. *Water Research* **166**, 115058.

First received 12 June 2020; accepted in revised form 16 November 2020. Available online 2 December 2020

 © 2021 The Authors

Control theory-based data assimilation for hydraulic models as a decision support tool for hydropower systems: sequential, multi-metric tuning of the controllers

Miloš Milašinović, Dušan Prodanović, Budo Zindović, Boban Stojanović and Nikola Milivojević

ABSTRACT

Increasing renewable energy usage puts extra pressure on decision-making in river hydropower systems. Decision support tools are used for near-future forecasting of the water available. Model-driven forecasting used for river state estimation often provides bad results due to numerous uncertainties. False inflows and poor initialization are some of the uncertainty sources. To overcome this, standard data assimilation (DA) techniques (e.g., ensemble Kalman filter) are used, which are not always applicable in real systems. This paper presents further insight into the novel, tailor-made model update algorithm based on control theory. According to water-level measurements over the system, the model is controlled and continuously updated using proportional–integrative–derivative (PID) controller(s). Implementation of the PID controllers requires the controllers' parameters estimation (tuning). This research deals with this task by presenting sequential, multi-metric procedure, applicable for controllers' initial tuning. The proposed tuning method is tested on the Iron Gate hydropower system in Serbia, showing satisfying results.

Key words | hydraulic model update, model-driven forecasting, near future forecasting, PID controller, PID controllers' tuning

Miloš Milašinović (corresponding author)
Dušan Prodanović
Budo Zindović
Faculty of Civil Engineering, Department for Hydraulic and Environmental Engineering,
University of Belgrade,
Belgrade,
Serbia
E-mail: *mmilasinovic@grf.bg.ac.rs*

Boban Stojanović
Faculty of Science, Department of Mathematics and Informatics,
University of Kragujevac,
Kragujevac,
Serbia

Nikola Milivojević
Water Institute Jaroslav Černi,
Belgrade,
Serbia

HIGHLIGHTS

- Uncertainty of the boundary and initial conditions affects model-driven forecasting.
- Data Assimilation is used to overcome these problems.
- Research presents potential of using novel, tailor-made, PID controllers-based data assimilation method for river hydraulic models update.
- Method could be used as a decision-support tool for hydropower systems control.
- Sequential, multi-metric tuning procedure has been introduced.

INTRODUCTION

The global tendency for reduction of greenhouse gas emissions switches focus from fossil to renewable energy sources. This tendency puts hydropower systems, as the most important manageable renewable source, in the first place, considering its availability and cost-effectiveness (Eurelectric 2011). Hydropower production depends on numerous dynamically changing factors, like extreme hydrological events (floods and droughts), environmental constraints, price fluctuation on energy markets, and

This is an Open Access article distributed under the terms of the Creative Commons Attribution Licence (CC BY-NC-ND 4.0), which permits copying and redistribution for non-commercial purposes with no derivatives, provided the original work is properly cited (http://creativecommons.org/licenses/by-nc-nd/4.0/).

doi: 10.2166/hydro.2021.078

conflicting goals of the stakeholders in charge. To use the water optimally, engineers and operators in charge of hydropower systems require different decision support tools (Ehteram *et al.* 2018). In many cases, these tools must provide forecasts of the hydrological data (water levels and river flows) on a daily basis. Hydrological forecasting is a challenging task, addressed by numerous researchers (e.g., Wu & Chau 2013; Taormina & Chau 2015; Fu *et al.* 2020). For the purpose of forecasting, monitoring systems and physically based models are used in model-driven forecasting procedures.

The quality of model-driven forecast directly depends on boundary and initial conditions. However, they are often affected by numerous sources of uncertainty (Bozzi *et al.* 2015; Ocio *et al.* 2017). These problems eventually lead to poor forecast results and inadequate hydropower system control operations. To overcome these problems, data assimilation (DA) methods are widely used for improving hydrological-hydraulic forecasting (Vrugt *et al.* 2006).

Various DA methods are applied in hydrological-hydraulic modeling. One of the widely used methods is ensemble Kalman filter (EnKF) (Evensen 1994, 2003). This method is often used for improved river flood forecasting (Madsen *et al.* 2003; Neal *et al.* 2007; Li *et al.* 2014; Barthélémy *et al.* 2017) but also for forecasting flows and overflows in urban drainage systems (Lund *et al.* 2019). In many hydrological-hydraulic applications of DA, measured water levels from monitoring systems are used to update the model (Romanowicz *et al.* 2006; Hostache *et al.* 2010; Jean-Baptiste *et al.* 2011; Rakovec *et al.* 2012). Some authors have successfully used streamflow observations too (Thirel *et al.* 2010; Dumedah & Coulibaly 2014; Randrianasolo *et al.* 2014; Sun *et al.* 2015). In more recent researches, there has been a tendency to use remotely sensed hydrological data, such as satellite images (Matgen *et al.* 2010; Giustarini *et al.* 2011; Yoon *et al.* 2012; García-Pintado *et al.* 2013; Andreadis & Schumann 2014; Garambois *et al.* 2020) or even crowdsourced data (Mazzoleni *et al.* 2017, 2018; Annis & Nardi 2019).

All these studies show high applicability of the standard DA methods, such as EnKF, in hydrological-hydraulic modeling. On the other hand, in most real-world applications, EnKF, and similar DA methods, struggle to perform within a reasonable time frame (Madsen & Skotner 2005), which makes them not favored by engineers and operators in charge of water systems. Additionally, mathematical apparatus, which is relatively complex, means these methods avoided in real-world applications. Owing to the necessity for improved water resources management and control, many decision support tools are required, especially for improved short-term forecasting. This enables improved (near) real-time operations in various water resources (Schwanenberg *et al.* 2015; Ahmad & Hossain 2019). This also makes DA an important decision support tool. Therefore, problems with real-world applications of standard DA methods are under the spotlight, creating a challenging task. To overcome these problems, researchers are searching for simplified and fast DA methods. Many of them are trying to develop tailor-made assimilation techniques, suitable for solving some specific problems. Madsen & Skotner (2005) developed a cost-effective assimilation method for river models. This method uses the simplified version of Kalman filtering, which certainly provides reduction of computational time, but requires the predefinition of the parameters used in a pre-processing phase. All these applications of ensemble-based DA methods or its derivatives use the direct approach for model updating. In hydraulic models, this means that water levels (stages) are directly corrected according to measured values. Sometimes, this procedure can cause 'shocks' in the model, disrupting the continuity equation (mass balance). Therefore, some researchers are trying to use a different and indirect model updating procedure. For example, Fava *et al.* (2020) presented a fast DA methodology for improved flood prediction by hydrological models in urbanized areas, using the rainfall input corrections instead of direct water level update. The methodology for indirect water level update in urban drainage systems, presented by Hansen *et al.* (2014), uses an approach of adding/subtracting correction flow to the system in the junctions where observations are available. This approach, with all details in the Methodology, has narrowed the scope of its application.

The general approach of the indirect model update procedure (Hansen *et al.* 2014) was the basis for the introduction of DA methodology for (near) real-time river model update where correction flows are calculated using control theory (Rosić *et al.* 2017a, 2017b; Milašinović *et al.* 2019). Development of this novel, control theory-based DA method and EnKF benchmark test was presented and verified by Milašinović *et al.* (2020). The study has proved the applicability of proportional–integrative–derivative (PID) controllers as a DA tool. Due to its potential as a time-efficient method, compared to the widely used EnKF method, it can be used as decision support tool for near real-time control of river hydropower systems. However, the study underlined the need of further investigation in the field of controllers' tuning.

Assigning the values to the controllers' parameters (tuning of the controllers) is one of the most important steps in any application of the PID controllers. Previous research used trial-and-error tuning rather than any systematical tuning procedure. For further and more general applications of this novel DA method, proper controllers' tuning procedure has to be proposed. Therefore, this research addresses tuning of the controllers by presenting sequential and heuristic multi-metric tuning procedure. The goal is not to determine a global optimal solution on the controllers' parameters, but to present a simple, engineering procedure to estimate the initial values and possible range of the controllers' parameters preserving the model's stability. The presented procedure shows satisfying results, and thus can be considered for initial tuning of the controllers.

MATERIALS AND METHODS

Overview of the methodology

Standard DA procedure (e.g., EnKF) corrects the state variable (e.g., water level) based on observation data, a model's sensitivity on state variables and compared model uncertainty and observation uncertainty. When model uncertainty is bigger than observation uncertainty, DA procedure puts more trust in observations, correcting the model output towards observed data, and vice versa. This approach is derived using the assumption that there is no dominant source of uncertainty. On the other hand, in many engineering problems, some simplified approaches can, or must be used.

Control theory-based DA is developed using the assumption that system state observations (water levels) are with much lower uncertainty than models and can be used for direct, deterministic model update (Hansen *et al.* 2014). Therefore, comparison of the model and observation uncertainty is skipped, which allows significant speed-up of the assimilation process (Milašinović *et al.* 2020). The second used assumption is that dominant sources of the model uncertainty are boundary and initial conditions, while model parameters are properly determined during model calibration. In most cases, inflow data are not directly measured, have high uncertainty, especially during flood events, and are with poor time resolution. Additionally, there is lack of the input data (e.g., rainfall) for large water systems due to inability to implement monitoring stations, with good spatial resolution, across an entire catchment area. Therefore, initial and boundary conditions are inadequate and could be blamed for discrepancy between calculated and observed water levels. Eventually, these problems lead to a bad estimation of reservoir state and hydropower system operation rules.

Considering these two assumptions, a novel control theory-based DA with simplified model update procedure is developed. This DA procedure treats the problem (water level discrepancy) as the lack or excess of water amount in the model. Hence, the model update procedure is operated by adding or subtracting the water from the model at the locations where observations are available. This reduces the water level discrepancies, improving the initial conditions for future forecasting. The process of adding/subtracting water from the model is controlled by PID controllers (hence it is named control theory-based DA). Figure 1 represents the general model update procedure in control theory-based DA.

PID controllers as data assimilation tool for hydraulic model

Model-driven forecasting requires physically based models. In this paper, diffusion wave hydraulic model is used, as proposed by Milašinović *et al.* (2019, 2020). This model is based on one-dimensional (1D) Saint-Venant's Equations (1) and (2). Diffusion wave model for 1D open channels (*DiffW1D*) uses a simplified dynamic equation (Equations (2)) derived from the full Saint-Venant's equations by neglecting convective acceleration

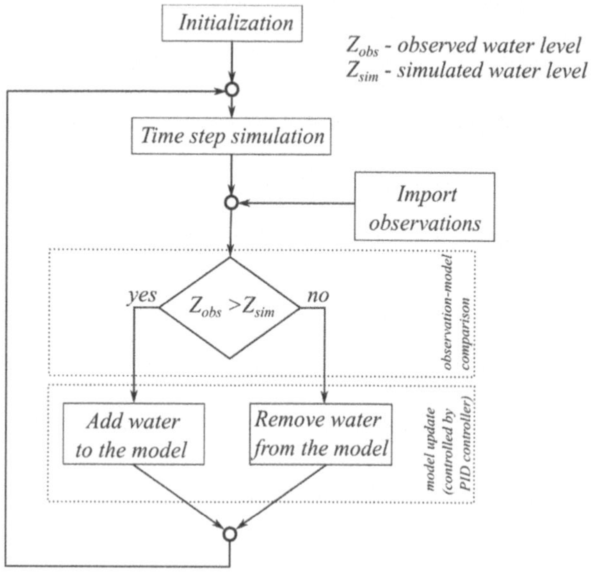

Figure 1 | Algorithm for control theory-based DA procedure.

(second term in Equation (2)):

$$\frac{\partial A}{\partial t} + \frac{\partial Q}{\partial x} = q \quad (1)$$

$$\frac{\partial Q}{\partial t} + \frac{\partial}{\partial x}\left(\beta \frac{Q^2}{A}\right) + gA\frac{\partial Z}{\partial x} + gn^2 \frac{Q|Q|}{A^2 \cdot R^{4/3}} = 0 \quad (2)$$

Here, x is space coordinate, t time, A cross-sectional area, Q discharge, q lateral inflow, Z local water surface elevation (water level), g acceleration due to gravity, R hydraulic radius calculated as the ratio between cross-sectional area and wetted perimeter, β velocity distribution coefficient, and n is Manning's roughness, calculated according to Costabile & Macchione (2012). Model domain discretization is presented in Figure 2(a). Numerical *DiffW1D* model uses a staggered numerical scheme where water levels and flows are calculated in alternating cross sections, as presented in Figure 2(b). Numerical model of the diffusion wave is given by Equations (3) and (4):

$$Z_i^{t+\Delta t} = Z_i^t - \frac{\Delta t}{B_i^t} \cdot \frac{Q_{i+1}^t - Q_{i-1}^t}{2\Delta x} \quad (3)$$

$$Q_{i+1}^{t+\Delta t} = \frac{\dfrac{Q_{i+1}^t}{g\left(\dfrac{A_{i+2}^{t+\Delta t} + A_i^{t+\Delta t}}{2}\right)\Delta t} - \dfrac{Z_{i+2}^{t+\Delta t} - Z_i^{t+\Delta t}}{2\Delta x}}{\dfrac{1}{g\left(\dfrac{A_{i+2}^{t+\Delta t} + A_i^{t+\Delta t}}{2}\right)\Delta t} + \dfrac{n^2}{\left(\dfrac{A_{i+2}^{t+\Delta t} + A_i^{t+\Delta t}}{2}\right)^2 \left(\dfrac{R_{i+2}^{t+\Delta t} + R_i^{t+\Delta t}}{2}\right)^{4/3}} \cdot |Q_{i+1}^t|} \quad (4)$$

where index i represents spatial location (i-th cross section), B is the top width of the cross section (water surface width), t in superscript denotes current time, Δx spatial resolution, and Δt is temporal resolution.

Procedure for adding/subtracting water to/from the model is based on adding/subtracting correction flows, in a form of lateral inflow, at the location where observations are available (assimilation point). Correction flows are calculated based on the difference between observed water levels and simulated water levels (error). Conversion of the errors to correction flows is conducted using the PID controllers. Correction flows are implemented as (fictive) lateral inflow elements in the continuity equation (Equation (3)) of the hydraulic model. A new form of continuity

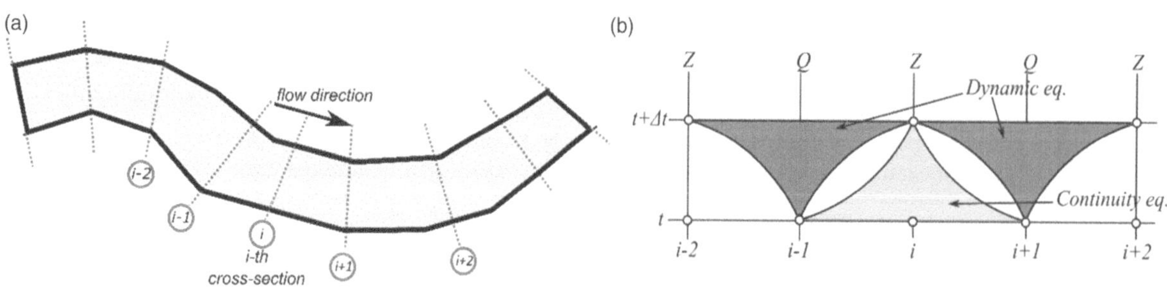

Figure 2 | (a) Model domain discretization and (b) numerical scheme using *DiffW1D* model (Milašinović et al. 2020).

equation is presented by Equation (5).

$$Z_i^{t+\Delta t} = Z_i^t - \frac{\Delta t}{B_i^t} \cdot \frac{Q_{i+1}^t - Q_{i-1}^t}{2\Delta x} + \frac{\Delta t}{B_i^t} \frac{Q_{corr}^t}{2\Delta x} \tag{5}$$

where Q_{corr} represents correction flow that reduces the water level discrepancy between observations and model. It is calculated using the PID controller's theory.

PID controller is a control loop feedback mechanism, where input in the next step is a function of the previous output (Astrom & Hagglund 1995). This mechanism is often used for real-time-control (RTC) of different process (e.g., RTC of hydraulic structures in urban drainage systems (Schütze *et al.* 2004), RTC of irrigation canals (Bonet *et al.* 2019), or RTC in water resources (Schwanenberg *et al.* 2015)). In this paper, the mechanism is used for on the fly hydraulic model update.

The PID controller's input is named as *error e(t)*, which is calculated as the difference between current value of the *process variable* (simulated water level – Z_{model}) and the *set-point* of the variable (observed water level – Z_{obs}), as stated in Equation (6). In many real-world problems the simulation time step Δt is much shorter than the observation time step Δt_{obs}. Therefore, in the period between two existing observations, the 'observed' water level Z^*_{obs} is calculated using linear interpolation. It can be assumed that the accuracy of interpolated 'observed' level decreases as the time interval from the last observation increases. To reduce the influence of the interpolated water levels Z^*_{obs}, multiplier C, named as attenuation factor (Figure 3), is used. In Equation (7), t_{prev_obs} is time of the previously available observation, t is current simulation time, t_{obs} is observation time, and Δt_{obs} is observation time step. Attenuation factor in Equation (7), will gradually turn off the PID controller (Figure 3) in periods between two measurements. This means that, as the model progress forward in time, in periods without measurements, smaller weight is given to the errors calculated using interpolated water levels.

The PID controller tends to reduce *error* using the *control variable*. The control variable used to reduce this error is lateral inflow $Q_{corr}(t)$ presented by Equation (8).

$$e(t) = [Z^*_{obs}(t) - Z_{model}(t)] \cdot C \tag{6}$$

$$C = \begin{cases} 1 & t = t_{obs} \\ \dfrac{\Delta t_{obs}}{t + \Delta t_{obs} - t_{prev\_obs}} & t \neq t_{obs} \end{cases} \tag{7}$$

$$Q_{corr}(t) = Q_{corr}^t = K_p \cdot e(t) + K_i \cdot \int_{t_0}^{t} e(t)dt + K_d \frac{de}{dt} \tag{8}$$

PID parameters are: K_p – proportional gain factor used to multiply the current error value, K_i – integrative gain factor used to add the influence of error history, and K_d – derivative gain factor used to adapt control to current trends in error change. Proportional gain produces an output based only on the current value of the error. High

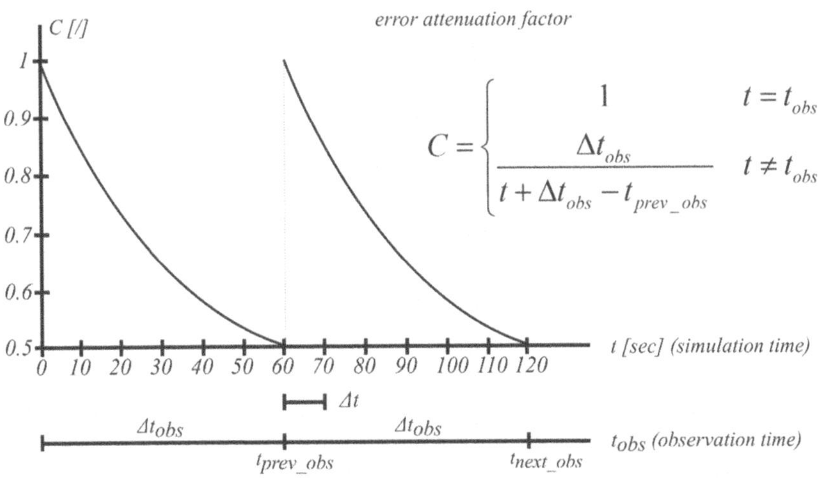

Figure 3 | Error attenuation factor (periodical, discontinuous, function) as a measure of uncertainty (minimal value of the attenuation factor depends on specific values of simulation time step Δt and observation time step Δt_{obs}) (Milašinović *et al.* 2020).

values of K_p cause big variations in controllers' output that can make the system unstable (extremely big water level oscillations causing the occurrence of physically impossible values of water depths). Low values (towards zero) of K_p avoid problems of an unstable system, but time needed for reaching the setpoint increases and, practically, makes system unable to reach the goal. Therefore, K_i, along with error integral, forms integrative gain, which is used to collect error history and its duration, to minimize them over time. This gain can significantly reduce time needed for reaching the setpoint. In some cases, in highly dynamic systems with rapid changes, the derivative component is included to estimate the error trend (what will be the error in the near future). However, the derivative component is sensitive in systems with high measurement noise and can enhance the controllers' instability. Hence, proper tuning of the PID controllers' gain factors must be provided. Finding optimal values of the controllers' parameters is a challenging task. In common applications of the PID controllers this is often conducted using some of the simple, heuristic methods (e.g., Ziegler & Nichols 1995; Skogestad 2004). However, these tuning methods, systematically presented by Ang *et al.* (2005), are mostly physical systems-oriented (in other words, they are proven for automated control of the real physical systems). Recommendations in these methods are derived from the specific field of applications and are unreliable when the PID controllers' theory is used in hydraulic modeling. Additionally, they are single metric or just visually based rather than multi-metric based.

Rather than using simple, heuristic tuning methods, this procedure can be done by optimization algorithms (Ou & Lin 2006; Altinten *et al.* 2008; Solihin *et al.* 2008; dos Santos Coelho 2009; Wang & Wei 2009; Chiha *et al.* 2012). The main problem in optimization algorithm usage (e.g., genetic algorithm) is the difficulty to find the global optimal solution (in reasonable time) when the number of variables increases (multiple controllers). The solution depends on the initial assumption on the variables' values (Deng *et al.* 2015). To prevent this problem, appropriate initial values of the controllers' parameters (K_p, K_i, and K_d) should be estimated. Therefore, simple tuning procedures are necessary to complete this step.

This paper presents sequential, heuristic, multi-metric tuning procedure for initial tuning of the controllers. It is a simple tuning procedure and convenient to start with. Hence, it should not be omitted. Further, fine tuning can be performed using the results of sequential multi-metric procedure as an initial point. Procedure is based on gradual increase of the gain factors of each component independently in three phases. Phase 1 is used to estimate the value of the proportional gain only (P controller) that provides fastest approaching to the prescribed level, with minimal overshooting or oscillations. In Phase 2, integrative gain is coupled (PI controller) with the value of proportional gain factor K_p which produced the best assimilation results according to assimilation quality indicators. In Phase 3, derivative gain is added (PID controller) to the values of K_p and K_i which showed the best results. K_d is increased to analyze if there is any improvement by adding the derivative gain in the controllers. Phases have to be performed in this order, because application of derivative gain as the first or coupling it first with proportional gain (PD controller) is not recommended (suggested by many researchers, e.g., Stanbury *et al.* (2017)). Additionally, almost 80% of PID controllers' applications have the derivative part switched off while the other 20% use it coupled with proportional and integrative gains (Ang *et al.* 2005), because inappropriate tuning of K_d can affect the system's stability. This is a solid reason to use the proposed phases in sequential tuning procedure.

Indicators for DA performance assessment

To assess the effect of the multi-metric tuning procedure, DA performance assessment (DA quality) indicators must be established. The aim of the PID controllers' tuning is to minimize difference between process value and the setpoint, oscillations in process error and settling time. For the application of PID controllers as a DA tool, four indicators are used in this paper: root mean square error *RMSE* (to represent difference between process value and the setpoint), amplitude of the process error *maxError* (to represent the oscillations), assimilation time ratio *AssimTRatio* (to represent the settling time), and total correction volume *CorrVol*, which is method-specific indicator, a product of PID controller-based assimilation (it represents the intensity of the controllers' 'intervention' into the hydraulic model).

Root mean square error

Usage of the root mean square error *RMSE* as DA quality indicator depends on the number of locations where direct model update is conducted (number of assimilation points). In this research, mean value of *RMSE* indicators for all observation locations is used (Equation (9)).

$$RMSE = \frac{1}{M}\sum_{i=1}^{M}\sqrt{\frac{1}{N}\sum_{j=1}^{N}(Z_{j,obs} - Z_{j,sim})^2} \quad (9)$$

where $Z_{j,obs}$ represents sample from the observed water level time series, $Z_{j,sim}$ represents sample from the simulated water level time series, N is the number of time samples in time series, and M is the number of locations used for calculation of *RMSE* indicator.

The lower bound of this metric is 0, which represents the ideal case showing the excellent assimilation performance.

Amplitude of the process error

During the assimilation process, the model is updated towards the measured values, but big oscillations of simulated water level are present, especially at the beginning of the assimilation process. Therefore, amplitude of water level oscillations (error amplitude) *maxError* is used to estimate DA quality (Equation (10)). Since there are several observation locations where DA performance has to be estimated, mean value of error amplitudes is used. The lower bound of this metric is 0, and DA performance is better when *maxError* tends to this value.

$$maxError = \frac{1}{M}\sum_{i=1}^{M} max(|e_i|) \quad (10)$$

where e_i is error time series for the i-th observation location and M is the number of those locations.

Total correction volume

Total correction volume *CorrVol* represents the total volume of water added or subtracted by controllers at the assimilation points. This indicator is method-specific, applicable only in PID controller-based assimilation. It represents the estimation of controllers' intervention into the model. When confidence in boundary conditions is high, the total correction volume tends to be zero. In this research, it is evaluated as a sum of total correction volumes added/subtracted from the model at all assimilation points (Equation (11)). This indicator also has the lower boundary set at 0.

$$CorrVol = \sum_{i=1}^{M_A}\left|\int_0^{t_{sim}} Q_{corr,i}(t)dt\right| \quad (11)$$

where M_A is the number of assimilation points, t_{sim} is the time period of the simulation when assimilation is performed, and Q_{corr} is the correction flow.

Assimilation time ratio

Assimilation time ratio *AssimTRatio* [/] represents the ratio of period in which process error exceeds the threshold to the total simulation time t_{sim}. When *AssimTRatio* ratio has value 0, it shows excellent DA performance (best value), while value 1 shows poor DA performance (worst value), where assimilation fails to correct the model towards measurements. To evaluate the *AssimTRatio*, error time series has to be transformed into the error duration curve (Figure 4). The threshold used in this paper is set to 0.05 m. As there are several observation locations where assimilation quality has to be estimated, *AssimTRatio* is calculated as the average of assimilation times for all locations.

Test case

The presented methodology is developed for the purpose of near real-time control of transboundary hydropower system Iron Gate 1 on the Danube river shared between Serbia and Romania. Everyday operations of the hydropower plant (HPP) require short-term forecasting of water levels (and river flows) in the vicinity of HPP. For improved forecasting (improved initial conditions) DA is required.

The test case of the hydropower system consists of hydropower plant Iron Gate 1 (IG1) and a 170 km-long upstream Danube river section (Figure 5), the lower half of the total reservoir's length. The domain is discretized by

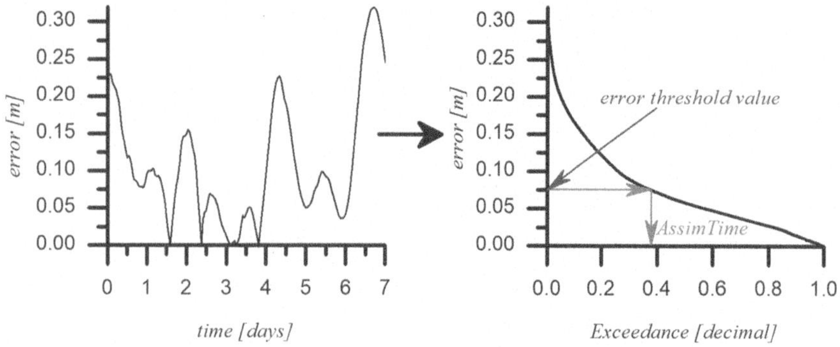

Figure 4 | Transformation of the error time series into the error duration curve for determining the *AssimTRatio* indicator.

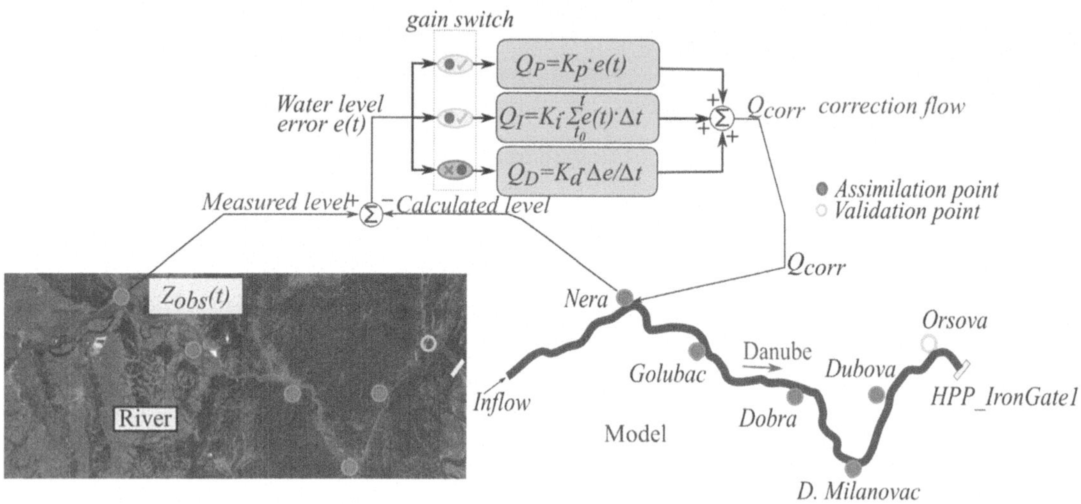

Figure 5 | Iron Gate hydropower system with 170 km-long Danube river section coupled with PID controllers for continuous model update.

189 cross sections, with the average distance of 900 m between. The river model is developed using the *DiffW1D* model with one upstream inflow, neglecting all tributaries. Tributaries are neglected due to lack of data, further supporting the assumption that dominant uncertainty source in the model is the inflow data. Six observation locations are available across the river section: Nera (132 km from IG1), Golubac (100 km from IG1), Dobra (74 km from IG1), D. Milanovac (47 km from IG1), Dubova (25.7 km from IG1), and Orsova (10.3 km from IG1). The first five locations are used as assimilation points (PID controllers are set at these locations), while Orsova is used as a validation point. The test case used for the analysis of the proposed DA methodology contains real river geometry.

Synthetic test case scenario is used for analysis in this paper. Observed water levels at six locations are generated using the 'true' inflow hydrograph (black line in Figure 6). The total inflow volume for a given 7 days is $2.163 \times 10^9\,m^3$. Then, the 'true' inflow is altered and changed to 'uncertain' inflow, or model driving inflow (dashed line with black circles in Figure 6). Difference between the 'true' and the 'uncertain' inflow is a sum of two sinusoidal functions with the amplitude of 1,000 m^3/s and frequencies of $\pi/(10{,}000\,s)$ and $\pi/(5{,}000\,s)$, which are arbitrarily selected. The scenario is adopted to emulate extreme conditions where model driving inflow ('uncertain' inflow) significantly differs from the 'true' inflow (in this study used to emulate 'true' water levels). This also represents the case of sampling the inflow data on poor timescale, which results in big discrepancies. It is also used to show the potential of using PID controllers as a DA method in a more dynamic environment (note that 'true' inflow is more

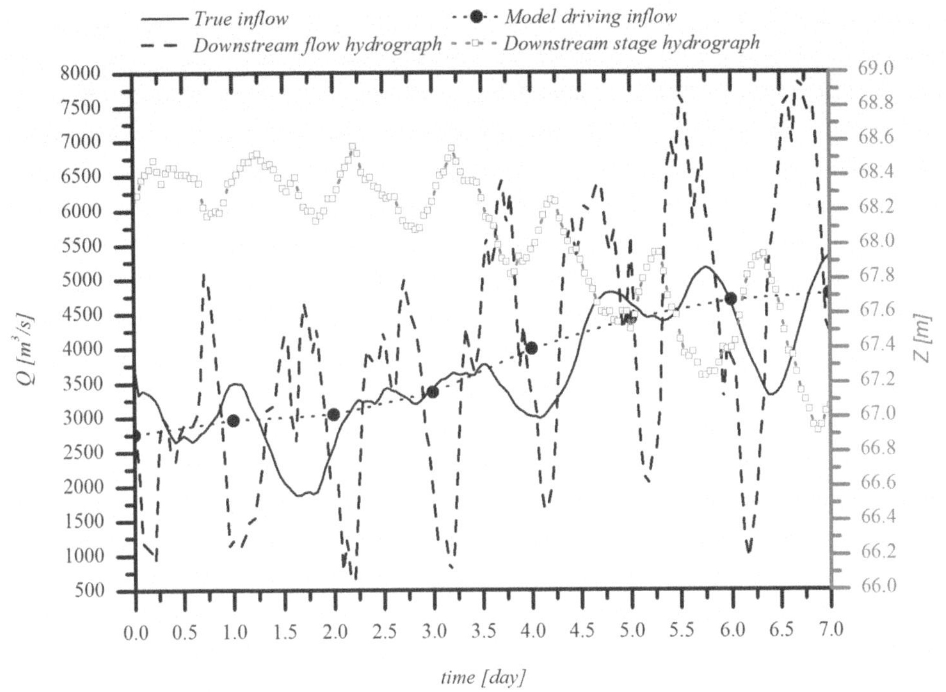

Figure 6 | Boundary conditions: true inflow, uncertain upstream inflow hydrograph (model driving inflow), downstream outflow hydrograph, and downstream stage hydrograph.

dynamic than model driving inflow). The altered inflow is set as the upstream boundary, with total volume of 2.268×10^9, with average 'uncertain' inflow which is larger than 'true' inflow for 104.7×10^6 m^3/s. At the downstream boundary, the measured outflow and stage (dashed line and dashed line with square markers in Figure 6) are used. Stage hydrograph is used to generate the true state, while outflow hydrograph is used as model driver. Water level observations are obtained with $\Delta t_{obs} = 1$ h, while computation time step Δt is set to 60 s, simulating real-world conditions where computation time step is, in most cases, smaller than observation time step. Total simulation time of the test case scenario is $t_{sim} = 7$ days.

Stage (water level) hydrograph or stage-discharge curve (rating curve) should be implemented as the downstream boundary condition for proper open channel hydraulic modeling. But, in most real-world cases, HPP operations are controlled by the discharge (when reservoir exists), not water level (stage) and stage-discharge curve is not easy to determine. Therefore, the outflow hydrograph is used as the downstream boundary condition. As can be seen in Figure 6, the outflow hydrograph reflects the daily power production in real-world HPP operations. Thus, this supports the assumption that boundary conditions create, dominantly, the model's uncertainty.

To analyze the influence of PID controllers' gains, sets of parameter values are used. Parameters are increased step-by-step by the order of magnitude. A set of identical controllers with the same values of parameters (K_p, K_i, and K_d) are used for all assimilation points. Tuning each controller separately requires thorough, independent, investigation, including multi-objective optimization, which will be the subject of a forthcoming research. This paper is focused on determining the initial values and bounds of these parameters, and good to start with.

Analysis is carried out in three phases. In Phase 1, only proportional gain is analyzed (integrative and derivative are switched off), by changing the K_p using the values of 10^n, where n is integer from $-3 \leq n \leq 3$. In Phase 2, K_p is fixed to the value that gave the best DA quality indicators in Phase 1, while K_i is changed using the values of 10^n ($-3 \leq n \leq 3$) and derivative gain is switched off. In the last, Phase 3, the K_d is changed using the values of 10^n ($-3 \leq n \leq 4$), K_p is fixed to the same value as used in Phase 2 and K_i is fixed to the value that gave the best results in Phase 2. The set of parameters' values could be done with more samples and it would probably show an even better solution. The goal of the presented sequential tuning procedure is not to determine the global optimal solution on the controllers' parameters but to estimate the initial values and bounds preserving the model's stability. Hence, the adopted values are good enough to present the tuning method potential.

RESULTS AND DISCUSSION

The base case scenario is the model's free run, without DA. Figure 7 shows the water levels obtained using the model in free run mode. The water levels obtained show the model's inability to reach the true state, which is most notable at the stations closer to the upstream boundary (Nera and Golubac). This originates from the bad boundary conditions being used and the fact that these stations are more affected by the upstream boundary condition. Stations closer to the downstream boundary are dominantly under the influence of the hydropower plant which practically mitigates all upstream model driving due to big reservoir volume. Using such a model for HPP control operations will lead to numerous problems. Hence, DA is required.

Initial tuning analyzed in this paper starts with Phase 1, where only proportional gain is used. Figure 8 shows the changes in DA quality indicators by increasing the K_p. Results are presented for five assimilation points (black bars) and one validation point (Orsova, gray bars). Please note that *CorrVol* is zero at validation point since there is no controller here to add/subtract the correction flow. *RMSE* values are pretty much the same for the values of K_p smaller than 100. *RMSE* values vary around 0.3 m for assimilation points and around 0.37 m for validation point. The value of *maxError* has the same trend, varying around 0.63 m for assimilation points and 0.77 m for validation point. Setting the value of K_p at 100, *RMSE* and *maxError* jump to values around 2.5 and 5, respectively, while increasing the K_p to 1,000 makes the model unstable. This means that the controller's operations cause the model to become first unstable (large water levels oscillations) and then to produce physically impossible solutions. In such situations, *RMSE* and *maxError* are assigned a value of 100, reflecting the poor performance of DA. Also, the *CorrVol* presented

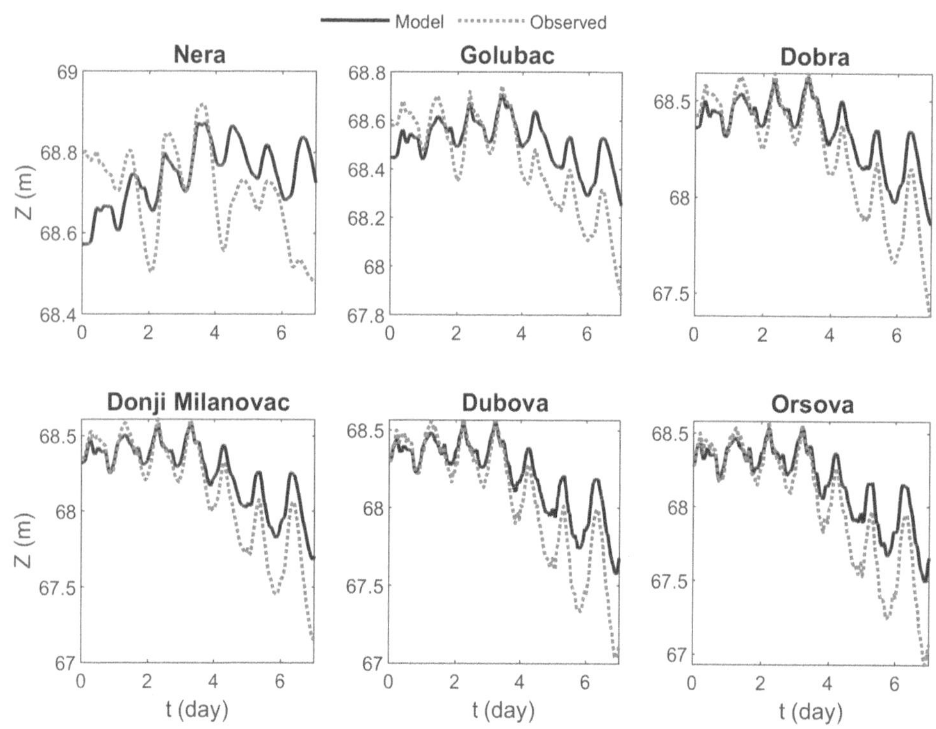

Figure 7 | Comparison of the observed water levels and water levels obtained by the model free run simulation.

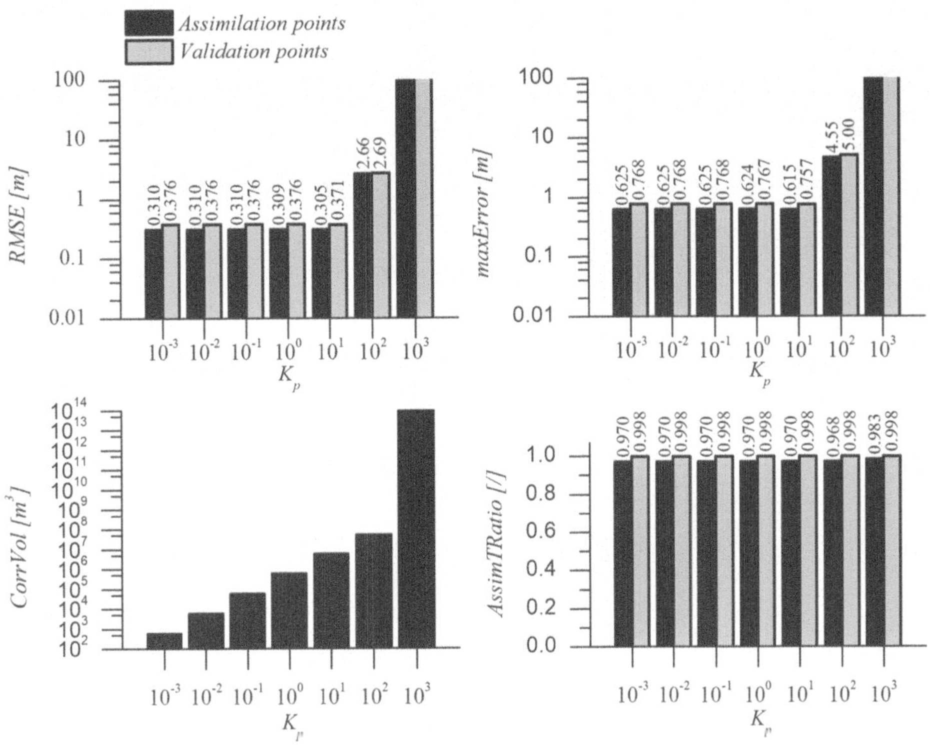

Figure 8 | Change of the DA quality indicators by increasing the K_p ($K_i = 0$, $K_d = 0$).

in Figure 8 increases for a few orders of magnitude when K_p is increased to 1,000.

Small values of the *RMSE* and *maxError* for the values of K_p below 100 could lead to the wrong conclusion that values around 10 can produce satisfying results. Therefore, *AssimTRatio* has also to be analyzed. This indicator shows that process error (water level difference) is not below the threshold set at 0.05 m for more than 97% of simulation time. This indicates that using the proportional gain only is not a good option for the application of the controllers in the hydraulic model. Therefore, Phase 2 is required.

In Phase 2, integrative gain is switched on. Values of K_i are changed between 10^{-3} and 10^3. Integrative gain is coupled with the proportional gain factor value of $K_p = 10$. Figure 9 shows changes of DA quality indicators with the increase of K_i. *RMSE* and *maxError* values are better than using the proportional gain only. *RMSE* is less than 0.25 for assimilation points and less than 0.3 for the validation point, for listed K_i values below 100. In these cases, *maxError* is below 0.5 for assimilation points and below 0.6 for the validation point. K_i set to 10 gives by far the best results for the *RMSE* and *maxError*. *RMSE* is 0.02 m for the assimilation points, and 0.07 m for the validation point. Value of *maxError* is 0.074 m for assimilation points and 0.187 m for the validation point. Further increase of K_i makes the model unstable (in Figure 8 values of *RMSE* and *maxError* are limited to 100). Total correction volume increases by increasing K_i, but not significantly as was the case when only K_p was used. Increasing K_i shows the increase in *CorrVol* but with a tendency to reach the maximum of the total correction volume. When *AssimTRatio* is analyzed, by far the best results are obtained when K_i is set to 10. In this case, *AssimTRatio* is 0.075 for assimilation points, which means that process error has the value over the threshold for only 7.5% of the simulation time. This is an extreme improvement from Phase 1. This value is 0.48 for the validation point, which is bigger than for the assimilation points, yet much better than in Phase 1.

Even though Phase 2 shows significant improvement in DA quality indicators, Phase 3 is used to determine whether the addition of derivative gain could give some further improvement. Factor K_d is included with increasing values from 10^{-3} to 10^4 (Figure 10), coupled with constant values of $K_p = 10$ and $K_i = 10$. Increasing the value of K_d up to

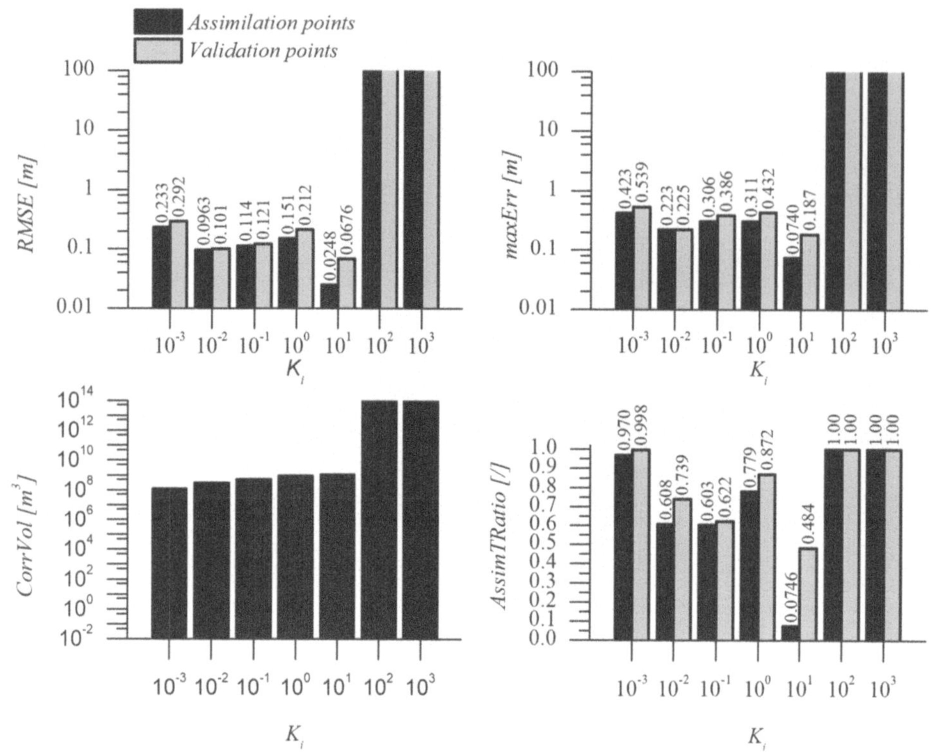

Figure 9 | Change of DA quality indicators by increasing K_i ($K_p = 10$, $K_d = 0$).

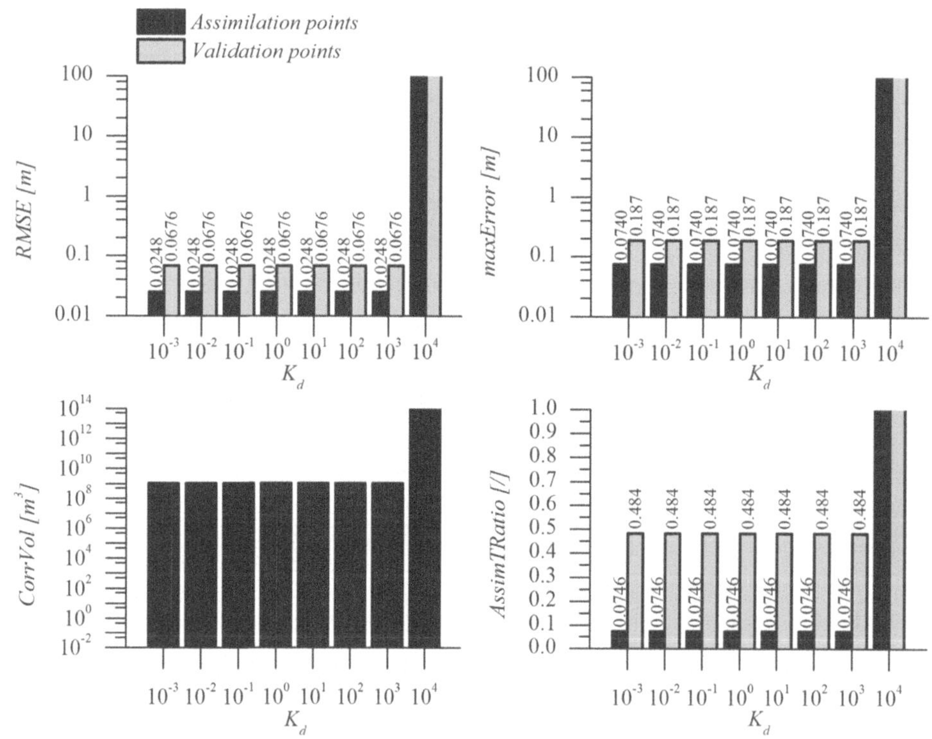

Figure 10 | Change of DA quality indicators by increasing K_d ($K_p = 10$, $K_i = 10$).

1,000 shows no improvement in the DA quality indicators. All DA performance indicators have the same values as obtained with the best solution from Phase 2. This can be explained by the fact that properly tuned PI controllers reduce process error to values near 0, so derivatives *de/dt* in Equation (8) have extremely low values. Multiplication of these values by K_d gives low addition in correction flow thus, practically, having no impact on the assimilation process. Effects of the derivative gain can be seen only if K_d has extreme values, such as 10,000. It increases the impact to the correction flow. In this case study, usage of the derivative gain creates problems, because the model becomes too sensitive to derivative gain output, resulting in the model's instability. This can be explained by the fact that derivative gain output (and total correction flow accordingly) creates highly dynamic input to the model (extremely different from the natural events). These events cannot be propagated by the model, which creates the model's instability. This is not unexpected considering other applications of the PID controllers (80% of the applications omit the derivative gain). Therefore, the multi-metric tuning procedure used in this case study shows that adding the derivative gain gives no benefits. This outcome is case specific, and it should not be taken for granted. Instead, the main contribution of the sequential tuning procedure is showing the potential to narrow the search space for optimal controllers' parameters. It could improve fine tuning of the controllers using the multi-objective optimization, which will be analyzed in a forthcoming research.

Figure 11 shows the levels of the model simulation coupled with the PI controllers, where proportional and integrative gains' factors, K_p and K_i, are set to 10. In this case, the model will reach the true state in the first half of a day, which gives the improved water level state at the end of the 7th day needed for the short-term forecasting. Figure 12 shows the correction flow hydrographs for all five assimilation points. When correction flow hydrograph at assimilation point 1 (Nera) is integrated to estimate the correction volume at this point, the total volume of 132.1×10^6 m^3 is obtained. Total volume difference between true inflow hydrograph (Figure 6) and the one used to drive the model in the assimilation process is 104.7×10^6 m^3. This shows that controller(s) located near the uncertain inflow can reconstruct the 'missing' inflows (flow hydrograph – Nera in Figure 12). The rest of the total correction volume *CorrVol* is the result of usage of inappropriate downstream boundary condition in simulation mode (which happens often). Controllers react to water level difference between

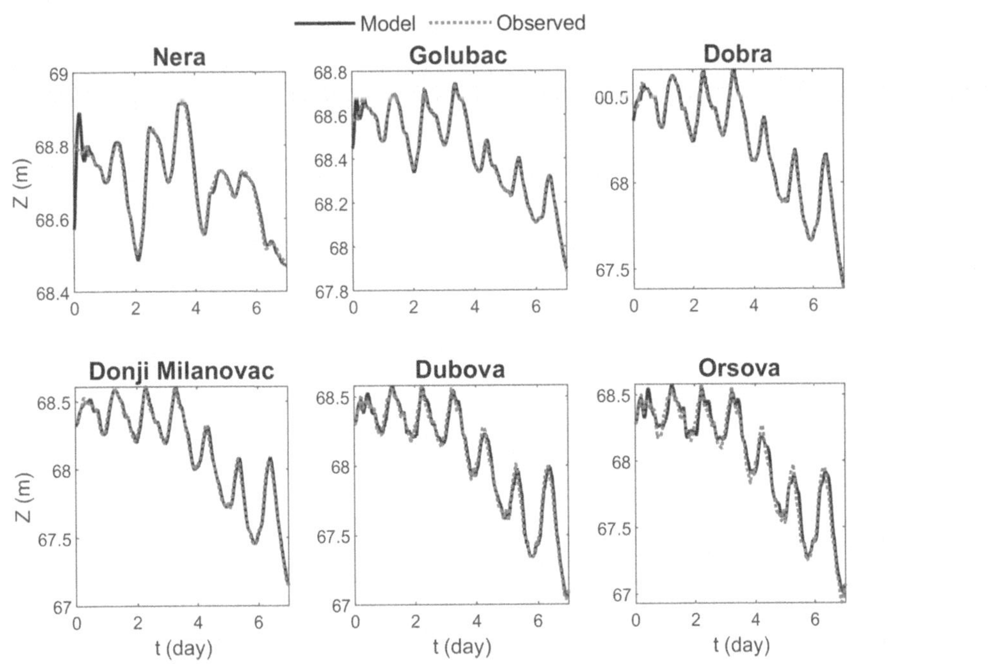

Figure 11 | Comparison of the observed water levels and water levels obtained by the model coupled with PI controller ($K_p = 10$, $K_i = 10$).

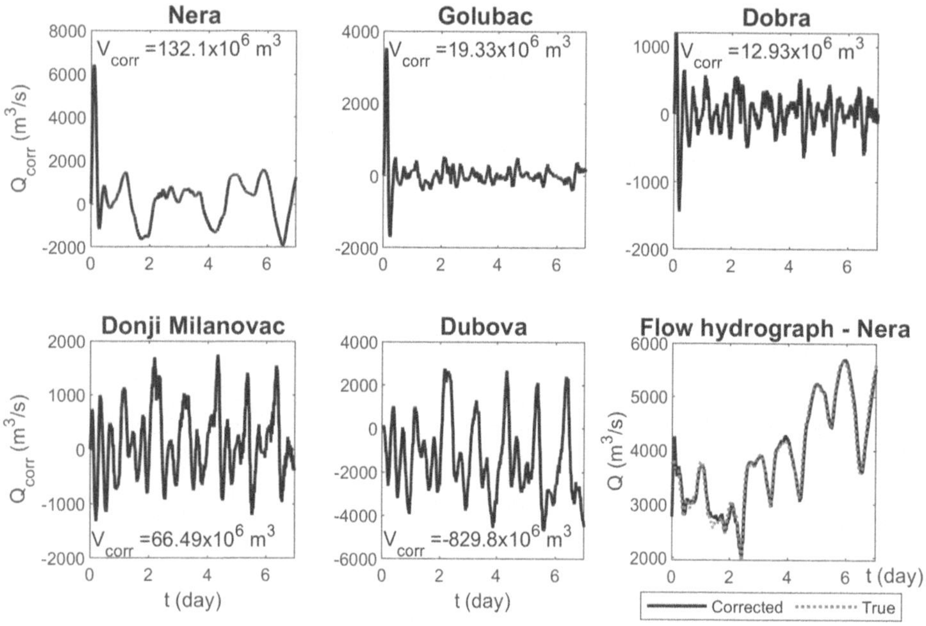

Figure 12 | Correction flows added (positive values) and subtracted (negative values) at five assimilation points (Nera, Golubac, Dobra, Donji Milanovac, and Dubova) by PI controllers ($K_p = 10$, $K_i = 10$) and flow hydrograph downstream of Nera (true inflow corrected by Nera PI controller).

true and simulated state even when simulated levels are the product of bad boundary condition. This is even more visible at downstream controllers. They are dominantly under the influence of the hydropower plant operations which makes them more sensitive to the faults in downstream boundary condition. Correction volumes for each station separately (Figure 12) show significant difference between Dubova station and the others. At this station the controller subtracted 829.8×10^6 m^3, which is enormous compared to other stations. This indicates that downstream boundary condition strongly affects the results. Therefore, PID controllers have the potential to be used for detection of problems which are the result of bad boundary conditions applied, beside estimated uncertain inflows. This can be a very significant decision support tool, and it will be analyzed in a future research.

Figure 12 also shows that some controllers supply the high correction flows at the beginning of the assimilation. Since the simulation model needs some time to respond to those corrections, during this short period the integrative part will accumulate the error causing water flow to overshoot. To overcome this problem at the beginning of the assimilation, hot start for integral error should be adopted as a common practice. This would increase the impact of the integrative gain from the assimilation start.

CONCLUSIONS

This paper presents further insight into the application of control theory-based data assimilation for 1D river hydraulic models used as decision support tool for hydropower production control. PID controllers are applied as a DA tool, in a form of simple (fictive) lateral inflow elements that reduce the water level difference between the model and the observations. The methodology is applied on the real-world case study, HPP Iron Gate with a 170 km-long Danube section with a synthetic test case scenario. The general goal of this analysis is to show the capability of PID controllers to properly update the model state and produce the improved initial conditions for the forecasting. Sequential, multi-metric tuning procedure for the controllers is introduced. It is performed in three phases, considering the DA quality indicators: *RMSE*, *maxError*, *CorrVol*, and *AssimTRatio*. In Phase 1, controllers are used in a form of P controllers. Multi-metric assessment of the tuning shows that that this type of controller is not able to produce satisfying results based on all indicators. In Phase 2, controllers are upgraded by adding the integrative gain (PI controllers). This upgrade produced highly improved results of model updating, based on all indicators. It also showed that tuning of the controllers' parameters must be carefully carried out. Inadequate tuning of the controllers can result in model instability. In Phase 3, PI controllers are upgraded to the full form of PID controllers by adding the derivative gain. Multi-metric assessment can show the necessity of adding the derivative gain. If the metrics show no improvement in Phase 3, derivative gain can be switched off. According to the results obtained in this research, the following specific conclusions can be derived:

- PID controllers can be applied as a DA tool for 1D open channel hydraulic model update.
- Usage of only one DA quality indicator, such as *RMSE*, cannot guarantee the proper implementation (tuning) of the controllers.
- Other DA quality indicators should be used as well, especially *AssimTRatio*, in combination with *RMSE*.
- Multi-metric assessment of controllers' parameters, used in this research, cannot result in the global optimum (K_p, K_i, and K_d). This procedure provides parameters' values good to start with, and this procedure should not be omitted, because it can significantly narrow the search space for finding the global optimal solution.
- More precise tuning (fine tuning) should be conducted using the multi-objective optimization algorithms. The presented sequential tuning procedure should be used to determine initial values of the controllers' parameters.

Based on the results and previous specific conclusions, control theory-based DA justifies its application in large-scale open channel hydraulic models. However, full application of this methodology requires further testing on the influence of numerical model complexity on the system's performance and procedure for fine tuning of the controllers (each controller individually and all parameters simultaneously), which will be the subject of forthcoming research.

DATA AVAILABILITY STATEMENT

All relevant data are included in the paper or its Supplementary Information.

REFERENCES

Ahmad, S. K. & Hossain, F. 2019 A web-based decision support system for smart dam operations using weather forecasts. *Journal of Hydroinformatics* **21** (5), 687–707.

Altinten, A., Ketenanlioglu, F., Erdogan, S., Hapoglu, H. & Alpbaz, M. 2008 Self-tuning PID control of jacketed batch polystyrene reactor using genetic algorithm. *Chemical Engineering Journal* **138** (1–3), 490–497.

Andreadis, K. M. & Schumann, G. J.-P. 2014 Estimating the impact of satellite observations on the predictability of large-scale hydraulic models. *Advances in Water Resources* **73**, 44–54. http://dx.doi.org/10.1016/j.advwatres.2014.06.006.

Ang, K. H., Chong, G. & Li, Y. 2005 PID control system analysis, design, and technology. *IEEE Transactions on Control Systems Technology* **13** (4), 559–576.

Annis, A. & Nardi, F. 2019 Integrating VGI and 2D hydraulic models into a data assimilation framework for real time flood forecasting and mapping. *Geo-Spatial Information Science* **22** (4), 223–236. https://doi.org/10.1080/10095020.2019.1626135.

Astrom, K. J. & Hagglund, T. 1995 *PID Controllers: Theory, Design and Tuning*, 2nd edn. Instrument Society of America, Research Triangle Park, NC, USA.

Barthélémy, S., Ricci, S., Rochoux, M. C., Le Pape, E. & Thual, O. 2017 Ensemble-based data assimilation for operational flood forecasting – on the merits of state estimation for 1D hydrodynamic forecasting through the example of the 'Adour Maritime' River. *Journal of Hydrology* **552**, 210–224.

Bonet, E., Gómez, M., Yubero, M. T. & Fernández-Francos, J. 2019 Gorosobo simplified: an accurate feedback control algorithm in real time for irrigation canals. *Journal of Hydroinformatics* **21** (6), 945–961.

Bozzi, S., Passoni, G., Bernardara, P., Goutal, N. & Arnaud, A. 2015 Roughness and discharge uncertainty in 1D water level calculations. *Environmental Modeling and Assessment* **20** (4), 343–353.

Chiha, I., Noureddine, L. & Borne, P. 2012 Tuning PID controller using multiobjective ant colony optimization. *Applied Computational Intelligence and Soft Computing* **1**, 1–7.

Costabile, P. & Macchione, F. 2012 Analysis of one-dimensional modelling for flood routing in compound channels. *Water Resources Management* **26** (5), 1065–1087.

Deng, Y., Liu, Y. & Zhou, D. 2015 An improved genetic algorithm with initial population strategy for symmetric TSP. *Mathematical Problems in Engineering* **2015** (3), 212794. https://doi.org/10.1155/2015/212794.

dos Santos Coelho, L. 2009 Tuning of PID controller for an automatic regulator voltage system using chaotic optimization approach. *Chaos, Solitons and Fractals* **39** (4), 1504–1514. http://dx.doi.org/10.1016/j.chaos.2007.06.018.

Dumedah, G. & Coulibaly, P. 2014 Integration of an evolutionary algorithm into the ensemble kalman filter and the particle filter for hydrologic data assimilation. *Journal of Hydroinformatics* **16** (1), 74–94.

Ehteram, M., Mousavi, S. F., Karami, H., Farzin, S., Singh, V. P., Chau, K.-W. & El-Shafie, A. 2018 Reservoir operation based on evolutionary algorithms and multi-criteria decision-making under climate change and uncertainty. *Journal of Hydroinformatics* **20** (2), 332–355.

Eurelectric. 2011 *Renewables Action Plan (RESAP)*.

Evensen, G. 1994 Sequential data assimilation with a nonlinear quasi-geostrophic model using Monte Carlo methods to forecast error statistics. *Journal of Geophysical Research* **99** (C5), 10143. http://doi.wiley.com/10.1029/94JC00572.

Evensen, G. 2003 The ensemble Kalman filter: theoretical formulation and practical implementation. *Ocean Dynamics* **53** (4), 343–367.

Fava, M. C., Mazzoleni, M., Abe, N., Mendiondo, E. M. & Solomatine, D. P. 2020 Improving flood forecasting using an input correction method in urban models in poorly gauged areas. *Hydrological Sciences Journal* **65** (7). http://dx.doi.org/10.1080/02626667.2020.1729984.

Fu, M., Fan, T., Ding, Z., Salih, S. Q., Al-Ansari, N. & Yaseen, Z. M. 2020 Deep learning data-intelligence model based on adjusted forecasting window scale: application in daily streamflow simulation. *IEEE Access* **8**, 32632–32651.

Garambois, P. A., Larnier, K., Monnier, J., Finaud-Guyot, P., Verley, J., Montazem, A.-S. & Calmant, S. 2020 Variational estimation of effective channel and ungauged anabranching river discharge from multi-satellite water heights of different spatial sparsity. *Journal of Hydrology* **581**, 124409. https://doi.org/10.1016/j.jhydrol.2019.124409.

García-Pintado, J., Neal, J. C., Mason, D. C., Dance, S. L. & Bates, P. D. 2013 Scheduling satellite-based SAR acquisition for sequential assimilation of water level observations into flood modelling. *Journal of Hydrology* **495**, 252–266.

Giustarini, L., Matgen, P., Hostache, R., Monranari, M., Plaza, D., Pauwels, V. R. N., De Lannoy, G. J. M., De Keyser, R., Pfister, L., Hoffmann, L. & Savenije, H. H. G. 2011 Assimilating SAR-derived water level data into a hydraulic model: a case study. *Hydrology and Earth System Sciences* **15** (7), 2349–2365.

Hansen, L. S., Borup, M., Møller, A. & Mikkelsen, P. S. 2014 Flow forecasting using deterministic updating of water levels in distributed hydrodynamic urban drainage models. *Water (Switzerland)* **6** (8), 2195–2211.

Hostache, R., Lai, X., Monnier, J. & Puech, C. 2010 Assimilation of spatially distributed water levels into a shallow-water flood model. Part II: use of a remote sensing image of Mosel River. *Journal of Hydrology* **390** (3–4), 257–268. http://dx.doi.org/10.1016/j.jhydrol.2010.07.003.

Jean-Baptiste, N., Malaterre, P. O., Dorée, C. & Sau, J. 2011 Data assimilation for real-time estimation of hydraulic states and unmeasured perturbations in a 1D hydrodynamic model. *Mathematics and Computers in Simulation* **81** (10), 2201–2214. http://dx.doi.org/10.1016/j.matcom.2010.12.021.

Li, X.-L., Lü, H., Horton, R., An, T. & Yu, Z. 2014 Real-time flood forecast using the coupling support vector machine and data assimilation method. *Journal of Hydroinformatics* **16** (5), 973–988.

Lund, N. S. V., Madsen, H., Mazzoleni, M., Solomatine, D. & Borup, M. 2019 Assimilating flow and level data into an urban drainage surrogate model for forecasting flows and overflows. *Journal of Environmental Management* **248**, 109052. https://doi.org/10.1016/j.jenvman.2019.05.110.

Madsen, H. & Skotner, C. 2005 Adaptive state updating in real-time river flow forecasting – a combined filtering and error forecasting procedure. *Journal of Hydrology* **308** (1–4), 302–312.

Madsen, H., Rosbjerg, D., Damgård, J. & Hansen, F. S. 2003 Data assimilation in the MIKE 11 flood forecasting system using Kalman filtering. *Water Resources Systems – Hydrological Risk, Management and Development* **281**, 75–81.

Matgen, P., Montanari, M., Hostache, R., Pfister, L., Hoffmann, L., Plaza, D., Pauwels, V. R. N., De Lannoy, G. J. M., De Keyser, R. & Savenije, H. H. G. 2010 Towards the sequential assimilation of SAR-derived water stages into hydraulic models using the particle filter: proof of concept. *Hydrology and Earth System Sciences* **14** (9), 1773–1785.

Mazzoleni, M., Verlaan, M., Alfonso, L., Monego, M., Norbiato, D., Ferri, M. & Solomatine, D. P. 2017 Can assimilation of crowdsourced data in hydrological modelling improve flood prediction? *Hydrology and Earth System Sciences* **21** (2), 839–861.

Mazzoleni, M., When, U., Alfonso, L., Norbiato, D., Monego, M., Ferri, M. & Solomatine, D. 2018 Exploring the influence of citizen involvement on the assimilation of crowdsourced observations: a modelling study based on the 2013 flood event in the Bacchiglione catchment (Italy). *Hydrology and Earth System Sciences* **22** (1), 391–416.

Milašinović, M., Zindović, B., Rosić, N. & Prodanović, D. 2019 PID Controllers as Data Assimilation Tool for 1D Hydrodynamic Models of Different Complexity. In: *Proceedings of the 5th International Conference SimHydro 2019*, Nice, France.

Milašinović, M., Prodanović, D., Zindović, B., Rosić, N. & Milivojević, N. 2020 Fast data assimilation for open channel hydrodynamic models using control theory approach. *Journal of Hydrology* **584**, 124661. https://doi.org/10.1016/j.jhydrol.2020.124661.

Neal, J. C., Atkinson, P. M. & Hutton, C. W. 2007 Flood inundation model updating using an ensemble Kalman filter and spatially distributed measurements. *Journal of Hydrology* **336** (3–4), 401–415.

Ocio, D., Le Vine, N., Westerberg, I., Pappenberger, F. & Buytaert, W. 2017 The role of rating curve uncertainty in real-time flood forecasting. *Water Resources Research* **53** (5), 4197–4213.

Ou, C. & Lin, W. 2006 Comparison between PSO and GA for parameters optimization of PID controller. In: *2006 IEEE International Conference on Mechatronics and Automation, ICMA 2006*, Luoyang, Henan, China, pp. 2471–2475.

Rakovec, O., Weerts, A. H., Hazenberg, P., Torfs, P. J. J. F. & Uijlenhoet, R. 2012 State updating of a distributed hydrological model with ensemble Kalman filtering: effects of updating frequency and observation network density on forecast accuracy. *Hydrology and Earth System Sciences* **16** (9), 3435–3449.

Randrianasolo, A., Thirel, G., Ramos, M. H. & Martin, E. 2014 Impact of streamflow data assimilation and length of the verification period on the quality of short-term ensemble hydrologic forecasts. *Journal of Hydrology* **519**, 2676–2691. http://dx.doi.org/10.1016/j.jhydrol.2014.09.032.

Romanowicz, R. J., Young, P. C. & Beven, K. J. 2006 Data assimilation and adaptive forecasting of water levels in the River Severn catchment, United Kingdom. *Water Resources Research* **42** (6), 1–12.

Rosić, N., Jaćimović, N., Prodanović, D. & Stojanović, B. 2017a Data assimilation for operational reservoir management on the Danube River. In: *7th International Conference on Information Society and Technology ICIST 2017*, Kopaonik, Serbia, pp. 210–213.

Rosić, N., Prodanović, D., Stojanović, B. & Obradović, D. 2017b Near real time data assimilation of numerical simulation model for Danube river from Novi Sad to Iron Gate I, test results. *Vodoprivreda* **49** (288), 253–261 (in Serbian).

Schütze, M., Campisano, A., Colas, H., Schilling, W. & Vanrolleghem, P. A. 2004 Real time control of urban wastewater systems – where do we stand today? *Journal of Hydrology* **299** (3–4), 335–348.

Schwanenberg, D., Becker, B. P. J. & Xu, M. 2015 The open real-time control (RTC)-tools software framework for modeling RTC in water resources sytems. *Journal of Hydroinformatics* **17** (1), 130–148.

Skogestad, S. 2004 Simple analytic rules for model reduction and PID controller tuning. *Modeling, Identification and Control* **25** (2), 85–120.

Solihin, M. I., Wahyudi, Kamal, M. A. & Legowo, A. 2008 Objective Function Selection of GA-Based PID Control Optimization for Automatic Gantry Crane. *Proceedings of the International Conference on Computer and Communication Engineering 2008*, ICCCE08: Global Links for Human Development. Kuala Lumpur, Malaysia, pp. 883–887.

Stanbury, P. F., Whitaker, A. & Hall, S. J. 2017 *Instrumentation and Control*. In: *Principles of Fermentation Technology*. Elsevier, Oxford, UK, pp. 487–536.

Sun, L., Nistor, I. & Seidou, O. 2015 Streamflow data assimilation in SWAT model using extended Kalman filter. *Journal of Hydrology* **531**, 671–684. http://dx.doi.org/10.1016/j.jhydrol.2015.10.060.

Taormina, R. & Chau, K. W. 2015 ANN-based interval forecasting of streamflow discharges using the LUBE method and MOFIPS. *Engineering Applications of Artificial Intelligence* **45**, 429–440. http://dx.doi.org/10.1016/j.engappai.2015.07.019.

Thirel, G., Martin, E., Mahfouf, J.-F., Massart, S., Ricci, S., Regimbeau, F. & Habets, F. 2010 A past discharges assimilation system for ensemble streamflow forecasts over France – part 1: description and validation of the assimilation system. *Hydrology and Earth System Sciences* **14** (8), 1623–1637. http://www.hydrol-earth-syst-sci.net/14/1623/2010/.

Vrugt, J. A., Gupta, H. V., Nualláin, B. & Bouten, W. 2006 Real-time data assimilation for operational ensemble streamflow forecasting. *Journal of Hydrometeorology* **7** (3), 548–565. http://journals.ametsoc.org/doi/abs/10.1175/JHM504.1.

Wang, Y. & Wei, B. 2009 A new particle swarm optimization based auto-tuning of PID controller. *Journal of Information and Computational Science* **6** (1), 219–226.

Wu, C. L. & Chau, K. W. 2013 Prediction of rainfall time series using modular soft computing methods. *Engineering Applications of Artificial Intelligence* **26** (3), 997–1007.

Yoon, Y., Durand, M., Merry, C. J., Clark, E. A., Andreadis, K. M. & Alsdorf, D. E. 2012 Estimating river bathymetry from data assimilation of synthetic SWOT measurements. *Journal of Hydrology* **464–465**, 363–375. http://dx.doi.org/10.1016/j.jhydrol.2012.07.028.

Ziegler, J. G. & Nichols, N. B. 1995 Optimum settings for automatic controllers. *InTech* **42** (6), 94–100.

First received 4 May 2020; accepted in revised form 14 January 2021. Available online 22 February 2021

Visualisation of the combinatorial effects within evolutionary algorithms: the compass plot

Qi Wang IWA, Miaoting Guan, Wen Huang, Libing Wang, Zhihong Wang, Shuming Liu IWA and Dragan Savić IWA

ABSTRACT

Applications of evolutionary algorithms (EAs) to real-world problems are usually hindered due to parameterisation issues and computational efficiency. This paper shows how the combinatorial effects related to the parameterisation issues of EAs can be visualised and extracted by the so-called compass plot. This new plot is inspired by the traditional Chinese compass used for navigation and geomantic detection. We demonstrate the value of the proposed compass plot in two scenarios with application to the optimal design of the Hanoi water distribution system. One is to identify the dominant parameters in the well-known NSGA-II. The other is to seek the efficient combinations of search operators embedded in Borg, which uses an ensemble of search operators by auto-adapting their use at runtime to fit an optimisation problem. As such, the implicit and vital interdependency among parameters and search operators can be intuitively demonstrated and identified. In particular, the compass plot revealed some counter-intuitive relationships among the algorithm parameters that led to a considerable change in performance. The information extracted, in turn, facilitates a deeper understanding of EAs and better practices for real-world cases, which eventually leads to more cost-effective decision-making.

Key words | combinatorial effects, evolutionary algorithms, parameterisation, visualisation, water distribution systems

Qi Wang IWA
Miaoting Guan
Wen Huang
Libing Wang
Zhihong Wang
School of Civil and Transportation Engineering,
Guangdong University of Technology,
Guangzhou 510006,
China

Shuming Liu IWA (corresponding author)
School of Environment,
Tsinghua University,
Beijing 100084,
China
E-mail: *shumingliu@tsinghua.edu.cn*

Dragan Savić IWA
KWR Water Research Institute,
3430 BB Nieuwegein,
The Netherlands
and
Centre for Water Systems,
University of Exeter,
Exeter EX4 4QF,
UK

HIGHLIGHTS

- A new tool, called the compass plot, is proposed for visualising the combinatorial effects related to evolutionary algorithms.
- The compass plot helps identify key factors and corresponding values for a given problem.
- Applications to the optimal design of water distribution systems provide new insights into NSGA-II and Borg.

This is an Open Access article distributed under the terms of the Creative Commons Attribution Licence (CC BY 4.0), which permits copying, adaptation and redistribution, provided the original work is properly cited (http://creativecommons.org/licenses/by/4.0/).

doi: 10.2166/hydro.2020.073

GRAPHICAL ABSTRACT

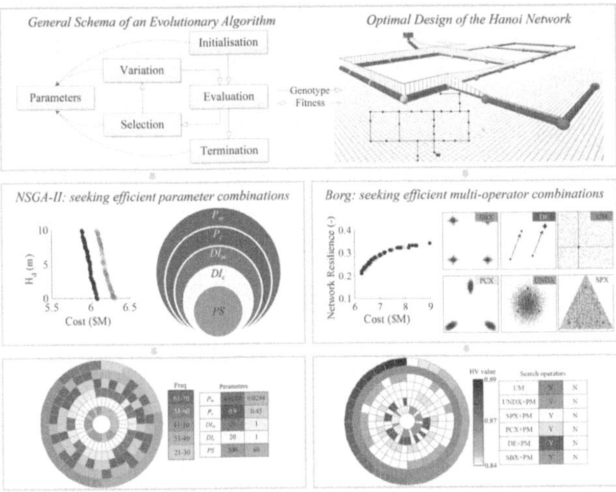

INTRODUCTION

Using evolutionary algorithms (EAs), especially multi-objective evolutionary algorithms (MOEAs), for the optimal design of water distribution systems (WDSs) is a prominent and active field in the domain of hydroinformatics (Mala-Jetmarova *et al.* 2018). The root of the complexity associated with a WDS design problem is mainly due to its highly combinatorial nature; thus, it is impossible to visit the entire search space systematically. There exist varying types of problem formulations, including single-objective, also known as the least-cost design problems (Savic & Walters 1997); multi-objective, i.e., concerning the trade-off among a set of goals usually competing with each other (Halhal *et al.* 1997); and even many-objective, i.e., considering four or more objectives simultaneously (Fu *et al.* 2013). In recent years, MOEAs have received increasing interest from the hydroinformatics community (Kollat & Reed 2006; Reed *et al.* 2013; Sweetapple *et al.* 2014; Mortazavi-Naeini *et al.* 2015; Wang *et al.* 2015; Salazar *et al.* 2016; Liu *et al.* 2017) because they provide a viable solution to complex design problems in a heuristic way and ensure a good approximation to the Pareto-optimal front.

Despite the success of MOEAs widely reported in the literature, the level of understanding of these methods, especially for scholars and practitioners who simply take MOEAs as the search engine, has generally lagged. This is particularly the case when facing the highly sensitive parameterisation issues associated with MOEAs (Mala-Jetmarova *et al.* 2015; Walker & Craven 2018b). An MOEA is a complex methodology, also known as a black-box. Its behaviour is significantly influenced by the settings of parameters, which interact in highly non-linear ways (De Jong 2007; Phan *et al.* 2020). Identification of significant parameters and their efficient values play a key role in successful applications of MOEAs to challenging problems in the real world, which are typically associated with complex constraints and huge search spaces. In such a case, it is beneficial to understand which parameters are dominant factors requiring fine-tuning and in which directions (or scales) those parameters should be adjusted. Note that the fine-tuning process is usually problem-specific, which implies that a group of efficient settings of parameters for one case may be less efficient for another. Consequently, a sophisticated method of analysing parameters' sensitivity is necessary, especially when it is computationally intensive to solve a problem. However, this aspect is beyond the scope of this paper.

Currently, researchers in the hydroinformatics community rely mostly on the trial-and-error approach to analyse the sensitivity of the parameters or simply follow the recommended settings (Wang *et al.* 2019b). One reason for such a gap lies in the lack of effective tools, preferably a visualisation method, for understanding the hidden mechanism of MOEAs responsible for their behaviour and success. That is, how their parameterisation can impact on the success of solving the problem at hand.

Visualisation tools for diagnosing whether or not an evolutionary algorithm works are available in the literature (Hart & Ross 2001). These tools are useful in choosing operators and parameters. In contrast, evidence of relevant work related to the visualisation of MOEAs is limited. Previous studies mainly focus on visualising the population in the objective space (Fu *et al.* 2013) and tracking the dynamic variations of the population by performance indicators (Zheng *et al.* 2016). The former enables a set of non-dominated solutions to be presented to decision-makers, followed by the cost-benefit analysis to determine the final design for implementation. The latter depicts the real-time execution of MOEAs from the convergence and diversity perspectives of the population. Most recently, some attempts to visualise algorithm performance, especially considering the design optimisation of WDSs, have been reported (Johns *et al.* 2018; Walker & Craven 2018a, 2018b), in which the exploration and exploitation aspects of MOEAs can be visually conveyed to non-expert users. Later, an extended and promising visualisation approach has been presented to benchmark different parameterisations for the challenging multi- and many-objective suites of test functions (Walker & Craven 2020).

On the other hand, since around 2007, there is a trend of embedding a multi-operator search strategy in advanced MOEAs, especially in an adaptive manner (Vrugt & Robinson 2007; Moral & Dulikravich 2008; Hadka & Reed 2013; Wang *et al.* 2017). Those algorithms, usually known as hybrid MOEAs, can mitigate the curse of the 'No Free Lunch Theorem' (Wolpert & Macready 1997) to some extent by effectively utilising the advantages of diversified search operators. Among hybrid MOEAs, Borg is probably the most famous and successful method, which has been tested on many challenging optimisation problems associated with water resources management (Reed *et al.* 2013). However, it is still unclear why several search operators are used together and how their combinations impact on optimisation performance. In essence, it is another form of parameterisation issue – deciding whether a specific operator is employed as a parameter.

This paper seeks to narrow the knowledge gap between experts of MOEAs and non-expert users by developing a visualisation method that can intuitively present the complex combinatorial effects related to MOEAs. The key question is how a visualisation tool can assist in identifying the dominant factor(s) and appropriate setting of algorithm parameters for a given problem. It will be shown that the compass plot, which stems from the ancient Chinese ingenuity, can help deal with the above question.

METHODOLOGY

Conceptual origin

Feng Shui, also known as Chinese geomancy, is a Chinese art that is based on the belief that how things are arranged within a room can affect one's life (e.g., luck and fortune). The creation of the compass plot was mainly inspired by the device widely used in this domain. Figure 1(a) shows the Chinese Feng Shui compass, which is a tool used by geomancers to confirm the right directions and positions in a room. It also serves as a protection appliance to stop bad luck or potential harm. A simpler but similar idea also exists in the 40-Year calendar, as shown in Figure 1(b). This delicate widget can correctly indicate

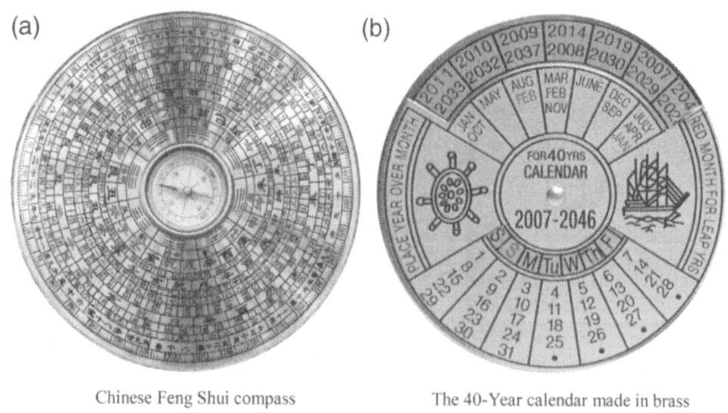

Chinese Feng Shui compass The 40-Year calendar made in brass

Figure 1 | The inspiration originated from other domains: (a) Chinese Feng Shui compass and (b) the 40-Year calendar made in brass. Please refer to the online version of this paper to see this figure in colour: http://dx.doi.org/10.2166/hydro.2020.073.

the days, weeks, and months for 40 years. Rotating the brass dial tells the corresponding dates given the year and month.

A Chinese Feng Shui compass is usually composed of Tianchi (the centric compass), an inner disc, and an outer square tray (deliberately omitted from Figure 1(a)). The inner disc can be rotated to align with the centric compass. There are many concentric annuli carved on the inner disc, one annulus is called one layer, and it represents a specific dimension of information. Each layer is divided into a number of equal-sized slices, and each slice is filled with a unique character. There may be varying numbers of slices in different layers. Since the principle of Feng Shui has a close connection with the eight diagrams (eight combinations of three whole or broken lines formerly used in divination in ancient China), a layer is divided into at least eight slices. Due to different types of the Feng Shui compass, the number of layers can be as few as five annuli or as many as 52 annuli. Thus, the number of slices increases rapidly, as new layers are added. In the extreme case, the outmost layer of the Feng Shui compass contains 384 slices. When a geomancer confirms the direction of things in the room being analysed with the Feng Shui compass, he/she can interpret the situation more easily (hierarchically and quantitatively) and provide comprehensive explanations to guide his/her client(s) in furniture placement.

As mentioned above, Feng Shui containing multi-dimensional information can be recorded and presented in a series of concentric annuli. In other words, we can use such an annular object to illustrate any type of phenomena that are influenced by different factors. Below, we will explain the development of the compass plot based on the ideas that originated from the ancient Chinese way of thinking.

Functional design

Inspired by the Chinese Feng Shui compass, we developed a six-step procedure to generate and utilise the proposed compass plot for analysing the combinatorial effects associated with multiple factors. Without loss of generality, here we refer to the independent variables considered under a specific situation as 'impact factors,' and the dependent variables influenced by these factors as 'outcome performance.' As shown in Figure 2, the procedure has six steps

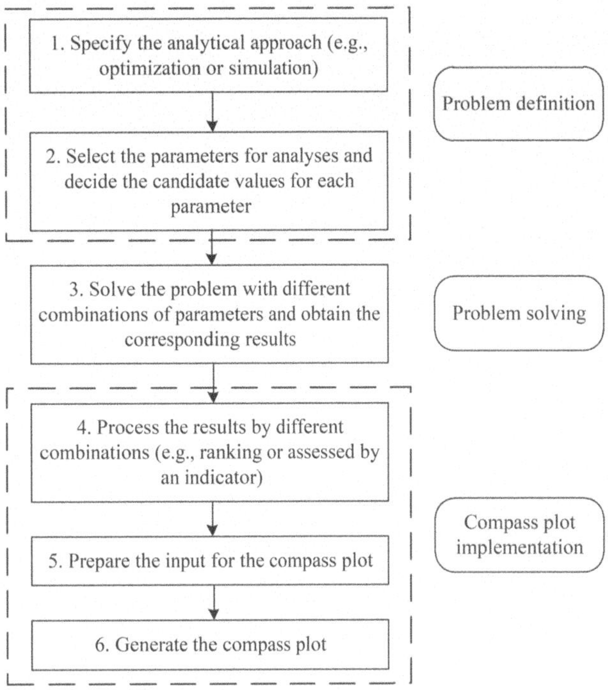

Figure 2 | The flowchart for the generation of the compass plot.

that are categorised into three stages (shown in rounded rectangles). At the first stage, the analytical approach (e.g., optimisation or simulation) is chosen when a well-structured problem is formulated (Step 1). This involves investigating the effects of different combinations of 'impact factors' on the 'outcome performance.' Then, the key impact factors are identified, and their candidate values are determined (Step 2). At the second stage, the problem is solved by the analytical approach specified in Step 1, and the corresponding results are obtained for each combination of impact factors (Step 3). At the last stage, in which the compass plot is generated and interpreted, results gained at the previous step are evaluated either based on the raw result values or by using a suitable performance indicator (Step 4). Next, the input for producing the compass plot is prepared, mainly including the specification of colours for 'impact factors' and 'outcome performance' and the layout (i.e., number and order of annuli) for the compass plot (Step 5). Finally, the compass plot is generated via the pseudo-code detailed below, and the important information is extracted based on the patterns observed in both the radial and annular directions of the compass plot (Step 6).

Function: Generation of the Proposed Compass Plot via Concentric Pie Charts

Input: colour data for impact factors and their corresponding outcome performance

Output: compass plot for a given problem

1 retrieve the key information: the number of combinations and the number of annuli;

2 **for** each annulus (from big to small) **do**

3 create an axis and set its position and size;

4 draw the pie chart and divide it into a number of equal-sized slots;

5 fill each slot with colours specified in the data for the current annulus;

6 adjust the position and size of the next concentric axis;

7 save the compass plot to the specified directory;

8 **return** the success or failure message;

Software implementation

For the sake of rapid prototyping, the implementation of the proposed compass plot (i.e., Steps 4–6) was carried out on the MATLAB® platform. Before generating the compass plot, it is necessary to process the raw data and yield the data for the colour specification. Firstly, the results are ranked, i.e., various combinations of impact factors and the corresponding performance, in the descending order of the outcome performance. Secondly, the colour settings are specified for each annulus and the slices denoting different candidate values. As a general rule-of-thumb, we prefer the vivid colours for inner annuli that indicate different impact factors and the shades of grey colours for the outmost annulus that illustrate the outcome performance (darker grey means better performance). For different candidate values, varying shades of the same colour are used. The colour data for each annulus are stored in an individual cell array, and the colour data for each slice are expressed in the RGB colour values, which is a vector of 1-by-3 with elements encoding the shades of the red, green, and blue colours, respectively. Finally, the proposed compass plot is rendered via a series of concentric annuli (pie charts in MATLAB®) with tailored positions and sizes (see the above pseudocode for detail).

Interpretation of the compass plot

The merit of the proposed compass plot lies in the fact that the patterns identified in both the radial and annular directions reveal essential messages. In the radial direction, the first few slices (depending on the scope of the top grey level in the outermost annulus) in the counter-clockwise direction (the default order in a pie chart produced by MATLAB®) demonstrate the combinations of factors that have a great impact on outcome performance. Furthermore, within these slices, the most significant factor(s) and its corresponding value(s) can be visualised if one candidate value appears frequently in a certain annulus, especially if it happens consecutively. Conversely, the insensitive factor(s) and/or its corresponding value(s) are more likely to be scattered around the annulus.

APPLICATIONS

Case study

Despite a wide range of benchmark networks available in the literature (Wang *et al.* 2015), we simply chose the Hanoi network (HAN), which is a widely used case study in the research community of WDS for demonstration purposes (Fujiwara & Khang 1990). Figure 3 depicts the layout of the HAN. It consists of 34 pipes, which form three basic loops. Water is supplied to a flat region (elevation equal to zero) comprised of 31 demand nodes from one reservoir with a fixed head of 100 m. The required minimum pressure head at each node is 30 m. There are six commercially available pipe sizes, and their corresponding unit costs are listed in Table 1. As a result, the search space of the HAN problem contains $6^{34} \approx 2.87 \times 10^{26}$ discrete combinations.

Problem formulation

Two kinds of problem formulations for WDS design purposes were considered in this paper. In the first definition, we converted the least-cost design to the bi-objective optimisation by transforming the constraints of the minimum required heads at nodes into the second objective, so that an MOEA can be applied directly to handle the least-cost

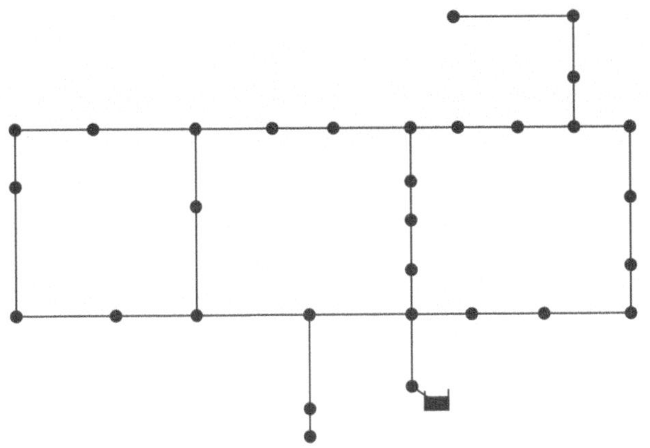

Figure 3 | The layout of the HAN.

Table 1 | Unit costs of the available pipe sizes for the HAN network

Diameter (mm)	304.8	406.4	508.0	609.6	762.0	1,016.0
Cost ($/m)	45.73	70.40	98.39	129.33	180.75	278.28

design problem. Such a relaxed form of definition may be practical in dealing with real-world problems under a tight budget (Muhammed *et al.* 2017), as a marginal violation of the minimum head requirement can be much cheaper than the strictly feasible solutions. In the second definition, we took both the capital cost and the network reliability into consideration, which is a widely used formulation adopted by many studies (Raad *et al.* 2010; Wang *et al.* 2015; Liu *et al.* 2017). The network resilience index (Prasad & Park 2004), denoted as I_n, was selected as a surrogate indicator for the network reliability because I_n has been proven to have a strong correlation with the mechanical and hydraulic reliability of WDSs among a wide range of surrogate reliability measurements (Raad *et al.* 2010; Liu *et al.* 2017; Wang *et al.* 2019a).

Definition 1: a relaxed form of the least-cost design of WDSs

The first definition is aimed at minimising both the capital cost and the total head deficit across the network. This transforms the hard constraints considered in the least-cost design of WDSs to soft ones. The mathematical definition of the relaxed least-cost design is given in Equation (1):

$$\begin{cases} \text{minimise } Cost = \sum_{i=1}^{np} C(D_i)L_i \\ \text{minimise } H_d = \sum_{j=1}^{nn} (H_j^{\min} - H_j), \quad \text{if } H_j < H_j^{\min} \\ \text{subject to } D_i \in \{A\}, \quad i = 1, 2, \ldots, np \end{cases} \tag{1}$$

where np is the total number of pipes in a network; nn is the total number of nodes in a network; $C(D_i)$ is the unit cost of pipe i with a diameter size of D_i; L_i is the length of pipe i; H_j is the total head at node j; $H_j^{\min}$ is the minimum head required at node j; H_d is the total head deficit across nodes; and A is the set of available pipe sizes in the local market.

Definition 2: the multi-objective design of WDSs

The second definition is aimed at minimising the capital cost and maximising I_n, approaching the trade-off between cost and reliability. The mathematical definition of the bi-objective design is given in Equation (2):

$$\begin{cases} \text{minimise } Cost = \sum_{i=1}^{np} C(D_i)L_i \\ \text{maximise } I_n = \dfrac{\sum_{j=1}^{nn} C_j Q_j (H_j - H_j^{\min})}{\left[\sum_{k=1}^{nr} Q_k H_k + \sum_{m=1}^{npu} \left(\frac{P_m}{\gamma}\right)\right] - \sum_{j=1}^{nn} Q_j H_j^{\min}} \\ \text{subject to } D_i \in \{A\}, \quad i = 1, 2, \ldots, np; \quad H_d = 0 \end{cases} \tag{2}$$

$$C_j = \frac{\sum_{i=1}^{np_j} D_i^j}{np_j \times max\{D_i^j\}} \tag{3}$$

where C_j is the uniformity of node j which is defined in Equation (3); Q_j is the demand at node j; nr is the number of reservoirs; Q_k and H_k are outflow and head at reservoir k, respectively; npu is the number of pumps; P_m is the power supplied by pump m; γ is the specific weight of water; np_j is the number of pipes connected to node j; and D_i^j is the diameter of pipe i connected to node j.

Scenarios

Two scenarios were demonstrated in this paper to show how the proposed compass plot can assist in exploring the combinatorial effects associated with the optimal design of WDSs by MOEAs. The first scenario considers how the parameterisation of the well-known NSGA-II (Deb *et al.* 2002) impacts on the quality of least-cost solutions under Definition 1. The second scenario investigates how the composition of search operators within Borg, which is an advanced MOEA (Hadka & Reed 2012, 2013), impacts on the convergence and diversity of the Pareto fronts under Definition 2. Note that the applications can also be extended to other fields in which MOEAs are used as the optimisation tools.

Scenario 1: parameterisation of NSGA-II

NSGA-II is arguably the most popular MOEA in the water research community, which can effectively handle a wide range of multi-objective optimisation problems. After a random initialisation process, the standard NSGA-II implements a fast non-dominated sorting approach to ranking solutions primarily based on the Pareto dominance relationship. Solutions with higher rankings survive and are selected to reproduce. For those solutions that are non-dominated to each other (i.e., with the same ranking), a secondary sorting criterion, known as the crowding distance, is used to discriminate solutions further (i.e., solutions that reside in a less crowded area are preferred). During the selection process, a binary tournament selection is carried out to establish the mating pool (i.e., parents). NSGA-II applies the Simulated Binary Crossover (SBX) and the Polynomial Mutation (PM) with designated probabilities to create children from parents (Deb *et al.* 2002). The implicit elitism strategy ensures that the best solutions ever found in the search history are always retained in the population. This allows a new population to be derived from the combination of parents and their offspring via the fast non-dominated sorting approach.

The literature contains many papers applying NSGA-II for WDS design tasks (Wang *et al.* 2013; Basupi & Kapelan 2015; Creaco *et al.* 2016; Zheng *et al.* 2016; Beygi *et al.* 2019). However, most previous studies followed the trial-and-error approach to determine the proper settings of NSGA-II's parameters (Wang *et al.* 2019b). This approach may not be sufficient to reveal the complex relationships among various parameters. Therefore, we selected NSGA-II as the representative MOEA to study the relationship between its parameters and the quality of solutions under Definition 1.

Specifically, we chose five key parameters of NSGA-II, including the population size (*PS*), the probability of SBX (P_c), the probability of PM (P_m), and the distribution indices of SBX and PM (denoted as DI_c and DI_m, respectively). For the sake of simplicity, we picked two candidate values for each parameter, so that there were 32 parameter combinations in total (i.e., $2^5 = 32$). Those two values for each parameter were contrasting enough to show the impact of the parameter value on the results. For *PS*, the minimum and the maximum values were 60 and 300, respectively. The recommended values for P_m and P_c were adopted (Deb *et al.* 2002), which were 1/34 (i.e., the inverse of the numbers of decision variables) and 0.9, respectively. A much higher P_m and a much lower P_c were also selected, and they were 1/17 (i.e., two times the recommended value for P_m) and 0.45 (i.e., half the recommended value for P_c), respectively. The boundary values of 1 and 20 were set for both DI_m and DI_c. A distribution index of 1 is small enough to guarantee the diversity of the offspring, and a distribution index of 20 is large enough to make the offspring focus on the area near their parents.

Parameterisation of NSGA-II was investigated by solving the HAN problem with the 32 combinations of NSGA-II's key parameters. For each combination of parameters, 100 individual runs with different random seeds for initialisation were carried out. For each run, the number of function evaluations (NFEs) was 600,000, as it ensured a sufficient convergence for the HAN problem.

Although the trade-off between *Cost* and H_d can be obtained, the quality of solutions obtained by each combination of parameters was evaluated via the frequency of locating the best-known least-cost solution for the HAN problem in 100 runs (denoted as *Freq*). For example, if a specific parameter combination manages to identify the best-known solution in 50 runs out of 100, the corresponding *Freq* is 0.5. According to Zheng *et al.* (2017), the currently best-known least-cost solution for the HAN problem is \$6.0811M.

Scenario 2: synergistic effect of the multi-operator search strategy in Borg

Borg is based on the steady-state MOEA framework in which only one solution is generated per iteration

(Laumanns *et al.* 2002). Besides the population, it stores the best solutions found so far in an external archive. In each iteration of the main loop, Borg generates one child using randomly selected parents. For a recombination operator requiring k parents, one parent is chosen uniformly at random from the archive. The remaining $k-1$ parents are discovered from the population using tournament selection. The child resulting from the recombination process is then evaluated by the objective functions, then it is considered for acceptance in both the population and the archive.

Borg consists of seven search operators, including SBX, Differential Evolution (DE), Parent-Centric Crossover (PCX), Simplex Crossover (SPX), Unimodal Normal Distribution Crossover (UNDX), Uniform Mutation (UM), and PM (Hadka & Reed 2013). During each iteration, Borg implements the so-called auto-adaptive multi-operator recombination process. In particular, it uses only one group of search operators for offspring reproduction, by either UM only or one crossover operator (including DE) followed by PM. The successful operators, which contribute more solutions to the archive, are rewarded by increasing their probabilities of being selected for recombination. On the other hand, no operators are completely discarded during the execution of optimisation.

Borg also implements several advanced strategies concurrently, including ε-dominance, ε-progress (a measure of convergence speed), adaptive population sizing, and randomised restart (Hadka & Reed 2013). It proves to be a flexible and robust hybrid MOEA by covering a large area of high-performing parameterisations (Hadka & Reed 2012; Reed *et al.* 2013). However, previous studies did not focus on how the different combinations of search operators can affect its performance. Therefore, in this scenario, we focused on the synergistic effect of the multi-operator search strategy in Borg under Definition 2.

We considered all the possible combinations of the default seven operators. As mentioned above, Borg uses only one group of search operators for recombination per iteration and PM plays a subordinate role. Besides, UM is essential during the restart process and thus cannot be removed. Consequently, Borg has a total of 32 different combinations, that is using only UM, all default operators, and all the possible combinations of five crossover alike operators (i.e., $C_5^1 + C_5^2 + C_5^3 + C_5^4 = 30$). For each combination of operators, 50 individual runs were implemented with the same NFEs as used for Scenario 1.

The quality of solutions obtained by each combination of search operators was assessed concerning diversity and convergence of the Pareto fronts. The hypervolume (Zitzler & Thiele 1999) indicator (denoted as HV) was used to measure the quality of Pareto fronts in a comprehensive sense. The range of HV is between zero and one, and a larger value of HV indicates better performance. The median value of HV over 50 runs was selected to represent the performance of a particular combination of search operators. Also, the usage of each operator on average over the optimisation history was monitored.

RESULTS AND DISCUSSION

Scenario 1: parameterisation of NSGA-II

Figure 4 shows the compass plot that demonstrates the impact of the parameterisation of NSGA-II on the least-cost design of the HAN problem. As mentioned above, there are 32 combinations of the five key parameters; thus, the compass plot has six annuli (five parameters and one *Freq* value) and is divided into 32 slices accordingly. Each inner annulus represents a specific parameter. We use red, yellow, green, blue, and purple to denote PS, DI_c, DI_m, P_c, and P_m, respectively. A lighter colour means the smaller candidate value for a specific parameter. The *Freq* values, which appear in the outermost annulus, are illustrated in shades of grey, and the values are categorised into five levels, as shown in the legend of Figure 5.

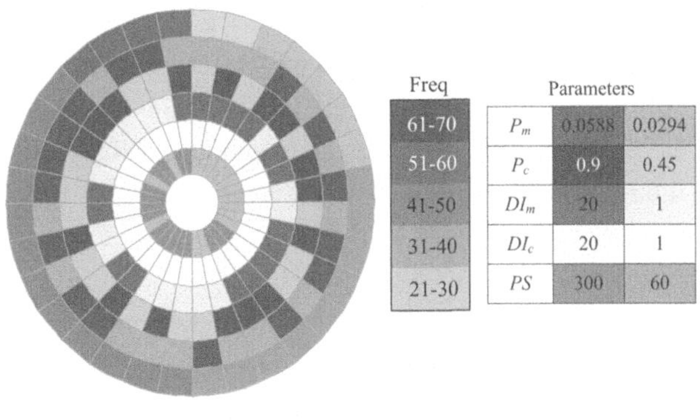

Figure 4 | Parameterisation of NSGA-II applied to the HAN design problem. Please refer to the online version of this paper to see this figure in colour: http://dx.doi.org/10.2166/hydro.2020.073.

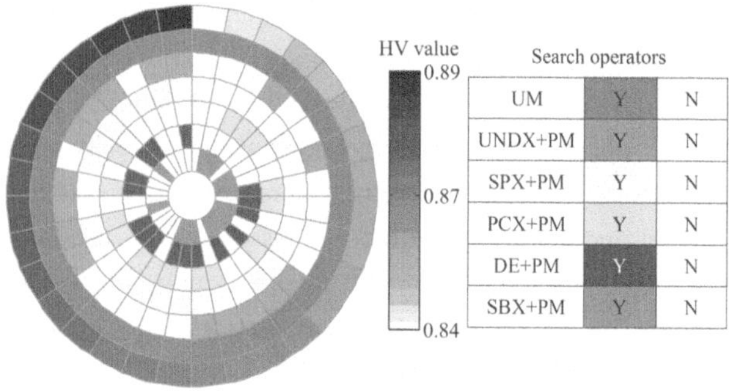

Figure 5 | Combinatorial effects of search operators within Borg according to HV (the coloured slice in each annulus indicates that a specific search operator is used; otherwise, a blank slice is shown). Please refer to the online version of this paper to see this figure in colour: http://dx.doi.org/10.2166/hydro.2020.073.

Additionally, the whole compass plot is sorted in the descending order of the *Freq* values counter-clockwise. Each slice contains five coloured slots and one grey slot illustrating a particular combination of parameters and its corresponding performance. For instance, the first slice shows that the combination with $PS = 300$, $DI_c = 1$, $DI_m = 20$, $P_c = 0.9$, and $P_m = 1/34$ gained the best performance with a *Freq* of 68%.

As mentioned in the last subsection of Methodology, a significant parameter has particular patterns in the compass plot. Specifically, a colour patch (slice) repeating in an annulus aligned with the scope of the high-performing grey levels (problem-specific), preferably in a consecutive manner, represents an efficient candidate value selected for the corresponding parameter. This pattern indicates that the parameter with the selected value can ensure superior performance no matter how the other parameters are set. On the contrary, a secondary parameter tends to disperse, encircling its annulus.

PS appears to have a dominant impact on the performance of NSGA-II. Larger values of *PS* can significantly improve the chance of identifying the best-known solution for the HAN problem in a single run. Within the top half combinations (16 in total) that managed to identify the best-known solution at a *Freq* higher than 50%, there exist 13 combinations with a *PS* of 300. All 16 combinations with a *PS* of 300 ensure a *Freq* higher than 40%. In contrast, using a *PS* of 60 is more likely to result in a *Freq* of less than 40%.

DI_m and DI_c turn out to be more critical parameters compared with P_c and P_m. This is probably surprising and inspiring as most previous studies using NSGA-II did not fine-tune DI_m and DI_c (Wang *et al.* 2019b). When both parameters were set to 1, all the related combinations (eight in total) found the best-known solution at a *Freq* higher than 40%. Seven combinations out of eight were able to achieve a *Freq* higher than 50%, and three combinations had a *Freq* higher than 60%. In particular, for the best combination with DI_m and DI_c set to 1 (i.e., the second slice counter-clockwise), P_c and P_m were equal to 0.45 and 1/17, which are far from the recommended values in the literature (Deb *et al.* 2002).

Scenario 2: synergistic effect of the multi-operator search strategy in Borg

Figure 5 shows the compass plot that demonstrates the impact of the auto-adaptive multi-operator strategy in Borg on the multi-objective design of the HAN problem. We considered 32 combinations of the six groups of search operators (UM alone and five crossover alike operators followed by PM); thus, the compass plot has seven annuli and is divided into 32 slices accordingly. Each inner annulus represents a group of operators (see the legend in Figure 5). We use red, orange, yellow, green, blue, and purple to denote UM, UNDX + PM, SPX + PM, PCX + PM, DE + PM, and SBX + PM, respectively. A slot shown in colour means a specific group of operators was used; otherwise, it is shown as blank. As mentioned above, UM plays an essential role in Borg; thus, it was included in all combinations of operators (i.e., the complete red annulus). The *HV* values, which appear in the outermost annulus, are illustrated in shades of grey, and the colour values are scaled to show the differences more clearly. Similarly to Figure 4, the whole compass plot is sorted in the descending order of the *HV* values counter-clockwise. Each slice containing six coloured/blank slots and one grey slot illustrates a particular combination of search operators and its corresponding performance. For instance, the first slice shows the combination of DE + PM, UNDX + PM, and UM gained the best performance with an *HV* of 0.89.

Although the variations of *HV* values are not significant for the HAN problem, the best performance was not gained by Borg with the default combination of all seven operators. Quite the opposite: this combination ranks 21st among the 32 combinations. This observation is somewhat counter-intuitive as users who are non-experts in MOEAs are inclined to follow the default settings with application to problems they are interested in. Note that the differences in terms of *HV* can be increased when larger design problems are considered. This implies that using the default combination of operators in Borg can be suboptimal or even misleading. Figure 5 also reveals that UM, UNDX + PM, and SPX + PM have a more profound impact on Borg's performance compared with the other groups of operators.

On the other hand, it is necessary to investigate the overall usage of each group of operators over 50 independent runs because Borg applies only one group per iteration for the recombination process. It is interesting to know whether a frequently used operator is an essential ingredient in Borg. As given in Supplementary Table S1, SBX + PM and SPX + PM were frequently used, with the overall usage across 50 runs and all possible combinations equal to 54 and 40%, respectively. In contrast, UM and UNDX + PM were less used (both around 16%). This observation generally corresponds to the situation when the default combination of operators in Borg was employed (21st in the *HV* ranking). This reveals that SBX + PM and SPX + PM were able to contribute more solutions to the archive compared with UM and UNDX + PM.

However, the best performance was achieved by the combination of UM, UNDX + PM, and DE + PM. Even in this combination, DE + PM was superior to its counterparts. It seems to be paradoxical that the combination relying only on UM and UNDX + PM ranks 9th among the 32 combinations, but the combination of UM, SBX + PM, and SPX + PM ranks only 27th. Moreover, in the top half of combinations, only four cases rely on the group of SBX + PM, but nine cases rely on the groups of UM and UNDX + PM. The other groups of operators consistently dominated the usage of UM and/or UNDX + PM. In other words, strong groups of operators in Borg are not as essential as expected, while the weak groups of operators are indispensable.

The phenomenon that operators with a minor contribution to the archive have a significant impact on the performance of Borg can be attributed to the mechanism of the auto-adaptive multi-operator strategy. Because no operators are lost during the execution of Borg, the contribution from the weak groups, probably at the early stage of optimisation, can still boost the search by other operators. The weak groups of operators, which are employed occasionally afterwards, can also take advantage of the contributions by other groups. Additionally, the paradox mentioned above emphasises that the tailored selection of search operators for a given type of problem (e.g., design of WDSs) can be beneficial as proved in Wang *et al.* (2017).

CONCLUSIONS

This paper proposes an innovative visualisation tool for investigation of the challenging combinatorial effects commonly found in many fields. The tool is called the compass plot and inspired by the traditional Chinese Feng Shui compass. We demonstrate how the compass plot can be used to extract the key messages effectively by identifying the patterns in both the radial and annular directions. This new approach can help identify the impact factors and the corresponding values that have a more significant impact on a given problem in an intuitive way. Two scenarios related to the applications of MOEAs to the optimal design of the HAN are considered, including the parameterisation of NSGA-II and the auto-adaptive multi-operator strategy in Borg.

The compass tool applied to the least-cost design of the Hanoi problem reveals that *PS* is the most critical parameter of NSGA-II, followed by DI_m and DI_c. In contrast, P_c and P_m, which are generally regarded as key parameters by most previous studies, play a secondary role. More interestingly, when applied to investigate the synergistic effect of the auto-adaptive multi-operator strategy in Borg, this tool identifies the paradox hidden inside the advanced Borg framework. That is, weak operators with a minor contribution to the archive have a significant impact on the performance of Borg. In contrast, strong operators are not as essential as expected. As a result, for a specific type of problem such as the multi-objective design of WDSs, using the default combination of operators can be suboptimal or even misleading. The key messages identified by the proposed compass plot contribute new knowledge to the research community; thus, it can guide users who rely on these MOEAs for coping with more challenging problems in the real world.

Note that the compass plot is not restricted to EAs that are closely connected with optimisation. It can also be useful to demonstrate the hidden and complex interactions among various parameters associated with the calibration of simulation models where a set of best calibration parameters is needed. Despite the value of the compass plot, it is worth noting that users can still suffer the 'dimensional explosion,' in which many impact factors and their candidate values (e.g., more than two) are involved. It can lead to computational overheads that are too high to bear or far more complex patterns in the compass plot to interpret. In such a case, users have to resort to alternatives for eliminating less important factors and limiting the number of candidate values. The compass plot is envisaged, however, to benefit the hydroinformatics community in deciding the effective values for key parameters for a given type of problem. An online version of such a tool, which provides real-time feedback on the efficiency of parameterisation, may also contribute to the dynamic or interactive parameter settings to further boost the power of MOEAs.

DATA AVAILABILITY STATEMENT

All relevant data are available from an online repository or repositories (http://dx.doi.org/10.17632/cgh6bpzjfs.1).

REFERENCES

Basupi, I. & Kapelan, Z. 2015 Flexible water distribution system design under future demand uncertainty. *Journal of Water Resources Planning and Management* **141** (4), 04014067.

Beygi, S., Tabesh, M. & Liu, S. 2019 Multi-objective optimization model for design and operation of water transmission systems using a power resilience index for assessing hydraulic reliability. *Water Resources Management* **33** (10), 3433–3447.

Creaco, E., Franchini, M. & Todini, E. 2016 The combined use of resilience and loop diameter uniformity as a good indirect measure of network reliability. *Urban Water Journal* **13** (2), 167–181.

Deb, K., Pratap, A., Agarwal, S. & Meyarivan, T. 2002 A fast and elitist multiobjective genetic algorithm: NSGA-II. *IEEE Transactions on Evolutionary Computation* **6** (2), 182–197.

De Jong, K. 2007 Parameter setting in EAs: a 30 year perspective. In: *Parameter Setting in Evolutionary Algorithms* (F. G. Lobo, C. F. Lima & Z. Michalewicz eds.), pp. 1–18. Spinger, Berlin, Heidelberg.

Fu, G., Kapelan, Z., Kasprzyk, J. & Reed, P. 2013 Optimal design of water distribution systems using many-objective visual analytics. *Journal of Water Resources Planning and Management* **139** (6), 624–633.

Fujiwara, O. & Khang, D. 1990 A two-phase decomposition method for optimal design of looped water distribution networks. *Water Resources Research* **26** (4), 539–549.

Hadka, D. & Reed, P. 2012 Diagnostic assessment of search controls and failure modes in many-objective evolutionary optimization. *Evolutionary Computation* **20** (3), 423–452.

Hadka, D. & Reed, P. 2013 Borg: an auto-adaptive many-objective evolutionary computing framework. *Evolutionary Computation* **21** (2), 231–259.

Halhal, D., Walters, G., Ouazar, D. & Savic, D. 1997 Water network rehabilitation with structured messy genetic algorithm. *Journal of Water Resources Planning and Management* **123** (3), 137–146.

Hart, E. & Ross, P. 2001 GAVEL – a new tool for genetic algorithm visualization. *IEEE Transactions on Evolutionary Computation* **5** (4), 335–348.

Johns, M., Walker, D. J., Keedwell, E. & Savic, D. 2018 Interactive visualisation of water distribution network optimisation. In: *The 13th International Conference on Hydroinformatics* (G. L. Loggia, G. Freni, V. Puleo & M. D. Marchis eds.). EasyChair, Palermo, Italy, pp. 995–1003.

Kollat, J. B. & Reed, P. M. 2006 Comparing state-of-the-art evolutionary multi-objective algorithms for long-term groundwater monitoring design. *Advances in Water Resources* **29** (6), 792–807.

Laumanns, M., Thiele, L., Deb, K. & Zitzler, E. 2002 Combining convergence and diversity in evolutionary multiobjective optimization. *Evolutionary Computation* **10** (3), 263–282.

Liu, H., Savic, D., Kapelan, Z., Creaco, E. & Yuan, Y. 2017 Reliability surrogate measures for water distribution system design: comparative analysis. *Journal of Water Resources Planning and Management* **143** (2), 04016072.

Mala-Jetmarova, H., Barton, A. & Bagirov, A. 2015 Sensitivity of algorithm parameters and objective function scaling in multi-objective optimisation of water distribution systems. *Journal of Hydroinformatics* **17** (6), 891–916.

Mala-Jetmarova, H., Sultanova, N. & Savic, D. 2018 Lost in optimisation of water distribution systems? a literature review of system design. *Water* **10** (3), 307.

Moral, R. J. & Dulikravich, G. S. 2008 Multi-objective hybrid evolutionary optimization with automatic switching among constituent algorithms. *AIAA Journal* **46** (3), 673–681.

Mortazavi-Naeini, M., Kuczera, G., Kiem, A. S., Cui, L., Henley, B., Berghout, B. & Turner, E. 2015 Robust optimization to secure urban bulk water supply against extreme drought and uncertain climate change. *Environmental Modelling & Software* **69**, 437–451.

Muhammed, K., Farmani, R., Behzadian, K., Diao, K. & Butler, D. 2017 Optimal rehabilitation of water distribution systems using a cluster-based technique. *Journal of Water Resources Planning and Management* **143** (7), 04017022.

Phan, H. D., Ellis, K., Barca, J. C. & Dorin, A. 2020 A survey of dynamic parameter setting methods for nature-inspired swarm intelligence algorithms. *Neural Computing & Applications* **32** (2), 567–588.

Prasad, T. D. & Park, N.-S. 2004 Multiobjective genetic algorithms for design of water distribution networks. *Journal of Water Resources Planning and Management* **130** (1), 73–82.

Raad, D. N., Sinske, A. N. & van Vuuren, J. H. 2010 Comparison of four reliability surrogate measures for water distribution systems design. *Water Resources Research* **46**, W05524.

Reed, P. M., Hadka, D., Herman, J. D., Kasprzyk, J. R. & Kollat, J. B. 2013 Evolutionary multiobjective optimization in water resources: the past, present, and future. *Advances in Water Resources* **51**, 438–456.

Salazar, J. Z., Reed, P. M., Herman, J. D., Giuliani, M. & Castelletti, A. 2016 A diagnostic assessment of evolutionary algorithms for multi-objective surface water reservoir control. *Advances in Water Resources* **92**, 172–185.

Savic, D. A. & Walters, G. A. 1997 Genetic algorithms for least-cost design of water distribution systems. *Journal of Water Resources Planning and Management* **123** (2), 67–77.

Sweetapple, C., Fu, G. & Butler, D. 2014 Multi-objective optimisation of wastewater treatment plant control to reduce greenhouse gas emissions. *Water Research* **55**, 52–62.

Vrugt, J. A. & Robinson, B. A. 2007 Improved evolutionary optimization from genetically adaptive multimethod search. *Proceedings of the National Academy of Sciences* **104** (3), 708–711.

Walker, D. J. & Craven, M. J. 2018a Toward the online visualisation of algorithm performance for parameter selection. In: *International Conference on the Applications of Evolutionary Computation* (K. Sim & P. Kaufmann eds.). Springer, Parma, Italy, pp. 547–560.

Walker, D. J. & Craven, M. J. 2018b Visualising the operation of evolutionary algorithms optimising water distribution network design problems. In: *International Conference on Hydroinformatics* (G. L. Loggia, G. Freni, V. Puleo & M. D. Marchis, eds.). EasyChair, Palermo, Italy, pp. 2250–2258.

Walker, D. J. & Craven, M. J. 2020 Identifying good algorithm parameters in evolutionary multi- and many-objective optimisation: a visualisation approach. *Applied Soft Computing* **88**, 105902.

Wang, Q., Savić, D. A. & Kapelan, Z. 2013 Hybrid metaheuristics for multi-objective design of water distribution systems. *Journal of Hydroinformatics* **16** (1), 165–177.

Wang, Q., Guidolin, M., Savić, D. & Kapelan, Z. 2015 Two-objective design of benchmark problems of water distribution system via MOEAs: towards the best-known approximation to the true Pareto front. *Journal of Water Resources Planning and Management* **141** (3), 04014060.

Wang, Q., Savić, D. A. & Kapelan, Z. 2017 GALAXY: A new hybrid MOEA for the optimal design of water distribution systems. *Water Resources Research* **53** (3), 1997–2015.

Wang, Q., Huang, W., Wang, L. B., Zhan, Y. S., Wang, Z. H., Wang, X., Zhan, X. Y. & Liu, S. M. 2019a How are various surrogate indicators consistent with mechanical reliability of water distribution systems: from a perspective of many-objective optimization. *Water* **11** (8), 1669.

Wang, Q., Wang, L., Huang, W., Wang, Z., Liu, S. & Savić, D. A. 2019b Parameterization of NSGA-II for the optimal design of water distribution systems. *Water* **11** (5), 971.

Wolpert, D. H. & Macready, W. G. 1997 No free lunch theorems for optimization. *IEEE Transactions on Evolutionary Computation* **1** (1), 67–82.

Zheng, F., Zecchin, A. C., Maier, H. R. & Simpson, A. R. 2016 Comparison of the searching behavior of NSGA-II, SAMODE, and Borg MOEAs applied to water distribution system design problems. *Journal of Water Resources Planning and Management* **142** (7), 04016017.

Zheng, F., Zecchin, A. C., Newman, J. P., Maier, H. R. & Dandy, G. C. 2017 An adaptive convergence-trajectory controlled ant colony optimization algorithm with application to water distribution system design problems. *IEEE Transactions on Evolutionary Computation* **21** (5), 773–791.

Zitzler, E. & Thiele, L. 1999 Multiobjective evolutionary algorithms: a comparative case study and the strength Pareto approach. *IEEE Transactions on Evolutionary Computation* **3** (4), 257–271.

First received 30 April 2020; accepted in revised form 26 October 2020. Available online 10 November 2020

© 2021 The Authors

Flood mitigation in coastal urban catchments using real-time stormwater infrastructure control and reinforcement learning

Benjamin D. Bowes, Arash Tavakoli, Cheng Wang, Arsalan Heydarian, Madhur Behl, Peter A. Beling and Jonathan L. Goodall

ABSTRACT

Flooding in coastal cities is increasing due to climate change and sea-level rise, stressing the traditional stormwater systems these communities rely on. Automated real-time control (RTC) of these systems can improve performance, and creating control policies for smart stormwater systems is an active area of study. This research explores reinforcement learning (RL) to create control policies to mitigate flood risk. RL is trained using a model of hypothetical urban catchments with a tidal boundary and two retention ponds with controllable valves. RL's performance is compared to the passive system, a model predictive control (MPC) strategy, and a rule-based control strategy (RBC). RL learns to proactively manage pond levels using current and forecast conditions and reduced flooding by 32% over the passive system. Compared to the MPC approach using a physics-based model and genetic algorithm, RL achieved nearly the same flood reduction, just 3% less than MPC, with a significant 88× speedup in runtime. Compared to RBC, RL was able to quickly learn similar control strategies and reduced flooding by an additional 19%. This research demonstrates that RL can effectively control a simple system and offers a computationally efficient method that could scale to RTC of more complex stormwater systems.

Key words | real-time control, reinforcement learning, smart stormwater systems, urban flooding

Benjamin D. Bowes
Arash Tavakoli
Cheng Wang
Arsalan Heydarian
Madhur Behl
Peter A. Beling
Jonathan L. Goodall (corresponding author)
Department of Engineering Systems and Environment,
University of Virginia, P.O. Box 400747,
Charlottesville, VA 22904,
USA
E-mail: *goodall@virginia.edu*

HIGHLIGHTS

- Reinforcement learning (RL) creates policies for real-time coastal stormwater system control.
- RL's ability to mitigate flooding and manage ponds is compared to a passive system, model predictive control, and rule-based control.
- RL was more efficient than model predictive control using a physics-based model and genetic algorithm.
- RL's ability to mitigate flooding exceeded the passive system and rule-based control.

This is an Open Access article distributed under the terms of the Creative Commons Attribution Licence (CC BY-NC-ND 4.0), which permits copying and redistribution for non-commercial purposes with no derivatives, provided the original work is properly cited (http://creativecommons.org/licenses/by-nc-nd/4.0/)

doi: 10.2166/hydro.2020.080

INTRODUCTION

As the frequency and intensity of storms increase due to changes in climate, the ability of existing stormwater infrastructure to mitigate urban flooding is being increasingly stressed (Mynett and Vojinovic 2009; Berggren *et al.* 2012; Neumann *et al.* 2015; Wuebbles *et al.* 2017; Mounce *et al.* 2020). In coastal cities, gravity-driven stormwater systems are critical for flood management, but their functionality is also being reduced by sea-level rise (Sadler *et al.* 2020). These stressors, combined with the flat, low elevation topography of many coastal cities, mean that these communities are already experiencing increased flooding during high tide events (Sweet and Park 2014; Moftakhari *et al.* 2015, 2017).

Advances in urban hydroinformatics (Mynett and Vojinovic 2009), including smart stormwater systems (Kerkez *et al.* 2016), provide promising tools and approaches to improve stormwater system performance in coastal communities. In the smart stormwater system approach, existing stormwater systems are retro-fitted with real-time sensors and actuators (e.g., remotely controlled valves and pumps) to allow active monitoring and control. Active control is a cost-effective way to more efficiently use the existing capacity of stormwater infrastructure (Jose Meneses *et al.* 2018). In addition, active control can allow a system to function as a whole, which can be much more effective than a piece-wise capital improvement of passive infrastructure systems (Wong and Kerkez 2018).

Key to the effectiveness of active systems is the use of real-time control (RTC) (Schwanenberg *et al.* 2015; Kerkez *et al.* 2016; Mounce *et al.* 2020). RTC uses streaming sensor data (i.e., current rainfall and retention pond depths) to approximate the current system states. The system state can then be used to inform changes to control assets (e.g., valves, pumps) that adapt the behavior of the system to current or forecast conditions. The decisions on when and what structures to control, and how to change them in order for a system to meet certain objectives (e.g., minimize flooding, maintain certain flow conditions), are based on a control policy. In a smart stormwater system, control policies map system states, such as water levels in a pipe network, to actions that need to be taken in order to meet management objectives (Sadler *et al.* 2019). In current practice, control policies are often predefined simple heuristics, such as opening a valve when a storage pond reaches a certain depth (level control or feedback control) and may be based on the experience of a human operator (García *et al.* 2015; Abou Rjeily *et al.* 2018). This heuristic approach may be effective in simple systems (i.e., a system with only a few controlled assets); however, it requires that control actions are predefined for all scenarios and becomes increasingly difficult as the number of assets grows and/or more external factors start influencing the system.

Heuristic control can be improved by incorporating some aspects of feedback control (i.e., system observations) and feed-forward control (i.e., forecasts and predicted system states) (Schwanenberg *et al.* 2015), termed rule-based control (RBC) in this research. RBC can be implemented in stormwater systems based on watershed characteristics and forecast rainfall data in order to meet flooding, water quality, and/or channel protection objectives (Marchese *et al.* 2018). By continuously monitoring retention pond depths and rainfall forecasts, inflow from storm events can be estimated using simple rainfall–runoff models and used to inform control decisions. For example, if a storm event has been forecast, the available volume in a pond can then be adjusted based on the estimated inflow from the storm event. This adjustment is made by actuating a valve at the pond's outlet that can be opened to drain water until the necessary storage volume in the pond is reached. RBC is intuitive to end-users and can be effective for controlling individual stormwater system components (OptiRTC and Geosyntec Consultants Inc. 2017). However, as system complexity increases (e.g., releases of water need to be coordinated if multiple ponds drain to the same downstream location in order to prevent flooding), RBC becomes more difficult to implement.

Two approaches for finding control policies for RTC beyond simple heuristics and RBC include model predictive control (MPC) and reinforcement learning (RL). MPC uses a process model to evaluate various control strategies for a specified control horizon (Camacho and Bordons 2007; García *et al.* 2015). It has been successfully applied to large-scale water systems for multi-objective optimization

(Schwanenberg *et al.* 2015) and for stormwater system control (Sadler *et al.* 2019, 2020). Sadler *et al.* (2019) examined the effectiveness of MPC for a tidally influenced stormwater system and successfully found control policies that reduced flooding compared to the passive system. This particular formulation of MPC used a physics-based stormwater system model optimized with a genetic algorithm. Using high-performance and cloud-based computing, the authors were able to speed up this computationally expensive MPC formulation. However, as a limitation of this approach, the authors indicate that for large real-world stormwater systems, such MPC techniques could be prohibitively slow for RTC. When scaled to part of a real-world stormwater system, Sadler *et al.* (2020) were able to run MPC for three simulated actuators and determined that RTC could help reduce flooding for future sea-level rise scenarios.

RL is a category of machine learning that aims to learn from trial-and-error experience by interacting with an environment (Sutton and Barto 2018). An RL agent does not have known answers to learn from but instead tries to maximize the amount of reward it can receive from its environment by taking certain actions. One of the differences between RL and MPC is that RL can learn control policies offline, which moves the computational burden to prior to taking any actions. The use of RL in water resources engineering has been compared with Dynamic Programming (DP) for multi-objective reservoir management (Lee and Labadie 2007; Castelletti *et al.* 2013, 2014; Pianosi *et al.* 2013; Delipetrev *et al.* 2017). These studies used tabular RL methods, which can be more computationally efficient than DP but are constrained to simple systems (i.e., a small number of potential system states and actions) and demonstrated that a physics-based model could act as the environment for RL agents to train on. Recent advances in deep learning have allowed RL to use neural networks as function approximators to overcome the limitations of tabular RL. For instance, Mullapudi and Kerkez (2018) demonstrated control of a stormwater system with a Deep Q-Network (DQN) (Mnih *et al.* 2015). By throttling valves, they were able to make control decisions for a discrete action space (a limitation of DQN). Their work also highlighted how rewards can be shaped for real-time stormwater control through deep RL and illustrated how RTC with RL can increase a stormwater system's effective capacity. However, their DQN agent was only reactive to the current conditions of the stormwater system; effective RTC strategies should be based on both the current conditions and forecasts. Additionally, discretizing control actions may not always provide the most efficient policy. Further research is needed, therefore, to determine if RL control can be further refined using a continuous action space that allows any valve position to be used and create forecast-based predictive control policies.

In this paper, RL is used to create control policies for RTC of a tidally influenced stormwater system. In addition to presenting the first work exploring RL for RTC of tidally influenced stormwater systems, we advance on prior work in smart stormwater by exploring the Deep Deterministic Policy Gradient (DDPG) RL algorithm to control valves over a continuous action space, allowing the agent to learn more refined control policies. Additionally, forecasts of rainfall and tide are included as part of the state information received by the RL agent, allowing the agent to learn proactive control strategies. The RL agent's performance is compared to (i) a passive, gravity-driven system (ii) MPC (as implemented by Sadler *et al.* (2019)), and (iii) RBC. Through this comparison, we illustrate the applicability of RL for RTC of a simulated coastal urban stormwater system.

METHODOLOGY

In this section, the simulated stormwater system is introduced, a hypothetical scenario for categorizing the impact of flood events is detailed, and each RTC method is described. The stormwater system is similar to that used by Sadler *et al.* (2019) and is the same for all scenarios (RL, Passive, MPC, RBC), except that retention ponds in the passive system have weirs at a fixed elevation to maintain a depth of water in the ponds while the ponds for the RTC scenarios have a controllable valve at the bottom of the pond side. An overview of the methodology used to compare the performance of an RL-controlled stormwater system with the other scenarios is provided in Table 1. Open-source code for these scenarios is available on github (Bowes 2020b). SWMM simulation files and data are available as a resource on HydroShare (Bowes 2020a).

Table 1 | Overview of methodology for stormwater system control scenarios

Method	Valve	Training/optimization	Control policy	Testing
RBC system	Controllable valve at 0 m	N/A	Fixed for simulation duration	Test on 2010–2019 data
MPC system	Controllable valve at 0 m	Online optimization with genetic algorithm	Adjusted valve positions for specified time horizon based on modeled scenarios and objective functions	Test on first week of August 2019 data (due to computational cost)
RL system	Controllable valve at 0 m	Offline training on August 2019 data for 197,000 control steps	Learned policy relating states to actions	Test on 2010–2019 data
Passive system	Fixed weir at 0.61 m	N/A	N/A	Test on 2010–2019 data

Stormwater simulation

Stormwater system simulations are carried out using the U.S. Environmental Protection Agency's Stormwater Management Model (SWMM), version 5. The hypothetical stormwater system used in this study is a simple model inspired by conditions within an urban catchment located in Norfolk, Virginia, USA. It consists of two subcatchments, two storage units (retention ponds), and pipes going to the system outfall that discharges to a tidally influenced waterbody (Figure 1). In the passive scenario, each pond has a weir at a fixed elevation and cannot be completely emptied. Infiltration and evaporation are excluded in this simple system as the focus is on flood control. Ponds in the RTC scenarios have been retro-fitted by removing the weirs and adding controllable valves (orifices) at the bottom of the pond side. This allows the full pond volume to act as active storage that can be adapted for different storm events. Input to this system is rainfall, which falls onto a subcatchment and is transformed into runoff that flows into a storage unit. Flow out of the storage units can be regulated by the valves; both ponds drain directly into a single node before flowing through two pipe segments to the outfall. The tail water condition at the outfall is influenced by the tide level. At high tide, the tail water condition can cause sea level to partially block the outfall, backing up stormwater drainage. The physical parameters of the SWMM model can be found in Table 2.

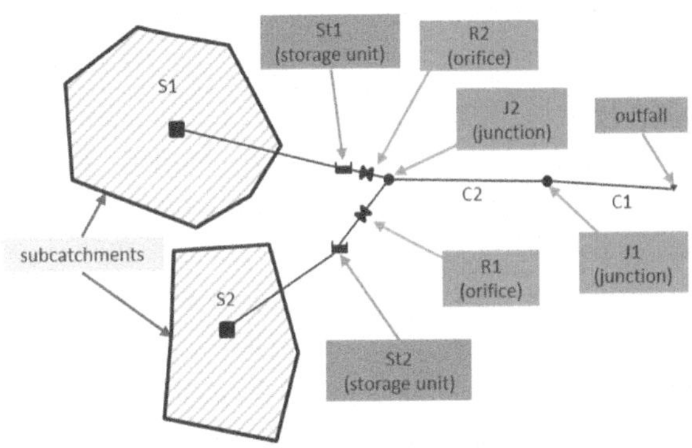

Figure 1 | SWMM simulation schema.

In this SWMM model, flooding can be caused by (i) rainfall, (ii) high tide, or (iii) a combination of rainfall and tide. Flooding caused by these factors can be in the form of the ponds over-topping or the downstream node J1, which can be thought of as a roadway storm drain, surcharging. Ponds can over-top if the subcatchment runoff volume for a storm event exceeds current pond capacity and inflow is greater than outflow. Flooding at the downstream node can occur if the flow from the ponds is not regulated and coordinated by adjusting the two valves. Node J1 can also flood if tidal conditions at the outfall are preventing the normal flow of water from the system or causing backflow if the tide is especially high.

The pyswmm (McDonnell *et al.* 2020) Python wrapper for SWMM is used for step-by-step running of simulations as needed for the RTC methods. Each control scenario can be updated once every 15 min (an adjustable control time step).

SWMM input data

Input data for the SWMM simulation come from observed data for stations in Norfolk, Virginia (Figure 2). Rainfall data are collected at a 15-min timestep by the Hampton

Table 2 | Properties of SWMM model

Subcatchment properties				
Name	Area (hectares)	Width (km)	Slope (%)	Impervious (%)
S1	32	0.46	0.5	30
S2	20	0.61	0.5	45
Storage unit properties				
Name	Surface area (m^2) (constant)	Initial depth (m)	Max depth (m)	Bottom Elev (m)
St1	6,039	0.61	1.41	0.91
St2	4,645	0.61	1.41	0.91
Pipe properties				
Name	Length (m)	Diameter (m)	Roughness	
C1, C2	122	0.3	0.01	
Node properties				
Name	Max depth (m)	Bottom Elev (m)		
J2	1.5	0.91		
J1	0.6	0.30		
Outfall	NA	0		
Orifice properties				
Name	Height (m)	Discharge coefficient		
R1 and R2	0.61	0.65		

Roads Sanitation District; rainfall data for subcatchment S1 is from gauge Rain1 and data for subcatchment S2 is from gauge Rain2. Rainfall data are cleaned through a number of checks. First, any values over the 1000-year, 15-min rainfall for Norfolk (59.18 mm) are assumed to be erroneous and removed. Next, missing data in each rainfall time series are replaced; if both rain gauges are missing values at the same time stamp, both get zero. Otherwise, if one station is missing a value but the other is not, the missing value is replaced with the known value from the other station. Observed tide level data come from the National Oceanic and Atmospheric Administration Sewells Point gauge and are measured at 6-min intervals. For use as an SWMM boundary condition, tidal data are resampled to an hourly interval and referenced to the North American Vertical Datum of 1988 (NAVD88). All of the observed data are for the period between 1 January 2010 and 6 November 2019 and are divided into individual months to make simulation run times tractable.

Forecasts were created from the observed data for use in the various control methods, therefore assuming perfect knowledge of future events. A single forecast, in this case, is an array of values representing the rainfall or tide measurement over the next n time steps. For example, a 24-h forecast of 15 min rainfall would contain 96 values. In future work, noise could be added to these forecasts to explore how RL (or any other RTC method) handles uncertainty, but this is beyond the scope of this research (for future directions, see Hartono and Hashimoto (2007) and van Andel *et al.* (2008, 2014)).

Flood event classification

In order to quantify the flooding impact, a hypothetical scenario is developed from physical data for Norfolk. In this scenario, the subcatchments considered are residential neighborhoods where any flooding of the ponds will impact roadway intersections. The downstream node J1 is considered a storm drain at a roadway intersection; flooding at this node will make the intersection impassable if the depth of the flood water is above a certain threshold. For this hypothetical scenario, 0.2 m of roadway flooding slows traffic considerably, the

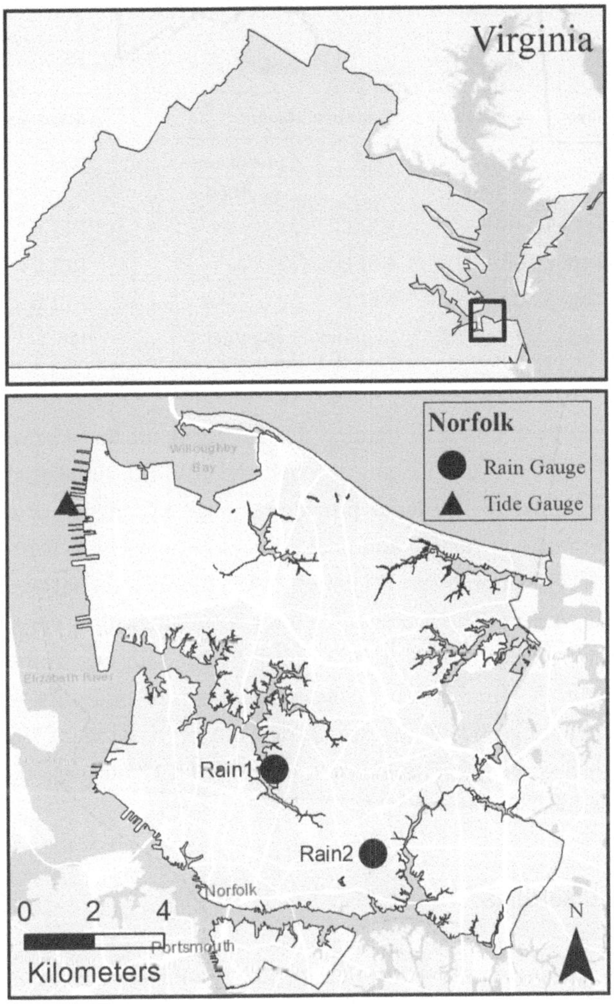

Figure 2 | Gauge locations in Norfolk, VA, USA.

threshold for safe passenger vehicle passage is 0.3 m (Pregnolato *et al.* 2017), and 0.4 m is the limit for safe emergency vehicle passage. The relationship between flood volume and depth was developed from digital elevation data from Norfolk as described in Appendix B. The number, volume, and duration of these flood events are used as an additional metric for quantifying flooding along with the total volume.

Implementing RL in stormwater systems

In RL, an agent learns to optimize its behavior by interacting with its environment (Sutton and Barto 2018). The environment is usually modeled as a Markov Decision Process: $\langle \mathcal{S}, \mathcal{A}, P, r, \gamma \rangle$, where $\mathcal{S}$ is the state space, $\mathcal{A}$ is the action space, $P(s'|s, a)$ is the stochastic probability of transitioning to a new state s' after taking action a at the current state s, $r(s, a, s')$ is the reward function, and $\gamma \in [0, 1]$ is the discount factor that weighs the importance of short-term and long-term reward. The RL agent's goal is to find an optimal policy that maximizes the expected discounted return:

$$G_t = r_t + \gamma r_{t+1} + \gamma^2 r_{t+2} + \cdots = \sum_{\infty}^{k=0} \gamma^k r_{t+k} \tag{1}$$

where $r_t = r(s_t, a_t, s_{t+1})$.

In this paper, the states $\mathcal{S}$ are defined as the current depths and rate of flooding (if any) at the ponds and downstream nodes (St1, St2, J1), the current valve positions (R1, R2), the sum of the 24 h rainfall forecast (F) for each subcatchment, and the mean value of the 24 h tide forecast. These values are gathered from the SWMM simulation at each control time step. The actions $\mathcal{A}$ that the RL agent can take at any step are to close or open any valve to any degree. Finally, the reward r the RL agent receives in this system is based on how well the agent prevents flooding and maintains certain target pond water depths. It is defined as:

$$r = \begin{cases} -\Sigma(\text{flooding}) & F > \delta \\ -\text{J1}_{\text{flooding}} - (|\text{St1}_{\text{depth}} - \text{target}| + |\text{St2}_{\text{depth}} - \text{target}|) & F = 0 \end{cases} \tag{2}$$

where flooding is the flooding rate at each particular node (St1, St2, J1) and δ represents a forecast rainfall threshold (>0 in this case); target is the target water depth for the storage ponds (St1 and St2). In this relatively simple stormwater system, the target depth is 0.61 m for both ponds. In a real system, different ponds would most likely have different target depths; this can be taken into account in the RL implementation by having different target depths for each pond in the reward function.

As an example, consider a case where the agent is in a specific state s in $\mathcal{S}$ (e.g., the water depth in a specific pond is 1.0 m) and takes an action a in $\mathcal{A}$ (e.g., completely opens the valve) with a probability given by the policy $\pi(a|s)$. The agent will then transition to a new state s' with a probability of $P^a_{s,s'} = P(s'|s, a)$ (e.g., the water depth in a specific pond is 0.75 m) and receives a reward $r(s, a, s')$.

The value of this action depends on the reward that the agent receives and the discounted value of all the future rewards if the agent follows the policy afterwards. Using a discount factor γ, the value of a future reward of x after n steps is $x\gamma^{n-1}$. The expected discounted return when starting in state s, then taking action a, and following π is called the Q-value function:

$$\begin{aligned} Q^{\pi}(s, a) &= E[G_t|s, a] \\ &= r(s, a, s') + \gamma \sum_{s' \in \mathcal{S}} P^{a}_{s,s'} \sum_{a' \in \mathcal{A}} \pi(a'|s') Q^{\pi}(s', a'), \end{aligned} \quad (3)$$

where the second equation is known as the Bellman equation (Sutton and Barto 2018).

By having the optimal Q-values, one can find the optimal policy by finding the specific actions in each state that give the maximum Q-value. However, due to the curse of dimensionality, this tabular type of Q-learning is limited to problems with relatively small state and action spaces. Recent advances in deep learning have been applied to RL to overcome this problem by using deep neural networks to approximate value functions instead of storing them in tables (Mnih *et al.* 2015).

In order to have an RL agent that can set the valves to any position over a continuous action space, the Deep Deterministic Policy Gradients (DDPG) (Lillicrap *et al.* 2015) actor-critic algorithm is used. DDPG uses deep neural networks to approximate a policy and the difference between policies, the gradient, is used to update the agent. In this case, the agent consists of two parts: an actor which represents the policy, and a critic which estimates the q-value of actions taken by the actor. During the training process, the actor is fed information on the state of the stormwater system and outputs the actions to be taken. These actions, along with the state information, are used as input to the critic. The actions and q-value estimates output from the critic are used to update the agent.

The keras-rl (Plappert 2016), openai gym (Brockman *et al.* 2016), and Tensorflow (Abadi *et al.* 2016) python packages are used to implement the DDPG algorithm for this research. Each part of the DDPG agent, the actor and the critic, is composed of a deep feed-forward neural network (Table 3). The hyperparameters of each neural network are determined by trial and error (Maier *et al.* 2010). Through experimentation, it can be found that training the RL agent on the August 2019 dataset and looping through the SWMM simulation approximately 100 times provided enough experience of a wide range of rainfall and tidal events for the agent to learn from. This month has a total of 256.54 mm of rainfall over seven events. The average tide level is 0.16 m with a maximum value of 1.01 m from Tropical Storm Erin late in the month. A visualization of these data is given in Figure 4. RL training and testing are carried out on a standard PC with 8 cores, 16GB RAM, and an NVIDIA Quadro P2000 Graphical Processing Unit.

Table 3 | DDPG RL agent architecture and hyperparameter settings

	Actor		Critic	
NN layer	**Neurons**	**Activation**	**Neurons**	**Activation**
Input	Current state s	N/A	Current state s and action a	N/A
Hidden 1	16	RELU	32	RELU
Hidden 2	16	RELU	32	RELU
Hidden 3	8	RELU	32	RELU
Output	1 [R1, R2]	Sigmoid	1 [q-value]	Linear

MPC settings

The swmm_mpc software developed by Sadler *et al.* (2019) is used to implement MPC for comparison with RL; readers are referred to this paper for full details on the MPC implementation. Briefly, swmm_mpc uses SWMM as a process model and an evolutionary algorithm to search for a control policy. At each time step in an SWMM simulation, swmm_mpc runs many variations of the SWMM simulation in order to determine which control actions minimize an objective function for a specified time horizon. In this case, the objective function is based on the amount of flooding and deviations from target water level depths as:

$$\text{MPC objective function} = \alpha(\boldsymbol{a} \cdot \boldsymbol{v}(\boldsymbol{u}, \boldsymbol{x})) + \beta(\boldsymbol{b} \cdot \boldsymbol{d}(\boldsymbol{u}, \boldsymbol{x})) \quad (4)$$

where $\boldsymbol{v}$ and $\boldsymbol{d}$ are one-dimensional vectors of flood volumes at each node and deviations from target depths, respectively. The two-dimensional vectors $\boldsymbol{u}$ and $\boldsymbol{x}$ represent the control policies for all controls and the system states, respectively. The user-defined parameters and their definitions are given in Table 4. The scalar multipliers α and β are overall weights for the cost of flooding and the cost of water-level deviations. These will be adjusted in order to optimize the MPC control.

Because of the computational expense of running swmm_mpc, where many variants of the SWMM model have to be executed at each simulated time step to find the best control actions, a high-performance computer (HPC) was used to run the software. The HPC computational resources consisted of 28 cores with a CPU speed of 2.4 GHz, an Intel Xeon processor, and 128 GB RAM.

Rule-based control

RBC was implemented based on documented industry-standard methods (OptiRTC and Geosyntec Consultants Inc. 2017; Wright and Marchese 2017; Marchese *et al.* 2018). In practice, this type of control uses forecasts of rainfall and watershed characteristics to inform the control of valves on stormwater assets (wet/dry ponds, bioswales) in order to meet flood control, water quantity, and/or quality objectives.

Because the current research is done on a simulated system, the expected flood volume from a forecast of rainfall, if any, is used to control the level of water in an individual pond. For example, if a forecast storm event is expected to cause 1,000 m^3 of flooding, the pond's outlet valve would open before the storm in order to drain out a corresponding volume of water plus a 20% safety factor. After the pond's depth is drawn down by the appropriate level, the valve can be closed to retain the incoming stormwater, which helps improve water quality. In this way, storm runoff should not flood the pond and will be retained to prevent flooding downstream. After a storm event, water can be held in the pond for a specified settling period (24 h in this case) and then slowly released (over 24 h) to bring the pond back to its standard operating range. Outside of storm events, rules can also be in place to maintain the pond level within the operating range or maintain certain flow conditions. The exact control rules and their hierarchy, as implemented in this research, are detailed in Figure 3.

Table 4 | MPC cost function parameters

Parameter (description)	Value
α (overall flood weight)	Scalar
a (individual node flood weight [St1, St2, J1, J2])	[1, 1, 1, 1]
β (overall deviation weight)	Scalar
b (individual deviation weight [St1, St2, J1, J2])	[1, 1, 0, 0]
Target depths (m)	[0.61, 0.61, NA, NA]

RESULTS

Comparison of RL and passive system

A comparison of the RL agent's policy against the passive system shows that the agent can learn to effectively control valves to maximize its reward. As indicated in Figure 4, the training data show valves are opened when rainfall is in the forecast, allowing additional storage space in the retention ponds. After a storm is over, valve positions are adjusted again in order to maintain a pond depth close to the target of 0.61 m. Following this policy allows the RL agent to reduce total flood volume for this month by approximately 70% (5,936 vs. 19,957 m^3) compared to the passive system. For example, in the first storm of August 2019, both valves (R1 and R2) are opened to drain water in response to the rainfall forecast (a detailed figure with this comparison is available in Appendix A, Figure 1). However, due to the difference in rainfall on the subcatchments, R2 closes earlier than R1 in order to maintain the target depth. Directly after storm events, the RL agent tries to balance returning the ponds to the target depth and preventing flooding downstream at J1. Because the RL agent also needs to maintain the target pond depths compared to the passive system, there is an increase in the number of events at the downstream node, despite these being minor events.

Applying the policy learned on the August 2019 training data to the test sets shows that this RL agent has learned a policy that works well in many other conditions (Appendix A, Figure 2). Compared to the passive system, total flooding was reduced by RL in 85 of the 120 months of data (71%). In particular, this policy works well on test sets with similar (e.g., August 2018) or larger (e.g., September 2016) amounts of flooding than the August 2019 training set (the mean total

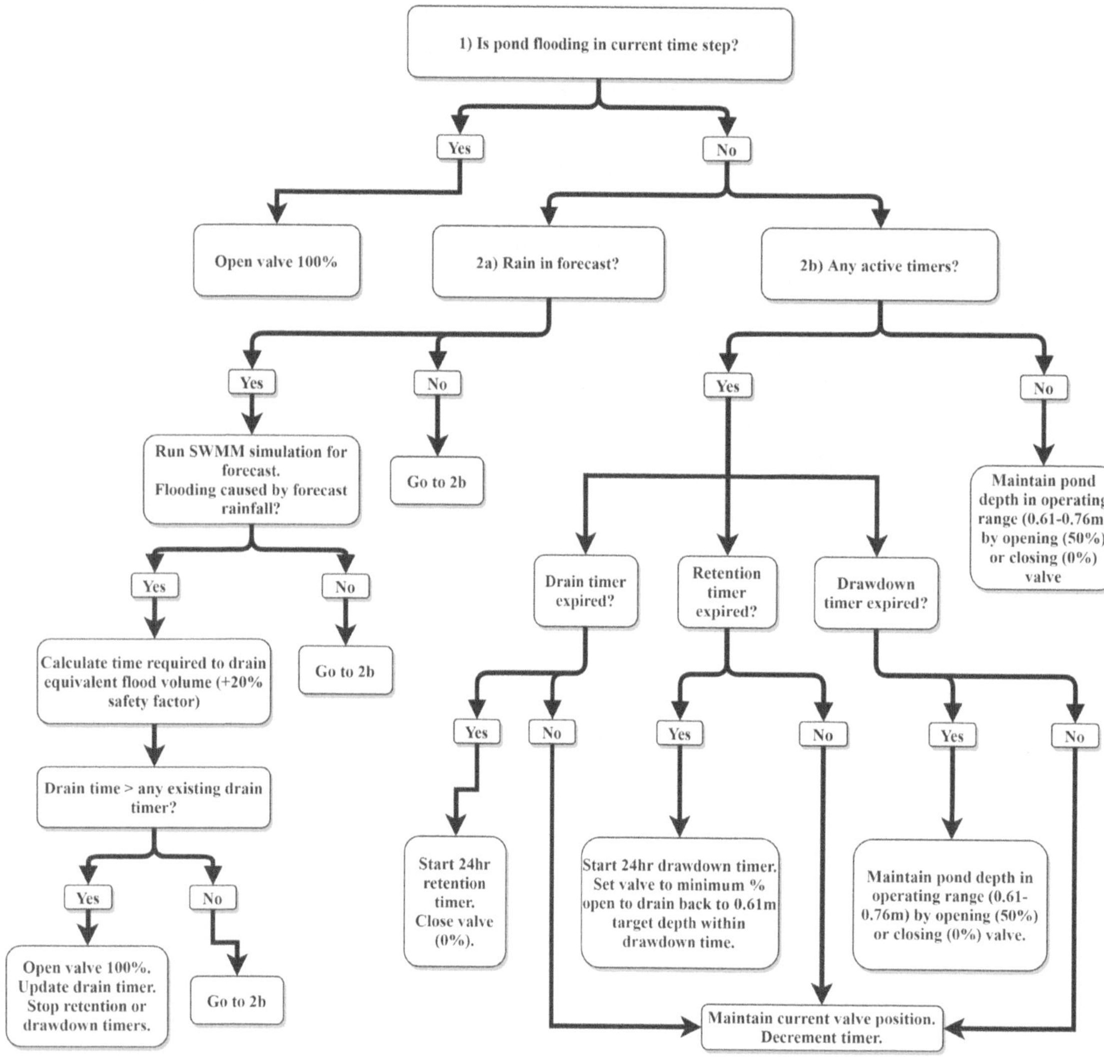

Figure 3 | Rule-based control hierarchy and settings.

flood volume for these months was 4,278 m^3). In a few cases (such as April 2019 or September 2017), the RL policy increases the amount of flooding. These are months with little or no flooding (mean of 606 m^3) and the agent has learned to respond to rainfall events in a manner that is not ideal for these months with less flood risk. The agent's performance on these months can be improved by increasing the threshold value for rainfall forecasts used in the conditional reward. For example, if the conditional reward threshold is increased from 0 to 1.3 mm of rain, the agent's performance on months with low flooding is improved (Appendix A, Figure 2, 'RL: 1.3'). However, this is at the expense of performance on the larger storms. Overall, the agent trained with the 1.3 mm threshold had lower flooding than the agent trained with the 0.0 mm threshold in 28 months (23% of the data), but increased flooding in the remaining 92 months (77%).

Comparison of RL and MPC

In order to make computational expense tractable, the MPC setup from Sadler *et al.* (2019) was only run using data from the first week in August 2018. However, due to the nature of this MPC formulation (online optimization using a genetic algorithm), computational times are still high. Finding an MPC policy for this week of data took almost 50 h (2 days,

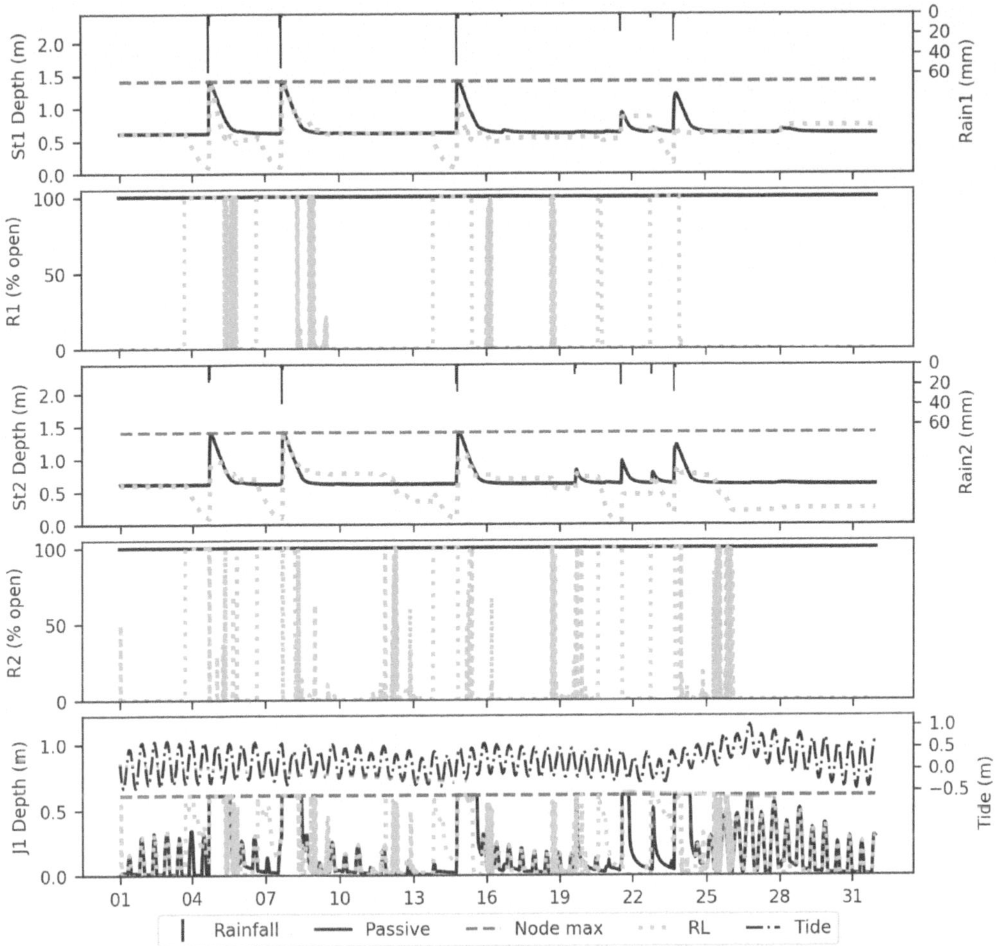

Figure 4 | Comparison of RL controlled and passive system performance on August 2019 training data.

1 h, and 48 min); computational time for each 15 min simulated control step is 3.9 min. This is tractable for a simple system but is partly a function of the simulation length and would increase with system complexity (Sadler *et al.* 2019).

To investigate MPC's performance for this specific dataset, several combinations of objective function weights were tried in order to prevent flooding and maintain target storage pond depths (Table 5). The best performing of these combinations is having the flood weight set to 1 and the deviation weight set to 10 (MPC4). This result was unexpected given that an even weighting seems like it would provide the best balance of flood mitigation and pond depth maintenance. Further, this MPC formulation was the only one in which both ponds were not kept empty for the dry periods in the simulation. A visualization of the policies carried out by MPC and RL shows that while MPC

Table 5 | MPC trials and performance comparison with the passive and RL systems for the first week in August 2019

Model	Alpha (overall flood weight)	Beta (overall deviation weight)	Control horizon (h)	Total flood volume (m^3)	Accumulated deviation (m)
Passive	N/A	N/A	N/A	13,586	126.3
MPC1	1,000.0	0.5	1.0	2,767	620.2
MPC2	0.75	0.25	1.0	2,582	614.5
MPC3	1.0	1.0	1.0	2,714	591.5
MPC4	1.0	10.0	0.5	3,929	439.2
RL	N/A	N/A	N/A	4,058	234.5

modulated orifice R2 and kept pond St2 close to the target depth, orifice R1 was slightly open and static for much of the time (Figure 5). This allowed pond St1 to essentially

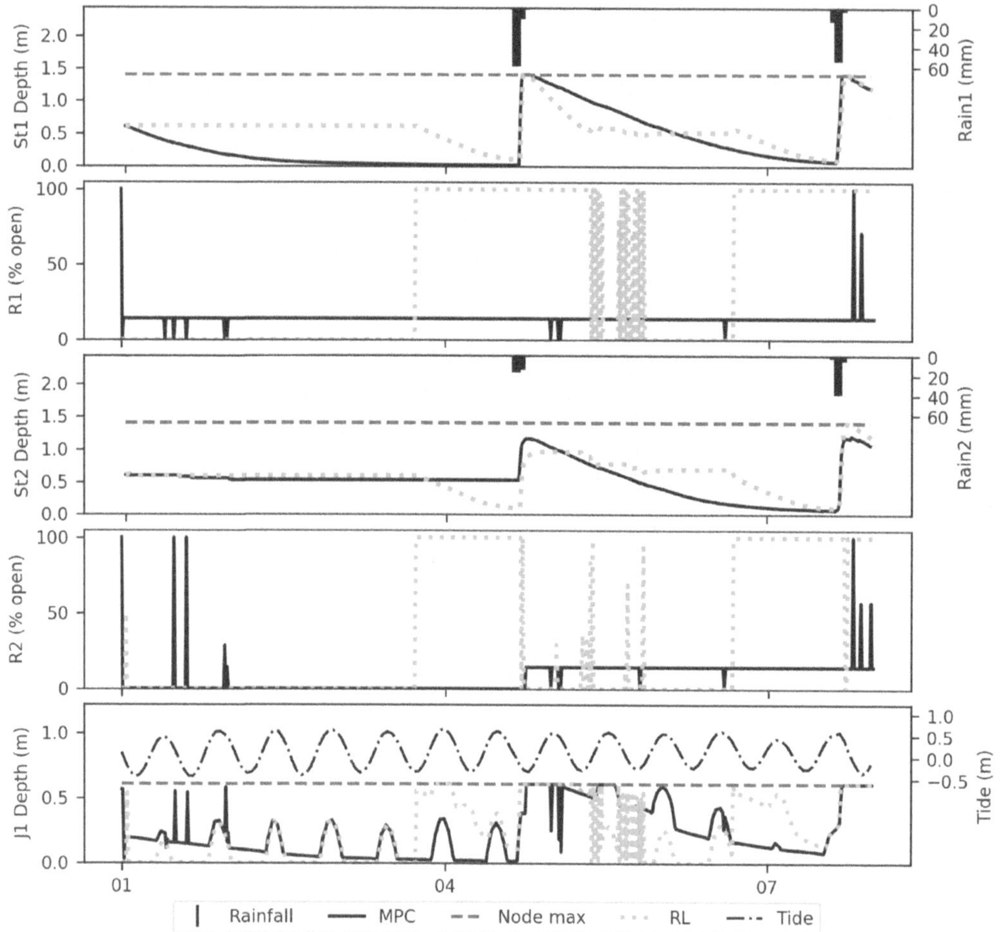

Figure 5 | RL and MPC control policies and states for the first week of August 2019.

empty during dry periods, which is an undesirable behavior. In contrast, RL was better at maintaining the target depth before the first storm and between the storm events.

Comparison of RL and RBC

RBC results were generated for the same monthly datasets used in the RL training and testing. RBC was able to reduce flooding by 57% for the month of August 2019 compared to the passive system (8,540 vs. 19,957 m^3). This method of control is also able to extend the retention time of stormwater in the ponds and maintain target pond depths during dry periods or small rainfall events. Because this RBC is based on rainfall forecasts (with perfect knowledge of future events), it is able to drawdown ponds prior to a storm event based on the expected flood volume (Figure 6).

The RL agent's performance on the August 2019 training data is similar to RBC but has an advantage in that the entire system state is used to inform control decisions, as opposed to using only the depth in the individual ponds (Figure 7). Because of this increased system knowledge and flexibility in its valve control settings, the RL agent was able to reduce flooding by 30% over RBC (5,936 vs. 8,540 m^3). The RBC logic is based on individual pond depths; conditions at other parts of the system (e.g., flooding downstream or tidal influence on the outfall) are only considered indirectly if they impact pond depth.

Over all the months of data, RBC reduced flooding over the passive system in 45 months (38%) (Figure 8). However, these were months with large total flood volumes (mean

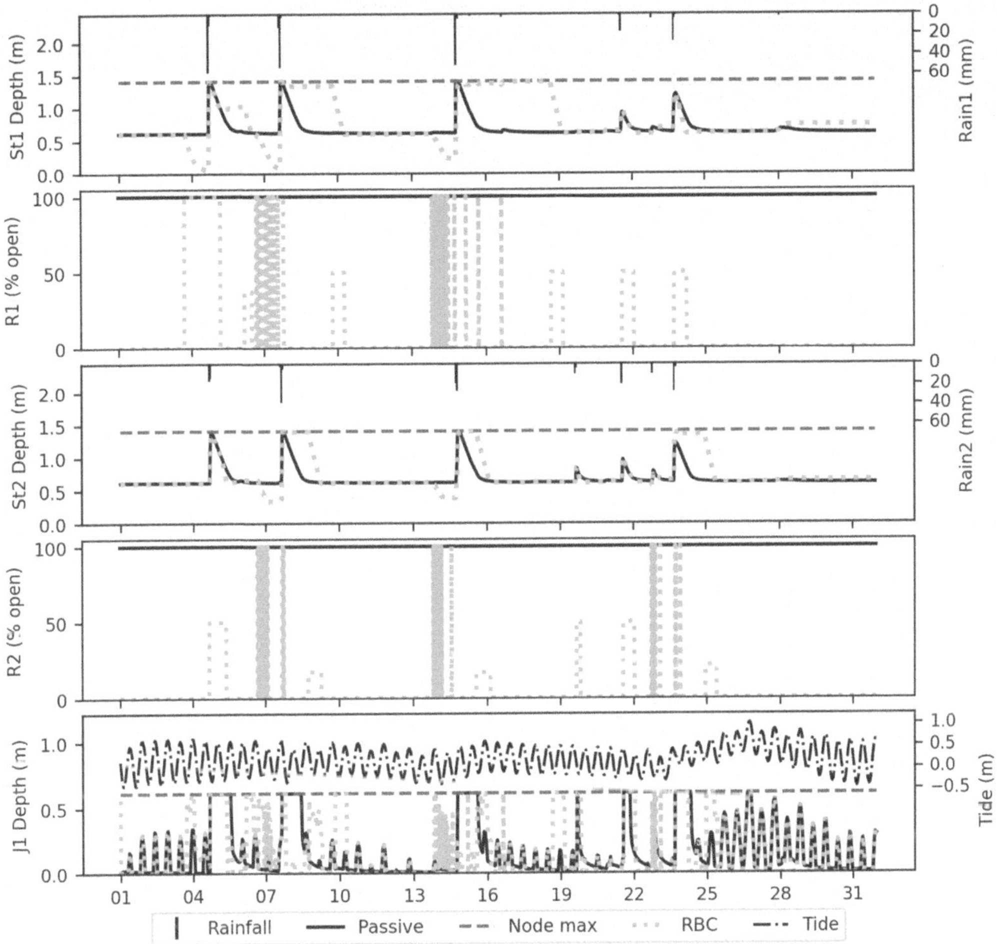

Figure 6 | Comparison of RBC and passive system performance on August 2019.

total flood volume of 8,559 m^3). Similar to RL, RBC performs less well on the months with little or no flooding of the passive system (e.g., February 2019, March 2014, March 2019) and a few months with more flooding (e.g., September 2018, October 2012). In comparison with RL, RBC had more or equal flooding in 101 months (84% of data, with a mean increase in flooding of 60%) and reduced flooding in the remaining 19 months (16% of data, with a mean decrease in flooding of 25%).

Examining the month of July 2010 shows a situation where RBC outperforms RL (see Appendix A, Figure 3). For this month, the difference in flooding between RL and RBC is relatively small (17,844 vs. 16,311 m^3, respectively) and both reduced flooding compared to the passive system (21,997 m^3), but RBC better maintained the target pond depths. This example illustrates the difficulty in shaping rewards for RL; because the conditional reward function is based on the rainfall forecast, very small amounts of rainfall will cause the agent to only be rewarded for preventing flooding. As mentioned in the RL/Passive system results, the rainfall threshold used in the reward function can influence this behavior. However, using a threshold value of >0 mm lets the agent keep pond St1 higher than the 0.61 m target depth. Because RBC maintained the target pond depths better than RL in this case, RBC did not have as much water to drain out of the ponds before the large storm event at the end of the month.

Examining the months where the passive system had lower total flooding than RBC helps illustrate its limitations. For these 75 months, the mean total flooding was 80 m^3. For

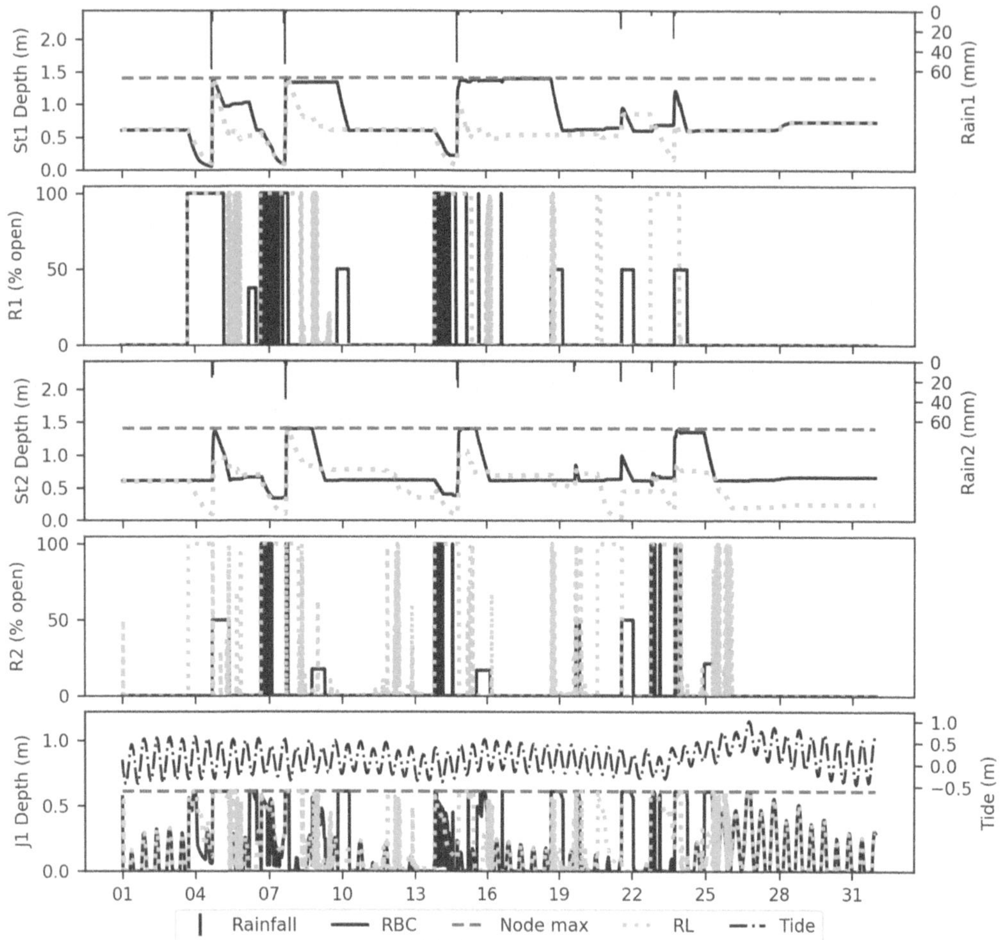

Figure 7 | Comparison of RL and RBC system performance on August 2019.

example, in March 2014, RBC increased total flooding of the system by nearly 5.5 times over the passive system (2,396 vs. 439 m^3MG) (see Appendix A, Figure 4).

Flood event classification results

Based on the flood event analysis at the two storage ponds, RL had the fewest flood events (St1: 22, St2: 26), followed by the passive system (St1: 41, St2: 42). RBC had the greatest number of pond flooding events (St1: 56, St2: 59). In terms of maximum flood volume for a single event, RL had the lowest, followed by RBC, and the passive system at pond St1 (27,558, 28,883, 30,094 m^3, respectively) and pond St2 (26,119, 27,331, 27,596 m^3). The mean single event flood volume showed a similar pattern at pond St1 (719, 719, 1,060 m^3) and pond St2 (871, 871, 1,136 m^3) for the RL, RBC, and passive systems, respectively. Flood event duration at the two ponds was similar for the three scenarios, with RL having the lowest mean duration (St1: 0.45 h, St2: 0.47 h) followed by RBC (St1: 0.51 h, St2: 0.53 h) and the passive system (St1: 0.73 h, St2: 0.66 h).

Results of flood event classification for downstream node J1 (a hypothetical roadway storm drain inlet) are shown in Figure 9. RL had the lowest number of flood events classified at the 0.2 and 0.3 m thresholds but the highest for the 0.4 m threshold (Figure 9(a)). RBC had more flood events for the 0.2 and 0.3 m thresholds than RL or the passive system, but the lowest for the 0.4 m threshold. This was expected, as these control rules manage the two ponds individually and not as a unified system considering downstream conditions. Flood volume for the 0.2 and 0.3 m thresholds was similar across the three systems

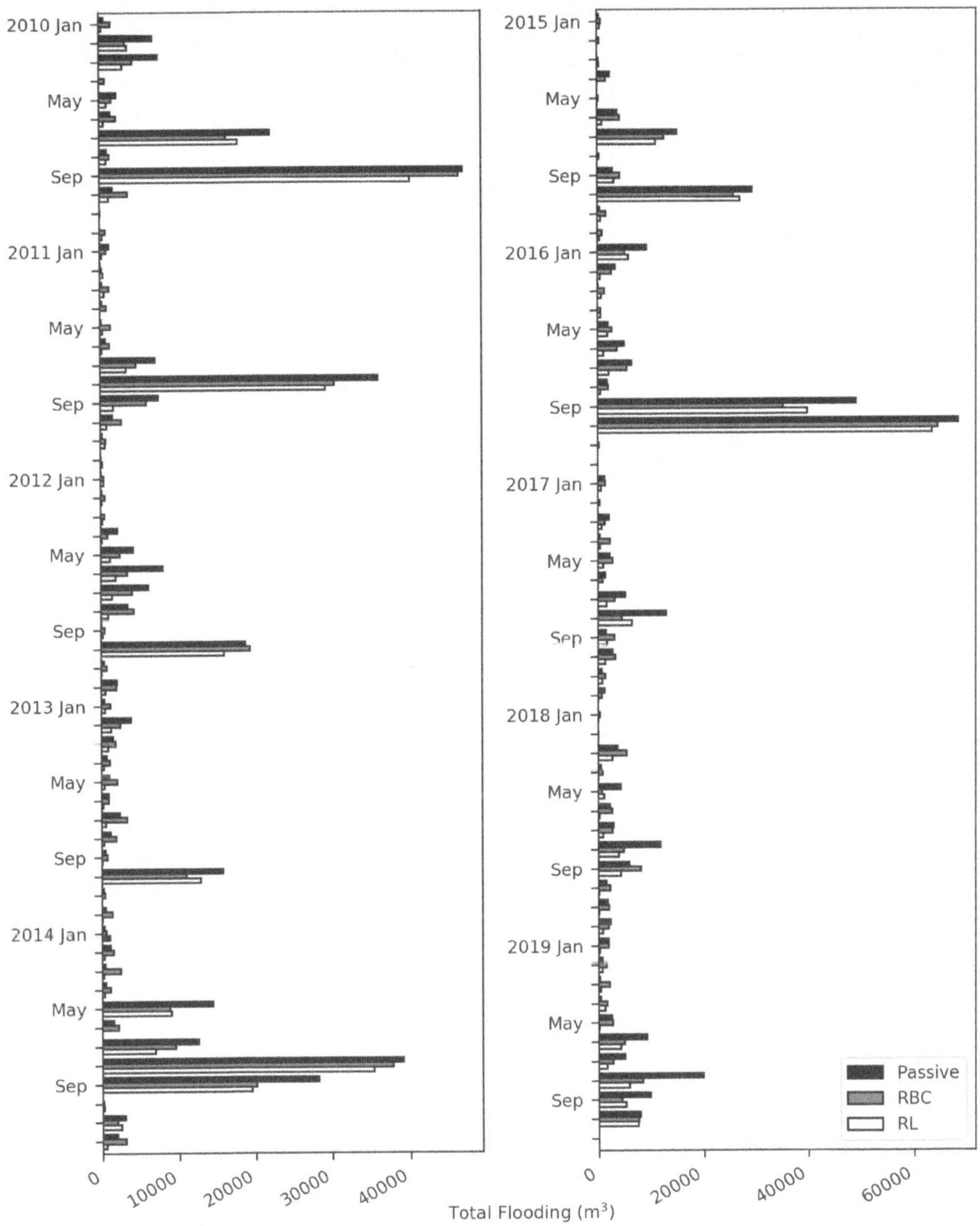

Figure 8 | Comparison of RL, RBC, and passive system performance on all months of data.

(Figure 9(c)). At the 0.4 m threshold, however, the two RTC methods (RL and RBC) had lower minimum, maximum, and mean flood volumes than the passive system.

Flood event duration for node J1 was lowest for the passive system, followed by RBC and RL (Figure 9(b)). This result makes sense when viewed in context with the flood volumes coming from the upstream ponds. RL prevented flooding at the ponds by routing more water downstream to node J1. RL had a similar mean flood duration to the passive systems, but a 60% higher maximum, indicating that RL allowed more low volume, but long duration flood events at J1 in order to reduce flooding at the two storage ponds. This behavior is influenced by the reward function; if the reward function was based on whether or not a node was flooding instead of the rate of flooding, the agent may have learned a different trade-off for managing flooding between the

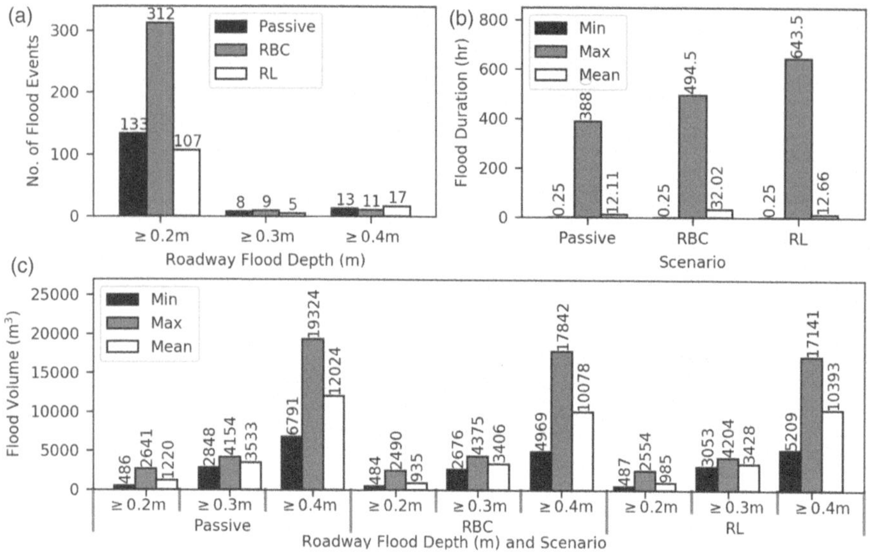

Figure 9 | Number (a), duration (b), and volume (c) of flood events at downstream node J1. Flood volumes at node J1 were categorized as causing $\geq$0.2, 0.3, or 0.4 m of water depth on the roadway.

three nodes. Due to the lack of system coordination, RBC had longer duration flood events at J1 than the passive system, but at lower volumes. In the passive system, the short duration, but a high volume of flood events at node J1 shows that without control, this system is flashy, which is a challenge in many urban systems.

DISCUSSION

One important aspect of any RTC application is the computational cost, both in terms of when the computation needs to happen and the time needed to compute a policy. With the RL algorithm used here, the agent learns offline on a training dataset. Once learned, the RL policy can be quickly applied. This formulation of MPC, in contrast, uses a genetic algorithm to perform on-line optimization (i.e., the best control actions are not known until the time that they need to be implemented). In this research, RL was able to learn a policy in approximately 34 min using 1 month of data on a standard desktop computer. Once developed, testing the RL policy simply requires running an SWMM simulation, passing the system state at each control time step through the agent, and implementing the resulting control actions (this takes 9 s for the August 2019 training data). Running MPC with the genetic algorithm and physics-based model as implemented in Sadler *et al.* (2019) required access to a HPC and took almost 50 h for 1 week of simulation. Because optimization is on-line, additional testing of MPC on other datasets would take a similar amount of time and would increase as the complexity of the system increases (Sadler *et al.* 2019). In practice, MPC may only need to run using the available forecast data (e.g., 18, 24, 36 h), not an entire week, reducing the computational burden (Sadler *et al.* 2020). Additionally, other formulations of MPC could use a different process model than SWMM (for instance using a state-space model learned from observed data) which could dramatically reduce MPC's computational cost (Balchen *et al.* 1992; Cigler *et al.* 2013; Corbin *et al.* 2013; Li *et al.* 2013; Behl *et al.* 2014). The training time for RL is also dependent on system complexity but needs further research to determine the feasibility and limitations for larger, more complex stormwater systems.

The RBC logic used in this research, like RL, can be considered an offline policy. Instead of an RL agent learning the control policy by interacting with the system, a human operator must understand the system well enough to formulate the rules. The growing adoption of Internet of Things (IoT) sensors for monitoring water levels provides the data needed to create control rules. In practice, the amount of time required to create these rules, and their quality, is dependent on factors like the availability of data on the physical watershed characteristics and the complexity of the system to be controlled. For the relatively simple system used in this research, and real single ponds (OptiRTC and Geosyntec Consultants Inc. 2017; Marchese *et al.* 2018), developing control rules is feasible. Ensuring coordinated and effective system-level control, however, will become increasingly difficult as complexity increases. As an example, the rule for valve position when trying to maintain the target depth was originally to completely open the valve. Through simulation, it was found that this often caused increased flooding at the downstream node. Adjusting that rule to only open the valve 50% when maintaining the target depth helped eliminate downstream flooding but is system specific and most likely not optimal. Adding a depth sensor in the downstream pipe as an additional factor in the control rules would be possible in this case; with enough time and IoT sensors, it may be possible to create control rules considering system-wide performance. However, there could be many such factors in a real urban stormwater system and accounting for each one and their interactions under different flow conditions will quickly become unmanageable. RL has an advantage here because the relationships between components of the system do not have to be known or stated explicitly, but can be learned. The disadvantage of RL, however, is that it is much less transparent than RBC in terms of how and why certain control decisions are made.

While this paper has explored RTC with RL, MPC, and RBC, there are other methods from the field of control theory that could be applied to stormwater systems. Wong and Kerkez (2018) provide an elegant example by using a linear quadratic regulator to manage storage pond depths in urban headwater catchments. This uses a state-space model as a linear representation of a watershed and performs control with a feedback controller. Another key contribution of this work is the ability to optimize the location of control structures and show that the entire system does not have to be controlled to achieve system-wide benefits. The state-space representation used by Wong and Kerkez (2018) or the discrete time dynamic system shown in Schwanenberg *et al.* (2015) could be used in RL or MPC as a replacement to the more computationally expensive SWMM model to speed up control of larger systems, but the full dynamics of the system represented in physics-based models may be lost.

Groundwater could contribute a significant amount to retention ponds that are being actively controlled, especially in coastal cities with high groundwater tables like Norfolk, Virginia, that respond quickly to storm events (Bowes *et al.* 2019). For a retention pond in Norfolk, we have estimated that groundwater would contribute approximately 0.16 m or 11% of the pond's volume per hour if the pond is completely emptied (see Appendix C for details on these approximations). Considering the storm event of 4–5 August 2019, the RL agent lowers the depth of water in the simulated ponds by almost 0.61 m over a 24-h period (Figure 1). Over that time, groundwater would have contributed an additional 0.71 m or 50% of the pond's total volume. This is not currently reflected in the SWMM simulations but has important implications in practice. While the additional inflow would most likely not change the general policy learned by the RL agent (i.e., lowering depths before a storm and maintaining depths otherwise), a larger valve may be needed to drain the ponds more quickly or the agent may need a longer forecast in order to drain the ponds prior to a storm event. Additionally, evaporation should be included in these simulations before being applied to real-world systems.

When implementing any of the RTC methods presented in this research, the method's interpretability will influence its adoption and use by decision makers. While RBC is easy to understand and highly transparent, MPC is less so, and RL is the least transparent. The control policies created by RL, while effective, can cause the system to make decisions that are non-intuitive to a human operator. Therefore, fully automating smart stormwater systems with RL may not be advisable at this time until more testing and safety controls can be put in place. However, RL could assist human operators in determining control policies and support decision making, for example, as part of a recommendation system (Solomatine and Ostfeld 2008). RL-based policies should continue to be trained with new data as it becomes available to increase confidence that the RL policies will produce desirable outcomes. This study shows, however, that even with a single month's worth of training data, RL shows great potential for determining effective control policies.

CONCLUSIONS AND FUTURE WORK

This research has explored the application of an RL agent for real-time stormwater system control where both rainfall and tidal level can impact flooding and retention pond depths in the system. In contrast to previous work, this paper used a continuous action space to create more refined control policies, by implementing the DDPG RL algorithm. A conditional reward structure based on the rainfall forecast and inclusion of forecasts in the system state allowed the RL agent to learn proactive control strategies. The performance of RL was compared to a passive system as well as two other RTC methods: MPC and RBC.

Results of this research show that both RL and RBC can improve stormwater system performance compared to the passive system. Using a control policy developed from a single month of rainfall and tide data, RL reduced total flood volume by 32% over the passive system for the 2010–2019 data. RBC, while only controlling ponds individually, still reduced total flood volume by 13% compared to the passive system. Additionally, this research showed that RL was able to learn to balance flooding throughout the system to maximize the conditional reward and meet the control objectives of mitigating flooding and maintaining target pond water levels. When implemented using the SWMM physics-based model, as described in Sadler *et al.* (2019, 2020), MPC was too computationally expensive to run for more than a small portion of the datasets. In this research, RL provided an 88× speedup in the creation of control policies compared to MPC.

Although the simple stormwater system, which is inspired by conditions in the coastal city of Norfolk, Virginia, demonstrates that RL can outperform other methods, more complex systems will face different computational burdens that could be a barrier to using such methods in real-time. This needs to be explored through future research testing RL on real-world systems. In addition, an alternative implementation of MPC using a state-space model, instead of the SWMM model used here, could dramatically reduce the computational cost for this control method. Lastly, the feasibility of using, and potentially combining, any of the RTC methods for decision support to enhance stormwater system performance should be investigated.

In order to move this work toward implementation within real-world systems, it may be valuable to explore more complex reward functions than the one used in this study. For example, it may be better to base the reward on different variables beyond flood volume and pond water depth. It may be the case that costs due to valve operation and drainage of ponds in specific cases are higher than a small amount of flooding that does not have a societal impact. Additionally, the flow rates and velocities in the system may have additional restrictions to consider in the reward function (e.g., maintaining certain flow conditions for water quality or stream biota health or preventing flow velocities that can cause soil erosion). More complex reward functions can be explored in future work to account for more complex situations, move towards the control of real systems, and integrate specific characteristics of valves and ponds in real-world systems. Finally, this paper trained an RL agent on a single month of data. In future research, the sensitivity of the algorithm to the amount and diversity of the data during training should be investigated. This will help construct a trade-off analysis between the amount of data needed, the training time required, and the accuracy of the predictive models needed for the training procedure.

ACKNOWLEDGMENTS

This work was funded as part of two National Science Foundation grants: Award #1735587 (CRISP-Critical, Resilient Interdependent Infrastructure Systems and Processes) and Award #1737432 (SCC-IRG Track 1: Overcoming Social and Technical Barriers for the Broad Adoption of Smart Stormwater Systems). We acknowledge HRSD for continued access to their high-quality data and the City of Norfolk for information regarding their stormwater retention ponds.

DATA AVAILABILITY STATEMENT

All relevant data are available from an online repository or repositories. The code repository link is https://github.com/UVAdMIST/swmm_rl. The data repository link is http://www.hydroshare.org/resource/e2d21c9224ab4aefaf1a5b6394b270b1.

REFERENCES

Abadi, M., Agarwal, A., Barham, P., Brevdo, E., Chen, Z., Citro, C., Corrado, G. S., Davis, A., Dean, J., Devin, M., Ghemawat, S., Goodfellow, I., Harp, A., Irving, G., Isard, M., Jia, Y., Jozefowicz, R., Kaiser, L., Kudlur, M., Levenberg, J., Mané, D., Monga, R., Moore, S., Murray, D., Olah, C., Schuster, M., Shlens, J., Steiner, B., Sutskever, I., Talwar, K., Tucker, P., Vanhoucke, V., Vasudevan, V., Viégas, F., Vinyals, O., Warden, P., Wattenberg, M., Wicke, M., Yu, Y., Zheng, X. & Research, G. 2016 *TensorFlow: Large-Scale Machine Learning on Heterogeneous Distributed Systems*. arXiv preprint arXiv:160304467.

Abou Rjeily, Y., Abbas, O., Sadek, M., Shahrour, I. & Hage Chehade, F. 2018 Model predictive control for optimising the operation of urban drainage systems. *Journal of Hydrology* **566**, 558–565.

Balchen, J. G., Ljungquist, D. & Strand, S. 1992 State space predictive control. *Chemical Engineering Science* **47** (4), 787–807.

Behl, M., Nghiem, T. X. & Mangharam, R. 2014 Model-iq: uncertainty propagation from sensing to modeling and control in buildings. In: *2014 ACM/IEEE International Conference on Cyber-Physical Systems (ICCPS)*. pp. 13–24. IEEE.

Berggren, K., Olofsson, M., Viklander, M., Svensson, G. & Gustafsson, A.-M. 2012 Hydraulic impacts on urban drainage systems due to changes in rainfall caused by climatic change. *Journal of Hydrologic Engineering* **17** (1), 92–98.

Bowes, B. D. 2020a *Monthly SWMM Input Files for Real-time Control Simulation*. http://www.hydroshare.org/resource/e2d21c9224ab4aefaf1a5b6394b270b1

Bowes, B. D. 2020b *UVAdMIST/swmm_rl: Repo for SWMM RL code*. https://github.com/UVAdMIST/swmm_rl

Bowes, B. D., Sadler, J. M., Morsy, M. M., Behl, M. & Goodall, J. L. 2019 Forecasting groundwater table in a flood prone coastal city with long short-term memory and recurrent neural networks. *Water* **11** (5), 1098.

Brockman, G., Cheung, V., Pettersson, L., Schneider, J., Schulman, J., Tang, J. & Zaremba, W. 2016 *OpenAI Gym*.

Camacho, E. F. & Bordons, C. C. 2007 *Model Predictive Control*. Springer, London, New York.

Castelletti, A., Pianosi, F. & Restelli, M. 2013 A multiobjective reinforcement learning approach to water resources systems operation: Pareto frontier approximation in a single run. *Water Resources Research* **49**, 3476–3486.

Castelletti, A., Yajima, H., Giuliani, M., Soncini-Sessa, R. & Weber, E. 2014 Planning the optimal operation of a multioutlet water reservoir with water quality and quantity targets. *Journal of Water Resources Planning and Management* **140** (4), 496–510.

Cigler, J., Gyalistras, D., Široky, J., Tiet, V. & Ferkl, L. 2013 Beyond theory: the challenge of implementing model predictive control in buildings. In *Proceedings of 11th Rehva world congress*, Clima, Vol. 250.

Corbin, C. D., Henze, G. P. & May-Ostendorp, P. 2013 A model predictive control optimization environment for real-time commercial building application. *Journal of Building Performance Simulation* **6** (3), 159–174.

Delipetrev, B., Jonoski, A. & Solomatine, D. P. 2017 A novel nested stochastic dynamic programming (nSDP) and nested reinforcement learning (nRL) algorithm for multipurpose reservoir optimization. *Journal of Hydroinformatics* **19** (1), 47–61.

García, L., Barreiro-Gomez, J., Escobar, E., Téllez, D., Quijano, N. & Ocampo-Martinez, C. 2015 Modeling and real-time control of urban drainage systems: a review. *Advances in Water Resources* **85**, 120–132.

Hartono, P. & Hashimoto, S. 2007 Learning from imperfect data. *Applied Soft Computing Journal* **7** (1), 353–363.

Jose Meneses, E., Gaussens, M., Jakobsen, C., Steen Mikkelsen, P., Grum, M. & Vezzaro, L. 2018 Coordinating rule-based and system-wide model predictive control strategies to reduce storage expansion of combined urban drainage systems: the case study of Lundtofte, Denmark. *Water* **10** (76), 1–15.

Kerkez, B., Gruden, C., Lewis, M., Montestruque, L., Quigley, M., Wong, B., Bedig, A., Kertesz, R., Braun, T., Cadwalader, O., Poresky, A. & Pak, C. 2016 Smarter stormwater systems. *Environmental Science and Technology* **50**, 72677273.

Lee, J.-H. & Labadie, J. W. 2007 Stochastic optimization of multireservoir systems via reinforcement learning. *Water Resources Research* **43** (11), 1–16.

Li, P., O'Neill, Z. & Braun, J. 2013 Development of control-oriented models for model predictive control in buildings. In *ASHRAE Trans*, Intelligent Buildings Operations Workshop, Boulder, CO.

Lillicrap, T. P., Hunt, J. J., Pritzel, A., Heess, N., Erez, T., Tassa, Y., Silver, D. & Wierstra, D. 2015 Continuous control with deep reinforcement learning. In *International Conference on Learning Representations*. p. 14.

Maier, H. R., Jain, A., Dandy, G. C. & Sudheer, K. P. 2010 Methods used for the development of neural networks for the prediction of water resource variables in river systems: current status and future directions. *Environmental Modelling and Software* **25**, 891–909.

Marchese, D., Johnson, J., Akers, N., Huffman, M. & Hlas, V. 2018 Quantitative comparison of active and passive stormwater infrastructure: case study in Beckley, West Virginia. *Proceedings of the Water Environment Federation* **2018** (9), 4298–4311.

McDonnell, B. E., Ratliff, K. M., Tryby, M. E., Wu, J. J. X. & Mullapudi, A. 2020 *PySWMM: The Python Interface to Stormwater Management Model (SWMM)*.

Mnih, V., Kavukcuoglu, K., Silver, D., Rusu, A. A., Veness, J., Bellemare, M. G., Graves, A., Riedmiller, M., Fidjeland, A. K., Ostrovski, G., Petersen, S., Beattie, C., Sadik, A., Antonoglou, I., King, H., Kumaran, D., Wierstra, D., Legg, S. & Hassabis, D. 2015 Human-level control through deep reinforcement learning. *Nature* **518**, 529–543.

Moftakhari, H. R., AghaKouchak, A., Sanders, B. F., Feldman, D. L., Sweet, W., Matthew, R. A. & Luke, A. 2015 Increased nuisance flooding along the coasts of the United States due to sea level rise: past and future. *Geophysical Research Letters* **42** (22), 9846–9852.

Moftakhari, H. R., AghaKouchak, A., Sanders, B. F. & Matthew, R. A. 2017 Cumulative hazard: the case of nuisance flooding. *Earth's Future* **5** (2), 214–223.

Mounce, S. R., Shepherd, W., Ostojin, S., Abdel-Aal, M., Schellart, A. N. A., Shucksmith, J. D. & Tait, S. J. 2020 Optimisation of a fuzzy logic-based local real-time control system for mitigation of sewer flooding using genetic algorithms. *Journal of Hydroinformatics* **22** (2), 281–295.

Mullapudi, A. & Kerkez, B. 2018 Autonomous control of urban storm water networks using reinforcement learning. *EPiC Series in Engineering* **3**, 1465–1469.

Mynett, A. E. & Vojinovic, Z. 2009 Hydroinformatics in multi-colours-part red: urban flood and disaster management. *Journal of Hydroinformatics* **11.3** (4), 166–180.

Neumann, J. E., Price, J., Chinowsky, P., Wright, L., Ludwig, L., Streeter, R., Jones, R., Smith, J. B., Perkins, W., Jantarasami, L. & Martinich, J. 2015 Climate change risks to US infrastructure: impacts on roads, bridges, coastal development, and urban drainage. *Climatic Change* **131** (1), 97–109.

OptiRTC and Geosyntec Consultants Inc 2017 *Water Quality Summary Report National Fish and Wildlife Foundation Smart, Integrated Stormwater Management Systems Anacostia River Watershed Water Quality Study*. Technical report.

Pianosi, F., Castelletti, A. & Restelli, M. 2013 Tree-based fitted Q-iteration for multi-objective Markov decision processes in water resource management. *Journal of Hydroinformatics* **15** (2), 258–270.

Plappert, M. 2016 *keras-rl*.

Pregnolato, M., Ford, A., Wilkinson, S. M. & Dawson, R. J. 2017 The impact of flooding on road transport: a depth-disruption function. *Transportation Research Part D: Transport and Environment* **55**, 67–81.

Sadler, J. M., Goodall, J. L., Behl, M., Morsy, M. M., Culver, T. B. & Bowes, B. D. 2019 Leveraging open source software and parallel computing for model predictive control of urban drainage systems using EPA-SWMM5. *Environmental Modelling & Software* **120**, 104484.

Sadler, J. M., Goodall, J. L., Behl, M., Bowes, B. D. & Morsy, M. M. 2020 Exploring real-time control of stormwater systems for mitigating flood risk due to sea level rise. *Journal of Hydrology* **583**, 1–10.

Schwanenberg, D., Becker, B. P. J. & Xu, M. 2015 The open real-time control (RTC)-tools software framework for modeling RTC in water resources systems. *Journal of Hydroinformatics* **17** (1), 130–148.

Solomatine, D. P. & Ostfeld, A. 2008 Data-driven modelling: some past experiences and new approaches. *Journal of Hydroinformatics* **10** (1), 3–22.

Sutton, R. S. & Barto, A. G. 2018 *Reinforcement Learning: An Introduction*, 2nd edn. The MIT Press, Cambridge, Massachusetts.

Sweet, W. V. & Park, J. 2014 From the extreme to the mean: Acceleration and tipping points of coastal inundation from sea level rise. *Earth's Future* **2** (12), 579–600.

van Andel, S. J., Price, R. K., Lobbrecht, A. H., van Kruiningen, F. & Mureau, R. 2008 Ensemble precipitation and water-level forecasts for anticipatory water-system control. *Journal of Hydrometeorology* **9** (4), 776–788.

van Andel, S. J., Price, R., Lobbrecht, A., van Kruiningen, F., Mureau, R. & Cordero, W. B. 2014 Framework for anticipatory water management: testing for flood control in the Rijnland storage basin. *Journal of Water Resources Planning and Management* **140** (4), 533–542.

Wong, B. P. & Kerkez, B. 2018 Real-time control of urban headwater catchments through linear feedback: performance, analysis, and site selection. *Water Resources Research* **54** (10), 7309–7330.

Wright, J. & Marchese, D. 2017 Briefing: continuous monitoring and adaptive control: the 'smart' storm water management solution. *Proceedings of the Institution of Civil Engineers – Smart Infrastructure and Construction* **170** (4), 86–89.

Wuebbles, D., Fahey, D., Hibbard, K., Dokken, D., Stewart, B. & Maycock, T. 2017 *Climate Science Special Report: Fourth National Climate Assessment, Volume I. Technical Report, U.S.* Global Change Research Program (USGCRP), Washington, DC.

First received 4 May 2020; accepted in revised form 16 September 2020. Available online 20 October 2020